METHODS IN MOLECULAR BIOLOGY

Series Editor:
John M. Walker
School of Life and Medical Sciences
University of Hertfordshire
Hatfield, Hertfordshire, AL10 9AB, UK

For further volumes:
http://www.springer.com/series/7651

RNA Interference and Cancer Therapy

Methods and Protocols

Edited by

Lekha Dinesh Kumar

Cancer Biology, CSIR-Centre for Cellular and Molecular Biology, Hyderabad, Telangana, India

Humana Press

Editor
Lekha Dinesh Kumar
Cancer Biology, CSIR-Centre for Cellular
and Molecular Biology
Hyderabad, Telangana, India

ISSN 1064-3745 ISSN 1940-6029 (electronic)
Methods in Molecular Biology
ISBN 978-1-4939-9222-5 ISBN 978-1-4939-9220-1 (eBook)
https://doi.org/10.1007/978-1-4939-9220-1

This Humana Press imprint is published by the registered company Springer Science+Business Media, LLC, part of Springer Nature.
The registered company address is: 233 Spring Street, New York, NY 10013, U.S.A.

Preface

Even after decades of conventional research, the war against cancer is still continuing. At present, even though chemotherapy is the only key approach for cancer treatment, it fails to improve patient mortality and compliance due to poor bioavailability and severe dose-dependent debilitating effects on healthy tissues. Biological therapies have not been strong enough to replace conventional therapy in spite of being used to augment chemotherapies in the case of certain types of cancers. In the recent past, RNA interference (RNAi) has emerged as a safe and powerful therapeutic tool which has ushered in a new hope to combat cancer. It is an attractive approach for silencing specific genes responsible for disease onset and progression and thus offers enough scope to be employed as a potential therapeutic tool. Several techniques and procedures, both in vitro and in vivo, have been developed and adopted for assessing RNAi-based strategies in cancer treatment. However, the major challenges facing the success of RNAi-based drugs include the difficulty in their targeted delivery to the primary site of the tumor, stability, penetration, and its effective accumulation to therapeutic levels. Though scientists have achieved this by using viral-based vectors, the stigma attached with the viral pathogenicity that might be acquired due to mutations prevents its exploitation. Therefore, efforts are on to achieve effectiveness of viral-based delivery without compromising on the safety issues by employing non-viral agents. The main drawback of non-viral agents is the inherent toxicity associated with high transfection efficiency. Different types of nanoparticles developed offer solutions to ligand-targeted delivery of RNAi molecules to tumor cells, thus preventing off-target effects. The first clinical trial with siRNA molecules as therapeutics showed that these novel molecules can be safely dosed to humans. This also shows the need for development of new and improved formulations to make this revolutionary therapeutic a reality. Thus, RNAi is the future biological therapeutic molecule with immense potential and promise to move from bench to bedside, even though progress in the field is at a slow pace. Several interesting and useful procedures and methodologies have been developed by researchers working in this field which should act as a repertoire of techniques for adoption and exploitation by those pursuing RNAi as a fertile strategy to develop the bio-drugs of future. It is appropriate, timely, and relevant to collate such protocols to catalyze this process. Therefore, an attempt has been made here to collate the protocols available in this field, and I strongly feel that this compilation of 26 procedures with its relevant rationale, tips, and background information provides a backpack guide to researchers, both novice and professional, who intend to discover and innovate newer means of using RNAi technology to combat cancer, the scourge of human health.

Chapter 1 describes the protocol to identify synergistic combinations from siRNA libraries using robotic screen. Chapter 2 gives a simple and straightforward in vitro protocol to evaluate candidate gene knockdown for cancer therapy. Chapter 3 outlines a protocol for the development of siRNA-based drugs against apoptotic factor and its successful delivery to orthotopic models of prostate cancer. Chapter 4 provides the protocol for quantification of individual siRNA strand loading to Ago 2, a crucial parameter that determines efficient silencing. Chapter 5 explains efficient targeted silencing by non-viral vectors which contribute to stable siRNA complex formation, specific intracellular uptake and release. Chapter 6 is about designing siRNA-encapsulating DNA nanosuitcases which conditionally release their cargo. Bio-drug in combination with chemo-drug as successful combinatorial therapy has been described in Chapters 7 and 8. Chapter 9 summarizes detailed protocol for preparation and targeted delivery of shRNAs using minicell (producing parent bacterial cells). Modified microRNAs have also been used to successfully target myeloid cells as narrated in Chapter 10. Different nanoparticle-based targeted deliveries including aptamers have been tried to combat shortcomings of viral-based deliveries, and the procedures adopted have been summarized in Chapters 11 through 17. As in the quote "the proof of the pudding is in the eating," the success of RNA interference technology in combating various cancers undoubtedly has to come from the preclinical studies. Chapters 18–26 demonstrate precisely this and thus prove that an RNAi-mediated bio-drug is a futuristic molecule which holds promise for cancer therapy.

My sincere thanks to all the authors for sharing their detailed protocols with expertise and experiences described in each chapter. Special thanks to the series editor, John M. Walker, Professor Emeritus in the School of Life Sciences at the University of Hertfordshire, for his guidance during the editing process. It was an absolute pleasure to work with the editor David C. Casey at Springer.

Hyderabad, Telangana, India ***Lekha Dinesh Kumar***

Contents

Contributors

DAVID J. ADAMS • *Illawarra Health and Medical Research Institute (IHMRI), University of Wollongong, Wollongong, NSW, Australia*

KWASI ADU-BERCHIE • *Department of Chemical Engineering and Materials Science, Michigan State University, East Lansing, MI, USA; School of Engineering and Applied Sciences, Harvard University, Cambridge, MA, USA*

REBAZ AHMED • *Department of Pathology, The University of Oklahoma Health Sciences Center, Oklahoma City, OK, USA; Graduate Program in Biomedical Sciences, The University of Oklahoma Health Sciences Center, Oklahoma City, OK, USA*

NARSIREDDY AMREDDY • *Department of Pathology, The University of Oklahoma Health Sciences Center, Oklahoma City, OK, USA; Stephenson Cancer Center, The University of Oklahoma Health Sciences Center, Oklahoma City, OK, USA*

PHILLIP A. ANGART • *Department of Chemical Engineering and Materials Science, Michigan State University, East Lansing, MI, USA; Office of Biotechnology Products, U.S. Food and Drug Administration, Silver Spring, MD, USA*

ANISH BABU • *Department of Pathology, The University of Oklahoma Health Sciences Center, Oklahoma City, OK, USA; Stephenson Cancer Center, The University of Oklahoma Health Sciences Center, Oklahoma City, OK, USA*

SYED SULTAN BEEVI • *Cancer Biology, CSIR-Centre for Cellular and Molecular Biology, Hyderabad, Telangana, India; Department of Cancer Biology, KIMS Foundation & Research Centre, KIMS Hospital, Secunderabad, India*

KATHERINE E. BUJOLD • *Department of Chemistry, McGill University, Montréal, QC, Canada*

GÜLAY BÜYÜKKÖROĞLU • *Department of Pharmaceutical Biotechnology, Faculty of Pharmacy, Anadolu University, Eskisehir, Turkey*

WEIBO CAI • *Department of Radiology, University of Wisconsin–Madison, Madison, WI, USA; Department of Medical Physics, University of Wisconsin–Madison, Madison, WI, USA*

REBECCA J. CARLSON • *Department of Chemical Engineering and Materials Science, Michigan State University, East Lansing, MI, USA; Harvard-MIT Program in Health Sciences and Technology, Cambridge, MA, USA*

CHRISTINA CHAN • *Department of Chemical Engineering and Materials Science, Michigan State University, East Lansing, MI, USA*

SZU-TING CHOU • *Five Prime Therapeutics, Inc., South San Francisco, CA, USA*

DANIEL F. COSTA • *Center for Pharmaceutical Biotechnology and Nanomedicine, Northeastern University, Boston, MA, USA; CAPES Foundation, Ministry of Education of Brazil, Brasília, DF, Brazil*

JESÚS M. DE LA FUENTE • *Instituto de Ciencia de Materiales de Aragón, CSIC-University of Zaragoza, Zaragoza, Spain; CIBER-BBN, Zaragoza, Spain*

KIRK DOUGLAS • *Department of Oncology, Karmanos Cancer Institute, Wayne State University, Detroit, MI, USA*

Anna A. Egorova • *Gynecology and Reproductology, D.O. Ott Research Institute of Obstetrics, Saint-Petersburg, Russia*
Hassan H. Fakih • *Department of Chemistry, McGill University, Montréal, QC, Canada*
Daniel P. Feldmann • *Department of Oncology, Wayne State University School of Medicine, Detroit, MI, USA*
Jianfeng Guo • *School of Pharmaceutical Sciences, Jilin University, Changchun, China*
Inbal Hazan-Halevy • *Laboratory of Precision NanoMedicine, School of Molecular Cell Biology and Biotechnology, Tel Aviv University, Tel Aviv, Israel; Department of Cell Research and Immunology, George S. Wise Faculty of Life Sciences, Tel Aviv University, Tel Aviv, Israel; Center for Nanoscience and Nanotechnology, Tel Aviv University, Tel Aviv, Israel; Cancer Biology Research Center, Tel Aviv University, Tel Aviv, Israel*
Justin D. Holmes • *Materials Chemistry and Analysis Group, School of Chemistry, Tyndall National Institute, University College Cork, Cork, Ireland; Centre for Research on Adaptive Nanostructures and Nanodevices (CRANN), Trinity College Dublin, Dublin 2, Ireland*
Kohzoh Imai • *The Institute of Medical Science, The University of Tokyo, Tokyo, Japan*
Ashok K. Jangid • *School of Nano Sciences, Central University of Gujarat, Gandhinagar, India*
Mehul Jivrajani • *Department of Pharmacology and Toxicology, B. V. Patel Pharmaceutical Education and Research Development (PERD) Centre, Ahmedabad, Gujarat, India; Faculty of Science, NIRMA University, Ahmedabad, Gujarat, India*
Steven Jones • *Department of Oncology, Wayne State University School of Medicine, Detroit, MI, USA*
Lei Kang • *Department of Nuclear Medicine, Peking University First Hospital, Beijing, China*
Maria Kavallaris • *Tumour Biology and Targeting Program, Lowy Cancer Research Centre, Children's Cancer Institute, UNSW Sydney, Sydney, NSW, Australia; Australian Centre for Nanomedicine, ARC Centre of Excellence in Convergent Bio-Nano Science and Technology, UNSW Sydney, Sydney, NSW, Australia*
Anton V. Kiselev • *Gynecology and Reproductology, D.O. Ott Research Institute of Obstetrics, Saint-Petersburg, Russia*
Marcin Kortylewski • *Department of Immuno-Oncology, Beckman Research Institute at City of Hope, Duarte, CA, USA*
Hitesh Kulhari • *School of Nano Sciences, Central University of Gujarat, Gandhinagar, India*
Lekha Dinesh Kumar • *Cancer Biology, CSIR-Centre for Cellular and Molecular Biology, Hyderabad, Telangana, India*
Andrew Lee • *International Institute for Nanotechnology, Northwestern University, Evanston, IL, USA; Department of Chemistry, Northwestern University, Evanston, IL, USA*
Dian-Jang Lee • *Department of Pharmacy, Center for NanoScience, Ludwig-Maximilians-Universität München, Munich, Germany; Nanosystems Initiative Munich (NIM), Munich, Germany*
Qixin Leng • *Department of Pathology, University of Maryland School of Medicine, Baltimore, MD, USA*
Guido Marcucci • *Hematologic Malignancies, Beckman Research Institute at City of Hope, Duarte, CA, USA*

MARIANNA A. MARETINA • *Gynecology and Reproductology, D.O. Ott Research Institute of Obstetrics, Saint-Petersburg, Russia; Saint-Petersburg State University, Saint-Petersburg, Russia*

JOSHUA A. MCCARROLL • *Tumour Biology and Targeting Program, Lowy Cancer Research Centre, Children's Cancer Institute, UNSW Sydney, Sydney, NSW, Australia; Australian Centre for Nanomedicine, ARC Centre of Excellence in Convergent Bio-Nano Science and Technology, UNSW Sydney, Sydney, NSW, Australia*

OLIVIA M. MERKEL • *Department of Oncology, Wayne State University School of Medicine, Detroit, MI, USA; Department of Pharmaceutical Sciences, Wayne State University, Detroit, MI, USA; Department of Pharmacy, Pharmaceutical Technology and Biopharmacy, Ludwig-Maximilians-Universität München, Munich, Germany*

CHAD A. MIRKIN • *International Institute for Nanotechnology, Northwestern University, Evanston, IL, USA; Department of Chemistry, Northwestern University, Evanston, IL, USA*

A. JAMES MIXSON • *Department of Pathology, University of Maryland School of Medicine, Baltimore, MD, USA; Greenebaum Cancer Center, University of Maryland School of Medicine, Baltimore, MD, USA*

ANUPAMA MUNSHI • *Stephenson Cancer Center, The University of Oklahoma Health Sciences Center, Oklahoma City, OK, USA; Department of Radiation Oncology, The University of Oklahoma Health Sciences Center, Oklahoma City, OK, USA*

MANISH NIVSARKAR • *Department of Pharmacology and Toxicology, B. V. Patel Pharmaceutical Education and Research Development (PERD) Centre, Ahmedabad, Gujarat, India*

DAN PEER • *Laboratory of Precision NanoMedicine, School of Molecular Cell Biology and Biotechnology, Tel Aviv University, Tel Aviv, Israel; Department of Cell Research and Immunology, George S. Wise Faculty of Life Sciences, Tel Aviv University, Tel Aviv, Israel; Department of Materials Sciences and Engineering, Iby and Aladar Fleischman Faculty of Engineering, Tel Aviv University, Tel Aviv, Israel; Center for Nanoscience and Nanotechnology, Tel Aviv University, Tel Aviv, Israel; Cancer Biology Research Center, Tel Aviv University, Tel Aviv, Israel*

PHOEBE A. PHILLIPS • *Australian Centre for Nanomedicine, ARC Centre of Excellence in Convergent Bio-Nano Science and Technology, UNSW Sydney, Sydney, NSW, Australia; Pancreatic Cancer Translational Research Group, Lowy Cancer Research Centre, School of Medical Sciences, UNSW Sydney, Sydney, NSW, Australia*

DEEP POOJA • *Pharmacology and Toxicology Division, CSIR-Indian Institute of Chemical Technology, Hyderabad, Telangana, India*

KAMIL RAHME • *Department of Sciences, Faculty of Natural and Applied Sciences, Notre Dame University-Louaize, Zouk Mosbeh, Lebanon; Materials Chemistry and Analysis Group, School of Chemistry, Tyndall National Institute, University College Cork, Cork, Ireland*

RAJAGOPAL RAMESH • *Department of Pathology, The University of Oklahoma Health Sciences Center, Oklahoma City, OK, USA; Graduate Program in Biomedical Sciences, The University of Oklahoma Health Sciences Center, Oklahoma City, OK, USA; Stephenson Cancer Center, The University of Oklahoma Health Sciences Center, Oklahoma City, OK, USA; Stanton L. Young Biomedical Research Center, The University of Oklahoma Health Sciences Center, Oklahoma City, OK, USA*

SRINIVAS RAMISHETTI • *Laboratory of Precision NanoMedicine, School of Molecular Cell Biology and Biotechnology, Tel Aviv University, Tel Aviv, Israel; Department of Cell Research and Immunology, George S. Wise Faculty of Life Sciences, Tel Aviv University, Tel Aviv, Israel; Center for Nanoscience and Nanotechnology, Tel Aviv University, Tel Aviv, Israel; Cancer Biology Research Center, Tel Aviv University, Tel Aviv, Israel*

K. C. REMANT • *Department of Chemical and Material Engineering, Faculty of Engineering, University of Alberta, Edmonton, AB, Canada*

DANIEL ROSENBLUM • *Laboratory of Precision NanoMedicine, School of Molecular Cell Biology and Biotechnology, Tel Aviv University, Tel Aviv, Israel; Department of Cell Research and Immunology, George S. Wise Faculty of Life Sciences, Tel Aviv University, Tel Aviv, Israel; Center for Nanoscience and Nanotechnology, Tel Aviv University, Tel Aviv, Israel; Cancer Biology Research Center, Tel Aviv University, Tel Aviv, Israel*

ZACHARY T. ROSENKRANS • *Department of Radiology, University of Wisconsin–Madison, Madison, WI, USA; Department of Medical Physics, University of Wisconsin–Madison, Madison, WI, USA*

CAN SARISOZEN • *Center for Pharmaceutical Biotechnology and Nanomedicine, Northeastern University, Boston, MA, USA*

BEHIYE ŞENEL • *Department of Pharmaceutical Biotechnology, Faculty of Pharmacy, Anadolu University, Eskisehir, Turkey*

GEORGE SHARBEEN • *Pancreatic Cancer Translational Research Group, Lowy Cancer Research Centre, School of Medical Sciences, UNSW Sydney, Sydney, NSW, Australia*

ANTHONY F. SHIELDS • *Department of Oncology, Karmanos Cancer Institute, Wayne State University, Detroit, MI, USA*

RAMAKRISHNA SISTLA • *Pharmacology and Toxicology Division, CSIR-Indian Institute of Chemical Technology, Hyderabad, Telangana, India*

HANADI F. SLEIMAN • *Department of Chemistry, McGill University, Montréal, QC, Canada*

ALEXANDER H. STEGH • *Ken and Ruth Davee Department of Neurology, The Robert H. Lurie Comprehensive Cancer Center, The Northwestern Brain Tumor Institute, Northwestern University, Chicago, IL, USA; International Institute for Nanotechnology, Northwestern University, Evanston, IL, USA*

YU-LIN SU • *Department of Immuno-Oncology, Beckman Research Institute at City of Hope, Duarte, CA, USA*

PIOTR SWIDERSKI • *Department of Molecular Medicine, Beckman Research Institute at City of Hope, Duarte, CA, USA*

YOSHIFUMI TAKEI • *Department of Medicinal Biochemistry, School of Pharmacy, Aichi Gakuin University, Nagoya, Japan*

XING FEI TAN • *Department of Anatomy, Yong Loo Lin School of Medicine, National University of Singapore, Singapore, Singapore*

NAVEEN KUMAR TANGUDU • *Cancer Biology, CSIR-Centre for Cellular and Molecular Biology, Hyderabad, Telangana, India; Institute of Comparative Molecular Endocrinology, Ulm University, Ulm, Germany*

HIROAKI TANIGUCHI • *Clinical and Translational Research Center, Keio University, Tokyo, Japan; The Institute of Medical Science, The University of Tokyo, Tokyo, Japan*

WEI XUAN TEO • *Department of Anatomy, Yong Loo Lin School of Medicine, National University of Singapore, Singapore, Singapore*

Bindu Thapa • *Faculty of Pharmacy and Pharmaceutical Sciences, University of Alberta, Edmonton, AB, Canada*

Serena Tommasini-Ghelfi • *Ken and Ruth Davee Department of Neurology, The Robert H. Lurie Comprehensive Cancer Center, The Northwestern Brain Tumor Institute, Northwestern University, Chicago, IL, USA*

Vladimir P. Torchilin • *Center for Pharmaceutical Biotechnology and Nanomedicine, Northeastern University, Boston, MA, USA*

Claudia Tortiglione • *Istituto di scienze applicate e sistemi intelligenti "E.Caianiello", Consiglio Nazionale delle Ricerche, Pozzuoli, Italy*

Hasan Uludağ • *Faculty of Pharmacy and Pharmaceutical Sciences, University of Alberta, Edmonton, AB, Canada; Department of Chemical and Material Engineering, Faculty of Engineering, University of Alberta, Edmonton, AB, Canada; Department of Biomedical Engineering, Faculty of Medicine and Dentistry, University of Alberta, Edmonton, AB, Canada*

Christine Vauthier • *Institut Galien Paris Sud, UMR CNRS 8612, University of Paris-Sud, Chatenay-Malabry, France*

Vinod Kumar Verma • *Cancer Biology, CSIR-Centre for Cellular and Molecular Biology, Hyderabad, Telangana, India; Department of Cancer Biology, KIMS Foundation and Research Centre, KIMS Hospital, Secunderabad, India*

Daniel B. Vocelle • *Department of Chemical Engineering and Materials Science, Michigan State University, East Lansing, MI, USA*

Ernst Wagner • *Department of Pharmacy, Center for NanoScience, Ludwig-Maximilians-Universität München, Munich, Germany; Nanosystems Initiative Munich (NIM), Munich, Germany*

S. Patrick Walton • *Department of Chemical Engineering and Materials Science, Michigan State University, East Lansing, MI, USA*

Martin C. Woodle • *AparnaBio, Inc., Gaithersburg, MD, USA*

Leyuan Xu • *Section of Nephrology, Department of Internal Medicine, Yale University School of Medicine, New Haven, CT, USA*

Hu Yang • *Department of Chemical and Life Science Engineering, Virginia Commonwealth University, Richmond, VA, USA; Department of Pharmaceutics, Virginia Commonwealth University, Richmond, VA, USA; Massey Cancer Center, Virginia Commonwealth University, Richmond, VA, USA*

Evrim Yenilmez • *Department of Pharmaceutical Technology, Faculty of Pharmacy, Anadolu University, Eskisehir, Turkey*

George W. Yip • *Department of Anatomy, Yong Loo Lin School of Medicine, National University of Singapore, Singapore, Singapore*

Chapter 1

siRNA Library Screening to Identify Complementary Therapeutic Pairs in Triple-Negative Breast Cancer Cells

Bindu Thapa, KC Remant, and Hasan Uludağ

Abstract

The existence of tightly integrated cross talk through multiple signaling and effector pathways has been appreciated in malignant cells. The realization of the plasticity of such networks is stimulating the development of combinational therapy to overcome the limitations of one-dimensional therapies. Synergistic pairs of siRNAs or siRNA and drug combinations are the new frontiers in identifying effective therapeutic combinations. To elucidate effective combinations, we developed a versatile protocol to screen siRNA libraries in triple-negative breast cancer cell models. This protocol outlines the steps to identify synergistic combinations of siRNA-siRNA or siRNA-drug combinations using siRNA libraries via a robotic screen. By focusing on smaller functional siRNA libraries, we present methodologies to identify synergistic siRNA pairings against cancerous cell growth and molecular targets to augment the activity of pro-apoptotic TRAIL protein. Here, we summarize the critical steps to undertake such combinational target identification, emphasizing critical factors that affect the outcome of the screens. Our experience suggests that siRNA library screening is an efficient protocol to identify complementary therapeutic pairs of new or already-existing drugs. This protocol is simple, robust and can be completed within a 1-week working period.

Key words siRNA library screening, Triple-negative breast cancer cells, siRNA transfection

1 Introduction

The coupling between rapid human genome sequencing and techniques for high-throughput screening has revolutionized the studies of gene function and their role in different diseases. Silencing of individual genes predicted from the genome sequencing provides a clear-cut way to systematically probe the role of individual genes in different diseases. Noncoding RNAs (ncRNAs) such as microRNA (miRNA) and short interfering RNAs (siRNAs) that participate in RNA interference (RNAi) mechanism have been developed as a new line of therapeutics for cancer gene therapy. siRNA silences a targeted gene by inducing natural RNAi pathway, which results in degradation or translational blockage of a complementary messenger RNA (mRNA). siRNA treatments have been used for func-

Lekha Dinesh Kumar (ed.), *RNA Interference and Cancer Therapy: Methods and Protocols*, Methods in Molecular Biology, vol. 1974, https://doi.org/10.1007/978-1-4939-9220-1_1, © Springer Science+Business Media, LLC, part of Springer Nature 2019

tional genomics and to reveal the molecules involved in biochemical pathways [1]. The complex physiological changes associated with different diseases can be better understood by silencing specific genes associated with such diseases. Silencing with siRNA allows reversible deletion of individual participants in biochemical cascades, revealing their role and function in the investigated aspects of the diseases. Hence, siRNA library screens, in which large numbers of siRNAs have been compiled broadly (e.g., against genome-wide transcriptome) or with a specific focus (e.g., against apoptosis-regulating proteins), have been indispensable to identify therapeutic targets in different diseases [2].

Cancer is a particularly attractive disease for siRNA screens since aberrant changes in gene expression and/or regulation is the main cause of the disease and large numbers of aberrant transformations are likely to emerge in individual cancers. The outcome of siRNA screens can identify aberrant mediators that can serve as drug targets, in addition to providing specific siRNAs that can be employed in therapy. Genome-wide screens against breast cancers have been attempted to provide an unbiased approach for target identification [3, 4]. Assessing the outcomes of every possibility head-to-head can provide a more objective assessment of the relative importance of various targets, but handling large libraries is time-consuming, requires significant resources, and is more likely to lead to false-positive hits due to technical errors. Alternatively, we preferred to screen "focused" libraries, such as the libraries against apoptosis-regulating proteins [2, 5], protein kinases [2], phosphatases [6], and protein regulators of cell cycle [7] in malignant cell lines, since less resources are required especially for subsequent validation studies. The findings have typically revealed that individual target silencing altered the assessed feature of the malignant cells. Since the ultimate goal is to control unchecked growth, the screens have been most notably conducted to inhibit cellular growth (i.e., as a functional outcome). However, in most diseases, especially in cancer, where cellular transformation arises from the interplay or accumulation of multiple mediators, identifying and targeting single mediators may not be sufficient. Complex signaling network including redundancies, extensive cross talk, compensatory and neutralizing activities in addition to heterogeneity in the population of disease-causing cells, is responsible for the therapeutic limitations of monotherapy [8–10]. To this end, combination therapy comprising multiple therapeutic agents, which target multiple pathways, has been developing as a promising approach in cancer gene therapy [10]. Three major approaches to combinational therapy include (1) inhibiting specific targets by multiple strategies, (2) abolishing multiple components in a given pathway (to better eradicate a given pathway), and (3) interfering with multiple mechanisms in tumor growth and metastasis [11]. The combination of therapeutic agents that generate the synergism via complementary effects with minimal overlapping of toxicity spectrum is an ideal model in therapeutic intervention. This modality may further attenuate the side effects associated with the clinical doses of individual drugs by reducing the doses of individual component [12, 13]. Therefore, here, we established a standard protocol as a proof of concept to identify complementary therapeutic pair for cancer gene therapy using siRNA library screening in breast cancer cells. Triple-negative breast cancer MDA-MB-231 cells were used as a model of breast cancer, given the lower therapeutic response (with current drugs) in the case of triple-negative breast cancer.

2 Materials

2.1 Cell Culture and Seeding

1. Identity-authenticated triple-negative breast cancer cells, MDA-MB-231.
2. Tissue culture media: Dulbecco's Modified Eagle's Medium (DMEM) supplemented with 10% fetal bovine serum and 100 U/mL penicillin and 100 μg/mL streptomycin.
3. Hank's Balanced Salt Solution (HBSS).
4. Tissue culture plates: sterile standard T75 tissue culture flask for adherent cells, 96-well transparent tissue culture plates.
5. Instruments: Hemocytometer, bright-field microscope, and cell culture incubator (5% CO_2, 37 °C).

2.2 siRNA Library Screening with Drug

1. RNase-free sterile water.
2. Serum-free culture medium, DMEM.
3. siRNA library: siGENOME Human Apoptosis library, G-003905 (*see* **Note 1**).
4. Positive and negative control siRNAs (*see* **Note 2**).
5. Drug: recombinant human TRAIL.
6. Transfecting reagent: lipid-modified small molecular weight (MW 1200 Da) polyethyleneimine (*see* **Note 3**).
7. Microplate seals.
8. Instruments: plate centrifuge, Perkin Elmer JANUS automated liquid-handling system, and "WinPREP" software.

2.3 Final Read-Out Assay

1. 3-(4,5-dimethylthiazol-2-yl)-2,5-diphenyltetrazolium bromide (MTT).
2. Dimethyl sulfoxide (DMSO).
3. Instruments: microplate reader (to measure absorbance), multichannel pipette, syringe, and membrane filter.

3 Methods

The protocol we developed is semiautomated and optimized for triple-negative breast cancer cells lines and an apoptosis-related siRNA library which contains 446 siRNAs related to apoptosis events. Some of the steps need to be modified depending on cell types and siRNA libraries. The experimental design is divided into the following sections: cell culture and seeding, sample preparation, transfection and drug treatment, and final read-out assay as outlined in Fig. 1. *See* **Note 4** for additional information.

3.1 Cell Culture and Seeding

1. Aspirate 10 mL of pre-warmed cell culture medium into sterile 75-cm^2 cell culture flask inside laminar flow hood. Gently, swirl the flask to ensure even distribution of flask bottom surface. Collect pre-warmed cell culture medium into 15 mL centrifuge tube. Put these two mediums into the incubator to keep warm until they are used.

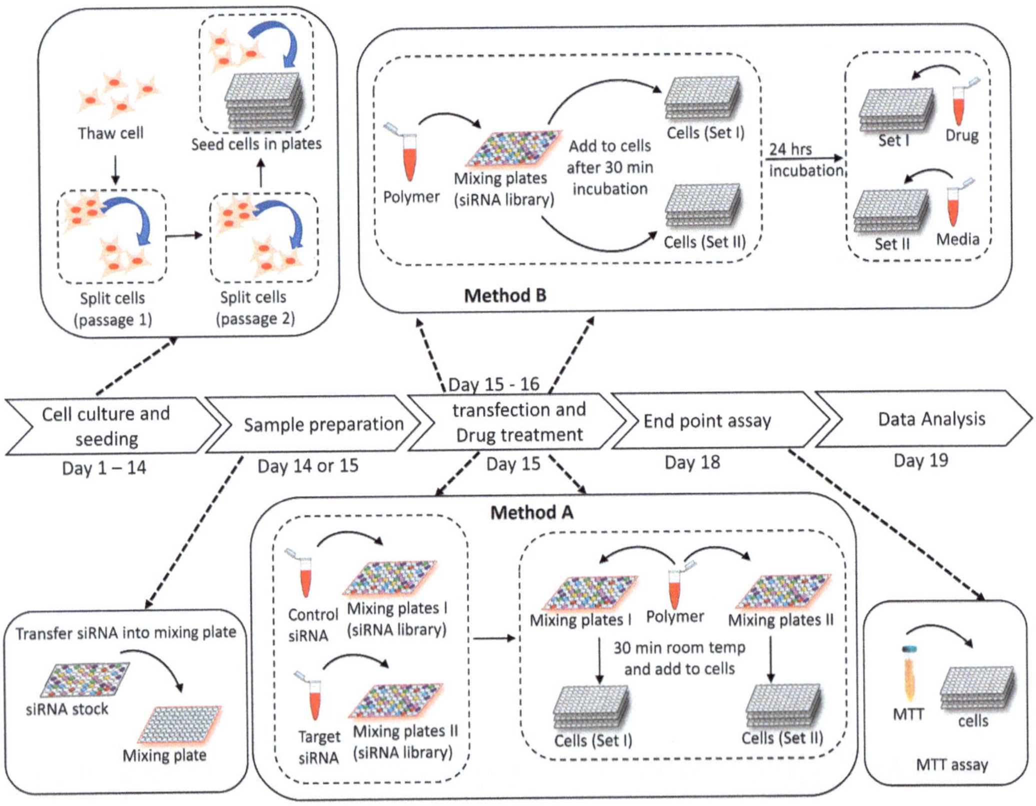

Fig. 1 Workflow and timeline for apoptosis siRNA library screening to identify complementary therapeutic pair of drug (TRAIL) or siRNA in triple-negative breast cancer cells (e.g., MDA-MB-231 cells). Total time required is about 3 weeks in which first 2 weeks are for cell thawing and subculturing. If cells are ready for testing, only 1 week is required to complete the library screening

2. Remove the frozen cell stock from liquid nitrogen, spray with 70% alcohol, and wipe. Open the cap of cryotube for 2 s, and close (to remove liquid nitrogen from cryotube) inside laminar flow hood. Immediately, thaw the cells in 37 °C water bath (*see* **Note 5**).
3. Pipette the cells from cryovial into centrifuge tube containing medium (prepared in **step 1**), and centrifuge at 600 rpm for 5 min to remove the cell freezing medium (*see* **Note 6**).
4. Remove supernatant, add 5 mL of complete DMEM, re-suspend the cells, and transfer into 75-cm^2 cell culture flask (prepared in **step 1**) with vented cap.
5. Place the flask in incubator at 37 °C, 5% CO_2 under humidified condition, and allow cells to attach. Change the medium after 24 h. Check the cells daily and allow it to grow until 80–90% confluence (*see* **Note 7**).
6. Once cells reach 80–90% confluent, aspirate the media, wash twice with of HBSS (~10 mL), and add 0.05% trypsin-EDTA (1 mL). Incubate it at room temperature until it starts to dislodge from the flask (~2 min).
7. Add 10 mL of complete DMEM to stop the enzymatic activity of trypsin-EDTA. Over-incubation with trypsin might digest the cells.
8. Collect the cells into a 15 mL centrifuge tube and centrifuge at 72 × *g* for 5 min.
9. Remove supernatant and re-suspend cell in 10 mL complete DMEM.
10. Pipette 10 mL of complete DMEM into 75 cm^2 cell culture flask.
11. Add 2 mL of cell suspension into it (1:5 dilutions).
12. Gently swirl the flask to distribute the cells throughout the flask, and allow them to grow.
13. Repeat **steps 6–12** at least two times before seeding into plates for screening, and proceed to **step 14** (*see* **Note 8**).
14. Once the cells reached the 80–90% confluence, aspirate the medium, wash HBSS, and trypsinize (as explained in **step 6**).
15. Add 10 mL of complete DMEM to stop trypsinization, pipette cell suspension into 15 mL centrifuge tube, and centrifuge at 600 rpm for 5 min.
16. Remove supernatant and re-suspend cells in 10 mL of complete DMEM by carefully pipetting the cell suspension up and down around 10–15 times to separate cell clumps. If required, mix cells from other flasks since the cells from single flask might not be enough (*see* **Note 9**).

17. Count the cells on a Neubauer cell-counting chamber.
18. Dilute cell suspension, which gives 5000 cells in 90 μL of medium (*see* **Note 10**).
19. Mix well, and pour cell suspension into sterile flat bottom rectangular container prepared for loading into 96-well plate through liquid-handling robot (*see* **Note 11**).
20. Gently pipette 90 μL of cell suspension into each well using liquid-handling robot. Gently shake the plates to ensure a uniform distribution of cells throughout the well surface (*see* **Note 12**).
21. Move cell-seeded 96-well plates into incubator at 37 °C, 5% CO_2 humidified atmosphere. Allow them to grow for 24 h. Do not overlay the plates inside incubator.
22. Repeat **steps 20** and **21** until all the required 96-well plates were seeded.
23. Incubate cells overnight and check them under microscope. If the cells reached ~40% confluence, proceed with transfection.

3.2 Sample Preparation

1. Calculate how much siRNA is needed per well using the formula:

$$\text{siRNA needed}\ (\mu L) = \frac{\text{Final volume}\ (\text{medium} + \text{compexes};\ \mu L) \times \text{dose of siRNA}\ (\text{nM})}{\text{Concentration of stock siRNA}\ (\text{nM})}$$

   Our siRNA stock concentration was 1 μM and the final volume was 100 μL. So, for 30 nM treatment concentration, 3 μL of siRNA was needed (*see* **Note 13**).
2. Thaw the stock siRNA library, and centrifuge to collect residual agents on surfaces. Pipette the required amount of siRNA from the stock siRNA plate (1 μM) into 96-well mixing plate (round bottom) using liquid-handling robot in a sterile environment. First and last columns of all the plates were left empty where the same amount of control siRNAs (positive and negative control) and blank (saline) were pipetted as shown in Fig. 2 (*see* **Note 14**).
3. Prepare polymer (lipid-modified 1.2 kDa polyethyleneimine) solution in serum free DMEM (without serum). Prepare extra (~10%) polymer solution to encounter the dead volume while pipetting. Concentration of polymer should be calculated in such a way that polymer to siRNA weight ratio becomes 6 in a total of 10 μL of complexes per well (*see* **Note 15**).

3.3 Transfection and Drug Treatment

Transfection procedure in this study was "forward" transfection (i.e., cell addition is followed by siRNA complex) using aliphatic lipid-grafted low-molecular-weight (1200 Da) polyethyleneimine

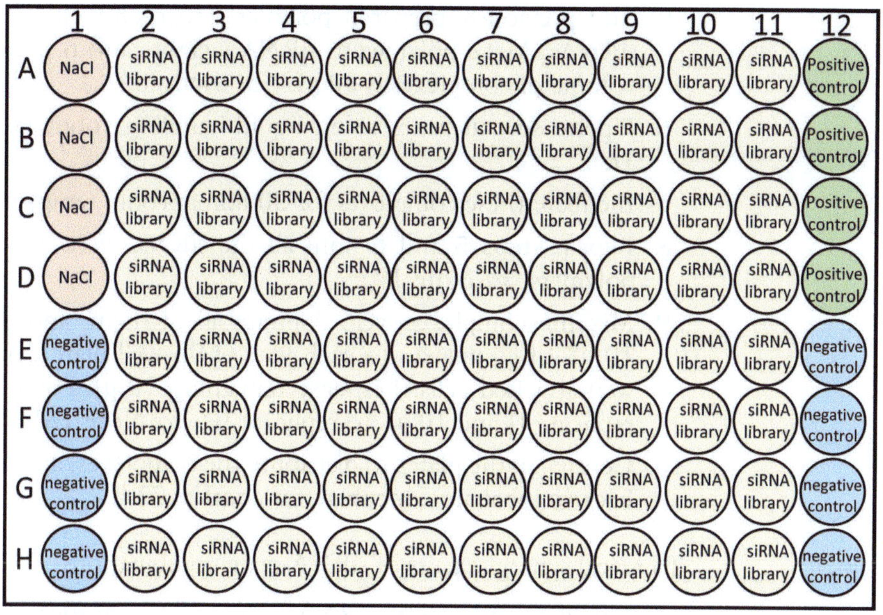

Fig. 2 Layout of 96-well mixing plate with siRNA printing

(PEI-L) as transfecting agent [14]. This library screening is intended to find out synergistic pairs of siRNA-siRNA molecules (Subheading 3.3.1, Fig. 1) or siRNA-drug molecules (Subheading 3.3.2, Fig. 1). In Method A, siRNAs were plated in 96-well mixing plate in two parallel sets of plates. A predetermined desired siRNA was then added to each well of one set of prepared siRNA plates, while a negative control (scrambled) siRNA was added to each well of the other set of siRNA plates. In Method B, drug (TRAIL) treatment was performed after 24 h of transfection to one set, while another set remained without drug (TRAIL). Screening was performed in triplicate wells. Hence, sufficient complexes should be prepared for addition to six wells for each siRNA treatment (with 10% excess volume to account for pipetting losses).

3.3.1 Method A

To reveal synergistic pairs of siRNA-siRNA combinations:

1. Take two sets of siRNA library plated in 96-well mixing plate (round bottom).
2. Add target siRNA (e.g., Mcl-1) to each well of one set, and control siRNA to each well of another set. The following procedure is the same for both sets. Label the mixing plates properly.
3. Mix well by pipetting and centrifuge them briefly.
4. Allow them to reach room temperature.

5. Add previously prepared polymer into each well containing siRNAs. Make polymer to siRNA weight ratio 6. Now, total volume of complexes is 30 μL (for triplicate).
6. Pipette several times using liquid-handling robot (*see* **Note 16**).
7. Incubate these complexes for 30 min at room temperature.
8. Repeat **steps 5** and **6** until all the siRNA were mixed with polymers (*see* **Note 17**).
9. Mix the complexes well. Aspirate all of them, and dispense 10 μL of complexes to each well of 96-well plate containing well-attached cells. Dispense remaining 10 μL complexes to replicate II and 10 μL to replicate III (*see* **Note 18**).
10. Gently tap the plates to distribute the complexes, and return it back to incubator. Incubate for another 72 h.

3.3.2 Method B

To reveal synergist pairs of siRNA-drug combinations:

1. Calculate and pipette siRNA solution enough for two sets of screening into round bottom 96-well mixing plate. Label the mixing plates properly.
2. Mix wells by pipetting and centrifuge them briefly.
3. Allow them to reach to room temperature.
4. Add previously prepared polymer solution into each well containing the siRNAs. Make polymer to siRNA weight ratio 6. Total volume of complexes would be 60 μL for the two sets of screening in triplicate (10 μL per well × 6 wells).
5. Pipette the complexes several times using liquid-handling robot (*see* **Note 16**).
6. Incubate the complexes for 30 min. at room temperature.
7. Repeat **steps 4** and **5** until all the siRNA were mixed with polymers (*see* **Note 17**).
8. Mix the complexes well. For Set 1, aspirate all of them, and dispense 10 μL of complexes to each well of 96-well plate containing well-attached cells. Dispense remaining 10 μL complexes to replicate II and 10 μL to replicate III. For Set 2, repeat the same procedure for this set as well (*see* **Note 18**).
9. Gently tap the plates to distribute the complexes well in the plate, and return it back to incubator and incubate for 24 h.
10. Prepare drug (TRAIL) solution in complete DMEM (*see* **Note 19**).
11. Add 20 μL of TRAIL solution into one set of cells (from **step 8**) treated with siRNA complexes (replicate I, II, and III). To have a proper control, add 20 μL of complete DMEM into

another set (from **step 8**) of cells with siRNA complexes (replicate I, II, and III).

12. Gently tap the plates to ensure proper mixing of drug, and return back to incubator. Incubate for another 48 h.

3.4 End Point Assay and Data Analysis

1. Prepare MTT solution (5 mg/mL) in pre-warmed HBSS, and filter through syringe membrane filter (pore size 0.2 μm) (*see* **Note 20**).
2. After 72 h of polymer/siRNA complexes treatment, add MTT solution to each well using multichannel pipette, and incubate at 37 °C for 1.5 h. Volume of MTT solution should be adjusted in such a way that final concentration becomes 1 mg/mL (*see* **Note 21**).
3. Check the plate for MTT crystal generation and remove the media from well. If the crystal did not appear in non-treated group, then incubate for an extra time (*see* **Note 22**).
4. Add 100 μL of DMSO to each well using multichannel pipette. Gently tap the plate to dissolve MTT crystal completely within a 10-min time window (*see* **Note 23**).
5. Read the absorbance at 570 nm at plate reader and proceed to data analysis (*see* **Note 24**).

3.5 Data Analysis and Selection of Hits

It is always not feasible to replicate the library screen; therefore, our confidence level in the identified targets from primary screening is low. Further validation is always needed to establish effective hits. We performed the screening experiments in triplicate to maximize the sensitivity along with increasing the confidence in hits.

1. To identify the effective siRNAs, calculate relative cell viability of treated group as a percentage of cell growth in non-treatment control group on a per plate basis using following formula:

$$\text{Relative cell viability} = \frac{\text{O.D. of siRNA or Drug treated well}}{\text{O.D. of Non} - \text{treated well}} \times 100\%$$

   In calculating the O.D. of wells, background O.D. (i.e., that of DMSO) should be reduced from the measured O.D.s.

2. After calculating relative cell viability of each well of the entire plate, calculate mean and standard deviation of triplicate wells (plates) using excel or other suitable software.
3. Calculate significance by student's two-tailed *t*-test (assuming equal variance) and *z*-score to identify the effective targets. A value of $p < 0.05$ was considered significant, and the outliers

were noted by selecting responses with $-1.96 < z < 1.96$. Calculate z value using the following formula:

$$z = \frac{xi - \mu}{s}$$

where xi is the percentage of cell growth compared to non-treatment cells for each well (relative cell activity), μ is the average, and s is the standard deviation of all xi in the whole plate.

4. The targets which satisfy the criteria (1) relative cell viability <70%, (2) z-score < -1.96, and (3) $p < 0.05$ were considered as hits in our screens.
5. Select the targets based on selection criteria mentioned above and prioritize the hits for validation. *See* **Notes 25** and **26** for detailed criteria to prioritize and validate selected targets.
6. Plot the correlation plots (between replicates) to assess the reproducibility, and assess the quality control of the library screen. *See* **Note 27** for details. *See* **Note 28** for anticipated results.

4 Notes

1. To this end, different siRNA libraries are commercially available, adding possibilities of focused screening toward specific class of cellular molecules, which are more cost-efficient and increase the chances of new drug target discovery. Although we used focused siRNA libraries in this protocol, the protocol can also be applicable to genome-wide screens with appropriate instrumentation. This protocol requires high-throughput liquid handler; therefore experimental costs will increase with the increase in experiment size. On the other hand, high-throughput screening instrumentation made the entire screen possible to set up, run, and evaluate in a reasonable time frame.
2. Scrambled siRNA is used as negative control siRNA to assess non-specific toxicity associated with siRNA, transfecting reagents and/or procedural steps. One can design or order it through different vendors. Negative control siRNA should have a chemical composition/size/architecture similar to the siRNA library members (e.g., a short 21 b.p. polynucleotide or long 27 b.p. DICER-substrate polynucleotide) that do not target any known transcript. Manufacturer-supplied scrambled siRNA should be also confirmed not to display any activity; despite best efforts, we still see some level of activity in separate assays when we use scrambled siRNAs. Keeping a low concentration/dose of siRNA is an effective way to minimize non-specific effects. Positive control siRNA, where strong effi-

cacy of the siRNA is well established, assures the accuracy of collective procedures during the screen, including transfection, incubation, and endpoint assays. Therefore, careful selection of positive control is important. We recommend finding a proper positive control for particular cell line before screening. Using different positive control with different siRNA libraries is recommended. Since our aim is to find out the best targets, which cause breast cancer cell death, we use siRNA-silencing CDC-20 (cell-cycle protein) as positive control, which had already shown significant cell death of breast cancer cells.

3. The siRNAs used here are not chemically modified, since such a modification makes library costs significantly higher, so that the screening will require an effective transfection reagent for intracellular delivery of siRNA. We rely on lipid-grafted low-molecular-weight (MW 1200 Da) polyethylenimines (PEIs) obtained from RJH Biosciences Inc., Edmonton, Canada, to undertake siRNA transfections in this protocol, while we found such PEIs to display an optimal efficacy over toxicity for a broad array of cells. Other transfection reagents could also be used in library screens. However, the choice of the polymer in our screens is facilitated by the availability of several polymeric analogues that allows us to match the performance of transfection reagent to the features of the cells, so that improved transfection efficiencies are obtained compared to generic transfection reagents commercially available [5, 15]. Polymeric delivery agent used for transfection is nontoxic and relatively easy to prepare. Unlike more cytotoxic liposomal reagents, the lack of toxicity by polymeric agents provides a clear advantage by maintaining normal physiology of the cells during testing. The availability of several analogues of the polymeric transfection reagents also enhances chances of success in specific cell types by employing an analogue with highest transfection efficiency in that cell type.

4. The protocol we developed is semiautomated and optimized for triple-negative breast cancer cells lines and an apoptosis-related siRNA library which contains 446 siRNAs related to apoptosis. For each targeted gene, a pool of four non-overlapping siRNAs were included in the treatment. Some of the steps will need to be modified while adopting to other cell types and siRNA libraries. In order to optimize this protocol for other cell lines, thorough testing of each individual step with negative and positive controls is recommended. While setting up large screens, advance planning and detailed scheduling are crucial for the success of the screening. Required materials, reagents, and equipment need to be checked beforehand. Listing of every required reagents and planning for extra

amount (overhead of 10%) is recommended. In this protocol, siRNA library screens are used to identify complementary pairs of therapeutic agents for cancer gene therapy. The synergism can be explored between the library members and a particular siRNA, a conventional chemotherapy drug or a protein drug used in cancer therapy. In order to identify complementary therapeutic pairs, two parallel siRNA library screenings are performed; in one screen, cells are treated with the library members alone (i.e., without a co-treatment), and, in a second screen, cells are treated with the combination of library members and a desired agent (e.g., TRAIL or a specific siRNA).

5. Cells should be thawed as fast as possible. While keeping in water bath, continuous shaking of the cell-containing cryotube by hand is recommended. Cells should be checked every half minute and kept in water bath until small portion of ice remains.
6. Centrifugal force and time should be optimized based on the size of the cells.
7. Cell culture condition such as metabolic activity, growth rate, and cell cycle is crucial for the transfection efficiency [16–18]. Cells prior to seeding should never reach the confluence state "plateau" phase, which may lower the metabolic activity in subsequent generations. Over-confluent cells will start to die and/or enter into senescence, which had reduced metabolic activity. This will substantially affect transfection efficiency. Beyond 80–90% of confluence, MDA-MB-231 cell becomes round and detaches from the flask. Therefore, cells should never grow more than 5–7 days. Cells with higher passage numbers (~20 to 25) are less metabolically active and should be avoided, and some passage (e.g., at least twice) should be allowed for frozen cells before employing in screening.
8. Cells should be maintained in several flasks such that sufficient cells are generated for seeding into 32 plates (96 well) as described in this Protocol. To avoid experimental variation, those flasks should be prepared from the cells of same batch and passage. This protocol was optimized for 96-well plate format, which was considered an optimum scale to screen large numbers of samples at a reasonable cost (i.e., small enough for reagent amounts) and reproducibility (i.e., sufficient cell numbers). However, 384-well culture plate can also be used. The number of cells and medium per well should be adjusted accordingly.
9. Pipette slowly and avoid aspirating air to prevent the bubble formation in the suspension.
10. Total cell suspension required for the entire plate should be prepared in a single container (500 mL bottle). Preparing in

separate container may result in variation in cell numbers between plates. Always calculate for 110 wells instead of 96 well, and prepare extra cell suspension. The cell culture condition and seeding density significantly influence the transfection efficiency and functional outcome of any siRNA library screening. The number of cells per well should be considered based on the size of well and type of cells and their growth speed. Cells should be seeded at such a density that the sham (non-treatment) group should grow exponentially but not to reach the over-confluence and plateau phase at the day of analysis. In addition, toxicity and transfection efficiency of polymer/siRNA complexes are closely related to cell density. As an example, polymer and siRNA concentrations could be more than optimal and generate non-specific effects if the cell density is low, as compared with confluent cell cultures. Efficiency of siRNA may be reduced in high cell density. Therefore, optimization of cell density to achieve optimum transfection without inducing any toxicity is crucial. Since cell seeding density does not necessarily translate to attached cell density, optimization is recommended based on cell culture condition, speed of cell growth, handling process, and passage number of the cell line. Sufficient dynamic range should be available to detect functional effects in both positive and negative directions. MDA-MB-231 cell is a fast-growing breast cancer cell; therefore, 5000 cells per well is optimal for 96-well plate in our case.

11. Programming of the robot should be done prior to cell suspension preparation. Label 96-well plate beforehand. Wipe workstation with 70% ethanol. Turn on lamina flow hood at least 15 min before experiment.
12. Avoid moving anything over cell suspension container. Put the lid as soon as seeding is complete to avoid any possible contamination.
13. Always prepare 10% extra volume in order to compensate loss during pipetting.
14. siRNA can be aliquated into mixing plate and stored at −20 °C beforehand to reduce work load at the day of transfection. It should be sealed tightly with aluminum plate sealer before storing and should be centrifuged before using. Each of the mixing plates was designed in such a way that positive and negative controls are accommodated. *See* **Note 2** for details on positive and negative control. Positive and negative control should be included in all plates. Needless to say, concentration and cell exposure time of both positive and negative control siRNAs should be identical to the siRNA library members.

15. Polymer solution should be prepared fresh. Optimization of transfection conditions (e.g., cell density, polymer and siRNA concentration, polymer to siRNA ratio, etc.) in particular cell culture plates that is used during screening is crucial. We found polymer: siRNA weight ratio of 6 optimal for siRNA transfection in MDA-MB-231 cells. This can vary depending on the type of cells and polymers or transfecting reagents. *See* **Note 3** for details.
16. Pipetting should be done gently. While mixing, pipette only fraction of mixture and discharge slowly. Repeat these processes at least five times. Avoid aspirating air, which creates bubbles.
17. This type of library screening contains more than one siRNA mixing plates. Hence, in order to provide same incubation time with all the siRNAs, it is recommended to have an interval of 5 min between two plates.
18. Before adding to cells, check for air bubbles in each well. If any air bubble is present, wait until it bursts or use pipette tips to burst it. The speed of complex mixing is critical to avoid any air bubble formation. Centrifugation of the complexes is not recommended. The triplicate samples were placed in three separate plates, rather than added consecutively in the same plate, to improve reproducibility and confidence in the obtained results.
19. We aim to reveal the complementary therapeutic targets of TRAIL in the library screening. TRAIL induces apoptosis in MDA-MB-231 cells. Since the effect of the TRAIL was more after 48 h than 24 h, we add TRAIL after 24 h of siRNA transfection, which allows total 48 h of incubation with TRAIL. For other drugs, time of treatment may vary and should be optimized before library screening. TRAIL protein loses its activity with repeated freeze-thaw cycles. Hence, it should be aliquoted into small volumes and stored at −80 °C according to manufacturer's instruction. Dilute TRAIL into complete medium just before adding to cells. Concentration of drugs used must be predetermined. Here, we used final concentration of 5 ng/mL of TRAIL. This will vary depending on drug and cell types. Always prepare an extra (around 10%) drug to compensate the pipetting error.
20. The MTT assay was used for assessing cellular growth, which provides immediate results without further analysis, optimal signal-to-noise ratios, and, more importantly, a read-out that is directly related to the desired clinical outcome in this therapeutic model. MTT must be dissolved in HBSS completely by vortexing and protected from light. Soluble versions (e.g., XTT) of the dye could be used to eliminate the organic solvent (DMSO) in the processing.

21. To synchronize the incubation time with MTT reagent, MTT reagent was added to each plate in 10 min intervals. In this way, sufficient time is provided to process and read the plate before the next plate is processed. This way addition of MTT solution and reading plate can be done parallel, and total incubation time for each plate would be the same. Incubation time with MTT reagent depends on the number of cells and metabolic activity of the cells and should be optimized depending on cell type used in screening.
22. If media is aspirated, then care must be taken not to touch the cells, and vacuum force should not be strong so that cells are lost. Alternatively, upside-down the cell culture plate to dump media, and gently tap on tissue paper to remove media completely. Presence of even small amount of media could affect the absorbance.
23. Check each well and make sure crystals are completely dissolved. If the crystals are not dissolved completely, it will give false results. Shake the plate in plate shaker for 30 s if undissolved crystals are visible.
24. Plate reader should be turned on at least 10 min before reading plate. The O.D. measured corresponds to the total metabolic activity (mitochondrial dehydrogenase activity) in the well and can be used as a measure of cell numbers. Cell growth by the treated cells was expressed as a relative percentage of the cells incubated with medium only (no treatment). It should be remembered that some agents can alter metabolic state of the cells (hence MTT signal) without affecting the cell numbers (growth), so that the possibility of false hits due to this complication should be considered.
25. Once hits were identified, multiple criteria should be utilized to prioritize the hits for validation. Druggability is one criteria to select if the aim is to find out the hits, which would represent the candidate drug targets. Function(s) of the identified targets can be obtained from the literature review if they are well known. In addition, different bioinformatics tools are available which can help to build functional network of hits and their interaction with other genes. Considering the expression and or mutational status of the hits focusing on cancer type in question is also important. Some databases such as the Cancer Genome Atlas are available for this analysis. In addition, it is not advisable to select the hits based on the ranking from the investigated functional outcome (i.e., growth inhibition). Hits with weaker performance but strong association to the disease of interest pathway should always be prioritized. It should be kept in mind that the performance of the identified siRNAs can always be increased by optimizing siRNA sequences and transfection conditions. However, we advise to

consider the hits with strong performance even though supporting evidences in disease of interest is missing. This may lead to identify the novel biomarker and to unforeseen insights into the biological pathway associated with disease.

26. Successful outcomes from siRNA library screening greatly depend on the optimization of the protocol. Experimental artifacts may be further exasperated from the use of high-throughput screening which can be minimized by optimizing each step in overall protocol. Since pooled siRNAs were used in this protocol, there is always the possibility of non-specific effects arising from individual siRNAs in the cocktail. At the same time, using pooled siRNA reduces the overall cost of siRNA library screening and increases the throughput of single screen. Finally, validation studies with independent siRNAs are needed; one can use the same source of siRNA for this purpose (e.g., from the same manufacturer of libraries) or prepare new siRNAs with different sequences against the same target. Either approach should lead to equivalent outcomes. The validation of identified targets with complementary assays is also accordingly required. *See* ref. 5 for follow-up validation studies in detail for this particular study.

27. Quality control throughout the experiment is very important. Performance of positive and negative controls throughout plates, analysis of standard deviation and *p*-value derived from the triplicates, and distribution of viability across the entire library and within plates are some parameters to assess the quality of outcomes. Exclusion of the outermost wells of each plate is also recommended to avoid edge effects (especially important if there is high evaporation from plates). For a successful screening, cells should be checked under microscope, while screen is in progress to uncover potential technical problems such as edge effect, plate-to-plate variability, or contamination. If the edge effect and other technical problems are obvious within a plate, then *B* score instead of *z*-score can be used to exclude the wells or row/columns affected by such technical problems. *B* score is relatively robust to outliers and can be calculated using open-source BioConductor bioinformatics software [19].

28. This protocol was developed to identify two types of complementary therapeutic pairs, one involving siRNA-siRNA combinations and one involving siRNA-drug combinations, for cancer therapy using focused siRNA libraries in triple-negative breast cancer cells (e.g., MDA-MB-231). Using this protocol, we identified synergistic combinations of therapeutic agents for protein-based anticancer drug TRAIL (Fig. 3). Optimization of experimental parameters such as cell density, concentration of therapeutic agents, and composition of com-

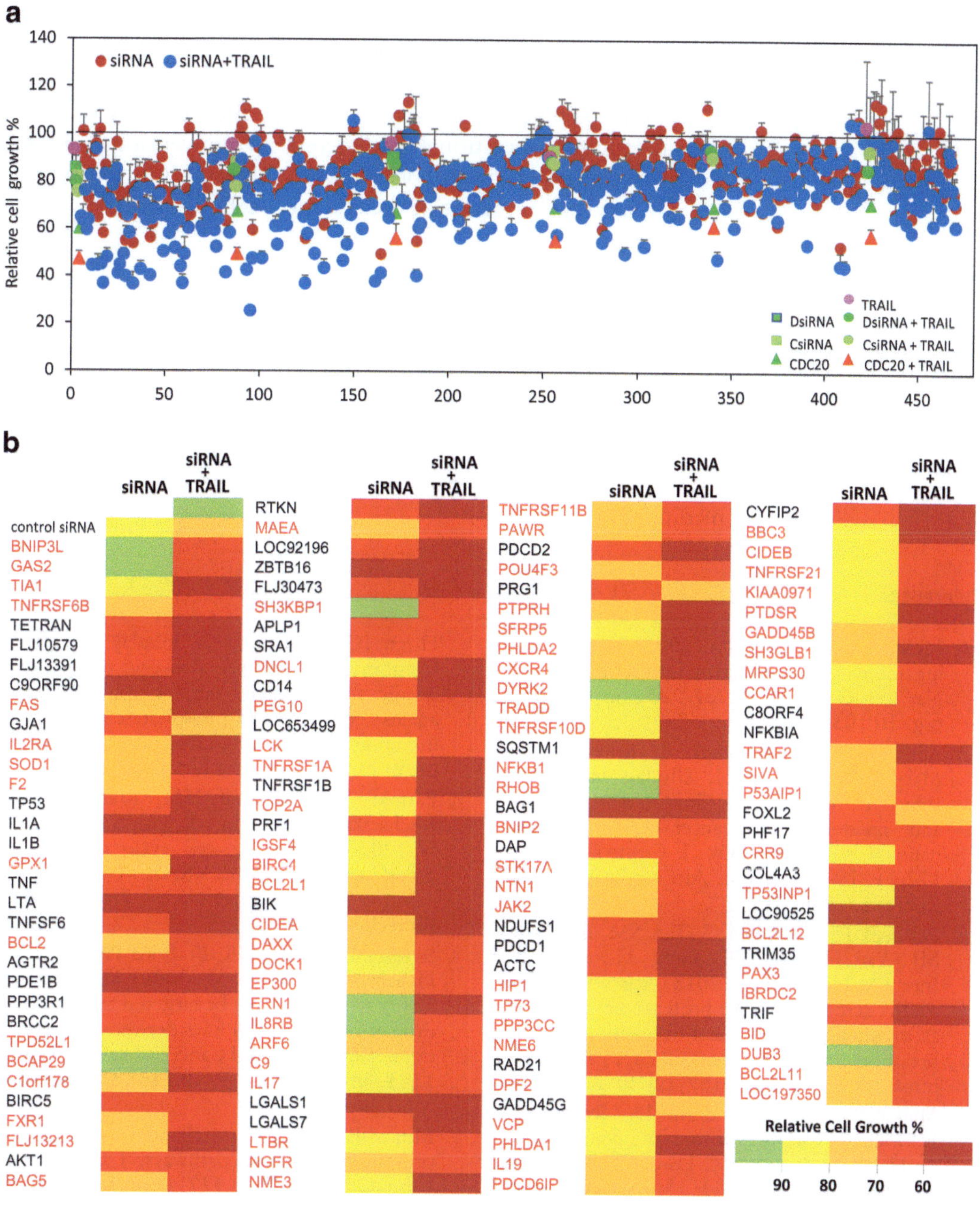

Fig. 3 (**a**) Human apoptosis siRNA library screen in MDA-MB-231 cells without and with TRAIL (5.0 ng/mL) treatment. The relative cell growth for treated cells was calculated as a percentage of cell growth of non-treated group. Final concentration of siRNA used for cell treatment was 30 nM. CDC-20 siRNA was used as positive control, and two negative control siRNA were used: DsiRNA (27-mer) and CsiRNA (21-mer). CDC-20 siRNA is 27-mer DICER-substrate polynucleotide; therefore, DsiRNA is used as its control. siRNAs from library are 21-mer polynucleotide; therefore, CsiRNA is used as its control. (**b**) Heat map shows the siRNAs that induced significant cell death (relative cell growth <70%) in MDA-MB-231 cells (without or with TRAIL). Many siRNAs, including BCL2L12, SOD1, BCL2L1, FLJ13391, NTN1, and FLJ13213, showed significant cell death in the presence of TRAIL. Figures are adapted from ref. 5

plexes (in particular polymer/siRNA ratio) are critical factors in performing this type of protocol. In siRNA-TRAIL combination study, post-transfection time for the addition of drug should be properly selected to get the optimum outcome. By fine-tuning operational parameters of conventional siRNA library screening protocols, we were able to identify promising siRNAs that can sensitize breast cancer cells to a drug (TRAIL) therapy.

Acknowledgments

B.T. was supported by Alberta Innovates Graduate Studentship. The authors RBKC and HU are the founder and shareholders in RJH Biosciences Inc. that are developing the lipopolymers for medical applications.

References

1. McManus MT, Sharp PA (2002) Gene silencing in mammals by small interfering RNAs. Nat Rev Genet 3(10):737–747. https://doi.org/10.1038/nrg908
2. Aliabadi HM, Maranchuk R, Kucharski C, Mahdipoor P, Hugh J, Uludag H (2013) Effective response of doxorubicin-sensitive and -resistant breast cancer cells to combinational siRNA therapy. J Control Release 172(1):219–228. https://doi.org/10.1016/j.jconrel.2013.08.012
3. Petrocca F, Altschuler G, Tan SM, Mendillo ML, Yan H, Jerry DJ, Kung AL, Hide W, Ince TA, Lieberman J (2013) A genome-wide siRNA screen identifies proteasome addiction as a vulnerability of basal-like triple-negative breast cancer cells. Cancer Cell 24(2):182–196. https://doi.org/10.1016/j.ccr.2013.07.008
4. Mendes-Pereira AM, Sims D, Dexter T, Fenwick K, Assiotis I, Kozarewa I, Mitsopoulos C, Hakas J, Zvelebil M, Lord CJ, Ashworth A (2012) Genome-wide functional screen identifies a compendium of genes affecting sensitivity to tamoxifen. Proc Natl Acad Sci U S A 109(8):2730–2735. https://doi.org/10.1073/pnas.1018872108
5. Thapa B, Bahadur Kc R, Uludag H (2018) Novel targets for sensitizing breast cancer cells to TRAIL-induced apoptosis with siRNA delivery. Int J Cancer 142(3):597–606. https://doi.org/10.1002/ijc.31079
6. Parmar MB, Meenakshi Sundaram DN, Kaur H, KC RB, Maranchuk R, Aliabadi HM, Hugh JC, Löbenberg R, Uludağ H (2018) Combinational siRNA delivery using hyaluronic acid modified amphiphilic polyplexes against cell cycle and phosphatase proteins to inhibit growth and migration of triple-negative breast cancer cells. Acta Biomater 66:294–309
7. Parmar MB, Aliabadi HM, Mahdipoor P, Kucharski C, Maranchuk R, Hugh JC, Uludağ H (2015) Targeting cell cycle proteins in breast cancer cells with siRNA by using lipid-substituted polyethylenimines. Front Bioeng Biotechnol 3:14
8. Smalley KS, Haass NK, Brafford PA, Lioni M, Flaherty KT, Herlyn M (2006) Multiple signaling pathways must be targeted to overcome drug resistance in cell lines derived from melanoma metastases. Mol Cancer Ther 5(5):1136–1144. https://doi.org/10.1158/1535-7163.MCT-06-0084
9. Muller R (2004) Crosstalk of oncogenic and prostanoid signaling pathways. J Cancer Res Clin Oncol 130(8):429–444. https://doi.org/10.1007/s00432-004-0570-y
10. K CR, Thapa B, Valencia-Serna J, Aliabadi HM, Uludag H (2017) Nucleic acid combinations: a new frontier for cancer treatment. J Control Release 256:153–169. https://doi.org/10.1016/j.jconrel.2017.04.029
11. Dancey JE, Chen HX (2006) Strategies for optimizing combinations of molecularly targeted anticancer agents. Nat Rev Drug Discov 5(8):649–659. https://doi.org/10.1038/nrd2089

12. Jia J, Zhu F, Ma X, Cao Z, Cao ZW, Li Y, Li YX, Chen YZ (2009) Mechanisms of drug combinations: interaction and network perspectives. Nat Rev Drug Discov 8(2):111–128. https://doi.org/10.1038/nrd2683
13. Kaelin WG Jr (2005) The concept of synthetic lethality in the context of anticancer therapy. Nat Rev Cancer 5(9):689–698. https://doi.org/10.1038/nrc1691
14. Hsu CY, Uludag H (2012) A simple and rapid nonviral approach to efficiently transfect primary tissue-derived cells using polyethylenimine. Nat Protoc 7(5):935–945. https://doi.org/10.1038/nprot.2012.038
15. Meenakshi Sundaram DN, Kucharski C, Parmar MB, Kc RB, Uludag H (2017) Polymeric delivery of siRNA against Integrin-beta1 (CD29) to reduce attachment and migration of breast cancer cells. Macromol Biosci 17(6). https://doi.org/10.1002/mabi.201600430
16. Brunner S, Sauer T, Carotta S, Cotten M, Saltik M, Wagner E (2000) Cell cycle dependence of gene transfer by lipoplex, polyplex and recombinant adenovirus. Gene Ther 7(5):401–407. https://doi.org/10.1038/sj.gt.3301102
17. Izumisawa T, Hattori Y, Date M, Toma K, Maitani Y (2011) Cell line-dependent internalization pathways determine DNA transfection efficiency of decaarginine-PEG-lipid. Int J Pharm 404(1–2):264–270. https://doi.org/10.1016/j.ijpharm.2010.11.017
18. Douglas KL, Piccirillo CA, Tabrizian M (2008) Cell line-dependent internalization pathways and intracellular trafficking determine transfection efficiency of nanoparticle vectors. Eur J Pharm Biopharm 68(3):676–687. https://doi.org/10.1016/j.ejpb.2007.09.002
19. Birmingham A, Selfors LM, Forster T, Wrobel D, Kennedy CJ, Shanks E, Santoyo-Lopez J, Dunican DJ, Long A, Kelleher D, Smith Q, Beijersbergen RL, Ghazal P, Shamu CE (2009) Statistical methods for analysis of high-throughput RNA interference screens. Nat Methods 6(8):569–575. https://doi.org/10.1038/nmeth.1351

Chapter 2

In Vitro Evaluation of Candidate Gene Targets for Cancer Therapy

Xing Fei Tan, Wei Xuan Teo, and George W. Yip

Abstract

Discovery and development of gene targets for cancer therapeutics are lengthy and highly costly processes. Identification and evaluation of candidate gene targets are of fundamental importance. RNA interference allows candidate genes to be specifically and effectively knocked down in cancer cells. This tool can be easily incorporated into a loss-of-function approach in the initial evaluation of candidate gene targets for cancer treatment prior to moving on to animal studies and clinical trials. This chapter describes a relatively simple and straightforward protocol that makes use of small interfering RNA to achieve knockdown of the candidate gene target and to evaluate the resultant effects on four aspects of cancer cell behavior: migration, invasion, proliferation, and adhesion.

Key words siRNA, Breast cancer, Migration, Invasion, Proliferation, Adhesion

1 Introduction

In recent years, RNA interference (RNAi) has proven to be an important and effective tool in cancer research. Through the introduction of small interfering RNA (siRNA) into cultured cells, the expression of specifically targeted proteins can be rapidly and efficiently suppressed via degradation of the corresponding mRNA transcripts [1]. This enables researchers to investigate the functional roles of different protein molecules in cancer cells, identify potential tumor biomarkers, and evaluate candidate therapeutic targets [2–7]. Thus, a loss-of-function approach using cancer cell lines provides a rapid and straightforward means for the initial evaluation of candidate gene targets for cancer therapy before moving on to animal studies and subsequent clinical trials. This chapter provides a protocol for using siRNA to achieve knockdown of the

Xing Fei Tan and Wei Xuan Teo contributed equally to this work.

Lekha Dinesh Kumar (ed.), *RNA Interference and Cancer Therapy: Methods and Protocols*, Methods in Molecular Biology, vol. 1974, https://doi.org/10.1007/978-1-4939-9220-1_2, © Springer Science+Business Media, LLC, part of Springer Nature 2019

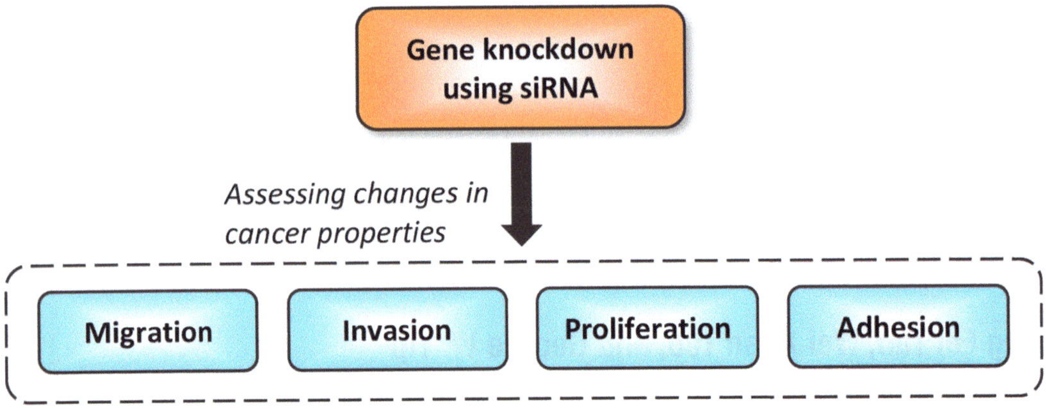

Fig. 1 Overview of the siRNA-mediated loss-of-function workflow to evaluate candidate gene targets for cancer therapy

candidate gene target and subsequent assessment of the effects on four aspects of cancer cell behavior: migration, invasion, proliferation, and adhesion (Fig. 1).

The migration assay can be used to measure the ability of breast carcinoma cells to move across a permeable Transwell membrane insert along a chemotactic gradient, as shown in Fig. 2 [8]. By coating the surface of the membrane insert with Matrigel matrix (which serves as a reconstituted basement membrane), Transwell chambers can be used to study the invasive characteristics of breast cancer cells [9]. Uncontrolled cell proliferation is one of the hallmarks of cancer [10, 11]. The MTS assay makes use of the 3-(4,5-dimethylthiazol-2-yl)-5-(3-carboxymethoxyphenyl)-2-(4-sulfophenyl)-2H-tetrazolium compound to measure metabolic activity of living cells, which serves as a proxy indicator of cell number and cell proliferation [12]. Cancer metastasis requires tumor cells to detach themselves from the primary site of the malignant growth and to attach themselves to the metastatic site. The cell adhesion assay can be used to examine the ability of breast cancer cells to adhere to different components of the extracellular matrix [13].

2 Materials

The protocol in this chapter has been optimized for gene target evaluation in human breast carcinoma MDA-MB-231 cells. For studies using other cancer cell lines, further optimization of the various experimental parameters may be required. These may include culture conditions, cell seeding density, siRNA concentration, and reagents for transfection.

2.1 Human Breast Cancer Cell Culture

1. Human breast carcinoma cells MDA-MB-231 (American Type Culture Collection, Manassas, VA, USA).

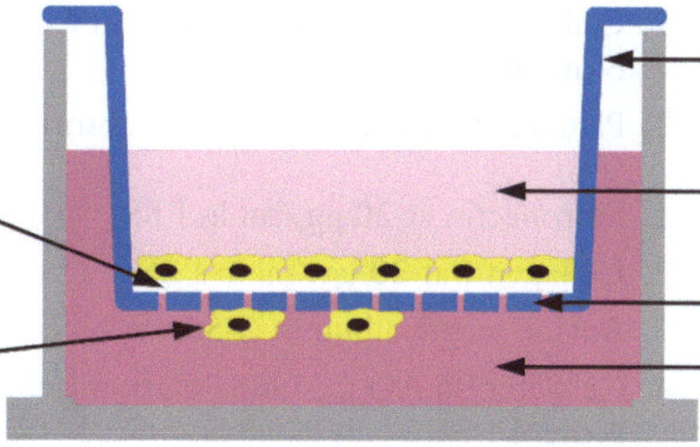

Fig. 2 Diagrammatic representation of a Transwell chamber used in a migration/invasion assay. Breast cancer cells were seeded onto the upper surface of the microporous membrane and allowed to pass through the micropores to the undersurface of the membrane. A chemotactic gradient was established across the membrane by putting culture medium containing a higher concentration of the chemoattractant (such as FBS) in the outer chamber and culture medium with no/lower concentration of the chemoattractant in the inner chamber. For the invasion assay, the upper surface of the microporous membrane was coated with a layer of Matrigel matrix

2. Complete medium consisting of Roswell Park Memorial Institute (RPMI) 1640 medium supplemented with 10% fetal bovine serum (FBS).
3. Cell culture incubator to provide a humidified environment maintained at 37 °C with 5% CO_2.
4. Cell culture grade 0.05% trypsin-0.02% ethylenediaminetetraacetic acid (EDTA) solution.

2.2 Transfection with siRNA

1. Silencer Select siRNA at 1 μM concentration (*see* **Note 1**).
2. Silencer Select nontargeting negative control siRNA at 1 μM concentration.
3. Oligofectamine Transfection Reagent.
4. Opti-MEM I Reduced Serum Medium.

2.3 Transwell Migration and Invasion Assays

1. Transwell polycarbonate membrane cell culture inserts: 6.5 mm diameter with pore size of 8.0 μm (*see* **Note 2**).
2. Matrigel matrix on ice (*see* **Note 3**).
3. Pipette tips and tubes pre-chilled to 4 °C (*see* **Note 3**).
4. Phosphate buffered saline (PBS) at 4 °C.
5. Methanol.
6. Crystal violet 0.5% (w/v) in 10% methanol.
7. Cotton buds.
8. Stereomicroscope with camera.

2.4 Proliferation Assay (MTS Assay)

1. CellTiter 96 AQ_{ueous} One Solution Cell Proliferation Assay (*see* **Note 4**).
2. Plate reader capable of reading absorbance at 490 nm.

2.5 Adhesion Assay

1. Fibronectin at 20 μg/ml in PBS.
2. Collagen I at 20 μg/ml in PBS.
3. Bovine serum albumin (BSA) at 1% concentration in PBS.
4. CellTiter 96 AQ_{ueous} One Solution Cell Proliferation Assay (*see* **Note 4**).
5. Plater reader capable of reading absorbance at 490 nm.

3 Methods

All steps are to be carried out in a biological safety cabinet using aseptic technique unless otherwise indicated. There should be strict adherence to biosafety and waste disposal regulations.

3.1 Transfection with siRNA

1. Seed MDA-MB-231 cells at a density of 1×10^5 cells per well in a six-well plate. Leave the cells in a cell culture incubator at 37 °C overnight (*see* **Note 5**).
2. Add 5 μl of 1 μM siRNA to 105 μl of Opti-MEM I in a sterile microcentrifuge tube labeled "A." Add 10 μl of Oligofectamine to 100 μl of Opti-MEM I in another tube labeled "B" (*see* **Note 6**). Allow the two tubes to stand for 5 min at room temperature.
3. Transfer the contents of tube A to tube B to form the transfection mixture. Gently mix the contents. Incubate the transfection mixture at room temperature for 20 min.
4. Remove the complete medium from the cultured cells.
5. Add 780 μl of Opti-MEM I to each well of MDA-MB-231 cells. Thereafter, add 220 μl of transfection mixture in a dropwise manner to the Opti-MEM I to form the transfection medium.
6. Rock the plate to mix the transfection medium. Allow the transfection to proceed in a cell culture incubator for 8 h.
7. Add 500 μl of 30% FBS to each well. Rock gently to mix, and continue the incubation for an additional 16 h.
8. Remove the transfection medium and replace it with fresh complete medium. Continue the culture until the cells are harvested for subsequent experiments (*see* **Note 7**).

3.2 Transwell Migration and Invasion Assays

1. Perform transfection with siRNA as described in Subheading 3.1.
2. On the following day, thaw an aliquot of Matrigel matrix in a 4 °C refrigerator overnight for the invasion assay (*see* **Note 3**). For the migration assay, hydrate the Transwell inserts by putting them into a 24-well plate containing 600 μl of serum-free medium per well (outer chamber). Add 200 μl of serum-free medium to each Transwell insert (inner chamber; Fig. 2). Allow hydration to take place overnight in a 4 °C refrigerator.
3. Dilute the Matrigel matrix to a concentration of 0.3 mg/ml using ice-cold PBS for the invasion assay. Add 100 μl of the diluted matrix to the upper surface of the polycarbonate membrane of each Transwell insert (*see* **Note 8**). Allow the Matrigel matrix to gel for 2 h at 37 °C in a cell culture incubator. The hydrated Transwell inserts for the migration assay should also be transferred to the incubator at this time.
4. Prepare a suspension of MDA-MB-231 cells from **step 1** in culture medium supplemented with 5% FBS. For the invasion assay, 3×10^4 cells are suspended in 200 μl of medium, whereas for the migration assay, 5×10^4 cells are suspended in 200 μl of medium (*see* **Note 9**).
5. Apply the cell suspension dropwise to the Matrigel matrix-coated insert for the invasion assay (*see* **Note 10**). Add 600 μl of culture medium containing 20% FBS to the well of the culture plate via the opening at the side of the insert (*see* **Note 11**). In the case of the migration assay, remove the non-coated Transwell insert from the 24-well plate. Aspirate residual fluid from the inside compartment of the insert (inner chamber). Transfer the Transwell chamber to a new culture plate containing 600 μl of culture medium with 20% FBS in the well (outer chamber). Add the cell suspension dropwise to the inner chamber.
6. Transfer the culture plate and Transwell inserts to the cell culture incubator for 24 h (*see* **Note 12**).
7. Aspirate the culture medium from the inner chamber, and rinse the outer surface of the Transwell insert with PBS.
8. Fix the cells in methanol for 15 min at room temperature.
9. Aspirate the methanol from the Transwell chamber, and rinse the outer surface with PBS.
10. Stain the cells with 0.5% crystal violet for 30 min at room temperature.
11. Aspirate the crystal violet from the Transwell chamber, and rinse the Transwell insert in a beaker of water.
12. Use a wet cotton bud to remove cells from the upper surface of the polycarbonate membrane by gentle twirling, taking care not to damage the membrane.

13. Rinse the Transwell insert in water again. Allow the insert to air dry at room temperature.
14. Use a stereomicroscope to examine the stained cells on the undersurface of the polycarbonate membrane. Take photomicrographs of five to ten randomly selected high-power fields. Count the cells in each field (*see* **Note 13**).

3.3 Proliferation Assay

1. Perform transfection with siRNA as described in Subheading 3.1 (*see* **Note 14**).
2. After 72 h, thaw an aliquot of the CellTiter 96 AQ_{ueous} One Solution Reagent. Dilute the reagent with complete medium at a 1:5 ratio (*see* **Note 15**).
3. Remove the culture medium from each well, and replace it with 1.8 ml of the diluted reagent per well (*see* **Note 16**).
4. Return the culture plate to the incubator for 1–4 h (*see* **Note 17**).
5. Measure the absorbance at 490 nm using a plate reader (*see* **Note 18**).

3.4 Adhesion Assay

1. Perform transfection with siRNA as described in Subheading 3.1.
2. Add 50 μl of 20 μg/ml fibronectin or 20 μg/ml collagen I to each well of a 96-well culture plate. Gently tap the side of the plate to ensure that the solution covers the growth area of each well completely (*see* **Note 19**).
3. Leave the culture plate to stand at 4 °C for 24 h.
4. Aspirate the fibronectin or collagen I solution from the wells.
5. Wash each well with 100 μl of PBS.
6. Add 50 μl of 1% BSA to each well for 1 h at room temperature.
7. Remove the BSA solution from the wells.
8. Rinse each well with 100 μl of PBS.
9. Allow the wells to dry overnight in an incubator.
10. Prepare a suspension of MDA-MB-231 cells from **step 1** in culture medium. Add 3×10^4 cells in 100 μl of culture medium to each fibronectin-coated well or 1×10^5 cells in 100 μl of culture medium to each collagen I-coated well.
11. Allow the cells to adhere for 30–45 min in a cell culture incubator (*see* **Note 20**).
12. Remove the culture medium from the samples in the "washed" group, and rinse each well gently with 100 μl of PBS twice to remove nonadherent cells (Fig. 3). Thereafter, add 100 μl of culture medium to each well.

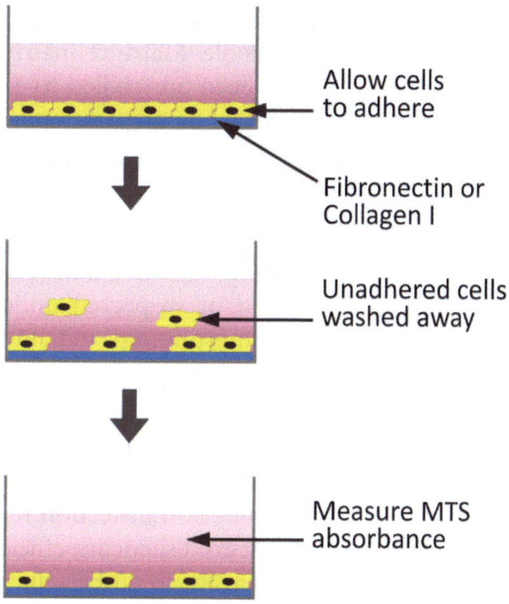

Fig. 3 Principle of the adhesion assay. Cancer cells were seeded into wells that were pre-coated with either collagen I or fibronectin and allowed to adhere. The non-adhered cells were then removed by gentle washing with PBS. The number of remaining adherent cells was quantified using the MTS assay

13. Add 20 μl of undiluted CellTiter 96 AQ_{ueous} One Solution Reagent to each well (*see* **Note 21**).
14. Return the culture plate to the incubator for 1–4 h, and then measure the absorbance at 490 nm as described in **steps 4** and **5** in Subheading 3.3.

4 Notes

1. One recommended approach to address concerns of off-target effects is to use at least two siRNA sequences that target different regions of the mRNA transcripts [14]. Diluted siRNA solutions should be aliquoted to minimize repeated freeze-thaw cycles.
2. Selection of the Transwell insert pore size will depend on the size and deformability of the cells used in the migration and invasion assays. Sufficient clearance is needed between the polycarbonate membrane and the bottom of the well of the culture plate to allow a chemoattractant gradient to be established (Fig. 2).
3. Matrigel matrix can be thawed overnight in a 4 °C refrigerator, aliquoted, and stored at −20 °C to avoid repeated freeze-thaw cycles. It is a liquid at 4 °C but begins to gel at 10 °C and

polymerizes rapidly between 22 °C and 35 °C. Thus, it is easier to handle Matrigel matrix using pipette tips and tubes that have been pre-chilled to 4 °C. Matrigel matrix should be kept on ice while preparation is made to coat the Transwell inserts.

4. Reagent is light sensitive.
5. Seeding density may need to be optimized for different cell lines in order to obtain good silencing efficiency. Before proceeding with the transfection on the following day, the cells should be inspected under a phase contrast microscope for their morphology and amount of cell death.
6. The stated volumes will produce sufficient transfection medium containing 5 nM siRNA for use in one well of a six-well plate. This can easily be scaled up based on the number of samples present in a given experiment. Determination of the optimal volumes of transfection reagent and siRNA needed to achieve the desired silencing efficiency for a chosen cell line and seeding density could be achieved by titration.
7. The silencing efficiency can be determined using quantitative polymerase chain reaction (PCR) to detect a change in mRNA transcript level or Western blot to measure a change in the amount of targeted protein. The effect of exogenous siRNA is transient; the knockdown efficiency decreases over time. Thus, it is useful to measure the silencing efficiency at the time point when cells will be harvested for downstream experiments. Using the protocol in this chapter, we have consistently achieved a knockdown efficiency of 70% or more for *GAPDH* in MDA-MB-231 cells on Day 3 of the experiment. Day 1 is defined as the day of adding transfection medium to the cultured cells.
8. Exercise precaution, and apply the Matrigel matrix rapidly before it starts to solidify to avoid unevenness of the gel layer, which could affect cellular invasion. If air bubbles are seen in the matrix, remove them by gently tapping the side of the chamber or burst them with a dry needle.
9. The optimal seeding density may need to be optimized, based on the migration and invasion properties of different types of cancer cells, the type of chemoattractant used, and the steepness of the chemoattractant gradient.
10. Applying the cell suspension dropwise to the center of the insert helps to reduce uneven seeding of cells (Fig. 4).
11. Make sure that no air bubbles are trapped under the insert.
12. Do not incubate for an extended period of time. The number of cells may increase due to mitotic division and be a confounding factor in analyzing the results of the migration and invasion assays.

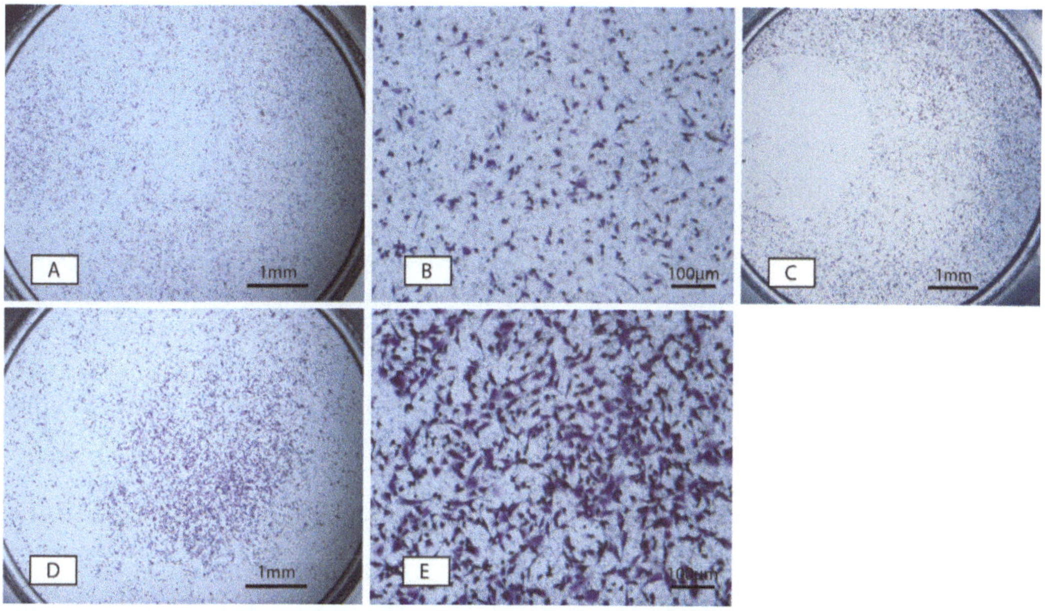

Fig. 4 Representative images of MDA-MB-231 breast cancer cells that were evenly seeded onto a Transwell membrane insert (**a**, **b**). The absence of cell clumping makes it easy to count the cells that have passed through the microporous membrane. In contrast, trapped air bubbles (**c**), uneven seeding (**d**), and/or cell clumping (**e**) may affect the quality and reliability of the data obtained from the assay

13. This can be done manually or with the help of image analysis software, such as the open source ImageJ program (available at https://imagej.nih.gov/ij/download.html) and Cell Counter plugin (https://imagej.nih.gov/ij/plugins/cell-counter.html).
14. If relevant, serum starvation to synchronize the cell cycle can be performed prior to starting the assay.
15. Depending on the doubling time of the cells, the proliferation assay can be done at an earlier or later time point. We find that the assay works well when the cells are 80–90% confluent. Do not let the cells become overcrowded as this will affect growth of the cells.
16. Set up an additional well, containing 1.8 ml of the diluted reagent without any cells, to serve as the blank. The absorbance value from this well represents the background absorbance.
17. If necessary, the plate can be covered in aluminum foil as the reagent is light sensitive.
18. The absorbance value increases with incubation time as a greater amount of formazan would be produced. Thus, if the absorbance value at the end of the first hour is low, the plate can be further incubated, and the absorbance measured again at the end of the second, third, or fourth hour.

19. Coat two wells for each sample in each experimental group, one for the "washed" group and the other for the corresponding "unwashed" group. The ratio of the number of cells in the "washed" group to the "unwashed" group provides a measure of cell adhesion.
20. The required incubation time may vary, depending on how rapidly the cells are able to adhere to the substrate used to coat the wells. In the case of MDA-MB-231 cells, their morphology changes from spherical to spindle shape once the cells have adhered.
21. Set up an addition well, containing 100 μl of culture medium and 20 μl of undiluted CellTiter 96 AQ_{ueous} One Solution Reagent without any cells, to serve as the blank.

Acknowledgment

We would like to thank S. L. Bay for her help in preparing the illustrations and C. X. Ng for contributing images for Fig. 3. Work in the laboratory of G. W. Yip is supported by Grants NMRC/CSA/0041/2012 and NMRC/CIRG/1436/2015 from the National Medical Research Council, Singapore.

References

1. Dana H, Chalbatani GM, Mahmoodzadeh H, Karimloo R, Rezaiean O, Moradzadeh A et al (2017) Molecular mechanisms and biological functions of siRNA. Int J Biomed Sci 13(2):48–57
2. Iravani O, Bay BH, Yip GW (2017) Silencing HS6ST3 inhibits growth and progression of breast cancer cells through suppressing IGF1R and inducing XAF1. Exp Cell Res 350(2):380–389
3. Schlabach MR, Luo J, Solimini NL, Hu G, Xu Q, Li MZ et al (2008) Cancer proliferation gene discovery through functional genomics. Science 319(5863):620–624
4. Ahmadzada T, Reid G, McKenzie DR (2018) Fundamentals of siRNA and miRNA therapeutics and a review of targeted nanoparticle delivery systems in breast cancer. Biophys Rev 10(1):69–86
5. Chen X, Mangala LS, Rodriguez-Aguayo C, Kong X, Lopez-Berestein G, Sood AK (2018) RNA interference-based therapy and its delivery systems. Cancer Metastasis Rev 37(1):107–124
6. Lai Y, Lim D, Tan PH, Leung TK, Yip GW, Bay BH (2010) Silencing the metallothionein-2A gene induces entosis in adherent MCF-7 breast cancer cells. Anat Rec (Hoboken) 293(10):1685–1691
7. Vijaya Kumar A, Salem Gassar E, Spillmann D, Stock C, Sen YP, Zhang T et al (2014) HS3ST2 modulates breast cancer cell invasiveness via MAP kinase- and Tcf4 (Tcf7l2)-dependent regulation of protease and cadherin expression. Int J Cancer 135(11):2579–2592
8. Roussos ET, Condeelis JS, Patsialou A (2011) Chemotaxis in cancer. Nat Rev Cancer 11(8):573–587
9. Hall DM, Brooks SA (2014) In vitro invasion assay using matrigel: a reconstituted basement membrane preparation. Methods Mol Biol 1070:1–11
10. Hanahan D, Weinberg RA (2011) Hallmarks of cancer: the next generation. Cell 144(5):646–674
11. Hanahan D, Weinberg RA (2000) The hallmarks of cancer. Cell 100(1):57–70
12. Cory AH, Owen TC, Barltrop JA, Cory JG (1991) Use of an aqueous soluble tetrazolium/formazan assay for cell growth assays in culture. Cancer Commun 3(7):207–212
13. Oskarsson T (2013) Extracellular matrix components in breast cancer progression and metastasis. Breast 22(Suppl 2):S66–S72
14. (2003) Whither RNAi? Nat Cell Biol 5(6):489–490

Chapter 3

siRNA-Based Drug Targeting Human Bcl-xL Against Cancers

Yoshifumi Takei

Abstract

The 2006 discovery of the process of RNA interference opened the door to application of this phenomenon for disease treatments. Short interfering RNA (siRNA) can be used to induce RNA interference, and this approach has generated much interest as a new type of nucleic acid-based drugs in humans and other mammals. However, despite the great potential of siRNA-based drugs in cancer therapy, some drawbacks of siRNAs, such as their instability in vivo and poor cellular uptake, remain unresolved. Here, we review the development of siRNA-based drugs targeting the human Bcl-xL gene, an anti-apoptotic factor overexpressed in many cancers, including prostate cancers. We also introduce a novel application of the biomaterial atelocollagen as a vehicle to functionally deliver tumor-specific siRNA molecules in nude mice. Here, we introduce an orthotopic tumor inoculation model in nude mice: the resulting orthotopic tumors more closely replicate the clinical conditions in humans—including the metastasis mode—than ectopic subcutaneously inoculated tumors.

Key words Short interfering RNA (siRNA), Cancer therapy, Bcl-xL, Anti-apoptotic factor, Prostate cancer, Atelocollagen, Anti-metastasis therapy

1 Introduction

RNA interference (RNAi) technologies are the most widely used in functional genomic studies [1–4]. The possibility of developing short interfering RNAs (siRNAs) for the treatment of numerous refractory diseases, such as viral infections, neurodegenerative disorders, and cancers, has been vigorously investigated [5, 6]. We developed an siRNA that potently inhibits the expression of human Bcl-xL [7]. Bcl-xL is an anti-apoptotic factor that is overexpressed in many cancers, including prostate cancers, to protect the cancer cells from apoptosis [8, 9] but which is expressed at much lower levels in normal tissues. Thus Bcl-xL is a suitable molecular target for the treatment of various cancers.

The largest obstacle to developing an siRNA-based drugs is the difficulty of delivering the siRNA molecule into the desired tissues/cells in vivo. In a previous work, we successfully developed a

Lekha Dinesh Kumar (ed.), *RNA Interference and Cancer Therapy: Methods and Protocols*, Methods in Molecular Biology, vol. 1974,
https://doi.org/10.1007/978-1-4939-9220-1_3,

systemic atelocollagen-mediated siRNA delivery method specifically for use in cancer therapy [10]. In this article, we introduce the method in detail.

2 Materials

2.1 siRNA

HPLC-purified grade siRNA is recommended (*see* **Note 1**). We used freeze-dried siRNA, which we reconstituted in RNase-free water (*see* **Note 2**) to prepare a stock solution (*see* **Notes 3** and **4**). The effect of the siRNA (RNAi effect) in the cell culture experiment (in vitro study) should be separately validated prior to the in vivo experiment. In this protocol, we used our originally designed siRNA against human Bcl-xL, which was described in our previous reports (ref. 7: Bcl-xL siRNA): 5′-GGUAUUGGUGAG UCGGAUCdTdT-3′ (sense), 5′-GAUCCGACUCACCAAUAC CdTdT-3′ (antisense). We also used a scrambled siRNA as a control. Bcl-xL siRNA-SCR, 5′-UAGGGGUUGCGACGUUUAG dTdT-3′ (sense), 5′-CUAAACGUCGCAACCCCUATdTdT-3′ (antisense). Bcl-xL siRNA-SCR had almost no effect on the target [7]. Our Bcl-xL siRNA almost perfectly knocked down the Bcl-xL gene in cells of the human adenocarcinoma cell line, PC-3, both in terms of the mRNA level via quantitative PCR and the protein level via specific ELISA [7]. A stabilized siRNA with chemical modifications could also be used if desired (*see* **Note 5**). An adequate aliquot of the siRNA stock to prevent freeze-thawing cycles should be maintained in a freezer (−80 °C). To monitor the delivery efficiency of the siRNA in vivo, Cy3-labeled siRNA can also be used.

2.2 Cell Culture

1. PC-3 cells and PC-3-Luc cells were maintained in Ham's F-12K (Kaighn's) (Invitrogen) with 10% fetal bovine serum (FBS).
2. Phosphate-buffered saline (PBS).
3. 0.25% Trypsin-EDTA in PBS.
4. Culture flasks or dishes and plastic pipettes.

2.3 Atelocollagen

We successfully developed an atelocollagen-mediated method for delivering tumor-specific siRNA molecules into the tumors of nude mice via a systemic administration route [7, 10]. Atelocollagen, which is commercially available from Koken Co., Ltd. (Tokyo), is derived from tropocollagen of calf dermal collagen (type I collagen). In order to prepare the atelocollagen, we removed telopeptides on both terminals via pepsin digestion to completely abolish the antigenicity [11–13]. Atelocollagen possesses basicity (positive charge) under physiological conditions, whereas siRNA and other nucleic acids have acidity (negative charge). Therefore, atelocollagen and siRNA are considered to form a complex by electrostatic binding. The nucleic acid complexed with atelocollagen can avoid degradation by nuclease(s) and can be efficiently transported in the bloodstream to tumor tissues and cells, where the siRNA/atelocollagen complex may be incorporated through endocytosis.

2.4 Animal Experiments

1. Male BALB/c athymic nude mice (9 weeks old; Japan SLC, Hamamatsu, Japan) were used.
2. Syringes with a 29-gauge needle (Terumo, Tokyo) were used for microinjecting the cancer cells into the prostate gland in nude mice. A disposable insulin syringe (Myjector 29G; Terumo Clinical Supply, Japan) with a fixed needle is convenient.
3. Somnopentyl (Kyoritsu Seiyaku Corp., Tokyo) was used for anesthetization.
4. Surgical silk sutures of size 5-0 were used (Alfresa Holdings, Tokyo).
5. Dissection and surgical instruments for use in mice are required.

3 Methods

3.1 Preparation of PC-3 Cells for Generating Inoculated Tumors in Nude Mice

1. Harvest PC-3 cells with Trypsin-EDTA in PBS when the culture is 80% confluent (*see* **Note 6**).
2. Resuspend the cells in Ham's F-12K (Kaighn's) without any FBS, and count the cell number.
3. Prepare the cell suspension in Ham's F-12K (Kaighn's) without any FBS (*see* **Note 7**).

3.2 Preparation of the PC-3 Orthotopic Tumor Model for Prostate in Nude Mice

1. Deeply anesthetize the nude mice (male BALB/c athymic nude mice, 9 weeks old) with Somnopentyl (5 mg/kg, i.p.) (*see* **Note 8**).
2. Expose the prostate gland (located underneath the bladder) after a lower midline abdominal incision.
3. Microinject the PC-3 cells (5×10^5 cells in 20 μl of F-12K medium without any serum) into the ventral lobe of the prostate (*see* **Note 9**).
4. Close the incision with 5-0 surgical silk sutures (*see* **Note 10**).
5. Place the mouse on a heat pad until awake.
6. Typical results are shown in Fig. 1.

3.3 Preparation of the Complex of siRNA and Atelocollagen

1. Mix an equal volume (e.g., 1.0 ml) of siRNA in phosphate-buffered saline (PBS) and atelocollagen solution (0.1%, e.g., 1.0 ml) at 4 °C (*see* **Note 11**).

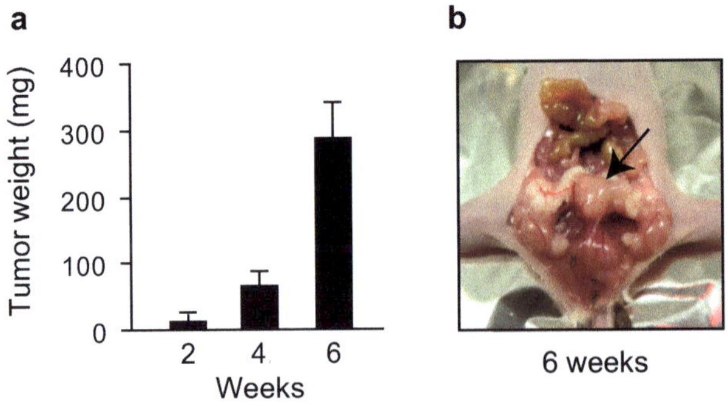

Fig. 1 PC-3 orthotopic tumor model in nude mice. (**a**) Tumor weight of PC-3 orthotopic tumors. PC-3 orthotopic tumors were harvested at 2, 4, and 6 weeks after the inoculation. The results are shown as the means ± SD ($n = 6$ tumors). (**b**) Typical macroscopic observation of a PC-3 orthotopic tumor (6 weeks after inoculation). Arrow: PC-3 orthotopic tumor

2. Incubate the mixture using a gentle rotator machine for 20 min in a cold room, and keep the complex at 4 °C until injection (*see* **Note 12**).

3.4 Tumor Therapy

1. Prepare nude mice with orthotopic PC-3 tumors on the prostate gland.
2. The start point of the therapy is 14 days after inoculation (Fig. 2a).
3. Deeply anesthetize the nude mice with Somnopentyl (5 mg/kg, i.p.).
4. Intravenously inject the complex of siRNA and atelocollagen into the tail veins of mice (siRNA dose, 50 μg/mouse) [Day 0].
5. Repeat the injection on Day 1 and Day 2 (Fig. 2a).
6. According to the injection schedule indicated in Fig. 2a, i.v. inject the complex a total of 12 times over the course of therapy.
7. The end point of the therapy is Day 28 (Fig. 2a).
8. On Day 28, sacrifice all of the mice treated, and remove and weigh the tumors.
9. Typical results are shown in Fig. 2b.

3.5 Evaluation of the RNAi Effect In Vivo

1. Prepare the tumors as indicated in the above section.
2. Lyse in lysis buffer for ELISA (Human Bcl-xL ELISA kit; R&D Systems).
3. For quantitative PCR, extract a total RNA. A TaqMan probe for human Bcl-xL (Probe ID, Hs00236329_m1) can be used. For the normalization assay, we used a TaqMan probe for

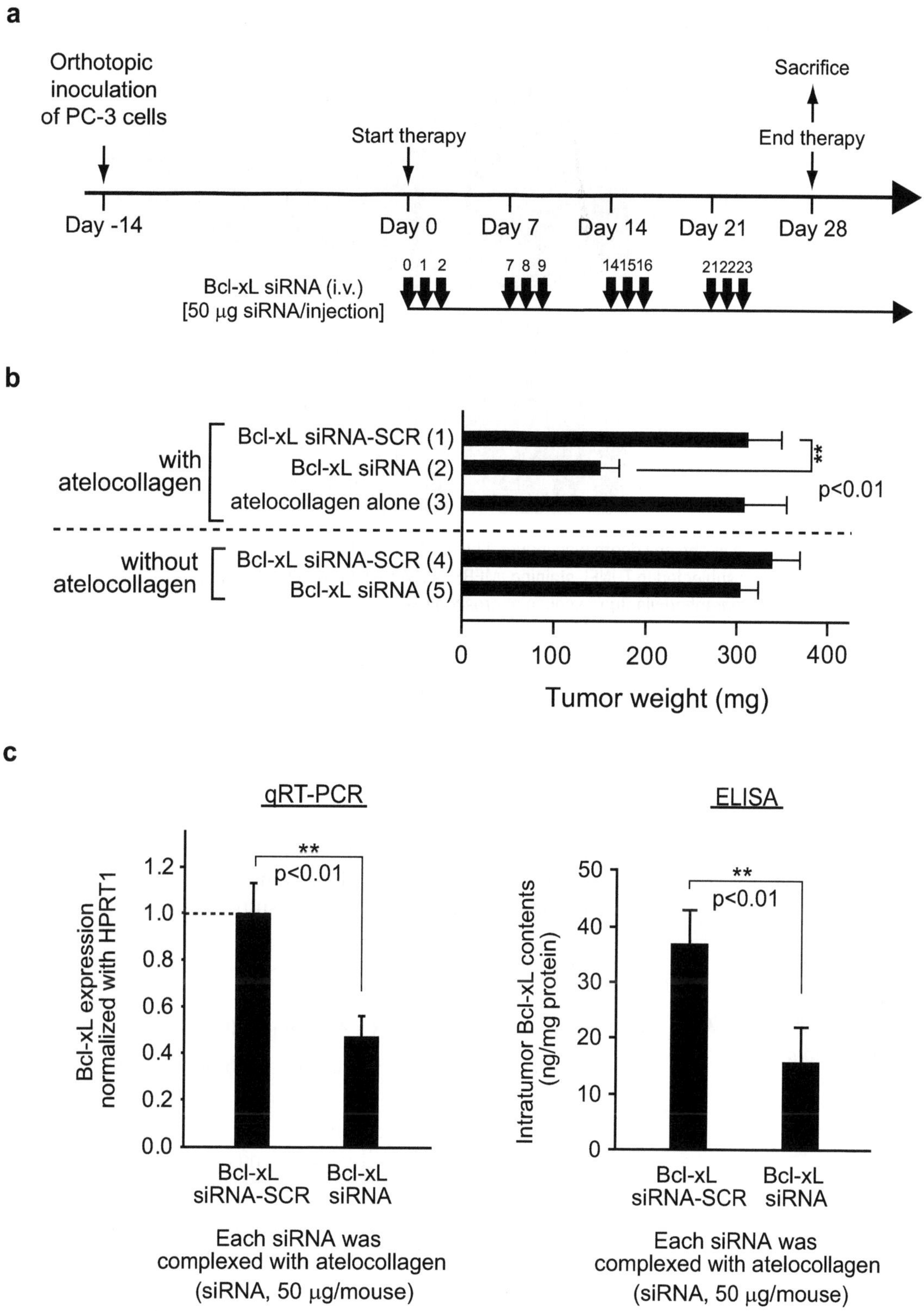

Fig. 2 Significant antitumor effect of the siRNA complexed with atelocollagen on PC-3 orthotopic tumors. (**a**) Experimental procedures. PC-3 cells were orthotopically inoculated into the prostate gland of nude mice, and 14 days later, Bcl-xL siRNA complexed with atelocollagen (50 μg siRNA per injection) was intravenously injected into the tail vein of each mouse (injection volume, 200 μl; final atelocollagen concentration, 0.05%). A total of 12 injections were administered. (**b**) All of the tumor weights were measured 28 days later, at the end point of the therapy. The results are shown as the means ± SD ($n = 6$ mice). $^{**}p < 0.01$. (**c**) The Bcl-xL expression levels determined by qRT-PCR and an ELISA using a specific TaqMan probe and antibodies. The results are the means ± SD ($n = 6$ mice). $^{**}p < 0.01$

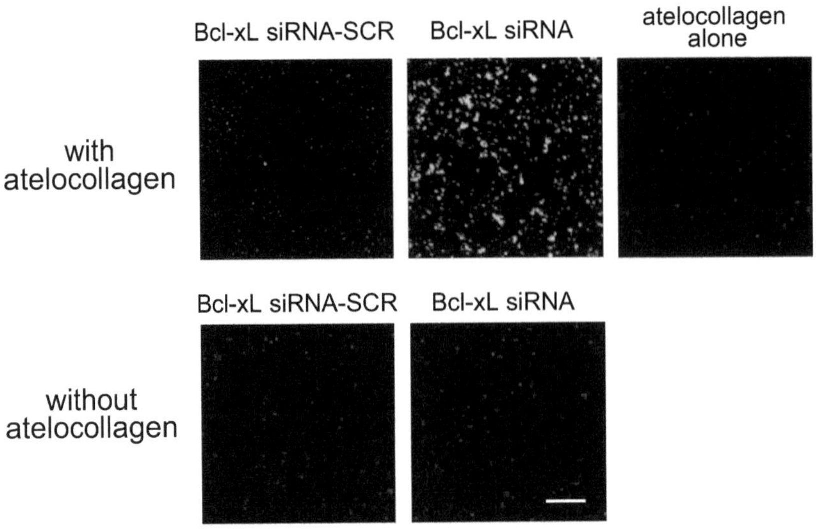

Fig. 3 Significant apoptotic cell death was induced by our therapy. Frozen sections from the tumors at the end point (Day 28) were subjected to TUNEL staining (MEBSTAIN Apoptosis Kit III; MBL, Nagoya, Japan). The stained sections were photographed via fluorescence microscopy. Bar, 100 μm

human hypoxanthine phosphoribosyltransferase 1 (HPRT1, Probe ID, Hs02800695_m1). Typical results are shown in Fig. 2c.

4. Determine the Bcl-xL expression levels in the tumors.
5. As in the case of Bcl-xL, a strong anti-apoptosis protein, observe the apoptosis-induced tumor cells in the treated tumors. The apoptosis-induced cells in the tumors were easily determined with a TUNEL-staining kit as described in our previous report [7, 10]. Typical results are shown in Fig. 3.

3.6 Examination of the Delivery of Cy3-Labeled siRNA into Tumors

1. Prepare nude mice with orthotopic PC-3 tumors on the prostate gland.
2. Prepare the complex of Cy3-labeled siRNA and atelocollagen.
3. Deeply anesthetize the nude mice with Somnopentyl (5 mg/kg, i.p.).
4. Intravenously inject the complex of Cy3-labeled siRNA and atelocollagen into the tail veins of mice (siRNA dose, 50 μg/mouse) [14, 15].
5. Twenty-four hours later, sacrifice the mice, excise the tumors, and imbed them in OCT compound to prepare the frozen sections [7, 10, 14, 15].
6. Observe the frozen sections under a confocal microscope to chase the red fluorescence (Cy3).

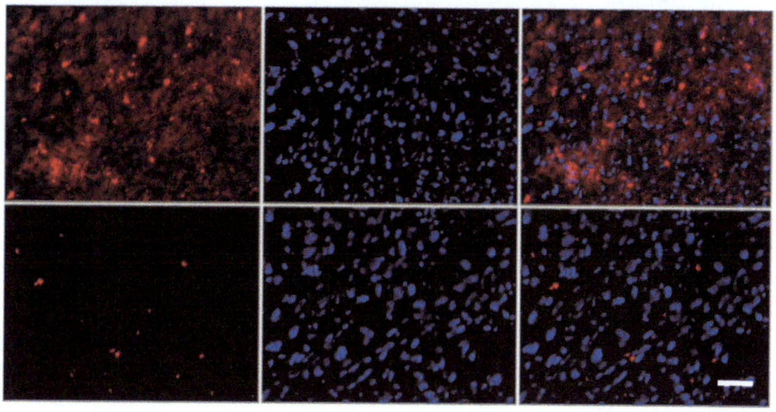

Fig. 4 Cy3-labeled Bcl-xL siRNA complexed with atelocollagen was delivered to PC-3 orthotopic tumors. Frozen sections were made from the tumors after the intravenous injection of Cy3-labeled Bcl-xL siRNA (50 μg/injection) with (upper) or without (lower) atelocollagen. DAPI staining was also performed. All of the sections were photographed by fluorescence microscopy. Bar, 50 μm

7. Stain nuclei with DAPI (4′,6-diamidino-2-phenylindole, for nuclear staining) to determine the total cell number on the section.
8. Determine the ratio of Cy3 fluorescence-positive cells (red) and DAPI-stained cells (blue). The ratio represents the delivery efficiency of the siRNA.
9. A typical observation under a confocal microscope is shown in Fig. 4.
10. The efficiency of siRNA delivery into tumors by this method was approximately 62%.

3.7 Evaluation of the Anti-metastatic Effects of Therapy

1. To evaluate the anti-metastatic effects of the therapy, PC-3 with luciferase expression (PC-3-LUC) can be used.
2. Prepare nude mice with orthotopic PC-3-LUC tumors on the prostate gland.
3. Perform the therapy (Bcl-xL siRNA with atelocollagen) according to the procedure shown in Fig. 2a.
4. At the end of the therapy (on Day 28), sacrifice the mice, and excise the liver (*see* **Note 13**).
5. Lyse liver tissues in lysis buffer, and centrifuge them (20,000 × *g*, 10 min at 4 °C).
6. Assay the supernatant with dual luciferase assay (Promega).
7. Normalize each luciferase activity with the protein concentration in samples to calculate RLUs (relative light units).
8. A typical result is shown in Fig. 5.

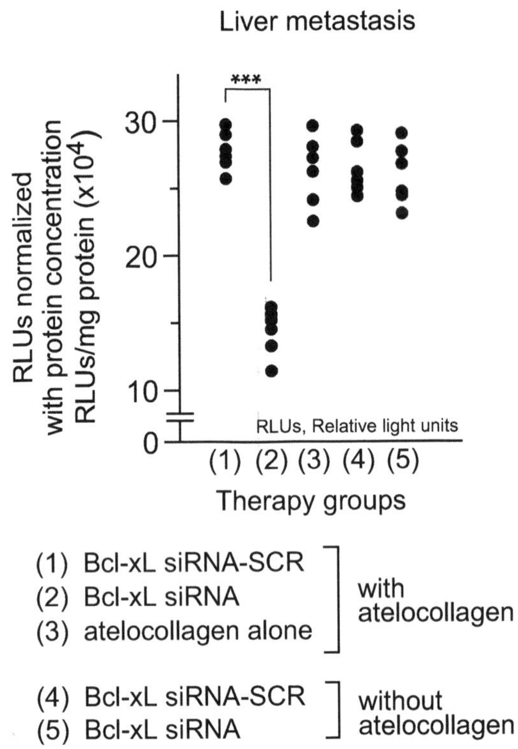

Fig. 5 Inhibition of liver metastasis by our proposed therapeutic method. The orthotopic tumors were prepared using PC-3-Luc cells, and 14 days later, the Bcl-xL siRNA/atelocollagen complex and other therapeutics were administered. The therapy was continued for 4 weeks, as indicated in Fig. 2a. On Day 28, all of the mice were sacrificed, and their livers were removed. The lysed samples from the livers were measured by a dual-luciferase reporter assay (Promega). The luciferase activity in relative light units (RLUs) was normalized to the total protein concentration. The results are the means ± SD ($n = 6$ mice). ***$p < 0.001$

4 Notes

1. The HPLC-purified siRNA was manufactured by Thermo Fisher Scientific K.K. (Dharmacon).
2. UltraPure™DNase/RNase-Free Distilled Water (Invitrogen).
3. An siRNA stock with high concentration is needed, because the injection volume of the siRNA into the xenografted tumor is quite limited (approximately 20–50 μl/tumor). Therefore, do not prepare a stock with lower concentration.
4. To handle the siRNA, we recommend working in an experimental environment for general RNA work and wearing a mask and gloves. Of course, the handling of siRNA must be done on a clean bench. RNase-free tubes and tips should be used.

5. siSTABLE-modified siRNA (Thermo Fisher Scientific K.K.).
6. Do not use confluent cultures (overgrowth cultures). The cells on logarithmic growth phase provide reproducible tumor formation.
7. Disperse the cells completely.
8. If necessary, diethyl ether can be used as an ancillary agent for inhalation anesthesia.
9. Do not allow the cell suspension to leak into the inoculated site.
10. Black-colored sutures are helpful to enhance visibility.
11. Do not mix the solution at room temperature. When the temperature is high (over 10 °C), the complex always forms gels.
12. Prepare the complex at the time of use. The final concentration of atelocollagen for systemic i.v. injection is 0.05%. The i.v. injection volume of the complex per mouse is 200 μl.
13. Organ(s)/tissue(s) other than the liver can also be examined if necessary.

References

1. Elbashir SM, Harborth J, Lendeckel W, Yalcin A, Weber K, Tuschl T (2001) Duplexes of 21-nucleotide RNAs mediate interference in cultured mammalian cells. Nature 411:494–498
2. Fire A, Xu S, Montgomery MK, Kostas SA, Driver SE, Mello CC (1998) Potent and specific genetic interference by double-stranded RNA in *Caenorhabditis elegans*. Nature 391:806–811
3. Carpenter AE, Sabatini DM (2004) Systematic genome-wide screens of gene function. Nat Rev Genet 5:11–22
4. Zheng L, Liu J, Batalov S et al (2004) An approach to genomewide screens of expressed small interfering RNAs in mammalian cells. Proc Natl Acad Sci U S A 101:135–140
5. Karagiannis TC, El-Osta A (2004) Mechanism of RNA interference, in vivo and potential clinical applications. Cancer Biol Ther 3:1069–1074
6. Ryther RCC, Flynt AS, Phillips JA III, Patton JG (2005) siRNA therapeutics: big potential from small RNAs. Gene Ther 12:5–11
7. Mu P, Nagahara S, Makita N, Tarumi Y, Kadomatsu K, Takei Y (2009) Systemic delivery of siRNA specific to tumor mediated by atelocollagen: combined therapy using siRNA targeting Bcl-xL and cisplatin against prostate cancer. Int J Cancer 125:2978–2990
8. Boise LH, González-García M, Postema CE, Ding L, Lindsten T, Turka LA, Mao X, Nuñez G, Thompson CB (1993) bcl-x, a bcl-2-related gene that functions as a dominant regulator of apoptotic cell death. Cell 74:597–608
9. Schott AF, Apel IJ, Nuñez G, Clarke MF (1995) Bcl-XL protects cancer cells from p53-mediated apoptosis. Oncogene 11:1389–1394
10. Yuan Y, Makita N, Cao D, Mihara K, Kadomatsu K, Takei Y (2015) Atelocollagen-mediated intravenous siRNA delivery specific to tumor tissues orthotopically xenografted in prostates of nude mice and its anticancer effects. Nucleic Acid Ther 25:85–94
11. Ochiya T, Takahama Y, Nagahara S, Sumita Y, Hisada A, Itoh H, Nagai Y, Terada M (1999) New delivery system for plasmid DNA in vivo using atelocollagen as a carrier material: the Minipellet. Nat Med 5:707–710
12. Takei Y, Kadomatsu K (2005) In vivo delivery technique of nucleic acid compounds using atelocollagen: its use in cancer therapeutics targeted at the heparin-binding growth factor midkine. Gene Ther Mol Biol 9:257–264
13. Fujimoto I, Takei Y (2014) Atelocollagen-mediated siRNA delivery: future promise for therapeutic application. Ther Deliv 5:369–371
14. Inaba S, Nagahara S, Makita N, Tarumi Y, Ishimoto T, Matsuo S, Kadomatsu K, Takei Y (2012) Atelocollagen-mediated systemic

delivery prevents immunostimulatory adverse effects of siRNA in mammals. Mol Ther 20:356–366

15. Ishimoto T, Takei Y, Yuzawa Y, Hanai K, Nagahara S, Tarumi Y, Matsuo S, Kadomatsu K (2008) Downregulation of monocyte chemoattractant protein-1 involving short interfering RNA attenuates hapteninduced contact hypersensitivity. Mol Ther 16:387–395

Chapter 4

Relative Quantification of siRNA Strand Loading into Ago2 for Design of Highly Active siRNAs

Phillip A. Angart, Kwasi Adu-Berchie, Rebecca J. Carlson, Daniel B. Vocelle, Christina Chan, and S. Patrick Walton

Abstract

In RNA interference (RNAi), silencing is achieved through the interaction of double-stranded small interfering RNAs (siRNAs) with essential RNAi pathway proteins, including Argonaute 2 (Ago2). Based on these interactions, one strand of the siRNA is loaded into Ago2 forming the active RNA-induced silencing complex (RISC). Optimal siRNAs maximize RISC activity against the intended target and minimize off-target silencing. To achieve the desired activity and specificity, selection of the appropriate siRNA strand for loading into Ago2 is essential. Here, we provide a protocol to quantify the relative loading of individual siRNA strands into Ago2, one factor in determining the capacity of a siRNA to achieve silencing activity and target specificity.

Key words siRNA, Ago2, RT-qPCR, Small RNA, Transfection, Immunoprecipitation, Stem-loop, HeLa

1 Introduction

Short interfering RNA (siRNA)-based therapeutics has the potential to treat a variety of diseases, including cancer [1, 2]. Utilizing the native eukaryotic pathway of RNA interference (RNAi), siRNAs initiate posttranscriptional gene silencing through sequence-specific targeting of mRNAs [2]. Upon entering the cytoplasm, the siRNA duplex is recognized by RNAi pathway proteins that load one of the siRNA strands into Argonaute 2 (Ago2), an endoribonuclease [3, 4]. The selected strand, known as the guide strand, along with Ago2, forms the mature, active RNA-induced silencing complex (RISC) [5]. Active RISC targets and cleaves mRNA transcripts complementary to the guide strand [6, 7].

Proper guide strand loading is essential to achieve efficient knockdown of the intended target. Considerable effort has been invested in determining the mechanism of strand loading in a variety of organisms [8]. These studies have highlighted particular

Lekha Dinesh Kumar (ed.), *RNA Interference and Cancer Therapy: Methods and Protocols*, Methods in Molecular Biology, vol. 1974, https://doi.org/10.1007/978-1-4939-9220-1_4,

siRNA sequence features that are important for selection of the guide strand [9–12]. Loading of the unintended strand of the siRNA duplex (i.e., the passenger strand) can lead to significant non-specific, off-target effects [13, 14]. Additionally, loading of the passenger strand sequesters Ago2, preventing loading of the guide strand and limiting the magnitude of silencing of the intended target. Thus, to achieve maximal activity and specificity of silencing, it is essential to design siRNAs such that only the intended guide strand is loaded into Ago2.

Methods for quantifying the strand loading of siRNAs have included indirect reporter assays [7, 9], as well as direct detection of siRNA loading through Northern blotting or Reverse Transcription-quantitative Polymerase Chain Reaction (RT-qPCR) of siRNA from Ago2 immunoprecipitation [15]. Stem-loop RT-qPCR was developed to increase the length of small RNAs, like siRNAs, in order to make them more amenable as qPCR templates [16]; since then, several other articles have discussed the details and optimization of RT-qPCR on siRNAs [17–20]. Another option for lengthening small RNAs is to use enzymatic poly(A) tailing. Since this method indiscriminately amplifies all RNAs, a forward primer containing locked nucleic acids is preferred to increase qPCR specificity [21]. Among direct detection methods, RT-qPCR-based assays have higher sensitivity than Northern blots while maintaining high specificity [22]. However, RT-qPCR-based assays also require thoughtful assay design and selection of appropriate controls. Here, we describe a method and the controls to investigate relative strand loading for siRNAs into Ago2 in cultured mammalian cells.

2 Materials

2.1 Cell Culture

1. 0.25% trypsin (1×), phenol red.
2. Versene solution.
3. DMEM, high glucose.
4. FBS.
5. Penicillin-streptomycin.
6. 6-well plate.
7. Culture medium: DMEM, high glucose; 10% FBS; 1% Pen-Strep.
8. Culture medium without antibiotics: DMEM, high glucose; 10% FBS.

2.2 Transfection

1. Opti-MEM I Reduced Serum Medium.
2. Lipofectamine 2000 (LF2K).

2.3 Cell Lysis

1. Complete Protease Inhibitor.
2. Lysis buffer: 0.5% Igepal CA630 (USB), 150 mM KCl, 25 mM Tris–HCl pH 7.5, 2 mM EDTA, 0.5 mM DTT (add immediately before use), 1× cOmplete Protease Inhibitor (add immediately before use).
3. 10× PBS pH 7.4, diluted to 1×.
4. Cell Lifters.
5. RNase-free water.

2.4 Bead Preparation and RNA Immunoprecipitation

1. Protein G Magnetic Beads.
2. Monoclonal anti-Ago2 antibody produced in rat, clone 11A9.
3. Wash buffer: 300 mM NaCl, 50 mM Tris–HCl pH 7.5, 5 mM $MgCl_2$, 0.05% Igepal CA630.
4. Quick Start Bradford Assay.
5. BSA standard.

2.5 RNA Purification

1. Direct-zol RNA MiniPrep with TRI Reagent.
2. 100% absolute ethanol.
3. Proteinase K.
4. tRNA from *E. coli* MRE 600.
5. Proteinase K buffer: 300 mM NaCl, 200 mM Tris–HCl pH 7.5, 25 mM EDTA, 2% SDS.
6. RNaseZap.

2.6 Preparation of Oligonucleotides for RT-qPCR

1. 5 M NaCl.
2. TE, pH 8.0.

2.7 Reverse Transcription

1. dNTP Mix.
2. Stem-loop primer.
3. SuperScript III Reverse Transcriptase (includes Superscript, DTT, First-Strand Synthesis Buffer).
4. SUPERaseIn RNase Inhibitor.

2.8 qPCR

1. iQ SYBR Green Mix.
2. Forward Primer (Custom Synthesis).
3. Universal Reverse Primer (Custom Synthesis).
4. 96-well plate.
5. Plate Seal.

3 Methods

The following methods detail the transfection of HeLa cells followed by the recovery and quantification of siRNA strands loaded into Ago2 (*see* **Note 1**). This protocol can be adapted to different siRNA sequences, delivery vehicles, or cell types. However, changing these variables may require optimization of RNA recovery conditions.

Assay outline:

Day 1: Plate cells.

Day 2: Transfect cells and conjugate antibody to magnetic beads.

Day 3: Lyse cells, immunoprecipitate Ago2, and perform organic extraction/ethanol precipitation.

Day 4: Suspend RNA in water and perform stem-loop RT-qPCR.

3.1 Assay Design (Samples, Standards, and Controls)

When comparing the two strands of one siRNA, no internal standard is needed; however, if there is a need to compare multiple strands from different siRNAs, an additional non-specific siRNA acting as an internal standard should be transfected, as presented here (*see* **Note 2**). The internal standard will account for variability in transfection, cell number, cell lysis, immunoprecipitation, and RNA recovery. The method presented here utilizes a set of standards for the quantification of siRNA; this method was chosen to account for the differences in reverse transcription and PCR amplification efficiencies (Fig. 1) (*see* **Note 3**). Standard curves are built by spiking in siRNA to a sample generated from mock-transfected cells, mimicking sample treatment conditions (Table 1). Additional controls were prepared to check for (1) non-specific amplification of contaminant DNA and (2) amplification of endogenous RNA.

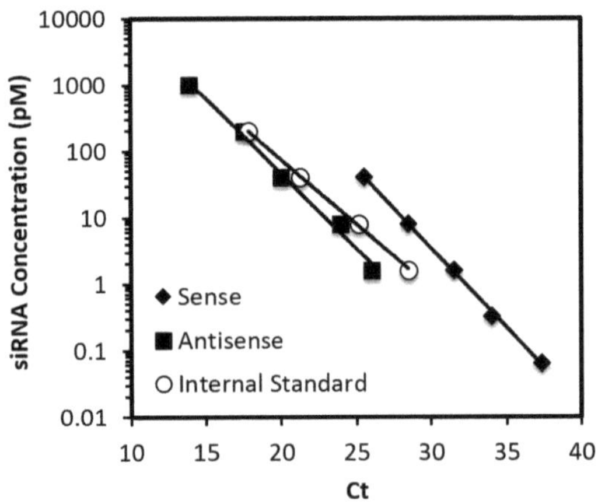

Fig. 1 Representative siRNA standard curves. Standards were generated by spiking in known quantities of siRNA into a mock-transfected lysate, followed by RT-qPCR. Amplification efficiencies were as follows: sense strand, 73.7%; antisense strand, 69.4%; and internal standard, 56.4%

Table 1
Assay design

	1	2	3	4	5	6	7	8	9	10	11
	Sample	IS	Sample standard curve	IS standard curve	Non-specific control	Non-specific control IS	–RT siRNA sample lysate	–RT IS sample lysate	–RT siRNA control lysate 1	–RT IS control lysate 1	–RT IS control lysate 2
Sample primers	+	–	+	–	+	–	+	–	+	–	–
IS primers	–	+	–	+	–	+	–	+	–	+	+
Sample lysate +siRNA +IS	+	+	–	–	–	–	+	+	–	–	–
Control lysate 1 –siRNA + IS	–	–	+	–	+	–	–	–	+	+	–
Control lysate 2 –siRNA –IS	–	–	–	+	–	+	–	–	–	–	+
RT	+	+	+	+	+	+	–	–	–	–	–
Additional siRNA			15 amol – 250 fmol	15 amol – 250 fmol		250 fmol					

qPCR samples and controls for an siRNA strand (referred to as siRNA) and an siRNA used as an internal standard (referred to as IS). Sample lysates were co-transfected with 10 nM siRNA (Column 1) and 10 nM IS (Column 2). Control lysates were generated lacking the siRNA (Control Lysate 1) and lacking both the siRNA and IS (Control Lysate 2). These lysates were utilized to generate the standard curves (Columns 3 and 4), which also serve as positive controls, and negative controls (Columns 5–11). The negative controls included the verification of RT specificity for siRNA (Column 5) and IS (Column 6). The verification products generated are the result of transcription in the sample lysate (Columns 7 and 8), in Control Lysate 1 (Columns 9 and 10), and in Control Lysate 2 (Column 11)

3.2 Stem-Loop RT-qPCR Primer Design

3.2.1 Outline for Primer Design

1. Determine the sequence of the siRNA strand. The passenger strand sequence for our siRNA will be used as an example: 5′-GGCCAGAAGGAUUUCAUUAUU-3′.
2. Find the reverse complement: 5′- AATAATGAAATCCTTC TGGCC-3′.
3. Generate the stem-loop primer sequence by combining the 44 nt stem-loop sequence, 5′-GTTGGCTCTGGTGCA GGGTCCGAGGTATTCGCACCAGAGCCAAC-3′ [17], with the first six nucleotides from **step 2**, 5′-GTTGGCTCT GGTGCAGGGTCCGAGGTATTCGCACCAGAG CCAAC***AATAAT***-3′.
4. Design the forward qPCR primer by taking the first 12–17 nucleotides of the siRNA sequence. Additional nucleotides can be added to the 5′ end, constituting a 5′ overhang, to raise the T_m to ~60 °C. We generated our forward primer by adding "GCGCG" to the first 15 nucleotides of our siRNA, resulting in the following sequence: 5′- **GCGCG**GGCCAGAAGGA TTTC-3′.
5. Design a universal reverse qPCR primer from sequences within the stem-loop. Primer T_m can be altered by changing the length of the primer from the 5′ terminus. We used the reverse primer: 5′-CCAGTGCAGGGTCCGAGGTA-3′ [19].
6. Confirm that qPCR primers do not form self-dimers, heterodimers, or stable hairpins and that the forward primer does not share complementarity with the stem-loop primer (*see* **Note 4**).

3.2.2 Expected Amplification Sequences (Fig. 2)

1. After reverse transcription, a 65 nt first strand cDNA results in the following sequence: 5′-GTTGGCTCTGGTGCAG GGTCCGAGGTATTCGCACCAGAGCCAACaata atgaaatccttctggcc-3′.

   The reverse complement of the siRNA sequence is in lowercase.
2. The second strand cDNA is expected to be 71 nt, consisting of the reverse complement of the first strand cDNA plus the additional six 5′ nt of the forward primer: 5′-**GCGCG ggccagaaggatttc**attattGTTGGCTCTGGTGC-GAATACCTCGGACCCTGCACCAGAGCCAAC-3′.

   The forward qPCR primer is in bold and siRNA sequence in lowercase.
3. The final PCR product is expected to be 63 nt with the following sequences:

   Forward strand:

5′-**GCGCGggccagaaggatttc**attattGTTGGCTCTGGTGC GAA<u>TACCTCGGACCCTGCACTGG</u>-3′

Reverse strand:

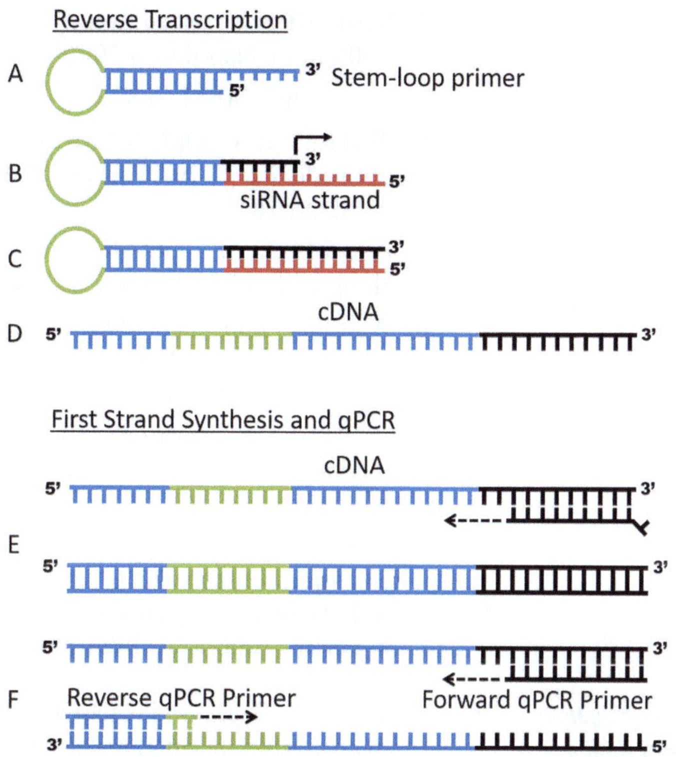

Fig. 2 Overview of stem-loop RT-qPCR amplification. The stem-loop primer is 44 nt with 6 nt complementary to the siRNA strand (**a**, **b**). Reverse transcription of the siRNA occurs from the 3′ end of the stem-loop primer adding 15 nt (**c**), making the final cDNA 65 nt (**d**). First strand synthesis occurs with the siRNA-specific primer (**e**), followed by PCR amplification (**f**)

5′-CCAGTGCAGGGTCCGAGGTATTCGCACC AGAGCCAACaataat**gaaatccttctggccCGCGC**-3′.

The forward qPCR primer and its reverse complement are in bold in the forward strand and reverse strand, respectively. The siRNA sequence is in lowercase. The reverse primer and its reverse complement are underlined in the reverse strand and the forward strand, respectively.

3.3 HeLa Cell Transfection

For a 6-well plate, prepare a 500 μL transfection solution containing 5.75 μg (2.3 μg/mL based on final well volume) of LF2K, 10 nM final concentration of siRNA of interest, 10 nM internal standard siRNA, and Opti-MEM to volume (*see* **Note 5**).

1. Maintain HeLa cells in culture medium (*see* **Note 6**).
2. Trypsinize and then suspend cells in culture medium without antibiotics at 175,000 cells/mL, and plate 2 mL/well in a 6-well plate for 24 h.
3. Confirm that the cells are >90% confluent before proceeding with transfection.

4. Verify siRNA concentration by measuring its absorbance at 260 nm by NanoDrop 2000 (*see* **Notes 7** and **8**).
5. Warm Opti-MEM to 37 °C.
6. Add 245 μL of Opti-MEM to a 1.7 mL centrifuge tube, and then add 2.5 μL of 10 μM siRNA and 2.5 μL of 10 μM internal standard.
7. Add 244.25 μL Opti-MEM to a new 1.7 mL centrifuge tube along with 5.75 μL of LF2K. Mix by gently pipetting.
8. Wait for 5 min, then add the diluted siRNA in the first centrifuge tube to the diluted LF2K solution in the second tube, mix by pipetting, and incubate for 20 min.
9. While the transfection solution is incubating, replace the medium in the 6-well plate with fresh culture medium without antibiotics.
10. After 20 min, mix the transfection solutions by pipetting several times before adding dropwise to each respective plate/well. Gently rock plates back and forth and side to side, avoiding circular motions. Return plates to incubator for 24 h.

3.4 Conjugation of Ago2 Antibody to Magnetic Protein G Beads

1. Transfer 25 μL of Protein G Magnetic Beads to a 1.7 mL microcentrifuge tube, making sure to achieve a homogeneous suspension prior to transfer (*see* **Note 9**).
2. Add 475 μL of 1× PBS. Place the tube on a magnetic separation rack for 1 min and then remove and discard the supernatant (*see* **Note 10**).
3. Suspend the beads in 1.5 mL of 1× PBS, and add 2.25 μL (3.375 μg) of Ago2 antibody.
4. Incubate overnight at 4 °C with end-over-end rotation.
5. Wash the beads 1× in 250 μL of wash buffer. Remove and discard the supernatant.
6. Resuspend the beads 1× in 250 μL of 1× PBS, and transfer them to a new 1.7 mL microcentrifuge tube.

3.5 Cell Lysis

1. Aspirate the medium, and wash the plate two times with a volume of ice-cold PBS equal to the volume of medium removed (2 mL per well for a 6-well plate), working with the plate on ice.
2. Add 500 μL of ice-cold lysis buffer, and rock to make sure the buffer wets the entire well.
3. Incubate on ice for 30 min.
4. Scrape cells and transfer lysate to a centrifuge tube.
5. Pellet debris at 16,000 × *g* for 30 min at 4 °C, and transfer to a fresh microcentrifuge tube (*see* **Note 11**).

6. Measure protein concentration by Quick Start Bradford Assay using a BSA standard. Lysates may need to be diluted to yield protein concentrations in the linear range of the standard curve and to minimize interference of the lysis buffer with quantification (*see* **Note 12**).

3.6 RNA Immunoprecipitation (RIP)

1. Remove 1× PBS from the prepared magnetic beads, and load 400 μg of lysate, diluted to 0.5 mL with lysis buffer, into the 1.7 mL centrifuge containing the beads.
2. Incubate at 4 °C for 3 h with end-over-end rotation.
3. Briefly centrifuge the tube, place it on a magnetic separator for 1 min, and remove the supernatant (*see* **Note 11**).
4. Wash the beads 3× in 250 μL of wash buffer. Each time, mix by pipetting up and down, and then place tube on magnetic separator for 1 min before removing and discarding the supernatant.
5. Wash the beads 1× in 250 μL 1× PBS, and transfer them to a new microcentrifuge tube (*see* **Note 11**).

3.7 RNA Purification

1. Place tube on magnetic separator.
2. Remove and discard the supernatant.
3. Suspend the beads in 140 μL Proteinase K Buffer with 6 μL of 20 mg/mL Proteinase K and 4 μL of 390 μg/mL tRNA (*see* **Note 13**).
4. Incubate beads at 60 °C for 15 min, mixing occasionally to bring thc bcads back into suspension.
5. Using the Zymo Research Direct-zol RNA Purification kit (*see* **Note 14**), add 3 volumes of TRI Reagent (450 μL), mix thoroughly, and freeze if not immediately performing reverse transcription (RT).
6. Add 1 volume of absolute ethanol and vortex.
7. Apply entire volume (600 μL) to Zymo-Spin Column, and centrifuge at 16,000 × *g* for 30 s.
8. Discard flow-through, apply remaining sample to column, and centrifuge again.
9. Wash spin column with 400 μL of RNA wash buffer.
10. Mix 75 μL DNA Digestion Buffer and 5 μL DNase I, then add to spin column, and incubate for 15 min.
11. Wash column with 400 μL RNA PreWash solution, and centrifuge for 30 s, discarding flow-through.
12. Wash column with 700 μL RNA wash buffer and centrifuge for 2 min.

13. Transfer column to a new centrifuge tube, and elute in 50 μL of RNase-free water preheated to 60 °C.
14. Proceed immediately to reverse transcription reaction or freeze recovered sample at −80 °C.

3.8 RT-qPCR Preparation

Suspend oligonucleotides in TE to a final concentration of 100 μM. Verify concentration using the NanoDrop, and adjust the concentration accordingly while making the first primer dilution.

3.8.1 Preparation of Stem-Loop Primer for RT

1. Fold the stem-loop primer by adjusting NaCl concentration to 50 mM (add 0.3 μL of 5 M NaCl to 30 μL of ~100 μM primer); then heat primer to 95 °C for 10 min; decrease to 75 °C for over 30 min; hold temperature at 75 °C, 68 °C, 65 °C, and 62 °C for 1 h at each temperature; hold at 60 °C for 2 h; and then decrease to 4 °C and hold.
2. Make a 200 nM stem-loop primer working stock by diluting 1:10 to 10 μM, verifying the primer concentration by measuring the absorbance at 260 nm (*see* **Note 7**), and then by diluting (1:10) to 1 μM, making adjustments to the dilution depending upon the measured concentration. Working primer stocks can be stored at −20 °C.

3.8.2 Preparation of qPCR Primers

Make a 10 μM working stock of each primer by first diluting primer 1:5 in TE, measuring the absorbance at 260 nm (*see* **Note 7**), and then diluting the primer ~1:2 (adjusted based on absorbance) to 10 μM. Store working stocks at −20 °C.

3.9 Reverse Transcription Reaction

1. A single reaction mix is prepared by adding the following components to a nuclease-free microcentrifuge tube:
   (a) 0.5 μL 10 mM dNTP mix.
   (b) 1 μL of 1 μM stem-loop RT primer (50 nM final concentration).
   (c) 4 μL 5× First-Strand Synthesis Buffer.
   (d) 2 μL 0.1 M DTT.
   (e) 0.2 μL SUPERaseIn RNase Inhibitor (20 U/μL).
   (f) 0.25 μL SuperScript III RT (200 U/μL).
   (g) 7.05 μL water.
   (h) 5 μL RNA sample (dilute 1:5 from purified RNA).
2. Prepare +RT and −RT master mixes without the RNA sample by scaling the volumes for a single qPCR reaction to the desired number of qPCR reactions, omitting the stem-loop primer, RNA sample, and 5 μL of water.
3. Prepare a second set of master mixes by adding the specific stem-loop primer. This will give four master mixes (+RT-IS, +RT-siRNA, −RT-IS, −RT-siRNA).

4. Assemble the RT reactions by adding 10 μL of the appropriate master mix to 5 μL of the corresponding 1:5 diluted RNA, according to Table 1, and 5 μL of water. For making standard curves or control samples requiring the addition of siRNA, replace the 5 μL of water with 5 μL of siRNA at the appropriate concentration (Table 1; Columns 3, 4, and 6).
5. Load thermal cycler and incubate as follows (*see* **Note 15**):
   (a) 1 Cycle:
      - 16 °C for 30 min.
   (b) 60 Cycles:
      - 30 °C for 30 s.
      - 42 °C for 30 s.
      - 50 °C for 1 s.
   (c) 1 Cycle:
      - 85 °C for 5 min.
      - 4 °C Forever.
6. Proceed immediately to qPCR or freeze at −20 °C.

3.10 qPCR Reaction

1. A single reaction mix is prepared by adding the following components:
   (a) 12.5 μL 2× iQ SYBR Green Mix.
   (b) 0.75 μL 10 μM Forward Primer.
   (c) 0.75 μL 10 μM Universal Reverse Primer.
   (d) 2 μL water.
   (e) 9 μL RT sample (cDNA diluted 1:5).
2. Prepare master mixes (without cDNA) by scaling the volumes for a single qPCR reaction to the desired number of qPCR reactions. Mix gently and spin down master mix.
3. Assemble the qPCR reaction by combining 11 μL of master mix with 9 μL of RT product.
4. Load real-time qPCR thermal cycler (BioRAD MyiQ) as follows:
   (a) 95 °C for 10 min.
   (b) 50 Cycles:
      - 95 °C for 15 s.
      - 60 °C for 1 min.
   (c) 72 °C for 5 min.
   (d) 95 °C for 1 min.
   (e) Melting curve: 0.5 °C per 10 s from 55 °C to 95 °C.

5. Verify the formation of a product with a T_m of ~84 °C. In addition, the C_t values of all samples and standards should be lower than those for every negative control (*see* **Note 16**).

4 Notes

1. The methods for cell culture can be scaled to increase signal by RT-qPCR as needed. Transfection of 250 pmol siRNA into 1 well from a 6-well plate of HeLa cells (350,000 cells/well plated 24 h prior to transfection) generated sufficient signal by RT-qPCR.
2. When investigating endogenous small RNAs (i.e., a miRNA), a miRNA with consistent expression in the tissue of interest is a suitable internal standard [23].
3. Analysis of qPCR data can be performed using either the $\Delta\Delta C_t$ method [24] or with the use of a known set of standards. Standards were used here because of the difference in reverse transcription and PCR efficiencies for different siRNA strands.
4. The OligoAnalyzer tool (https://www.idtdna.com/calc/analyzer) is useful for quickly checking primer hairpins, self-dimers, and heterodimers. If such structures do exist, depending on the issue, the reverse primer can be shifted or stem-loop primer sequence and reverse primer altered. If the forward primer forms a self-dimer or hairpin, accurate quantification may require increasing the qPCR annealing temperature or performing a pre-amplification step at a higher T_m followed by the normal qPCR reaction.
5. An siRNA concentration of 10 nM was chosen for these experiments, because it is the highest concentration we could confidently use without knowingly having any adverse effects on cell function [25].
6. Transfection efficiency can be highly variable. Maintaining cells in a metabolically active state, not exceeding a ~90% confluence, and maintaining a consistent LF2K:cell ratio will minimize transfection variance.
7. Because siRNAs are less than 100 nt, siRNA-specific extinction coefficients are recommended when measuring siRNA concentrations. Theoretical extinction coefficients can be calculated at https://www.idtdna.com/calc/analyzer.

$$\text{Concentration} = (\text{A260} \times \text{extinction coefficient}) / \text{path length}$$

8. Practice RNase-free technique, and use certified RNase-free reagents and consumables. This includes establishing a separate workspace that is cleaned with an RNase inhibitor (e.g., RNaseZap) prior to use and being cognizant of potential sources of RNase contamination. One common source of

contamination is the touching of the inside of a microcentrifuge cap when opening and closing microcentrifuge tubes; this can be avoided by using one hand to hold the tube and the second hand to open the cap. RNase-free technique is not critical for the manipulation of cells; however, RNase-free technique should be practiced for all lysis steps. It is recommended not to use DEPC-treated water as residual DEPC can interfere with enzymatic assays; suitable RNase-free water is available from commercial sources.

9. 25 μL of magnetic beads is more than enough to complex the amount of Ago2 antibody in our assay. 25 μL was chosen because it is the minimum quantity of beads that can be confidently manipulated throughout the immunoprecipitation.
10. When using a magnet for the manipulation of magnetic beads, ensure even suspension before applying samples. It is also good practice to briefly centrifuge the tube to remove any solution from the cap and then to resuspend the solution by pipetting.
11. Western blot analysis can be useful to verify the immunoprecipitation of Ago2 during assay development and troubleshooting. However, it is not practical to run a Western blot for every sample. Therefore, it is recommended to take samples of the diluted lysate, the Ago2-depleted lysate, and a fraction of the immunoprecipitated Ago2-Protein G Magnetic Beads (Fig. 3), and analyze these samples as necessary.

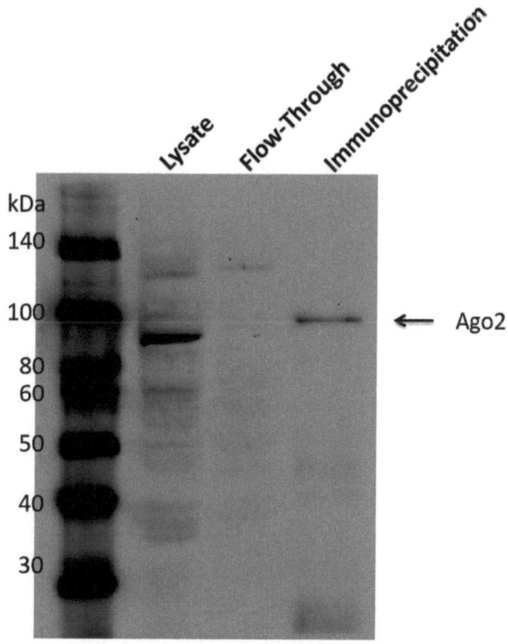

Fig. 3 Verification of Ago2 immunoprecipitation. Confirmation of Ago2 immunoprecipitation from HeLa cell lysates with an anti-Ago2 (11A9) antibody followed by Western blotting with an anti-Ago2 (C34C6) antibody

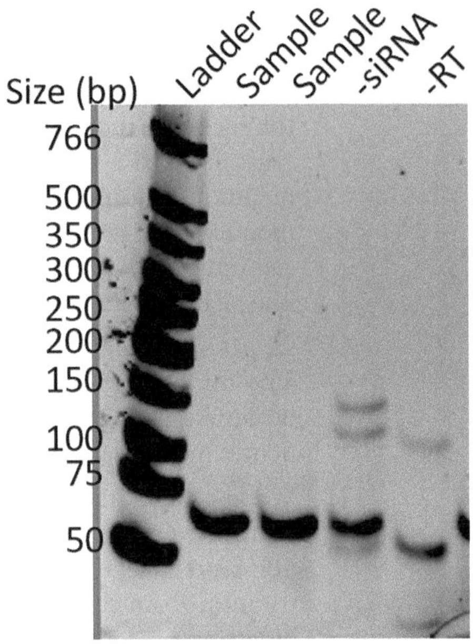

Fig. 4 Verification of PCR products. Representative qPCR products from Columns 1 (sample), 5 (siRNA), and 7 (RT), resolved on a 4–20% TBE gel and stained with SYBR Gold. Ladder is a low MW DNA ladder from NEB

12. A 1:5 dilution is a good starting dilution; for our experiments, protein concentrations after dilution were ~200 μg/mL.
13. Proteinase K and tRNA mixture should be added to extraction buffer immediately before use. The addition of tRNA to samples before RNA extraction from Ago2 improved the sensitivity of the detection, presumably by facilitating the efficiency of purification or RT.
14. The method of small RNA purification is important as some methods can lead to different precipitation efficiencies. In 2011, an article was retracted after the authors discovered bias in small RNA purification [26]. Multiple manufacturers have validated silica-based spin column purification methods as suitable for the unbiased purification of small RNAs.
15. A pulsed RT protocol [18] was chosen because greater separation was observed between sample and negative control C_t values; similar C_t values were observed for both pulsed RT and standard RT sample C_t values [16].
16. When using a new set of primers, it is useful to verify product formation by native gel electrophoresis. The hallmark of a specific product is a single band on a gel at the anticipated size. Negative control wells typically have several bands produced from non-specific amplification or primer dimers, possibly including a band of the anticipated size (Fig. 4).

Limiting the number of amplification cycles can also maximize the discrimination between specific and non-specific product formations.

References

1. Bobbin ML, Rossi JJ (2016) RNA interference (RNAi)-based therapeutics: delivering on the promise? Annu Rev Pharmacol Toxicol 56(1):103–122. https://doi.org/10.1146/annurev-pharmtox-010715-103633
2. Scherman D, Rousseau A, Bigey P et al (2017) Genetic pharmacology: progresses in siRNA delivery and therapeutic applications. Gene Ther 24(3):151–156. https://doi.org/10.1038/gt.2017.6
3. Yoda M, Kawamata T, Paroo Z et al (2010) ATP-dependent human RISC assembly pathways. Nat Struct Mol Biol 17(1):17–23
4. Liu JD, Carmell MA, Rivas FV et al (2004) Argonaute2 is the catalytic engine of mammalian RNAi. Science 305(5689):1437–1441
5. Rivas FV, Tolia NH, Song JJ et al (2005) Purified Argonaute2 and an siRNA form recombinant human RISC. Nat Struct Mol Biol 12(4):340–349
6. Martinez J, Patkaniowska A, Urlaub H et al (2002) Single-stranded antisense siRNAs guide target RNA cleavage in RNAi. Cell 110(5):563–574
7. Elbashir SM, Harborth J, Weber K et al (2002) Analysis of gene function in somatic mammalian cells using small interfering RNAs. Methods 26(2):199–213
8. Nakanishi K (2016) Anatomy of RISC: how do small RNAs and chaperones activate Argonaute proteins? Wiley Interdiscip Rev RNA 7(5):637–660. https://doi.org/10.1002/wrna.1356
9. Angart PA, Carlson RJ, Adu-Berchie K et al (2016) Terminal duplex stability and nucleotide identity differentially control siRNA loading and activity in RNA interference. Nucleic Acid Ther 26(5):309–317. https://doi.org/10.1089/nat.2016.0612
10. Schwarz D, Hutvagner G, Du T et al (2003) Asymmetry in the assembly of the RNAi enzyme complex. Cell 115(2):199–208
11. Noland CL, Ma E, Doudna JA (2011) siRNA repositioning for guide strand selection by human dicer complexes. Mol Cell 43(1):110–121
12. Sakurai K, Amarzguioui M, Kim D et al (2011) A role for human Dicer in pre-RISC loading of siRNAs. Nucleic Acids Res 39(4):1510–1525
13. Ozcan G, Ozpolat B, Coleman RL et al (2015) Preclinical and clinical development of siRNA-based therapeutics. Adv Drug Deliv Rev 87:108–119. https://doi.org/10.1016/j.addr.2015.01.007
14. Wittrup A, Lieberman J (2015) Knocking down disease: a progress report on siRNA therapeutics. Nat Rev Genet 16(9):543–552. https://doi.org/10.1038/nrg3978
15. Beitzinger M, Meister G (2011) Experimental identification of microRNA targets by immunoprecipitation of Argonaute protein complexes. In: Dalmay T (ed) MicroRNAs in development, vol 732. Humana, Totowa, NJ, pp 153–167
16. Chen CF, Ridzon DA, Broomer AJ et al (2005) Real-time quantification of microRNAs by stem-loop RT-PCR. Nucleic Acids Res 33(20):e179
17. Varkonyi-Gasic E, Wu R, Wood M et al (2007) Protocol: a highly sensitive RT-PCR method for detection and quantification of microRNAs. Plant Methods 3:12
18. Tang F, Hajkova P, Barton SC et al (2006) MicroRNA expression profiling of single whole embryonic stem cells. Nucleic Acids Res 34(2):e9. https://doi.org/10.1093/nar/gnj009
19. Kramer MF (2011) Stem-loop RT-qPCR for miRNAs. Curr Protoc Mol Biol. Chapter 15:Unit 15.10
20. Jung U, Jiang X, Kaufmann SH et al (2013) A universal TaqMan-based RT-PCR protocol for cost-efficient detection of small noncoding RNA. RNA 19(12):1864–1873. https://doi.org/10.1261/rna.040501.113
21. Benes V, Castoldi M (2010) Expression profiling of microRNA using real-time quantitative PCR, how to use it and what is available. Methods 50(4):244–249. https://doi.org/10.1016/j.ymeth.2010.01.026
22. Czimmerer Z, Hulvely J, Simandi Z et al (2013) A versatile method to design stem-loop primer-based quantitative PCR assays for detecting small regulatory RNA molecules. PLoS One 8(1):e55168. https://doi.org/10.1371/journal.pone.0055168
23. Peltier HJ, Latham GJ (2008) Normalization of microRNA expression levels in quantitative RT-PCR assays: identification of suitable reference RNA targets in normal and cancerous human solid tissues. RNA 14(5):844–852. https://doi.org/10.1261/rna.939908

24. Schmittgen TD, Livak KJ (2008) Analyzing real-time PCR data by the comparative C(T) method. Nat Protoc 3(6):1101–1108
25. Caffrey DR, Zhao J, Song Z et al (2011) siRNA off-target effects can be reduced at concentrations that match their individual potency. PLoS One 6(7):e21503
26. Kim Y-K, Yeo J, Ha M et al (2012) Retraction notice to: cell adhesion-dependent control of microRNA decay. Mol Cell 46(6):896

Chapter 5

VEGFA Gene Silencing in CXCR4-Expressing Cells via siRNA Delivery by Means of Targeted Peptide Carrier

Anna A. Egorova, Marianna A. Maretina, and Anton V. Kiselev

Abstract

Discovery of small interfering RNA as a tool for specific gene inhibition led to the development of new therapeutic strategy for the treatment of cancers. The efficacious delivery of therapeutic siRNAs into the cells is a crucial step in RNA interference (RNAi) application, but it remains challenging. Non-viral vectors can provide specific cellular uptake, stable siRNA complex formation, and intracellular siRNA release. Recently, we evaluated modular peptide carrier L1 bearing CXCR4 targeting ligand for its ability to condense siRNA and facilitate endosomal escape and VEGFA gene silencing in CXCR4-expressing endothelial and glioblastoma cells. The present chapter showcases the ability of L1 targeted peptide carrier to form complexes with siRNA and provide efficient VEGFA gene knockdown. We showed that siRNA delivery by means of L1 peptide carrier can result in significant decrease of VEGFA gene expression in A172 glioblastoma cells and in EA.hy 926 endothelial cells. Also, delivery of anti-VEGFA siRNA/peptide complexes led to significant inhibition of endothelial cell migration. Our results showed that L1 peptide carrier modified with CXCR4 ligand is a promising tool for targeted siRNA delivery into CXCR4-expressing cancer and endothelial cells.

Key words Cancer gene therapy, VEGFA, RNA interference, Peptide-based carrier, CXCR4, Non-viral delivery

1 Introduction

RNAi is a cellular pathway which uses sequence-specific small double-stranded RNAs (siRNA and miRNA) to target and subsequently cleave complementary mRNA [1]. Numerous studies confirm efficiency of RNAi-mediated knockdown of genes important for cancer cell growth, metastasis, angiogenesis, and chemoresistance, leading to tumor growth suppression [2]. However, development of RNAi-based therapeutics is hampered by delivery and cellular uptake issues. Small interfering RNAs are relatively high molecular weight polyanions which cannot cross cellular membrane easily. Different barriers of siRNA transport to cells must be overcome to gain therapeutic effect [3]. Uptake of siRNA can be greatly improved by means of various mostly cationic carri-

Lekha Dinesh Kumar (ed.), *RNA Interference and Cancer Therapy: Methods and Protocols*, Methods in Molecular Biology, vol. 1974, https://doi.org/10.1007/978-1-4939-9220-1_5,

ers capable of binding large anionic nucleic acids and deliver them into the cells. The most studied types of cationic carriers include polyethylenimine (PEI) and similar polymers, lipids, chitosans, dendrimers, and poly- and oligopeptides [4–9]. Modular proteins and peptides are promising group of siRNA carriers which have high level of biodegradability and consisting of functional modules for overcoming cellular transport barriers and providing targeted delivery [3].Targeted delivery of siRNA to cancer-associated cells can allow using siRNA at lower doses thus reducing nonspecific cytotoxic effects but increasing efficacy. Cell-targeted siRNA delivery can be achieved by modification of carriers with a ligand for specific receptor binding. A wide variety of ligands are available as targeting moieties and new ligand-receptor pairs were suggested to achieve specific delivery into cancer cells. One of the promising targets for specific delivery is chemokine receptor type 4 (CXCR4) which is a receptor of the stromal cell-derived factor-1 (CXCL12) [10–12]. CXCR4 is overexpressed in a variety of cancers including breast cancer, glioblastoma, cervical cancer, etc., but its expression in the adjacent normal tissue is rare that makes it an attractive for targeting [13, 14]. CXCL12 and CXCR4 is an important chemokine ligand/receptor pair, which plays a role in different biological processes including recruitment of endothelial progenitor cells during neoangiogenesis and chemoattraction of CXCR4-expressing normal or tumor stem cells thus mediating stem cell migration and metastasis development [15, 16]. In fact, CXCL12/CXCR4 axis is a master regulator of trafficking in both normal and cancer stem cells, which can contribute from 0.1% to 40% of tumor cell population. Existence of cancer stem cells fraction can explain why cancers recur even when existing treatments kill off up to 99% of a given tumor [17]. Thus, targeting of CXCR4-expressing cancer stem cells may help in complete eradication of tumors and prevent metastasis development. Involvement of CXCR4 into neoangiogenesis can also be exploited for the cancer treatment because tumors that fail to be vascularized do not progress. Inhibition of angiogenesis for cancer therapy is a well-known treatment approach, and RNAi is actively used for silencing of angiogenic factor's genes expression [18, 19]. Multiple factors regulate different steps of angiogenesis, but of particular importance is vascular endothelial growth factor A (VEGFA), which is involved in arterial destabilization, vascular permeability, migration, and proliferation of endothelial cells, including endothelial tube formation [20]. Thus, inhibition of VEGFA signaling by specific siRNA delivered by CXCR4-targeted carriers into endothelial cells is a promising approach to decrease cancer angiogenesis and tumor growth.

Several non-viral carriers conjugated with CXCR4 ligands have been reported to date [21–25]. Firstly, the receptor antagonist AMD3100 (plerixafor) has been utilized as a targeting moiety by Le Bon and co-authors for modification of PEI and cationic lipids, but the developed carriers provided specific DNA delivery to CXCR4-positive cells only when the nonspecific transfection pathways were minimized [21]. Later AMD3100 modification has been extensively studied, and new nucleic acid delivery systems have been proposed [14]. Developed carriers represent AMD3100-modified PLGA nanoparticles or polymeric AMD3100, and its analogues are modified with PEG or cholesterol which are able to simultaneously deliver nucleic acids and block CXCR4 to inhibit cancer cell invasion and metastasis [22, 26–28]. Second small molecule CXCR4 inhibitor—viologen (dialkylated 4,4′-bipyridinium salts)—also has been used as targeting moiety in viologen dendrimers which are promising class of gene delivery vehicles [25]. Other type of the targeting moieties is based on proteins and peptides which are able to bind the receptor. For example, anti-CXCR4 antibodies were used for liposomes modification to target metastatic breast cancer cells and to block their migration via RNAi-mediated silencing of lipocalin-2 migratory pathway. Simultaneously, developed lipoplexes inhibited migration along the CXCR4-CXCL12 axis [29]. CXCR4 receptor targeting by lipoplexes modified with peptide ligands has been demonstrated by Driessen and colleagues who used peptide analog 4F-benzoyl-TE14011. The developed lipoplexes efficiently delivered luciferase gene into rat glioma cells and endothelium of mesenteric arteries [12]. In our efforts to develop CXCR4-targeted delivery of genetic medicines, we designed group of ligands derived from N-terminus of CXCL12 and included it into the composition of modular peptide carriers [23, 24, 30]. Human glioblastoma and endothelial cells were successfully transfected by anti-VEGFA siRNA delivered by modular peptide carrier L1 that resulted in inhibition of VEGFA gene expression and protein production. Also VEGFA silencing led to significant decrease of endothelial cells migration [31]. L1 carrier consists of the KPVSLSYRSPSRFFESH targeting moiety derived from N-terminus of SDF-1 protein separated by two ε-aminocaproic acids (Ahx) from the siRNA binding peptide CHRRRRRRHC [23, 30].

In the present chapter, we describe modular peptide L1-mediated delivery of siRNA against in human glioblastoma (A172) and endothelial cells (EA.hy 926) as a model example.

2 Materials

1. Modular peptide carrier L1 (KPVSLSYRSPSRFFESH-Ahx-Ahx-CHRRRRRRHC) was synthesized with a solid-phase Boc chemistry and supplied as a dry powder. Quantities of 1–2 mg were dissolved in 0.1% TFA at 2 mg/ml and stored in small aliquots at −20 °C. The purity of the peptide, as determined by

high-performance liquid chromatography, should be in the range 90–95%.

2. Nuclease/protease-free water. Store desired amounts of aliquots at −20 °C.

2.1 Assembly of siRNA Duplexes

1. Manufacturer supplied sense and antisense strands of siRNAs; dissolved in nuclease/protease-free water and stored at −20 °C. The sense and antisense sequences of anti-VEGF siRNA 5′-GCG GAU CAA ACC UCA CCA Att-3′ and 3′-ttC GCC UAG UUU GGA GUG GUU-5′ target human VEGFA mRNA [32]. A non-silencing siRNA 5′-UUC UCC GAA CGU GUC ACG U-3′ and 3′-AAG AGG CUU GCA CAG TGC A-5′ served as a mock siRNA [33].
2. Thermocycler.
3. Aluminum plate.

2.2 Preparation of siRNA/Peptide Complexes

1. Synthesized modular peptide carrier L1; dissolved in 0.1% TFA at 2 mg/ml.
2. Hepes-buffered mannitol (HBM) (5% w/v mannitol, 5 mM Hepes, pH 7.5). Store at +4 °C temperature (*see* **Note 1**).
3. Vortexer.
4. siRNA duplexes dissolved in nuclease/protease-free water as per requirement of the experiment.

2.3 siRNA Binding Assay

1. Freshly prepared siRNA/peptide complexes in 1× HBM.
2. Sybr Green intercalating dye.
3. Plate spectrofluorimeter allowing fluorescence measurements.

2.4 In Vitro Transfection of siRNA/Peptide Complexes in Human Glioblastoma Cells and Endothelial Cells EA.hy 926

1. Human glioblastoma A172 cell line.
2. Dulbecco's Modified Eagle's Medium (DMEM). Store at +4 °C.
3. Fetal bovine serum (FBS) Store at −20 °C.
4. Gentamicin 40 mg/ml. Store at +4 °C.
5. Endothelial cells EA.hy 926 (a kind gift from Dr. Cora-Jean C. Edgell, the University of North Carolina, USA) are hybridoma of primary HUVEC (Human Umbilical Vein Endothelial Cells) and A549 cells (human lung adenocarcinoma epithelial cell line).
6. Dulbecco's Modified Eagle's Medium F12 (DMEM-F12). Store at +4 °C.
7. Fetal calf serum (FCS). Store at −20 °C.

8. Hypoxanthine-aminopterin-thymidine (HAT). Store at −20 °C.
9. Incubator under humidification and 5% CO_2 at +37 °C.

2.5 RNA Isolation

1. Trypsin-EDTA solution.
2. 1× PBS without Ca and Mg.
3. TRIzol reagent.
4. Chloroform.
5. Isopropanol.
6. 70% ethanol.
7. Standard laboratory equipment for the isolation of RNA (a fume hood, a centrifuge, tubes, pipettes, and tips).

2.6 cDNA Synthesis

1. cDNA synthesis kit.
2. Hexaprimers (15 OU/ml).
3. A thermal cycler for reverse transcription.
4. A spectrophotometer for RNA quantification.

2.7 Quantitative RT-PCR

1. The EVA Green master mix containing 10× EVA Green, dNTPs mix (2.5 mM), $MgCl_2$ (25 mM), Taq-polymerase (5 U/μl), nuclease/protease-free water.
2. Quantitative PCR was performed using VEGFA forward primer 5′-CAA CAT CAC CAT GCA GAT TAT GC-3′, reverse primer 5′-CCC ACA GGG ATT TTC TTG TCT T-3′, and endogenous reference gene β-actin was detected using forward primer 5′-TGC CGA CAG GAT GCA GAA G-3′, reverse primer 5′-GCC GAT CCA CAC GGA GTA CT-3′.
3. A real-time thermal cycler.

2.8 Scratch Migration Assay

1. Gentamicin 40 mg/ml. Store at +4 °C.
2. Endothelial cells EA.hy 926.
3. Dulbecco's Modified Eagle's Medium F12 (DMEM-F12) Store at +4 °C.
4. Fetal calf serum (FCS). Store at −20 °C.
5. Hypoxanthine-aminopterin-thymidine (HAT). Store at −20 °C.
6. Incubator under humidification and 5% CO_2 at +37 °C.
7. 0.2% crystal violet solution dissolved in 5% methanol.
8. AxioObserver Z1 microscope supplied with AxioVision software.

3 Methods

Use proper controls in all the experiments as required. Peptide solutions, siRNA solutions, nuclease protease-free water, and appropriate buffers are used at room temperature.

3.1 Assembly of siRNA Duplexes

1. Use supplied individual nucleic acid sense and antisense strands. Take appropriate concentrations of these sense and antisense strands to prepare duplexes of siRNA in nuclease protease-free water to get 100 μM stock solutions.
2. Preheat an aluminum plate at +70 °C in thermostat.
3. Heat the duplexes in a thermocycler at +90 °C in nuclease protease-free water for 3 min. Remove the duplexes from the thermocycler and let them anneal at slowly cooling aluminum plate (till room temperature) for 2 h.

3.2 Preparation of siRNA/Peptide Complexes

1. Take appropriate volumes in microlitres from peptide solution (2 mg/ml) to prepare complexes at various N/P ratios (*see* **Note 2**).
2. Take appropriate volumes in microliters from siRNA duplexes solution (100 μM) in 2× HBM buffer.
3. Add appropriate volumes of siRNA to the L1 peptide such that it matches the final concentration for the experiments and vortex for 30 s.
4. Incubate the mixture of L1 peptide and nucleic acids for 2 h at room temperature to ensure interpeptide disulfide bonds formation.

3.3 siRNA Binding Assay

Peptide binding to siRNA was monitored with SYBR Green fluorescence quenching method (Fig. 1).

1. Prelabel siRNA duplexes solution with 1× SYBR Green.
2. Prepare siRNA/peptide complexes at N/P ratios ranging from 0.1 to 5 in 96-well plate suitable for spectrofluorometric measurement.
3. Perform fluorescence measurement in scanning multilabel reader with 485 nm excitation and 590 nm emission.
4. SYBR Green displacement is calculated as (F − Ff)/(Fb − Ff), where Ff and Fb are the fluorescence intensities of SYBR Green in the absence and presence of siRNA, respectively.

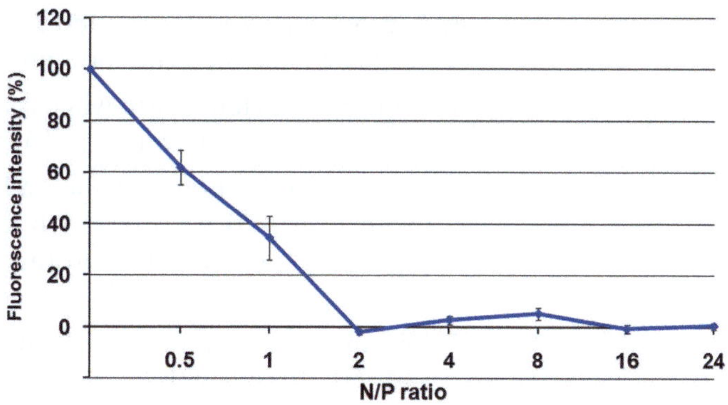

Fig. 1 Sybr Green quenching in siRNA/peptide L1 complexes. 100% of fluorescence intensity corresponds to free siRNA stained with SYBR Green. The data are shown as the mean ± SD

3.4 In Vitro Transfection of siRNA/Peptide Complexes in Human Glioblastoma Cells A172 and Endothelial Cells EA.hy 926

1. The cells are grown under cell culture conditions of humidified 5% CO_2 incubator at +37 °C. 24-well plates are used for the experiments.
2. One day before the experiment, seed A172 or EA.hy 926 cells at a density of 140,000 cells/well in 24-well plates in 1 ml of complete media in the incubator.
3. Prepare siRNA/peptide complexes at N/P ratios 8 and 16 as described above.
4. Wash each well of the cells two times with prewarmed (37 °C) DMEM (for A172) or DMEM-F12 (for EA.hy 926).
5. Add 1 ml of prewarmed (+37 °C) incomplete media-DMEM (for A172) or DMEM-F12 (for EA.hy 926).
6. Add siRNA/peptide complexes to each well to reach final siRNA concentration 200 nM.

   Incubate the cells at +37 °C for 4 h with siRNA/peptide complexes.
7. After 4 h of incubation, wash the cells with prewarmed (37 °C) DMEM (for A172) or DMEM-F12 (for EA.hy 926) two times.
8. Replace the incomplete media with 1 ml of prewarmed complete media.
9. Incubate the cells for 48 h.

3.5 RNA Isolation

1. After 48 h of incubation, wash the cells with 1× PBS without Ca and Mg two times.
2. Then dissociate the cells with trypsin-EDTA solution for 10 min at +37 °C.
3. Add 150 μl of 1× PBS without Ca and Mg while centrifuging them for 10 min at 200 × *g*.

4. Remove supernatant then add 125 μl of TRIzol, resuspend the cell pellet, and incubate for 5 min.
5. Add 25 μl of chloroform while centrifuging in pre-cooled centrifuge (+4 °C) for 15 min at 15,000 × *g*.
6. Take upper fraction and then add 63 μl of isopropanol and incubate at −70 °C at least 1 h for precipitation while centrifuging in pre-cooled centrifuge (+4 °C) for 20 min at 15,000 × *g*.
7. Discard supernatant and wash the RNA pellet with pre-cooled (+4 °C) 70% ethanol.
8. Dry the pellet for 1 h then dissolve in 50 μl of nuclease/protease-free water.

3.6 cDNA Synthesis

1. Mix 5 μl of RNA (600–900 ng), 1 μl of hexaprimers, and 15 μl of nuclease/protease-free water.
2. Incubate for 5 min at 70 °C.
3. For cDNA synthesis, prepare cDNA synthesis master mix by combining per reaction 2.5 μl reverse transcription buffer, 1 μl of 2 mM each dNTP, 0.5 μl of 200 U/μl M-MuLVRTase, and as according to the suppliers.
4. Incubate in a thermocycler in following conditions: 10 min at +25 °C, 60 min at +37 °C, 10 min at +70 °C.
5. Determine cDNA concentration by means of spectrophotometer.

3.7 Reverse Transcription-Quantitative Real-Time PCR (RT-qPCR)

1. Prepare a master mix for qPCR and simultaneously perform the qPCR reactions for the cDNA samples, a cDNA negative control, and a qPCR negative control. Master mix for qPCR should be prepared according to the manufacturer's instructions; e.g., for a single reaction, add the following components to a nuclease-free microcentrifuge tube:
2. 11.2 μl nuclease/protease-free water.
   2.5 μl 10× EVA Green I master mix.
   1 μl of the forward primer VEGFA (5 pmol/μl).
   1 μl of the reverse primer VEGFA (5 pmol/μl).
3. Prepare a similar master mix for qPCR using forward and reverse primers for β-actin expression analysis in the same samples as reference control.
4. Add 1.25 μg of sample cDNA to the each reaction tube with exception of the negative control.
5. Run the samples in parallel in real-time thermocycler at 94 °C for 10 min followed by 40 cycles of 95 °C for 20 s and 60 °C for 60 s.

6. Subsequent analysis was performed using the thermocycler software. All samples were measured three times, and a final result was inferred by averaging data. A standard curve of Ct value vs. log amount of standard was generated for VEGFA and β-actin genes using five times dilutions of cDNA (2; 1; 0.5; 0.25; 0.125 μg) prepared from total RNA extracted from untreated control cells.
7. Relative level of VEGFA gene expression is calculated by formula (Cv/Ca)/(Cv′/Ca′), where Cv and Ca are level of VEGFA and β-actin gene expression in the studied sample, whereas Cv′ and Ca′ are level of VEGFA and β-actin gene expression in the untreated control cells, respectively.

3.8 Scratch Migration Assay

1. For cell migration study seed 16,000 EA.hy 926 cells in 175 μl DMEM-F12 complete media per well of 96-well plate.
2. Prepare siRNA/peptide complexes as described above at N/P ratios 8/1 and 16/1 in quadruplicates (use 0.5 μg of siRNA per well).
3. Before transfection replace cell culture media with incomplete media without FCS.
4. Add siRNA/peptide complexes to the cells and incubate for 4 h then wash the cells by DMEM-F12 and scratch the cell monolayer with a 200–300 μl pipette tip (*see* **Note 3**).
5. Then wash the cells by 1× PBS without Ca and Mg, add culture medium containing 10% FCS, and photograph width of scratch (wound area) by AxioObserver Z1 microscope equipped with digital camera.
6. Incubate the cells for 24 h cells then wash by 1× PBS without Ca and Mg and stain in 100 μl of 0.2% crystal violet in 5% ethanol. Cells incubated in culture medium containing 10% FCS were used as positive control.
7. Stained cells should be dried and photographed by AxioObserver Z1 microscope. Three random fields should be registered.
8. EA.hy 926 cell migration during the wound repair is analyzed using ImagePro Plus 6.0 software. Number of cells (N) migrated to the wound area is counted. Cell density (ρ) is counted in area of 17,000 μm^2 (Fig. 2). Relative number of migrated cells is calculated by formula $(N/N') \times (\rho'/\rho)$, where N' is number of migrated cells in untreated control and ρ' is cell density in untreated control.

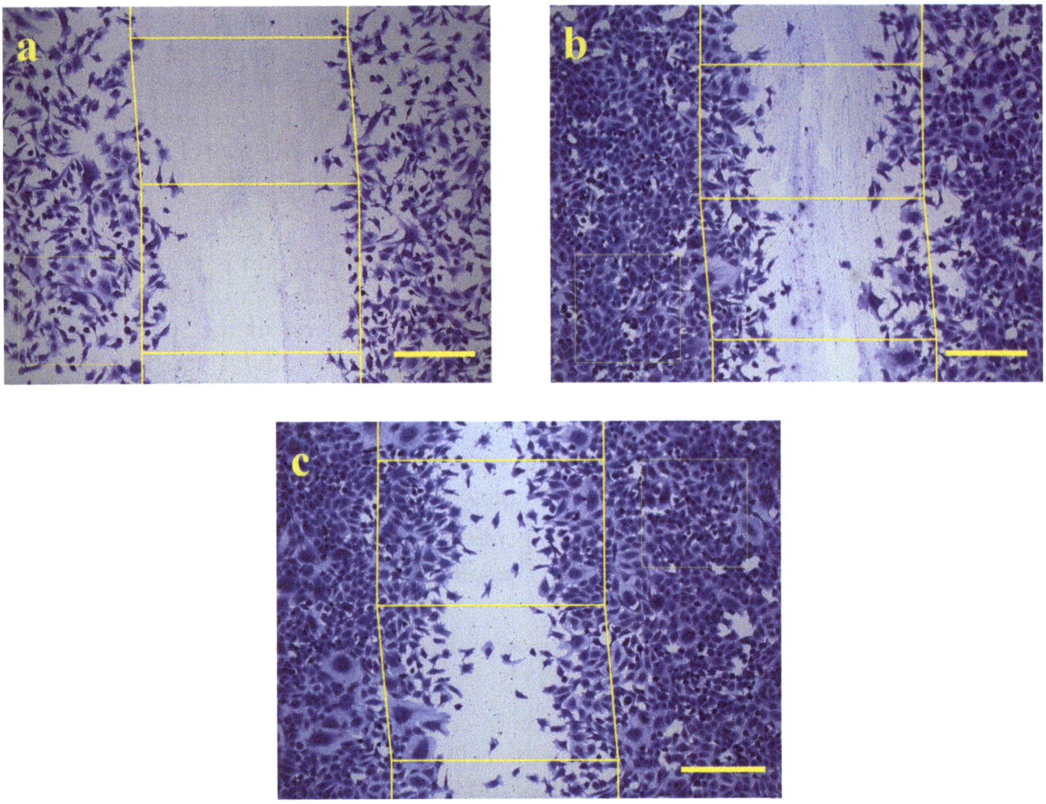

Fig. 2 Scratch migration assay on endothelial EA.hy 926 cells. (**a**) anti-*VEGFA* siRNA/peptide L1, N/P ratio 1/8; (**b**) mock siRNA/peptide L1, N/P ratio 1/8; (**c**) untreated control cells. Scale bar is 100 μm. The cell density was calculated in square area (17,000 μm^2)

4 Notes

1. Hepes-buffered mannitol (HBM) should be prepared as 4× stock solution (20% w/v mannitol, 20 mM Hepes, pH 7.5) because of low solubility of mannitol. Also it should be noted that storage at +4 °C temperature led to formation of crystals that can be dissolved by moderate heating. After the heating of HBM, it should be cooled up to room temperature.
2. N/P ratio means ratio between NH_2^+ and PO_4^- groups. This ratio firstly was suggested to formulate complexes of polylysine and DNA and means ratio between positive and negative charges [3]. Actually in peptides not only NH_2^+ groups are positive but guanidine group in arginine and imidazole group in histidine (only when pH is below 6.5) are also should be considered. In the present work, the molar N/P charge ratio of L1 peptide to negatively charged PO_4^- groups in DNA was calculated considering that 1 mg of cationic L1 peptide corresponds

to 2.18 nmol of positive charges, and also it should be noted that 1 mg of DNA has 3.08 nmol of negatively charged groups.

3. To make scratches of similar sizes, it is critical to minimize any possible variation caused by a difference in the width of the scratches.

Acknowledgments

This work was supported by Russian Science Foundation grant 14-15-00737. Marianna Maretina is supported by President of Russian Federation scholarship (SP-822.2018.4).

References

1. Ameres SL, Martinez J, Schroeder R (2007) Molecular basis for target RNA recognition and cleavage by human RISC. Cell 130:101–112. https://doi.org/10.1016/j.cell.2007.04.037
2. Tong AW, Zhang Y-A, Nemunaitis J (2005) Small interfering RNA for experimental cancer therapy. Curr Opin Mol Ther 7:114–124
3. Egorova AA, Kiselev AV (2016) Peptide modules for overcoming barriers of nucleic acids transport to cells. Curr Top Med Chem 16:330–342. https://doi.org/10.2174/1568026615666150812120755
4. Boussif O, Lezoualc'h F, Zanta MA et al (1995) A versatile vector for gene and oligonucleotide transfer into cells in culture and in vivo: polyethylenimine. Proc Natl Acad Sci 92:7297–7301. https://doi.org/10.1073/pnas.92.16.7297
5. Felgner JH, Kumar R, Sridhar CN et al (1994) Enhanced gene delivery and mechanism studies with a novel series of cationic lipid formulations. J Biol Chem 269:2550–2561
6. Huh MS, Lee EJ, Koo H et al (2017) Polysaccharide-based nanoparticles for gene delivery. Top Curr Chem 375:31. https://doi.org/10.1007/s41061-017-0114-y
7. Dufès C, Uchegbu IF, Schätzlein AG (2005) Dendrimers in gene delivery. Adv Drug Deliv Rev 57:2177–2202
8. Kiselev AV, Il'ina PL, Egorova AA et al (2007) Lysine dendrimers as vectors for delivery of genetic constructs to eukaryotic cells. Russ J Genet 43:593–600. https://doi.org/10.1134/S1022795407060014
9. Guryanov IA, Vlasov GP, Lesina EA et al (2005) Cationic oligopeptides modified with lipophilic fragments: use for DNA delivery to cells. Russ J Bioorganic Chem 31:18–26. https://doi.org/10.1007/s11171-005-0002-z
10. Juarez J, Bendall L, Bradstock K (2004) Chemokines and their receptors as therapeutic targets: the role of the SDF-1/CXCR4 axis. Curr Pharm Des 10:1245–1259. https://doi.org/10.2174/1381612043452640
11. Andre F, Soria JC, Assi H et al (2004) Expression of chemokine receptors by cancer cells. Bull Cancer 91(Suppl 4):S254–S256
12. Driessen WHP, Fujii N, Tamamura H, Sullivan SM (2008) Development of peptide-targeted lipoplexes to CXCR4-expressing rat glioma cells and rat proliferating endothelial cells. Mol Ther 16:516–524. https://doi.org/10.1038/sj.mt.6300388
13. Fulton AM (2009) Chemokine receptors in cancer. Humana, New York. https://doi.org/10.1007/978-1-60327-267-4
14. Wang Y, Xie Y, Oupicky D (2016) Potential of CXCR4/CXCL12 chemokine axis in cancer drug delivery. Curr Pharmacol Rep 2:1–10. https://doi.org/10.1007/s40495-015-0044-8
15. Kucia M, Reca R, Miekus K et al (2005) Trafficking of normal stem cells and metastasis of cancer stem cells involve similar mechanisms: pivotal role of the SDF-1-CXCR4 axis. Stem Cells 23:879–894. https://doi.org/10.1634/stemcells.2004-0342
16. Salcedo R, Oppenheim JJ (2003) Role of chemokines in angiogenesis: CXCL12/SDF-1 and CXCR4 interaction, a key regulator of endothelial cell responses. Microcirculation 10:359–370. https://doi.org/10.1038/sj.mn.7800200

17. Al-Hajj M, Clarke MF (2004) Self-renewal and solid tumor stem cells. Oncogene 23: 7274–7282. https://doi.org/10.1038/sj.onc.1207947
18. Chen S, Feng J, Ma L et al (2014) RNA interference technology for anti-VEGF treatment. Expert Opin Drug Deliv 11:1471–1480. https://doi.org/10.1517/17425247.2014.926886
19. Shubina AN, Egorova AA, Baranov VS, Kiselev AV (2013) Recent advances in gene therapy of endometriosis. Recent Pat DNA Gene Seq 7:169–178. https://doi.org/10.2174/18722156113079990021
20. Carmeliet P, Jain RK (2000) Angiogenesis in cancer and other diseases. Nature 407: 249–257
21. Le Bon B, Van Craynest N, Daoudi J-M et al (2004) AMD3100 conjugates as components of targeted nonviral gene delivery systems: synthesis and in vitro transfection efficiency of CXCR4-expressing cells. Bioconjug Chem 15:413–423. https://doi.org/10.1021/bc034220o
22. Li J, Zhu Y, Hazeldine ST et al (2012) Dual-function CXCR4 antagonist polyplexes to deliver gene therapy and inhibit cancer cell invasion. Angew Chem Int Ed Engl 51:8740–8743. https://doi.org/10.1002/anie.201203463
23. Egorova A, Kiselev A, Hakli M et al (2009) Chemokine-derived peptides as carriers for gene delivery to CXCR4 expressing cells. J Gene Med 11:772–781. https://doi.org/10.1002/jgm.1366
24. Egorova A, Bogacheva M, Shubina A et al (2014) Development of a receptor-targeted gene delivery system using CXCR4 ligand-conjugated cross-linking peptides. J Gene Med 16:336–351. https://doi.org/10.1002/jgm.2811
25. Li J, Lepadatu AM, Zhu Y et al (2014) Examination of structure-activity relationship of viologen-based dendrimers as cxcr4 antagonists and gene carriers. Bioconjug Chem 25:907–917. https://doi.org/10.1021/bc500191q
26. Li J, Oupický D (2014) Effect of biodegradability on CXCR4 antagonism, transfection efficacy and antimetastatic activity of polymeric Plerixafor. Biomaterials 35:5572–5579. https://doi.org/10.1016/j.biomaterials.2014.03.047
27. Wang Y, Li J, Chen Y, Oupický D (2015) Balancing polymer hydrophobicity for ligand presentation and siRNA delivery in dual function CXCR4 inhibiting polyplexes. Biomater Sci 3:1114–1123. https://doi.org/10.1039/C5BM00003C
28. Wang Y, Li J, Oupický D (2014) Polymeric plerixafor: effect of PEGylation on CXCR4 antagonism, cancer cell invasion, and DNA transfection. Pharm Res 31:3538–3548. https://doi.org/10.1007/s11095-014-1440-1
29. Guo P, You JO, Yang J et al (2014) Inhibiting metastatic breast cancer cell migration via the synergy of targeted, pH-triggered siRNA delivery and chemokine axis blockade. Mol Pharm 11:755–765. https://doi.org/10.1021/mp4004699
30. Kiselev A, Egorova A, Laukkanen A et al (2013) Characterization of reducible peptide oligomers as carriers for gene delivery. Int J Pharm 441:736–747. https://doi.org/10.1016/j.ijpharm.2012.10.020
31. Egorova A, Shubina A, Sokolov D et al (2016) CXCR4-targeted modular peptide carriers for efficient anti-VEGF siRNA delivery. Int J Pharm 515:431–440. https://doi.org/10.1016/j.ijpharm.2016.10.049
32. De Fougerolles A, Frank-Kamenetsky M, Manoharan M et al (2011) IRNA agents targeting VEGF. US Patent No. 7,919,473, 5 Apr 2011
33. Xu H, Jiang B, Meng L et al (2012) N-α-acetyltransferase 10 protein inhibits apoptosis through RelA/p65-regulated MCL1 expression. Carcinogenesis 33:1193–1202. https://doi.org/10.1093/carcin/bgs144

Chapter 6

Design Strategy to Access siRNA-Encapsulating DNA "Nanosuitcases" That Can Conditionally Release Their Cargo

Katherine E. Bujold, Hassan H. Fakih, and Hanadi F. Sleiman

Abstract

DNA nanotechnology enables the design and assembly of DNA nanostructures with unprecedented control over their size and shape. Additionally, the programmable base-pairing alphabet of DNA allows the incorporation of responsive units within these DNA nanostructures. Here, we describe a general design strategy to construct responsive DNA prisms that can encapsulate and selectively release an encapsulated siRNA upon recognition of an oligonucleotide trigger. This prismatic DNA scaffold design is adaptable and can encapsulate oligonucleotides of any length and type. Moreover, the prism can be made to respond to an oligonucleotide trigger of interest like a messenger RNA (mRNA) or a microRNA (miRNA), thus enabling dual or synergistic therapeutic strategies. We present an overview of the design strategy used to access these DNA nanostructures, followed by the steps involved in DNA sequence generation, assembly, and validation of this construct.

Key words DNA nanotechnology, DNA nanostructure design, DNA nanostructure sequence generation, Oligonucleotide drug encapsulation, Conditional release

1 Introduction

The last four decades have seen the birth and rise of DNA nanotechnology as an accessible method to synthesize monodisperse DNA nanostructures of any shape and size [1]. The founder of the field, Nadrian Seeman, inspired by the Holliday junctions that form in DNA homologous recombination, envisaged using rigid DNA junctions further assembled into well-defined extended arrays, to crystallize challenging biomolecules [2]. The field has since expanded and can be subdivided into three main approaches: DNA tile assembly [3], which uses rigid DNA tiles to produce extended patterned assemblies; DNA origami [4], where a long DNA backbone strand directs the assembly of hundreds to thousands of DNA strands; and supramolecular DNA assembly [5], which is our laboratory's specialty and creates highly functional

Lekha Dinesh Kumar (ed.), *RNA Interference and Cancer Therapy: Methods and Protocols*, Methods in Molecular Biology, vol. 1974, https://doi.org/10.1007/978-1-4939-9220-1_6, © Springer Science+Business Media, LLC, part of Springer Nature 2019

DNA-minimal nanostructures by imparting synthetic insertions onto them. Although the resulting DNA assemblies vary in size and topology, these three approaches share a common DNA design stage that seeks to program the nanoscale assembly of DNA.

We recently became interested in the design parameters required to assemble a DNA-minimal nanostructure that can selectively encapsulate and release a drug in a biological setting. For this purpose, we chose to investigate oligonucleotide drugs due to their potent and selective gene-silencing activity that is often plagued by poor pharmacokinetic profiles [6]. In a recent publication, we have demonstrated the design, assembly, and validation of a siRNA-encapsulating DNA prism that can conditionally release its cargo upon recognition of an oligonucleotide trigger [7]. This design allows for dual or synergistic therapy approaches if combined with mRNA or miRNA triggers.

In this book chapter, we describe a DNA nanostructure design strategy that allows the encapsulation and conditional release of a small interfering RNA (siRNA). The guiding principles behind the design of a responsive DNA prismatic scaffold are presented, followed by DNA sequence generation using a nucleic acid design software that minimizes secondary interactions. The assembly and conditional release properties of the DNA prism are then probed using polyacrylamide gel electrophoresis (PAGE). The cargo protection capabilities of the prism can be optimized using an enzymatic assay based on the site-specific cleavage of GAATTC by the EcoRI enzyme. Finally, the serum stability of the resulting constructs can be estimated from incubations in 10% fetal bovine serum (FBS) mixtures with cell culture media. In all, this protocol provides an overview of the steps involved in the design, assembly, validation, and optimization of functional DNA nanostructures.

2 Materials

2.1 DNA Computer-Aided Design

1. CANADA_2.0: All DNA sequence generations are carried using CANADA_2.0, a software that helps design DNA sequences for nanobiotechnology applications through minimization of secondary interactions [8] (*see* **Note 1**).
2. An oligonucleotide therapeutic against a chosen target. This protocol describes the encapsulation of a typical siRNA formed of a 19-base duplex with two-nucleotide overhangs at the 3′ end (*see* **Note 2**).
3. Two oligonucleotide sequences of at least 21 bases to be used as triggers for release. Alternatively, a single oligonucleotide sequence of at least 42 bases can be used (*see* **Note 3**). In both cases, the length of self-complementary interactions should be minimized, with melting temperatures below room temperature.

2.2 Polyacrylamide Gel Electrophoresis (PAGE)

Unless otherwise noted, all materials are kept at room temperature.

1. All steps involving DNA and related buffers are conducted using autoclaved ultrapure water, which was obtained from filtering deionized water at a sensitivity of 18 MΩ-cm at room temperature. When RNA was present, all solutions and buffers were prepared using diethylpyrocarbonate (DEPC)-treated ultrapure water. For this purpose, 0.1% DEPC was added to the water or buffer to be treated, thoroughly mixed, left to sit for 24 h, and autoclaved prior to use.
2. PAGE- or HPLC-purified DNA and RNA oligonucleotides with sequences and customizations are determined in the DNA design and computer-aided sequence generation section. Custom oligonucleotides can be purchased from many vendors or can be synthesized in-house using automated procedures [9]. All routinely used DNA oligonucleotides were stored in autoclaved ultrapure water at 4 °C. Otherwise, they were kept at −20 °C and fully thawed prior to usage. In contrast, RNA-containing oligonucleotides were kept frozen at −20 °C at all times in DEPC-treated water and only thawed when required.
3. A 96-well thermocycler to conduct thermal anneals.
4. A 20 × 20 cm gel vertical Hoefer 600 electrophoresis unit placed in a temperature-controlled water bath or an incubation chamber.
5. A fluorescence gel imaging system.
6. 10×TAMg buffer pH 8.0: 450 mM Tris and 76 mM magnesium chloride made up to 1 l with ultrapure water, pH 8.0 adjusted using acetic acid. 1×TAMg is prepared by diluting this buffer 1/10 with ultrapure water.
7. Native PAGE gels (6%): These can be obtained by polymerizing 25 ml of a 6% native PAGE solution (2.5 ml 10×TAMg and 3.75 ml of 40% acrylamide/bis-acrylamide 19:1 solution made up to 25 ml with ultrapure water) with 25 mg ammonium persulfate and 25 μl *N,N,N,N'*-tetramethylethylenediamine (TEMED).
8. GelRed DNA stain (Biotium) diluted to 1× in 1×TAMg buffer, prepared fresh every time, and stored in the dark.
9. Glycerine solution: 70% glycerol diluted in ultrapure water.
10. EcoRI endonuclease, stored at −20 °C.
11. Fetal bovine serum, stored at −20 °C.
12. Cell culture media, stored at 4 °C after opening.

3 Methods

3.1 DNA Prism Design and Considerations

This section describes the steps involved in the design of a cavity-spanning duplex inside a wireframe DNA prism with dimensions of 10 × 21 × 21 bases (Fig. 1a). These dimensions were chosen since they are integer multiples of the helical twist of DNA, which is 10.5 bases per turn and provide enough space in the cavity to fit a siRNA consisting in a 19-base duplex with two-nucleotide overhangs. All vertices of this prismatic DNA nanostructure are equipped with two to four bases of single-stranded spacers or hexaethylene glycol units to allow for flexibility and overall strain relief in the construct (Fig. 1a, highlighted in blue).

The selective release of the siRNA is mediated by two single-stranded triggers that are fully complementary to the red and orange regions on the prism scaffold (Fig. 1b). These two single-stranded triggers recognize the single-stranded regions next to the cargo strands and fully hybridized, thus releasing the cargo strand in a process called strand displacement. This is due to the favored increased thermodynamic stability of a fully hybridized duplex with the trigger strands compared to the cargo strands, which are only hybridized on a fraction of the 21 bases available for binding. The remainder of the prism provides a rigid double-stranded scaffold that surrounds the siRNA.

1. Determine the length of the siRNA duplex to be encapsulated. Typically, siRNAs consist in 19-base RNA duplexes with two-nucleotide DNA overhangs at the 3′ end [6].
2. Determine the nature of the spacer units. It is recommended to include spacers at each vertex for strain relief. Common spacers include 2–4 nucleotide regions or hexaethylene glycol units, where each unit is similar in size to a 4-nucleotide spacer with more conformational flexibility.
3. Determine the binding configuration of the cargo strand. The cargo strand consists of three domains: one middle domain that carries the sense strand of the siRNA and two binding domains that hybridize to the scaffold (Fig. 1a). It is recommended to choose a binding pattern where the hybridized strand and the middle section form an angle ≥90° to minimize strain, intermolecular binding of the cargo strand, and aggregation (Fig. 1c). However, this reduces the number of bases hybridized to the scaffold and the melting temperature of the overall assembly. It is recommended to use binding regions of 11 bases, which are expected to bind at approximately 90° angle and to use base modifications that enhance duplex stability, such as locked nucleic acids, on the DNA scaffold to mitigate this effect. However, if using only unmodified DNA bases, it is possible to hybridize the scaffold with up to 16

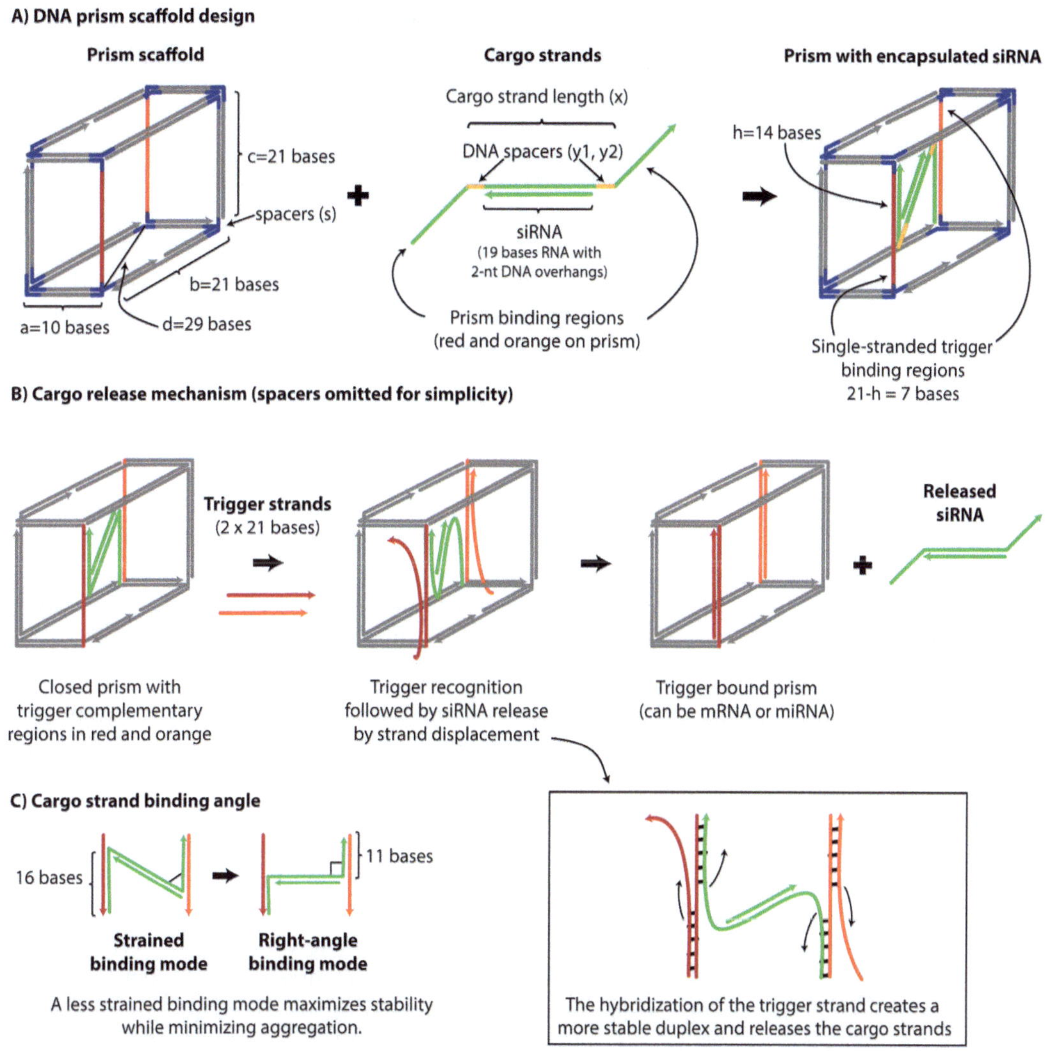

Fig. 1 Design of a siRNA-encapsulating DNA prism. (**a**) Prism scaffold design and dimensions. The cargo strands (green) hybridize to the red and orange regions on the prism and leave single-stranded regions for trigger recognition and strand displacement. (**b**) Cargo release mechanism. The two trigger strands recognize the single-stranded regions on the prism and displace the cargo strands to achieve full complementarity. (**c**) Cargo strand binding mode. The number of bases with which the cargo strands hybridize the DNA prism scaffold has an impact on its stability and propensity to aggregate

bases, with 14 bases displaying a good compromise between binding affinity and conformational strain.

4. Determine the approximate length of the cargo strand (x) using trigonometry (Fig. 1a). This step is required to identify the length of the spacers that will be placed on both sides of the siRNA to be encapsulated, allowing it to bind to the scaffold ($y1$ and $y2$, Fig. 1a). The length of the cargo strand (x) can be computed as follows for a prism with dimensions $10 \times 21 \times 21$ bases ($a \times b \times c$) where the cargo strands hybridize with 14

bases (h) and to which were added hexaethylene glycol spacers (equivalent to 4-nucleotide spacers) (s). First, calculate the size of the longest diagonal (d) across the base of the prism which should be equal to.

$$d = \sqrt{(a+s)^2 + (b+s)^2} = \sqrt{(10+4)^2 + (21+4)^2} \approx 29 \text{ DNA bases}$$

Then, compute the length of the cargo strand (x), which depends on the number of bases with which the cargo strands hybridize to the scaffold (h), the number of bases left unpaired on the prism for trigger recognition (21-h), and the diagonal of the base (d).

$$x = \sqrt{d^2 + (h-(21-h))^2} = \sqrt{29^2 + (14-(21-14)^2} \approx 30 \text{ DNA bases}$$

In the case of h = 14-base binding region, the length of the cargo strand could also be directly approximated as the length of the diagonal across the base since it adopts a conformation close to the right angle binding mode illustrated in Fig. 1c (*see* **Note 4**).

5. Based on the result from the previous step, determine the length of the DNA arms that will be added on each side of the sense strand to bind the scaffold. In this example, the prism cavity is expected to be able to fit a 30-base DNA strand. Many siRNA oligonucleotide drugs are composed of a 19mer RNA duplex, which is approximately equivalent in length to a 17mer DNA duplex (*see* **Note 5**), with a two-nucleotide DNA overhangs for a total length approximately equal to 19 DNA bases. The remaining 11 bases will be distributed equally on each side of the sense strand and act as spacers (Fig. 1a, *y*1, and *y*2, represented in yellow).

3.2 Computer-Aided Sequencing with CANADA_2.0

At this point, the main size parameters of the DNA prism have been determined. The next step consists in determining the sequences of each of its eight component strands using CANADA_2.0, a software that minimizes secondary interactions.

1. Build an input file for sequence generation. This file should be in text format and saved as input.txt in the same directory as the CANADA_2.0 program (*see* **Note 6** and example input file included as supporting information).
2. In the first section, divide each of the strands composing the prism into independent sets of sequences, one for each binding or single-stranded region (Fig. 2a–c). For each sequence to be generated define (*see* Fig. 2c for example):
   (a) An identifier: It should start with a letter and can contain only letters, numbers, and underscores. Note that the

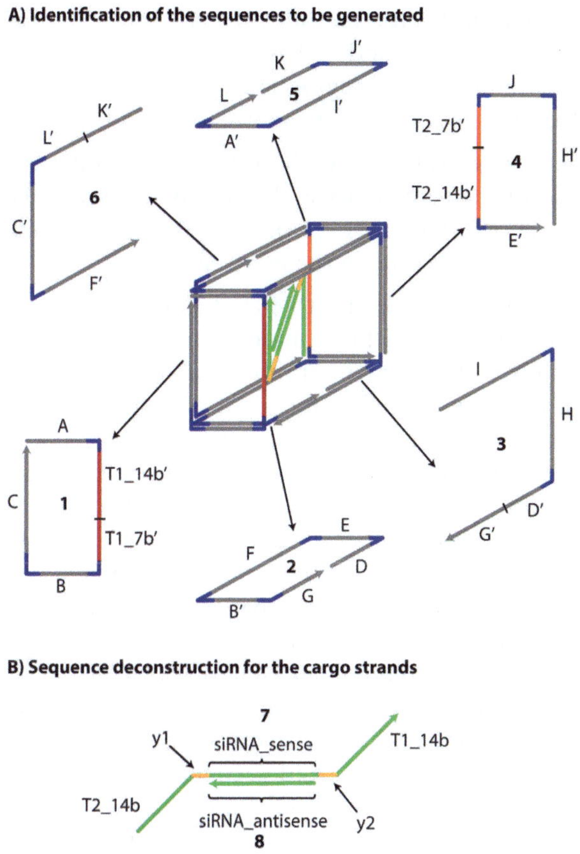

C) Example input for strand 4

```
// 1. Unknown sequence input:
SEQUENCE H {
length = 21;
GC_ratio = [0.3;0.7];
}
SEQUENCE J {
length = 10;
GC_ratio = [0.3;0.7];
}
SEQUENCE E {
length = 10;
GC_ratio = [0.3;0.7];
}

// 2. Fixed sequence input:
SEQUENCE T2_7b {
seq_mask = "CATACAG";
}
SEQUENCE T2_14b {
seq_mask = "AACTCCTAATCCAG";
}

// 3. Connect strands 5' to 3' (two at a time)
CONCAT complement(H), J;
CONCAT J, complement(T2_7b);
CONCAT complement(T2_7b), complement(T2_14b);
CONCAT complement(T2_14b), complement(E);

// 4. Define tolerated secondary interactions
POOL mypool {
n_uniqueness = 5;
violation_tolerance = 0;
}
DESIGNTOOL {
ID_in_table = true;      //show sequences in sequence.txt
}
```

Fig. 2 Dividing the prism strands for DNA sequence generation. (**a**) Deconstruction of each of the strands composing the prism into unique sequences, one for each independent binding region. (**b**) Deconstruction of the cargo strands into unique sequences and names. (**c**) Example input in CANADA_2.0 for strand 4

complementary strands to each of these sequences do not need to be defined in this section since the system will evaluate and generate them automatically.

(b) A length in bases based on the design considerations established above.

(c) The desired GC contents: In this example, the GC content was set between 0.3 and 0.7 to avoid forming sequences with only one type of base pair.

3. In the second section, add the fixed sequences for the siRNA and the two trigger strands from 5′ to 3′, along with their names (Fig. 2c). Input these sequences in DNA bases, replacing uracil (U) with thymine (T). Moreover, since the siRNA contains 2-nucleotide overhangs, only the 19 bases forming the core will be complementary. It is easier in this case to input these two items as two separate sequences, one for the sense strand and one for the antisense strand. Similarly, based on the

strand deconstruction in Fig. 2a, input the trigger strands as two separate pieces of 14 and 7 bases to account for the 14-base cargo strand binding region and the leftover 7-base single-stranded recognition site for the trigger strand.

4. In the third section, concatenate each of the eight component strands of the prism from 5′ to 3′ based on the sequence deconstruction above (Fig. 2a, c). Note that strands can only be concatenated two at a time in the input file.
5. In the fourth section, define the tolerated number of bases allowed for secondary interactions under n_uniqueness. This parameter ensures that sequences of this defined length occur only once and that their complement is not present, thus limiting secondary interactions to the number defined in the input file. Define also the number of times this rule can be broken in violation_tolerance. The violation_tolerance parameter allows for a defined number of sequences that will break the n_uniqueness rule after concatenation. For complex systems with many consecutive concatenated sequences, it is sometimes necessary to increase this parameter in order to generate sequences with a minimal number of secondary interactions. It is recommended to start with n_uniqueness = 2 and go progressively upward until a successful set of sequences is generated. Whenever possible, avoid secondary interactions that are more than half the length of a given sequence. For example, if a 7-base secondary interaction exists, it is preferable to place it on a 21-base binding region of the prism instead of a 10-base binding region. Whenever possible, keep the violation parameter at 0 or else at a maximum of 1.
6. Verify that the input.txt file compiles without error. It is likely the program will not be able to find a solution for very low n_uniqueness parameters. Increase the n_uniqueness by one every time the program fails to find a solution that minimizes the uniqueness parameter while keeping the violation tolerance under at 0 (*see* **Note 7**). Continue until the program converges. Alternatively, the violation_tolerance parameter can be increased to allow for more secondary interactions to be present in the system.
7. Regenerate strands 1–8 while adding spacers between their component sequences, based on Fig. 2a. We recommend thymine spacers since they form less secondary structures than other bases or hexaethylene glycol spacers due to their high conformational flexibility (*see* **Note 8**). Distribute the 11-base spacer calculated in **step 5** of Subheading 3.1 on each side of the siRNA, with $y1 = 6$ bases and $y2 = 5$ bases.
8. Verify secondary interactions using the OligoAnalyzer 3.1 tool from IDT (*see* **Note 9**). Manual insertion of spacers some-

times leads to undesired secondary interactions. Rerun the program if needed or change the spacer if possible.

9. Incorporate synthetic modifications to improve the properties of the DNA prism as desired.
   (a) For example, the cavity-spanning strand [7] can be stabilized using locked nucleic acids. Each insert can increase the melting temperature of a DNA duplex by 2–8 °C, which compensates for the low number of bases with which it hybridizes to the scaffold [10].
   (b) Protect single-stranded and spacer regions from enzymatic degradation by incorporating a phosphorothioate backbone to these regions. This modification prevents enzyme recognition and degradation of the DNA strand [11]. However, it lowers affinity, so it should be avoided in DNA duplexes whenever possible.
   (c) It is also possible to modulate the flexibility of the prism corners by comparing different lengths of T spacers and hexaethylene glycol units and measuring the melting temperature of the assembly. A more stable construct will display a higher melting temperature with a more cooperative melting transition.

3.3 DNA Cage Assembly, Validation, and Optimization Using Native PAGE

1. Determine the concentration of each of the strands to be used for prism assembly. Oligonucleotide concentrations can be determined using UV-visible spectroscopy from their absorbance at 260 nm. Concentrations are determined by dividing this value by the path length of the cell used for this measurement and the extinction coefficient of the strand (*see* **Note 9**), according to Beer's law.
2. Assemble the prism by mixing equimolar amounts of the strands forming the desired assembly in 1×TAMg buffer. Assembling the prism at a concentration range of 0.5–1 μM typically yields satisfactory results. To verify that all duplexes participate in the assembly as programmed, prepare the following samples where each of the strands composing the final product is added sequentially. This ensures the final product is assembled from all its component strands as designed (Fig. 3a, Lanes 1–8).

   Strand 2

   Strands 2 + 3

   Strands 2 + 3 + 5

   Strands 2 + 3 + 5 + 6

   Strands 2 + 3 + 5 + 6 + 4

   Strands 2 + 3 + 5 + 6 + 4 + 1

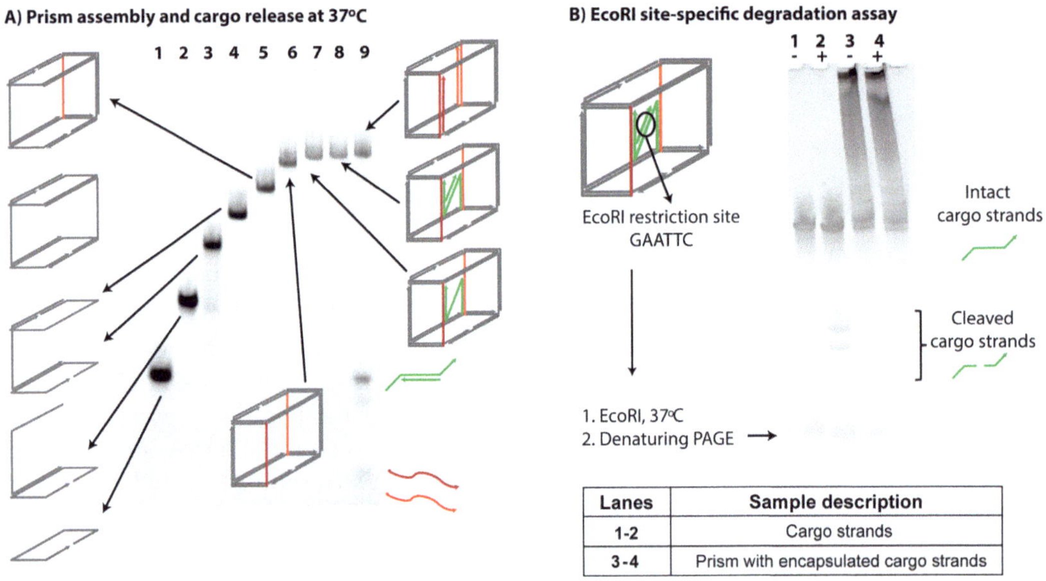

Lanes	Sample description
1-2	Cargo strands
3-4	Prism with encapsulated cargo strands

Fig. 3 Prism assembly, conditional cargo release, and stability to site-specific degradation. (**a**) Step-by-step assembly of the prism monitored using native PAGE (Lanes 1–8). The prism can selectively release its cargo when incubated with the trigger strands (Lane 9). (**b**) Placement on the prism of the GAATTC motif that is recognized and cleaved by the EcoRI restriction enzyme. Following incubation with the enzyme, the cargo strands alone show significant degradation (Lanes 1–2), while the prism can protect its cargo (Lanes 3–4)

Strands 2 + 3 + 5 + 6 + 4 + 1 + 7

Strands 2 + 3 + 5 + 6 + 4 + 1 + 7 + 8

3. Anneal from 95 °C to 4 °C using the following program: 5 min incubation at 95 °C, followed by 5 min per 1 °C from 80 °C to 4 °C. If not used immediately, store the assembled products at 4 °C.
4. Load the gel using an equal number of DNA bases per lane to ensure the contrast from one lane to another remains even. Since the final assembly contains the largest number of bases, it can be used to compute a normalization factor that will be used to determine the volume needed for samples 1–7 to achieve a similar staining. Additional 1×TAMg can be added to make all volumes equal and facilitate loading.
5. Add 10–20% glycerine to all samples to increase their density, so that they reach the bottom of the well.
6. Load on a 6% native polyacrylamide gel in 1×TAMg, and run at 250 V at room temperature. Alternatively, the gel can be run at 37 °C using a water bath to mimic physiological temperatures.

7. Stain the gel in a 1× GelRed solution in 1×TAMg for 20 min in the dark prior to imaging on a gel fluorescence imaging system.
8. Additionally, conditional release of the cargo strands can be tested by incubating a fully assembled cage with two equivalents of the trigger strands in 1×TAMg at room temperature for 5 min and loading this sample compared to a fully assembled cage (Fig. 3a, Lane 9).
9. The ability of the prism to protect its cargo can also be assessed using fully assembled DNA prisms that encapsulate cargo strands bearing a double-stranded GAATTC DNA motif [7]. This sequence is specifically cleaved by the EcoRI restriction enzyme, and the amount of enzymatic cleavage can be used as a measure of accessibility of the enzyme to the cargo strands, which in turn reflects on the cargo protection provided by the construct (Fig. 3c).
10. Finally, nuclease stability can be probed using fetal bovine serum (FBS) assays. FBS is a common additive in cell culture and contains a mix of enzymes known to degrade DNA. The lifetime of a given construct compared to single- or double-stranded cargo strands is indicative of the impact the 3D configuration of the prism on the protection of its cargo strands. Briefly, incubate DNA constructs of interest at 37 °C in a mixture of any given cell culture media of interest (these mimic physiological conditions) to which is added 10% FBS according to the procedure described by Conway et al. [12] In this procedure, aliquots are taken at timepoints of interest (0, 1, 2, 3, 4, 6, 8, 24, 48, and 72 h) and kept frozen at −20 °C in 1:1 sample: glycerine mixtures to inactivate the enzymes until analysis by PAGE.

4 Notes

1. CANADA_2.0 is available free of charge at http://ls11-www.cs.tu-dortmund.de/molcomp/downloads/. The program can be easily installed in a chosen directory; however, it lacks a graphical user interface and must be run from the command line using a text file as input. Examples of input files can be found at http://ls11-www.cs.tu-dortmund.de/molcomp/downloads/example_inputs.jsp.
2. In this design, the antisense strand, which is the mediator of biological activity, is left unmodified. However, the sense strand is flanked with DNA arms on both ends to bind the prism scaffold. A prior evaluation of the impact of this modification on gene silencing should be carried out using techniques such as

luciferase assays and Western blots to confirm the suitability of this encapsulation technique for the chosen target. In general, this method is very well suited to encapsulate most siRNAs or Dicer-substrate siRNA sequences. However, depending on the length of the oligonucleotide to be encapsulated, the size of the prism may have to be adjusted. Trigonometry can be used to estimate the prism cavity size. It is worth noting that the longer the duplexed region to be encapsulated, the higher the conformational rigidity, which may destabilize the assembly. Alternatively, a single-stranded oligonucleotide drug of similar length (antisense or microRNA) can be encapsulated, provided that adding single-stranded DNA extensions on both ends does not affect its activity.

3. For example, an mRNA sequence that contains at least two unique non-complementary regions could be used or two different miRNA sequences. It is also possible to use a single trigger sequence of more than 21 bases, yet less than 42 bases; however, this will introduce sequence symmetry into the design since the two trigger regions will have similar sequences. This may affect the self-assembly of the prism and lead to significant aggregation. This effect is highly sequence-dependent and can sometimes be mitigated using a slower anneal protocol around the melting temperature of the prism. Additionally, working at elevated temperatures, like 37 °C, as a mimic of physiological conditions, can melt some of these secondary interactions and help eliminate undesired products, provided they melt below the melting temperature of the prismatic scaffold with encapsulated cargo.
4. Note that this approximates the cavity size; it is recommended to synthesize the computed length ± 2 bases and to look at the stability of the resulting prisms before choosing a definite cargo strand length.
5. RNA duplexes have a pitch of 3.0 nm, which is slightly less than DNA duplexes, whose pitch is 3.4 nm. As a result, the 19-base duplex portion of the siRNA is expected to have a similar length to a 17-base DNA duplex.
6. CANADA_2.0 only recognizes files named input.txt as input. Save under another name after sequencing for record keeping. An example of input file with scrambled siRNA and trigger sequences is provided as supporting information.
7. Every time the CANADA_2.0 successfully runs, it generates a new sequence.txt file that overwrites the precedent one and provides a new solution to input.txt. Important sequences should therefore be saved under another name prior to running the program again.

8. In this example, *y*1 and *y*2 were not included in the sequence design and instead were manually replaced with 5 and 6 thymine spacers. This strategy was adopted because CANADA_2.0 does not run if one of the defined sequences is smaller or equal to the n_uniqueness factor.
9. Extinction coefficients, complementary interactions, and melting transitions for DNA and RNA strands can be computed using the OligoAnalyzer 3.1 tool from IDT technologies, which can be found at https://www.idtdna.com/calc/analyzer.

Acknowledgments

The authors would like to thank the Natural Science and Engineering Research Council of Canada, the Canada Research Chairs Program, the Fonds de recherche du Québec—Nature et technologies—the Canadian Institutes of Health, Prostate Cancer Canada, and FRQNT Center for Self-Assembled Chemical Structures for funding this work.

References

1. Seeman NC, Sleiman HF (2017) DNA nanotechnology. Nat Rev Mater 3:17068
2. Seeman NC (2003) DNA in a material world. Nature 421:427
3. Lin C, Liu Y, Rinker S, Yan H (2006) DNA tile based self-assembly: building complex nanoarchitectures. Chem Phys Chem 7(8): 1641–1647
4. Hong F, Zhang F, Liu Y, Yan H (2017) DNA origami: scaffolds for creating higher order structures. Chem Rev 117(20):12584–12640
5. McLaughlin CK, Hamblin GD, Sleiman HF (2011) Supramolecular DNA assembly. Chem Soc Rev 40(12):5647–5656
6. Deleavey GF, Damha MJ (2012) Designing chemically modified oligonucleotides for targeted gene silencing. Chem Biol 19(8): 937–954
7. Bujold KE, Hsu JCC, Sleiman HF (2016) Optimized DNA "Nanosuitcases" for encapsulation and conditional release of siRNA. J Am Chem Soc 138(42):14030–14038
8. Feldkamp U (2010) CANADA: designing nucleic acid sequences for nanobiotechnology applications. J Comput Chem 31(3):660–663
9. Beaucage SL, Iyer RP (1992) Advances in the synthesis of oligonucleotides by the phosphoramidite approach. Tetrahedron 48(12): 2223–2311
10. Koshkin AA, Singh SK, Nielsen P, Rajwanshi VK, Kumar R, Meldgaard M, Olsen CE, Wengel J (1998) LNA (Locked Nucleic Acids): synthesis of the adenine, cytosine, guanine, 5-methylcytosine, thymine and uracil bicyclonucleoside monomers, oligomerisation, and unprecedented nucleic acid recognition. Tetrahedron 54(14):3607–3630
11. Kibler-Herzog L, Zon G, Uznanski B, Whittier G, Wilson WD (1991) Duplex stabilities of phosphorothioate, methylphosphonate, and RNA analogs of two DNA 14-mers. Nucleic Acids Res 19(11):2979–2986
12. Conway JW, McLaughlin CK, Castor KJ, Sleiman H (2013) DNA nanostructure serum stability: greater than the sum of its parts. Chem Commun 49(12):1172–1174

Chapter 7

Combinatorial siRNA Polyplexes for Receptor Targeting

Dian-Jang Lee and Ernst Wagner

Abstract

As synthetic small interfering RNA (siRNA) against antitumoral gene targets show promise for cancer treatment, different siRNA delivery systems have sparkled intense investigations. To develop tumor-specific carriers for cytosolic and systemic siRNA delivery, our laboratory has recently generated folate-conjugated targeted combinatorial siRNA polyplexes based on sequence-defined oligomer platform compatible with solid-phase-supported synthesis. These polyplexes presented efficient siRNA-mediated gene silencing in folate receptor-expressing tumors in vitro and in vivo. In this chapter, we provide a brief background on the formulation design and detailed protocols to evaluate polyplex formation, gene silencing efficiency, and receptor-directed cell killing in cancer cells using targeted combinatorial siRNA polyplexes.

Key words siRNA delivery, Gene silencing, Polyplex, Receptor targeting

1 Introduction

Small interfering RNA (siRNA), as RNA interference (RNAi)-based therapeutics, shows promise for sequence-specific suppression of gene expression via mRNA degradation pathways [1]. Recently, numerous novel RNAi targets for malignancies have been identified, such as for breast cancer (WEE1) [2], glioma (PFKFB4) [3], multiple myeloma (GSK3) [4], and pancreatic cancer (CHK1) [5]. By silencing important gene targets in cancer cells, the siRNA therapeutics are showing impressive potential with several clinical trials that are underway [6]. One implicated oncological target gene is eglin 5 (EG5, also known as Kif11, KSP), a member of the Kinesin-5 subclass of kinesins, which is involved in the assembly and organization of the mitotic spindle apparatus, and is responsible for centrosome separation in cell division [7]. Thus, targeting siRNA against EG5 (siEG5) that corresponds to a sequence of the EG5 coding region conserved between human and mouse would represent an excellent strategy to hinder mitosis, leading to G2 arrest and eventually cell death of cancer cells [8–10].

Lekha Dinesh Kumar (ed.), *RNA Interference and Cancer Therapy: Methods and Protocols*, Methods in Molecular Biology, vol. 1974, https://doi.org/10.1007/978-1-4939-9220-1_7,

However, the challenging obstacles in siRNA delivery hinder its clinical applications. siRNA molecules are susceptible to nuclease destruction and cannot translocate across the cell membrane due to the highly charged backbone [11, 12]. During the last two decades, great efforts have been made in the development of various siRNA carriers [13]. Polymers [14–18], liposomes [19], inorganic materials [20], and hybrid systems [21–25] all showed encouraging improvements. Nevertheless, systemic delivery of siRNA directed to the tumor site remains a major limitation [26, 27]. Recently, by solid-phase-assisted synthesis, we have applied artificial oligoamino acids, such as succinoyl-tetraethylene pentamine (Stp) and succinoyl-pentaethylene hexamine (Sph) [28], in combination with natural α-amino acids, to generate sequence-defined cationic oligomers [29, 30]. These cationic oligomers and negatively charged siRNA can form polyplexes mainly due to electrostatic interaction [31, 32]. The oligomers provide excellent nucleic acid binding ability and endosomal buffering capacity and are also flexible for functional modifications [33]. Further studies have been undertaken to optimize such cationic carriers by diverse molecular topologies [34], incorporation of chemical motifs such as buffering histidines [35], fatty acids [36, 37], tyrosine trimers [38], or disulfide-forming units [12, 39]. It is also worth mentioning that PEGylated (polyethylene glycol-modified) siRNA polyplexes easily face the bottleneck of endosomal entrapment into the cytosol [40]. To facilitate endosomal escape, the synthetic peptide Inf7, previously designed as pH-specific analog of the influenza virus hemagglutinin subunit 2 (HA2) N-terminal fusion sequence [41–43], was covalently coupled to polyplexes or siRNA strand as endosomolytic agent [40, 44–46]. In addition, many different strategies have been investigated aiming to induce receptor-directed uptake in cancer cells [47–51]. To target transferrin (Tf) receptor-overexpressing tumors, the lipo-oligomers can be modified with PEG-linked Tf (for receptor targeting) and Inf7 peptide [46]. For that epidermal growth factor receptor (EGFR) is frequently upregulated in solid tumors, polyplexes are coupled with GE11 peptide, which lacks mitogenic activity in contrast to the natural epidermal growth factor (EGF), to combat EGFR-positive tumors [52]. Moreover, a broad variety of receptor-targeted experimental therapies have focused on folate receptor (FR), which has thus served as a golden standard for targeted drug delivery [45, 53, 54]. We previously reported folate-equipped PEGylated polyplexes that successfully exhibited FR-specific cellular uptake and gene silencing in vitro and in vivo [40]; yet due to small particle size, these polyplexes displayed a short circulation time of merely 15 min followed by renal excretion after systemic administration. Therefore, siRNA delivery systems were optimized based on post-PEGylation with folate ligand [55] or folate-conjugated lipid-based co-formulations [56] to improve the pharmacokinetic profile in FR-responsive tumor models.

As reviewed in the present work, the physicochemical properties of polyplexes were improved by combinatorial optimization of PEGylated folate-conjugated oligomer and 3-arm oligomer to generate targeted combinatorial polyplexes (TCPs). The 3-arm Stp oligomers ***386*** and ***689*** were selected for their strong siRNA binding ability, which provide a highly stable cationic core to compact siRNA [9, 29, 36, 50]. Additionally, PEGylated folate-conjugated Sph oligomers (4-arm) ***709*** and (2-arm) ***717*** were chosen for FR targeting and surface shielding [54]. These two types of oligomers were combined based on disulfide crosslinking. To achieve this, we activate the thiol groups of oligomer type #1 with 5,5′-dithiobis(2-nitrobenzoic acid) (DTNB) to produce a TNB-modified oligomer (Fig. 1a). The TNB-modified oligomer can be formulated with siRNA to form a binary siRNA polyplex, and then in the following incubation with the unmodified thiol-oligomer type #2, it would undergo a fast unidirectional coupling through disulfide formation with the free thiol groups of thiol-oligomer type #2. The combination of siRNA with both oligomers thus generate TCPs, which are composed of compact cationic core for siRNA binding, well-shielded PEG layer and folates as targeting ligands. Based on this approach, we developed four combinatorial formulations (TCP1, TCP2, TCP3, and TCP4, as in Table 1, Fig. 1b, c) (*see* **Note 1**) with favorable particle size and surface charge (Table 2) [57]. TCPs exhibit tumor-specific uptake of TCPs via FR-mediated endocytosis and mediate significant reporter gene silencing efficiency in the target cells, as 75–94% of gene expression is suppressed. TCPs loaded with siEG5-Inf7 (Inf7-modified EG5 siRNA) can further trigger profound EG5 gene suppression, leading to mitotic arrest and following FR-responsive tumor cell killing. Upon *i.v.* administration in mouse tumor model, TCP1 presented tumoral retention of siRNA and achieved EG5 gene silencing without noteworthy side effects [57].

In this chapter we describe a series of assays to successfully evaluate the performance of TCPs. At first, we introduce the preparation procedure of TCPs. Secondly, we show how agarose gel shift assay could be used to assign the polyplex formation between oligomers and siRNA. Thirdly, we describe how reporter cell lines can be employed to assess gene silencing potential of TCPs. Finally, we show how cell viability assays can be utilized to monitor cell killing effect mediated by EG5 gene knockdown.

2 Materials

2.1 Preparation of Targeted Combinatorial Polyplexes (TCPs)

1. For in vitro tests, oligomers **386**, **689**, **709**, and **717** and their corresponding TNB-modified oligomers (Table 1, Fig. 1) are dissolved with RNase-free Millipore water and stored 5 mg/mL in 50 μL of aliquots at −20 °C.

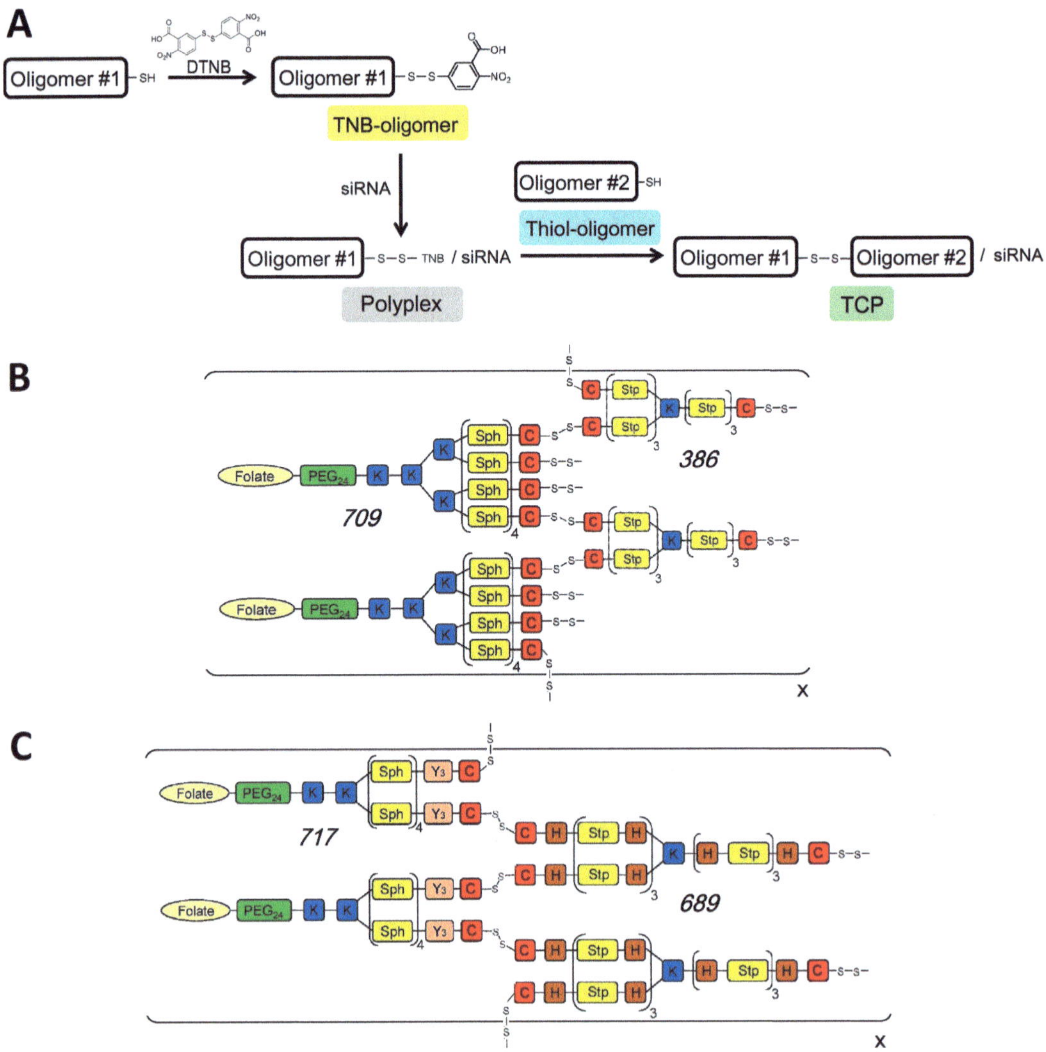

Fig. 1 Targeted combinatorial polyplexes (TCPs) with folate ligand for siRNA delivery. (**a**) The incubation of TNB-modified oligomer type #1 with siRNA forms a binary polyplex. Addition of thiol-oligomer type #2 results in fast unidirectional coupling of TNB-modified oligomer type #1 through disulfide formation with the free thiol groups of type #2. Together, siRNA, TNB-modified oligomer, and unmodified thiol-oligomer are formulated to produce a TCP. (**b**) TCP1 and TCP3 share the similar combined structure of 386 + 709 as $(709\text{-S-S-}386\text{-S-S})_x$. (**c**) TCP2 and TCP4 share the similar combined structure of 689 + 717 as $(717\text{-S-S-}689\text{-S-S})_x$. (adopted from [57] with permission from Elsevier)

2. The sequences of siRNA against eGFPLuc fusion protein (siGFP), siRNA against EG5 (siEG5), control siRNA (siControl), and endosomolytic peptide Inf7-modified siRNA (siGFP-Inf7, siEG5-Inf7 and siControl-Inf7) are shown in Table 3. We prepare 0.5 μg/μL in 50 μL of aliquots (dissolved in RNase-free Millipore water) at −20 °C (*see* **Note 2**).

Table 1
Targeted combinatorial polyplex (TCP) formulations. [TNB-modified oligomer/unmodified thiol-oligomer] molar ratio = 1:1

	TCP1	*TCP3*
TNB oligomer type #1	(4-arm ***873***): K-(PEG$_{24}$-Folate) K-[K-(Sph$_4$-C-TNB)$_2$]$_2$ (N/P 11.2)	(3-arm ***769***): TNB-C-Stp$_3$-K-(Stp$_3$-C-TNB)$_2$ (N/P 4.8)
Thiol oligomer type #2	(3-arm ***386***): C-Stp$_3$-K-(Stp$_3$-C)$_2$ (N/P 4.8)	(4-arm ***709***): K-(PEG$_{24}$-Folate)-K-[K-(Sph$_4$-C)$_2$]$_2$ (N/P 11.2)
Combined structure	(***709***-S-S-***386***-S-S)$_x$	
	TCP2	*TCP4*
TNB oligomer type #1	(3-arm ***770***): TNB-C-H-(Stp-H)$_3$-K-[(H-Stp)$_3$-H-C-TNB]$_2$ (N/P 7.4)	(2-arm ***874***): K-(PEG$_{24}$-Folate)-K-(Sph$_4$-Y$_3$-C-TNB)$_2$ (N/P 8.6)
Thiol oligomer type #2	(2-arm ***717***): K-(PEG$_{24}$-Folate)-K-(Sph$_4$-Y$_3$-C)$_2$ (N/P 8.6)	(3-arm ***689***): C-H-(Stp-H)$_3$-K-[(H-Stp)$_3$-H-C]$_2$ (N/P 7.4)
Combined structure	(***717***-S-S-***689***-S-S)$_x$	

The oligomer sequences are indicated (left to right) from C- to N-terminus. *C* cysteine, *H* histidine, *K* lysine, *Y* tyrosine, *S-S* disulfide crosslinking, *PEG$_{24}$* polyethylene glycol, *TNB* 5-thio-2-nitrobenzoic acid, *Stp* succinoyl-tetraethylene pentamine, *Sph* succinoyl-pentaethylene hexamine. K-(and K-[refer to branchings by α- and ε-amino modification of lysines. x hypothetic nanogel structures resulting from disulfide crosslinking in three dimensions within the TCP. (Reproduced from [57] with permission from Elsevier)

Table 2
Size and zeta potential of TCPs at N/P 16 measured by dynamic and electrophoretic light scattering [57]. [TNB-modified oligomer/unmodified thiol-oligomer] molar ratio = 1:1

Formulation	Size (nm)	Zeta potential (mV)
TCP1	103.5 ± 0.8	9.1 ± 0.2
TCP2	429.4 ± 53	7.93 ± 0.21
TCP3	208.8 ± 3.4	24.3 ± 0.3
TCP4	398.3 ± 63.5	8.1 ± 0.05

Table 3
siRNA sequences

siRNA	Target	Supplier	siRNA sequence
siGFP	GFP	Axolabs	AuAucAuGGccGAcAAGcAdTsdT (Sense) UGCUUGUCGGCcAUGAuAUdTsdT (antisense)
siControl		Axolabs	AuGuAuuGGccuGuAuuAGdTsdT (Sense) CuAAuAcAGGCcAAuAcAUdTsdT (antisense)
siGFP-Inf7	GFP	Axolabs	Inf7-ss-C6-AuAucAuGGccGAcAAGcAdTsdT (sense) UGCUUGUCGGCcAUGAuAUdTsdT (antisense)
siEG5- Inf7	EG5	Axolabs	Inf7-ss-C6-ucGAGAAucuAAAcu AAcudTsdT (sense) AGUuAGUUuAGAUUCUCGAdTsdT (antisense)
siControl-Inf7	–	Axolabs	Inf7-ss-C6-AuGuAuuGGccuGuAuuAGdTsdT (Sense) CuAAuAcAGGCcAAuAcAUdTsdT (Antisense)

Modified nucleotides: small letters, 2′methoxy; *dT* deoxythymidine, *s* phosphorothioate linkage. Inf7 peptide: GLFEAIEGFIENGWEGMIDGWYGC-amide

3. 20 mM HEPES buffered 5% glucose pH 7.4 (HBG), in 5 mL of aliquots at −20 °C.
4. 1.5 mL Eppendorf tubes.

2.2 siRNA Binding Assay

1. Agarose powder.
2. 1× TBE.
3. GelRed®.
4. Bromophenol blue. Store at 4 °C.
5. Electrophoresis system. The well comb should be large enough to form 25 μL of gel pockets, e.g., Sub-Cell GT System.
6. UV transilluminator, e.g., Dark Hood DH-40.

2.3 Cell Culture

1. Cell lines stably transfected with the firefly luciferase gene are used for siRNA silencing assay, and wild-type cell lines are used for MTT assay. The cell lines and corresponding culture medium are described in Table 4.
2. Trypsin (0.25%) and ethylenediaminetetraacetic acid (EDTA) (1 mM).
3. Phosphate-buffered saline (PBS).
4. 75 cm^2 cell culture flasks.
5. 96-well cell culture plates.

2.4 siRNA Silencing Assay

1. Cell culture 5× lysis reagent.
2. 10 mM luciferin solution: Weigh and add 48 mg of luciferin to 0.47 mL of 1 M glycylglycine (pH 8.0) and 10 mL of Millipore water. Then adjust with 1 M NaOH to pH 8.0. Finally bring

Table 4
Cell lines and corresponding culture media

Cell line	Cell culture medium
Human cervix carcinoma KB/ eGFPLuc cells	Folate-free RPMI 1640 medium, 10% fetal bovine serum, 4 mM stable glutamine, 100 U/mL penicillin, and 100 μg/mL streptomycin
Murine lung carcinoma M109 cells	Folate-free RPMI 1640 medium, 10% fetal bovine serum, 4 mM stable glutamine, 100 U/mL penicillin, and 100 μg/mL streptomycin
Human mammary adenocarcinoma MCF-7 cells	Dulbecco's Modified Eagle's Medium (DMEM), 20% fetal bovine serum, 100 U/mL penicillin, and 100 μg/mL streptomycin

to 16 mL with Millipore water and freeze in 500 μL of aliquots. Store below −20 °C.

3. Luciferase assay reagent (LAR): Weigh and add 50.8 mg of DTT and 27.8 mg of ATP to 2 mL of 1 M glycylglycine, 1 mL of 100 mM $MgCl_2$, 20 μL of 500 mM EDTA, and 0.5 mL of 42.6 mg/mL coenzyme A. Fill up to 100 mL with Millipore water. Adjust with 1 M NaOH to pH 8–8.5. Store at 4 °C (*see* **Note 3**).
4. Luminometer, e.g., Centro LB 960 plate reader luminometer.
5. Microplate 96 well, white. The multiwell plates must be compatible with the luminometer used.

2.5 MTT Assay for Evaluation of siEG5-Triggered Cell Killing

1. 3-(4,5-dimethylthiazol-2-yl)-2,5-diphenyltetrazolium bromide (MTT) and dimethylsulfoxide (DMSO) (*see* **Note 4**).
2. 96-well cell culture plates. The multiwell plates must be compatible with the assigned microplate reader.
3. Orbital shaker, e.g., Duomax 1030.
4. Microplate absorbance reader, e.g., SpectraFluor Plus microplate reader.

3 Methods

3.1 Preparation of Targeted Combinatorial Polyplexes (TCPs)

1. Thaw frozen aliquots of oligomer solution, siRNA (siGFP, siControl, siGFP-Inf7, siEG5-Inf7, or siControl-Inf7) solution, and HBG on ice.
2. Dilute 500 ng of siRNA in 10 μL of HBG.

3. The calculated amount of the two oligomers for the designated [TNB-modified oligomer/unmodified mercapto-form of oligomer] molar ratio (calculations *see* **Notes 5** and **6**) is separately diluted in 5 μL of HBG. The solution of the first TNB-modified oligomer is added to the siRNA solution and mixed by rapid pipetting (at least five times) to obtain 15 μL of binary siRNA polyplex solution (*see* **Note 7**). Incubate the siRNA polyplex solution in the closed Eppendorf reaction tube at room temperature for 30 min.
4. The solution of the second oligomer is added to the siRNA polyplex solution to obtain 20 μL of siRNA polyplex solution in total.
5. Allow the polyplex to form for 40 min exposure to air oxidation at room temperature in the closed Eppendorf tube (*see* **Note 8**).

3.2 siRNA Binding Assay

1. Prepare a standard 2.5% agarose gel.
   (a) Measure out 3.75 g of agarose powder.
   (b) Pour 3.75 g of agarose powder into a microwavable flask along with 150 mL of 1× TBE.
   (c) Microwave for 1–3 min (until the agarose is completely dissolved and rolling boil appears).
   (d) Add 15 μL of GelRed® for siRNA staining in agarose gels. Gently shake the flask to ensure homogenous mixing.
   (e) Let agarose solution cool down.
   (f) Pour the agarose into a gel tray with the well comb in place (*see* **Note 9**).
   (g) Let it at room temperature for 40 min until it has completely solidified.
   (h) Once solidified, transfer the agarose gel to the electrophoresis chamber and remove the well comb.
   (i) Add 1× TBE into the electrophoresis chamber until the gel is covered.
2. Prepare TCPs at different N/P 16 (*see* Subheading 3.1). For control experiments plain siRNA solution diluted in HBG is necessary.
3. Add 3 μL of bromophenol blue as loading buffer into each sample.
4. Carefully load samples into wells of the gel.
5. Perform the electrophoresis at 120 V for 40 min (*see* **Note 10**).
6. Record the result with a standard UV transilluminator. Such assay can validate siRNA is fully incorporated into polyplexes (Fig. 2a).

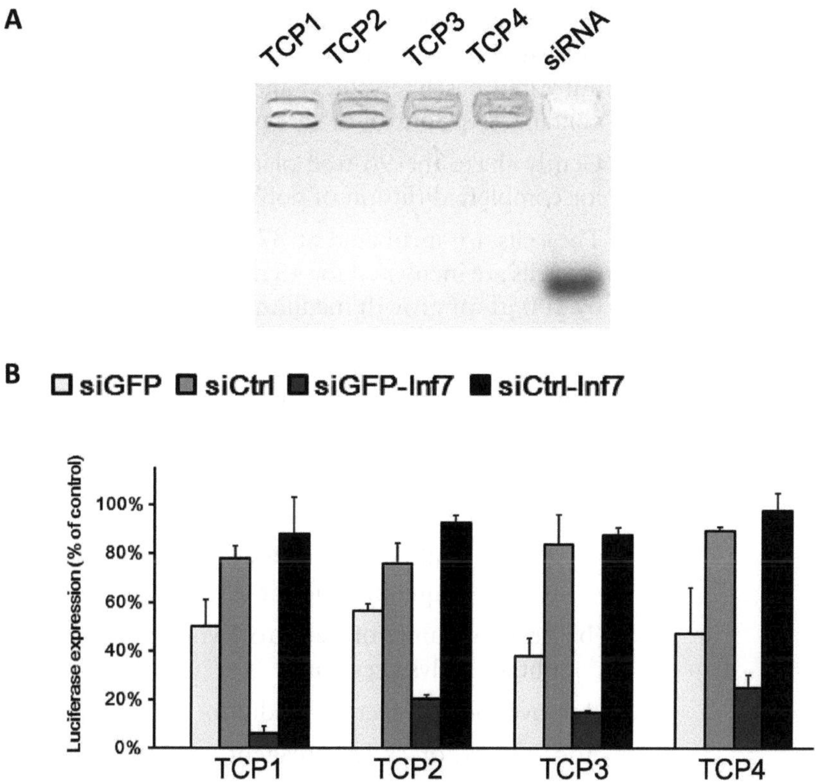

Fig. 2 Functional assessments for TCPs. (**a**) siRNA binding of TCPs at N/P 16 determined by agarose gel shift assay. (**b**) Gene silencing efficiency of TCPs in KB cells expressing eGFPLuc fusion protein (KB/eGFPLuc cells). The siRNA polyplexes are prepared at N/P 16 with different siRNA sequences: eGFP-targeted siRNA (siGFP), control siRNA (siCtrl), or siRNA chemically linked with the endosomolytic peptide Inf7 (siGFP-Inf7 or siCtrl-Inf7) (reproduced from [57] with permission from Elsevier)

3.3 Cell Culture and Treatment

1. FR-overexpressing (KB and M109) and FR-deficient (MCF-7) cells are used and cultured in the appropriate medium (Table 2). For siRNA silencing assay, KB cells stably transfected with the eGFPLuc gene (KB/eGFPLuc cells) are used. For MTT assay, FR-positive M109 cells and FR-negative MCF-7 are used. The cells are cultured in ventilated flasks inside incubators at 37 °C with 5% CO_2 in a humidified atmosphere. Cells are allowed to grow until 80% confluence and splitted when necessary.
2. 4 × 10^3 cells per well are seeded in 100 μL of medium using 96-well plates 24 h prior to the experiment.
3. Before the experiment, medium is replaced with 80 μL of fresh growth medium.
4. Treatments are carried out by treating cells with 20 μL of polyplex solution per well (*see* Subheading 3.1).

5. In parallel, it is imperative to run control experiments with untreated cells (as the gene silencing is measured against these untreated baseline values) and also treat cells with polyplexes containing siControl or siControl-Inf7.
6. Gently shake the 96-well plates horizontally for few seconds for complete diffusion of polyplexes.
7. The cells are incubated at 37 °C. For siRNA silencing assay, the cells are incubated for 45 min, then the medium is replaced by 100 μL of growth medium, followed by additional incubation for 48 h. For MTT assay, the cells are incubated for 48 h.

3.4 siRNA Silencing Assay

1. Cells are harvested and analyzed for gene silencing after incubation for 48 h with siRNA-containing polyplexes (siGFP, siControl, siGFP-Inf7 or siControl-Inf7) (*see* Subheading 3.3).
2. Prepare cell lysis buffer.
   (a) Thaw an aliquot of 5× lysis reagent at ambient temperature and equilibrate for at least 30 min.
   (b) Add 4 volumes of water to 1 volume of 5× lysis reagent to obtain 1× lysis reagent.
3. Remove the transfection medium.
4. Add 100 μL of 1× lysis reagent into each well to obtain cell lysate.
5. Measure the relative light units (RLU) of cell lysate by a plate-reading luminometer. We use Centro LB 960 plate reader luminometer, for the example below.
   (a) Thaw an aliquot of luciferin.
   (b) Mix 500 μL of luciferin with 10 mL of luciferase assay reagent (LAR), and apply the injector into the mixture.
   (c) The injector is programmed to add 100 μL of luciferin-LAR solution per well, then the well is read immediately.
   (d) Transfer 35 μL of cell lysate each well to white 96-well microplate.
   (e) Measure the light produced for a period of 10 s.
6. The relative light units (RLU) are presented as percentage of the luciferase gene expression obtained with untreated control cells. Example of successful gene silencing is shown in Fig. 2b.

3.5 MTT Assay for Evaluation of siEG5-Triggered Cell Killing

1. Cells are harvested and analyzed for metabolic activity after incubation with siRNA-containing polyplexes (siEG5-Inf7 or siControl-Inf7) for 48 h (*see* Subheading 3.3).
2. Thaw MTT solution.
3. Add 10 μL of MTT solution to the medium present in each well. Place the plate in the incubator at 37 °C for 2 h to allow formazan formation.

4. Remove the medium.
5. Place the plate at −20 °C for 1 h.
6. Add 100 μL of DMSO into each well.
7. Place the plate on an orbital shaker for 40 min to totally dissolve formazan.
8. The absorbance is measured at 590 nm (with 630 nm as reference wavelength) by microplate absorbance reader, e.g., SpectraFluor Plus microplate reader. The relative light units (RLU) of treated cells are presented as percentage compared to the untreated control cells. After incubating the cells with siEG5-Inf7 TCPs, the MTT assay reveals that TCPs present effective cell killing effect in FR-expressing M109 cells (Fig. 3a). Besides, siCtrl-Inf7 TCPs are not toxic to M109 cells. Meanwhile, siEG5-Inf7 TCPs are inactive in FR-negative MCF-7 cells (Fig. 3b). Therefore, these results clearly support that folate-conjugated TCPs enable efficient EG5 gene knockdown, followed by significant cell death in FR-positive cells.

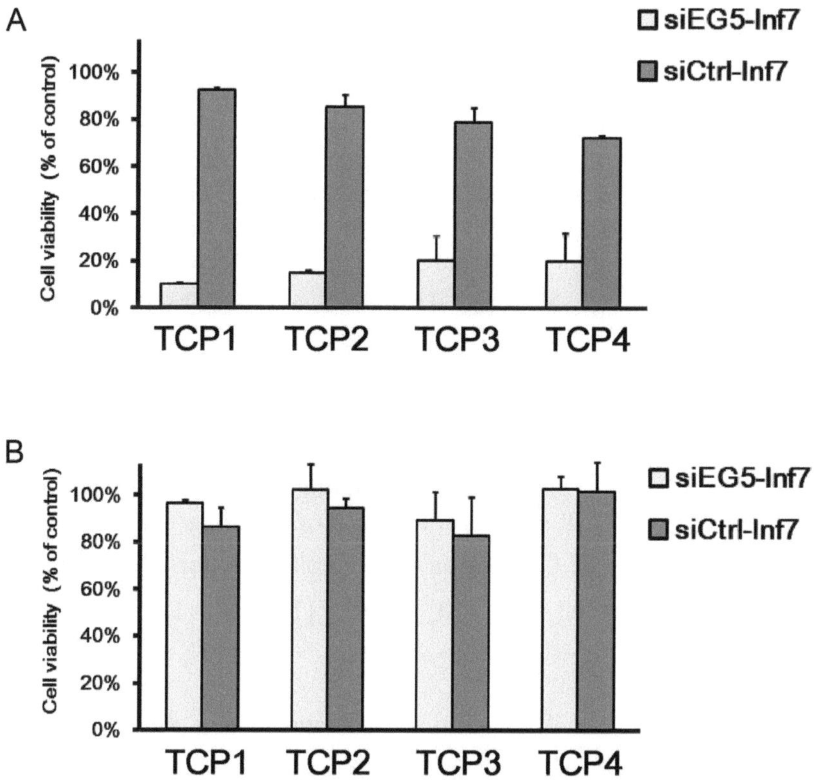

Fig. 3 Cell transfections with TCPs containing Inf7 peptide-modified EG5-targeted siRNA (siEG5-Inf7) or control siRNA (siCtrl-Inf7) at N/P 16. Cell viability of (**a**) FR-positive M109 cells and (**b**) FR-negative MCF-7 cells after transfection with TCPs (reproduced from [57] with permission from Elsevier)

4 Notes

1. TCP1 and TCP3 share the similar combined structure of ***386*** + ***709*** (Fig. 1a); TCP2 and TCP4 share the similar combined structure of ***689*** + ***717*** (Fig. 1b). Differences rise from the alterative TNB activations of oligomers. The equal molar (1:1) oligomer ratios and N/P 16 is determined as optimal condition for the TCP formulations.
2. To prevent freeze/thaw damage, we suggest preparing small volume of aliquots based on experimental design.
3. Light intensity is a measure of the rate of catalysis by luciferase and is therefore dependent upon temperature. The optimum temperature for luciferase activity is approximately room temperature (20–25 °C). It is important to fully equilibrate luciferase assay reagent to room temperature before measurements. The solution is stable for approximately 2 weeks at 4 °C.
4. We prepare MTT stock solution (5 mg/mL) in aliquots stored at −20 °C protected from light.
5. Calculations of the individual two oligomers used at N/P 16 in formation of TCPs at indicated molar ratios are made as shown here. Protonatable nitrogens (N) for the applied oligomers are calculated excluding histidine Ns (defined as unprotonated at pH 7.4) (*see* Table 3). The required molar amine amounts of oligomers at N/P 16 are calculated, which are 24.88 nmol nitrogens (N) for 500 ng of siRNA. For the subsequent calculation of the molar amount of each oligomer, general formulas are applied as follows:
   (a) $N/P_{total} = 16 = N/P_a + N/P_b$ ("a" and "b" stand for the two oligomers).
   (b) Total molar amount of nitrogen $M_{total} = N_a \times M_a + N_b \times M_b$, here N stands for the number of protonatable nitrogens, while M stands for the molar amount.
   (c) The molar ratio of oligomers $M_a/M_b = R_a/R_b$
   R stands for the ratio of each oligomer (1:1 in this case).
   (d) Calculation of the molar amount of each oligomer:
   $$M_a = 16 \times M_{total} \times N_a / (N_a + N_b \times R_b / R_a), \text{and}$$
   $$M_b = 16 \times M_{total} \times N_b / (N_a \times R_a / R_b + N_b).$$
   (e) The individual N/P ratio for each oligomer (at $N/P_{total} = 16$):

$$N/P_a = 16 \times N_a / (N_a + N_b \times R_b / R_a), \text{and } N/P_b = 16 \times N_b / (N_a \times R_a / R_b + N_b).$$

(f) As an example of TCP1 ***386/873*** at molar ratio 40:60, the individual N/P for ***386*** is $16 \times 29/(29 + 68 \times 60/40) = 3.5$, for ***873*** it is $16 \times 68/(29 \times 40/60 + 68) = 12.5$, and the molar amount M for ***386*** is $24.88 \times 29/(29 + 68 \times 60/40) = 5.45$ nmol for 500 ng of siRNA.

Please note that, depending on the selected molar ratios and TCPs, the final molar ratios of thiol SH /TNB will differ. At oligomer molar 1:1 ratio, SH/TNB are 3:4 for TCP1, 2:3 for TCP2, 4:3 for TCP3, and 3:2 for TCP4.

6. The amount of oligomer ($n_{oligomer}$, nmole) for the fixed amount of siRNA is calculated via N/P (protonatable nitrogens of the oligomer/phosphates of the siRNA) ratio:

$$n_{oligomer}\,(\text{nmole}) = \frac{m_{siRNA}\,(\text{ng})}{M_{siRNA}} \cdot P_{siRNA} \cdot \frac{N/P}{N_{oligomer}}$$

m_{siRNA}, weight of siRNA (ng); M_{siRNA}, molecular weight of siRNA; P_{siRNA}, amount of phosphates of siRNA; $N_{oligomer}$, amount of protonatable nitrogens of oligomer (*see* Table 5).

7. Avoid bubbles when pipetting.
8. Based on a pilot study monitoring disulfide formation, after 40 min of polyplex incubation time, approximately 60% of all free thiol groups are oxidized at N/P 6. The oxidation rate decreases with rising N/P ratios, reaching the lowest rate for plain oligomer.
9. Pour slowly to avoid bubbles which will disrupt the gel. Any bubbles can be pushed away from the well comb or toward the edges of the gel with a pipette tip.

Table 5
Number of protonatable nitrogens and molecular weight of oligomers. Protonatable nitrogens (N) for the applied oligomers are calculated excluding N of histidines

Formulation	Oligomer	Protonatable nitrogens (N)	Molecular weight
TCP1	*873*	68	10,793.8
	386	29	3955.2
TCP2	*770*	29	6629.9
	717	34	6670.5
TCP3	*769*	29	4546.7
	709	68	10,041.6
TCP4	*874*	34	7064.8
	689	29	6038.2

10. The negatively charged siRNA tends to diffuse throughout the gel in electrical field, whereas when cationic oligomer starts to bind to the siRNA it retards the siRNA diffusion.

Acknowledgments

The work presented in this article was supported by German Research Foundation (DFG) grant SFB1032 (Project B4) and German Excellence Cluster Nanosystems Initiative Munich.

References

1. Whitehead KA, Langer R, Anderson DG (2009) Knocking down barriers: advances in siRNA delivery. Nat Rev Drug Discov 8:129–138
2. Murrow LM et al (2010) Identification of WEE1 as a potential molecular target in cancer cells by RNAi screening of the human tyrosine kinome. Breast Cancer Res Treat 122:347–357
3. Goidts V et al (2012) RNAi screening in glioma stem-like cells identifies PFKFB4 as a key molecule important for cancer cell survival. Oncogene 31:3235–3243
4. Tiedemann RE et al (2010) Kinome-wide RNAi studies in human multiple myeloma identify vulnerable kinase targets, including a lymphoid-restricted kinase, GRK6. Blood 115:1594–1604
5. Azorsa DO et al (2009) Synthetic lethal RNAi screening identifies sensitizing targets for gemcitabine therapy in pancreatic cancer. J Transl Med 7:43
6. Haussecker D et al (2015) RNA interference. Drugging RNAi. Science 347:1069–1070
7. Valentine MT et al (2006) Eg5 steps it up! Cell Div 1:31
8. Judge AD et al (2009) Confirming the RNAi-mediated mechanism of action of siRNA-based cancer therapeutics in mice. J Clin Invest 119:661–673
9. Edinger D et al (2014) Gene silencing and antitumoral effects of Eg5 or Ran siRNA oligoaminoamide polyplexes. Drug Deliv Transl Res 4:84–95
10. Harborth J et al (2001) Identification of essential genes in cultured mammalian cells using small interfering RNAs. J Cell Sci 114: 4557–4565
11. Wagner E (2012) Polymers for siRNA delivery: inspired by viruses to be targeted, dynamic, and precise. Acc Chem Res 45:1005–1013
12. Klein PM, Wagner E (2014) Bioreducible polycations as shuttles for therapeutic nucleic acid and protein transfection. Antioxid Redox Signal 21:804–817
13. Kanasty R et al (2013) Delivery materials for siRNA therapeutics. Nat Mater 12:967–977
14. Boussif O et al (1995) A versatile vector for gene and oligonucleotide transfer into cells in culture and in vivo: polyethylenimine. Proc Natl Acad Sci U S A 92:7297–7301
15. De Smedt SC, Demeester J, Hennink WE (2000) Cationic polymer based gene delivery systems. Pharm Res 17:113–126
16. Pack DW et al (2005) Design and development of polymers for gene delivery. Nat Rev Drug Discov 4:581–593
17. Miyata K, Nishiyama N, Kataoka K (2012) Rational design of smart supramolecular assemblies for gene delivery: chemical challenges in the creation of artificial viruses. Chem Soc Rev 41:2562–2574
18. Lächelt U, Wagner E (2015) Nucleic acid therapeutics using polyplexes: a journey of 50 years (and beyond). Chem Rev 115:11043–11078
19. Sakurai Y et al (2013) Gene silencing via RNAi and siRNA quantification in tumor tissue using MEND, a liposomal siRNA delivery system. Mol Ther 21:1195–1203
20. Tang F, Li L, Chen D (2012) Mesoporous silica nanoparticles: synthesis, biocompatibility and drug delivery. Adv Mater 24:1504–1534
21. Kim HJ et al (2014) Precise engineering of siRNA delivery vehicles to tumors using polyion complexes and gold nanoparticles. ACS Nano 8:8979–8991
22. Pittella F et al (2014) Systemic siRNA delivery to a spontaneous pancreatic tumor model in transgenic mice by PEGylated calcium phosphate hybrid micelles. J Control Release 178:18–24

23. Heissig P et al (2016) DNA as tunable adaptor for siRNA polyplex stabilization and functionalization. Mol Ther Nucleic Acids 5:e288
24. Möller K et al (2016) Highly efficient siRNA delivery from core-shell mesoporous silica nanoparticles with multifunctional polymer caps. Nanoscale 8:4007–4019
25. Krzyszton R et al (2017) Microfluidic self-assembly of folate-targeted monomolecular siRNA-lipid nanoparticles. Nanoscale 9: 7442–7453
26. Scholz C, Wagner E (2012) Therapeutic plasmid DNA versus siRNA delivery: common and different tasks for synthetic carriers. J Control Release 161:554–565
27. Wagner E (2013) Biomaterials in RNAi therapeutics: quo vadis? Biomater Sci 1:804–809
28. Salcher EE et al (2012) Sequence-defined four-arm oligo(ethanamino)amides for pDNA and siRNA delivery: impact of building blocks on efficacy. J Control Release 164:380–386
29. Schaffert D et al (2011) Solid-phase synthesis of sequence-defined T-, i-, and U-shape polymers for pDNA and siRNA delivery. Angew Chem Int Ed 50:8986–8989
30. Lee DJ, Wagner E, Lehto T (2015) Sequence-defined oligoaminoamides for the delivery of siRNAs. Methods Mol Biol 1206:15–27
31. He D, Wagner E (2015) Defined polymeric materials for gene delivery. Macromol Biosci 15:600–612
32. Zhang P, Wagner E (2017) History of polymeric gene delivery systems. Top Curr Chem (Cham) 375:26
33. Reinhard S, Wagner E (2017) How to tackle the challenge of siRNA delivery with sequence-defined oligoamino amides. Macromol Biosci. 17:1 https://doi.org/10.1002/mabi.201600152
34. Scholz C, Kos P, Wagner E (2014) Comb-like oligoaminoethane carriers: change in topology improves pDNA delivery. Bioconjug Chem 25:251–261
35. Lächelt U et al (2014) Fine-tuning of proton sponges by precise diaminoethanes and histidines in pDNA polyplexes. Nanomedicine 10: 35–44
36. Frohlich T et al (2012) Structure-activity relationships of siRNA carriers based on sequence-defined oligo (ethane amino) amides. J Control Release 160:532–541
37. Reinhard S, Zhang W, Wagner E (2017) Optimized solid-phase-assisted synthesis of oleic acid containing siRNA nanocarriers. ChemMedChem 12:1464–1470
38. Troiber C et al (2013) Stabilizing effect of tyrosine trimers on pDNA and siRNA polyplexes. Biomaterials 34:1624–1633
39. Klein PM et al (2016) Precise redox-sensitive cleavage sites for improved bioactivity of siRNA lipopolyplexes. Nanoscale 8:18098–18104
40. Dohmen C et al (2012) Nanosized multifunctional polyplexes for receptor-mediated siRNA delivery. ACS Nano 6:5198–5208
41. Plank C et al (1994) The influence of endosome-disruptive peptides on gene transfer using synthetic virus-like gene transfer systems. J Biol Chem 269:12918–12924
42. Mechtler K, Wagner E (1997) Gene transfer mediated by influenza virus peptides: the role of peptide sequence. New J Chem 21: 105–111
43. Wagner E (1998) Effects of membrane-active agents in gene delivery. J Control Release 53:155–158
44. Klein PM et al (2015) Twin disulfides as opportunity for improving stability and transfection efficiency of oligoaminoethane polyplexes. J Control Release 205:109–119
45. Lee DJ et al (2016) Dual antitumoral potency of EG5 siRNA nanoplexes armed with cytotoxic bifunctional glutamyl-methotrexate targeting ligand. Biomaterials 77:98–110
46. Zhang W et al (2016) Targeted siRNA delivery using a lipo-oligoaminoamide nano-core with an influenza peptide and transferrin shell. Adv Healthc Mater 5:1493–1504
47. Martin I et al (2012) Solid-phase-assisted synthesis of targeting peptide-PEG-oligo|(ethane amino)amides for receptor-mediated gene delivery. Org Biomol Chem 10: 3258–3268
48. An S et al (2015) Peptide-like polymers exerting effective glioma-targeted siRNA delivery and release for therapeutic application. Small 11:5142–5150
49. Kos P et al (2015) Dual-targeted polyplexes based on sequence-defined peptide-PEG-oligoamino amides. J Pharm Sci 104:464–475
50. Kos P et al (2015) Histidine-rich stabilized polyplexes for cMet-directed tumor-targeted gene transfer. Nanoscale 7:5350–5362
51. Zhang W et al (2015) Combination of sequence-defined oligoaminoamides with transferrin-polycation conjugates for receptor-targeted gene delivery. J Gene Med 17: 161–172
52. Müller K et al (2016) EGF receptor targeted lipo-oligocation polyplexes for antitumoral siRNA and miRNA delivery. Nanotechnology 27:464001
53. de Bono JS, Ashworth A (2010) Translating cancer research into targeted therapeutics. Nature 467:543–549
54. He D et al (2016) Combinatorial optimization of sequence-defined oligo(ethanamino)amides

for folate receptor-targeted pDNA and siRNA delivery. Bioconjug Chem 27:647–659

55. Müller K et al (2016) Post-PEGylation of siRNA lipo-oligoamino amide polyplexes using tetra-glutamylated folic acid as ligand for receptor-targeted delivery. Mol Pharm 13: 2332–2345

56. Lee DJ et al (2017) Systemic delivery of folate-PEG siRNA lipopolyplexes with enhanced intracellular stability for in vivo gene silencing in leukemia. Bioconjug Chem 28:2393–2409

57. Lee DJ et al (2016) Tumoral gene silencing by receptor-targeted combinatorial siRNA polyplexes. J Control Release 244:280–291

Chapter 8

Synthesis of Doxorubicin and miRNA Stimuli-Sensitive Conjugates for Combination Therapy

Daniel F. Costa, Can Sarisozen, and Vladimir P. Torchilin

Abstract

Recent advances in combination therapy by using chemotherapeutic drugs and small noncoding RNAs have highlighted the need for optimization of such agents to allow their carriage in a single delivery system. This protocol details the synthesis of a doxorubicin prodrug, where a NHS coupling reaction was used to sensitize the drug to the proteolytic activity of tumor microenvironments. The design of a lipid-modified miRNA by an S-S coupling reaction is also described. Modification of both, doxorubicin and miRNA, facilitated their simultaneous incorporation into mixed micelles for use in combination therapy against tumor cells.

Key words Doxorubicin, miRNA, Prodrug, Stimuli-sensitive, Micelles

1 Introduction

The synthesis of drug conjugates and the chemical modification of genetic material have become essential steps in the optimization of drug delivery technologies. The aim to target a diseased tissue, to increase therapeutic efficacy, and to avoid off-target effects is closely related to the design of a system sensitive to biological processes accelerated or induced specifically at the target site [1, 2].

Abnormal protein expression in the extracellular matrix (ECM) of tissues is one of the key players in the development of human diseases. Cancer progression and invasiveness are effects directly related to the increased production of proteinases and the consequent breakdown of the ECM [3, 4]. Matrix metalloproteinases (MMPs) are among the most studied families of proteins responsible for the dynamic regulation of ECM composition. Recently, their activity has been considered to have potential as a trigger for drug delivery. Chemotherapeutic agents (doxorubicin, tumor necrosis factor, azademethylcolchicine, and others) have been modified with peptide sequences sensitive to cleavage by proteinases

Lekha Dinesh Kumar (ed.), *RNA Interference and Cancer Therapy: Methods and Protocols*, Methods in Molecular Biology, vol. 1974,
https://doi.org/10.1007/978-1-4939-9220-1_8,

which lead to successful drug delivery to target sites where the expression of MMPs is increased [5–8].

The intracellular microenvironment has also been shown to have great impact on the efficiency of drug delivery. Small noncoding RNAs have been chemically modified to optimize the design of RNA interference therapies using delivery systems with lipophilic cores, such as mixed micelles and solid lipid nanoparticles. In most of these cases, the encapsulation of the genetic material is achieved by coating it on the particle's surface. The cationic head of phospholipids used in their composition can be strategically modified to allow covalent attachment or ionic complexation with RNA molecules [9–11]. Lipid modification has improved payload delivery and transfection performance. An example was the use of a disulfide bond as linker between lipid and siRNAs. The sensitivity of the linker to a reductive biological environment drives the release of the genetic material into the cytoplasm of cancer cells with increased production of glutathione [12, 13].

This protocol details the steps for the synthesis of a doxorubicin prodrug sensitive to the proteolytic effect of MMPs as well as a miRNA conjugate cleaved in the presence of a highly reductive intracellular microenvironment. We were able to establish the concept of both stimuli-sensitive strategies and incorporate the conjugates in mixed micelles for combination therapy [14].

2 Materials

All solutions are prepared using analytical grade reagents and deionized water at room temperature. Additionally, miRNA-containing solutions are prepared using RNase-free water. Rinse glassware with RNaseZAP™ and RNase-free water prior to use. All plastic tubes and pipette tips must be RNase-free. Store reagents as recommended by the supplier. Typically, RNA samples are kept at −80 °C, and lipid stock solutions are stored at −20 °C. Waste disposal regulations must be followed to discard any material.

2.1 Synthesis of PEGylated MMP2-Sensitive Peptide (PEG2k-CLV)

1. PBS buffer, pH 7.4: NaCl 0.8% m/v; Na_2HPO_4 0.14% m/v; KH_2PO_4 24% m/v; KCl 20% m/v in water. Adjust pH with NaOH 0.1 N or HCl 0.1 N. Store at 4 °C.
2. MMP2-sensitive peptide solution: NH_2-GPLGIAGQ-COOH peptide 0.43% m/v in PBS buffer, pH 7.4.
3. PEG2k solution: MPEG-SVA-2000 1% m/v in PBS buffer, pH 7.4.

2.2 Synthesis of MMP2-Sensitive Doxorubicin Prodrug (PEG2k-CLV-Dox)

1. PEG2k-CLV solution: PEG2k-CLV peptide 2% m/v in dimethylformamide (DMF).
2. DCC solution: *N,N′*-dicyclohexylcarbodiimide (DCC) 3% m/v in DMF.

3. NHS solution: *N*-hydroxysuccinimide (NHS) 1.5% m/v in DMF.
4. Dox solution: doxorubicin (Dox) 0.5% m/v in DMF containing 0.4% v/v of triethylamine. Protect the solution from light (*see* **Note 1**).
5. Modified HEPES buffer, pH 7.4: NaCl 0.82% m/v; free acid HEPES 0.12% m/v; $CaCl_2$ 0.11% m/v in water. Adjust pH with NaOH 0.1 N or HCl 0.1 N. Store at 4 °C.

2.3 Synthesis of Sulfhydryl-Activated miRNA

1. miRNA solution: Modified miRNA (hsa-miR-34a-5p with 5′-disulfide-thiol modifier on the passenger strand) 180 μM in RNase-free water.
2. TCEP solution: tris (2-carboxyethyl) phosphine (TCEP) 4% v/v in RNase-free water.
3. NaAc solution: sodium acetate (NaAc) 25% m/v in RNase-free water.

2.4 Synthesis of GSH-Sensitive miRNA Conjugate (miRNA-S-S- DOPE)

1. Activated miRNA solution: sulfhydryl-activated miRNA 40 μM in PBS buffer, pH 7.4.
2. DOPE solution: 1,2-dioleoyl-*sn*-glycero-3-phosphoethanolamine-*N*-[3-(2-pyridyldithio)propionate] (DOPE-PDP) 0.4% m/v in methanol.
3. GSH solution: L-glutathione oxidized 3% m/v in RNase-free water. Make the solution immediately before use.

2.5 Preparation of Dual-Sensitive Mixed Micelles

1. PEG1k-PE solution: 1,2-distearoyl-*sn*-glycero-3-phosphoethanolamine-*N*-[methoxy(polyethylene glycol)-1000] (PEG1k-PE) 2.5% m/v in chloroform.
2. PEG2k-CLV-Dox solution: PEG2k-CLV-Dox prodrug 2% m/v in methanol.
3. miRNA-S-S-DOPE solution: miRNA 0.0066% m/v in RNase-free PBS buffer, pH 7.4 (*see* **Note 2**).

3 Methods

All methods are carried out at room temperature, unless otherwise specified. For the procedures where miRNA is used, rinse glassware with RNaseZAP™ and RNase-free water prior to use. All plastic tubes and pipette tips must be RNase-free. Waste disposal regulations must be followed to discard any material.

3.1 Synthesis of PEGylated MMP2-Sensitive Peptide (PEG2k-CLV)

1. Mix 1 mL of MMP2-sensitive peptide solution with 1 mL of PEG2k solution in an 8 mL glass vial. Pour N_2 into the system before closing the vial (*see* **Note 3**).
2. Stir reaction mixture at 4 °C for 12 h.

3. Transfer reaction mixture to a pre-treated dialysis tubing (MWCO 2000).
4. Purify the PEGylated peptide (PEG2k-CLV) by dialysis against 4 L of water for 4 h.
5. Replace the water and continue the dialysis for an additional 12 h.
6. Remove the solution from dialysis tubing, and filter it through a 0.20 μm PES filter into 10 mL tubes.
7. Freeze the solution at −80 °C for 2 h.
8. Freeze-dry the solution overnight (or until complete removal of water).
9. Store powder at −20 °C.
10. Check the presence of unreacted reagents by RP-HPLC (Fig. 1). Use a system equipped with a C18 column (250 × 4.6 mm, 5 μm) and DAD detector at λ = 214 nm. Chromatographic conditions: 0.1% TFA in acetonitrile (solvent A) and 0.1% TFA in water (solvent B). 0–15 min, 5–75% A; 15–16 min, 75% to 5% A; 16–20 min, 5% A. Flow rate: 1.0 mL/min at room temperature.

3.2 Synthesis of MMP2-Sensitive Doxorubicin Prodrug (PEG2k-CLV-Dox)

1. Mix 2 mL of PEG2k-CLV solution with 1 mL of DCC solution in an 8 mL amber glass vial. Stir reaction mixture for 10 min.
2. Add 1 mL of NHS solution to the reaction mixture. Pour N_2 into the system before closing the vial, and stir for 1 h.

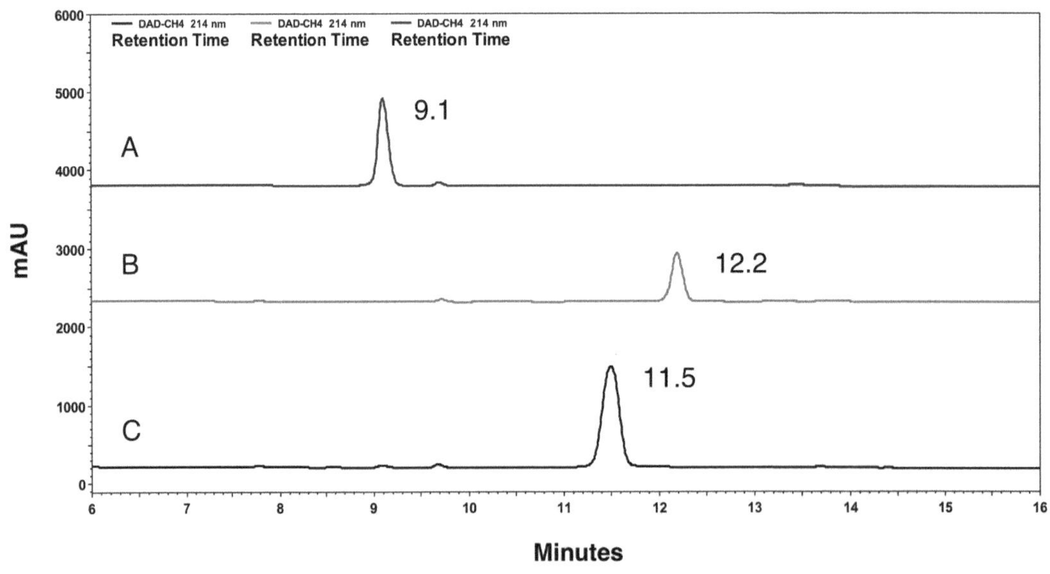

Fig. 1 Chromatograms of (**a**) MMP2-sensitive peptide, (**b**) MPEG-SVA-2000, and (**c**) purified PEGylated peptide

3. Add 2 mL of Dox solution to the reaction mixture. Pour N_2 into the system before closing the vial, and stir for 12 h protected from light.
4. Transfer reaction mixture to a pre-treated dialysis tubing (MWCO 2000).
5. Purify the Dox prodrug (PEG2k-CLV-Dox) by dialysis against 4 L of water for 2 h (*see* **Note 4**).
6. Transfer tubing to a glass beaker filled with 1 L of water-methanol (1:1 v/v), and continue dialysis for 12 h (*see* **Note 5**).
7. Transfer tubing to a beaker filled with 4 L of water and continue dialysis for 2 h.
8. Replace the water and continue the dialysis for an additional 12 h.
9. Remove the reaction mixture from dialysis tubing, and filter it through a 0.20 μm PES filter into 50 mL tubes.
10. Freeze the solution at −80 °C for 2 h.
11. Freeze-dry the solution overnight (or until complete removal of water).
12. Store powder at −20 °C.
13. Check the presence of unreacted reagents by RP-HPLC (Fig. 2). Use a system equipped with a C18 column (250 × 4.6 mm, 5 μm) and DAD detector at $\lambda = 485$ nm. Chromatographic conditions: 0.1% TFA in acetonitrile (solvent A) and 0.1% TFA in water (solvent B). 0–15 min, 5–75% A; 15–16 min, 75% to 5% A; 16–20 min, 5% A. Flow rate: 1.0 mL/min at room temperature.
14. Solubilize 0.5 mg of Dox prodrug in 1 mL of water-methanol (1:1 v/v), and measure fluorescence at λex/em = 485/590 nm in a Multi-Mode Microplate Reader.
15. Calculate drug loading by comparison with calibration curve of free Dox chloride in water-methanol (1:1 v/v) solution (*see* **Note 6**).
16. Check the sensitivity of Dox prodrug to enzymatic cleavage by HPLC.
    (a) Solubilize 0.5 mg of Dox prodrug in 100 μL of modified HEPES buffer pH 7.4.
    (b) Add 50 ng of human active MMP2 enzyme (MW 66 kDa) to the solution, and incubate at 37 °C for 3 h.
    (c) Analyze sample by RP-HPLC as described in **step 13** (*see* **Note 7**).

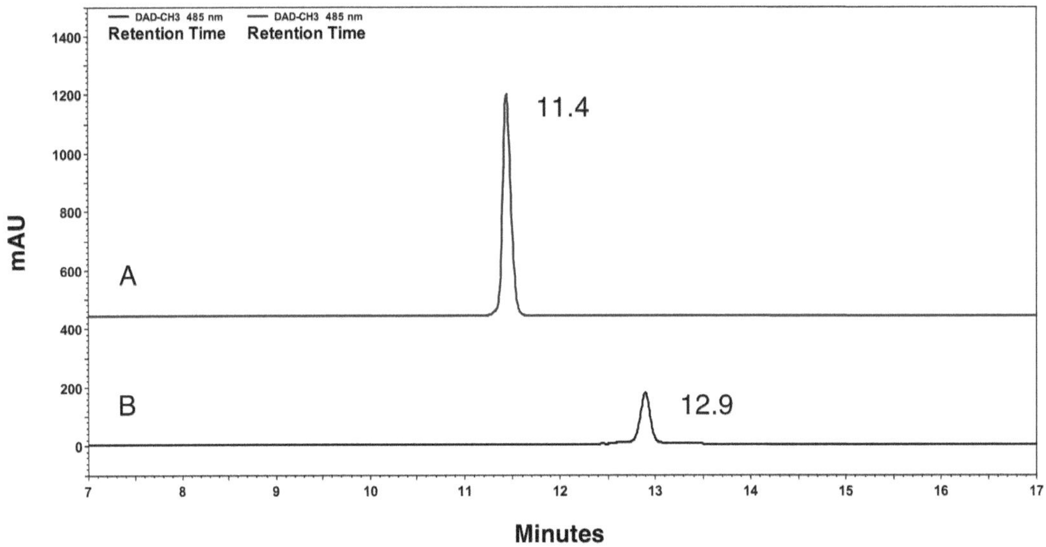

Fig. 2 Chromatograms of (**a**) doxorubicin chloride and (**b**) purified PEG2k-CLV-Dox prodrug

3.3 Synthesis of Sulfhydryl-Activated miRNA

1. Add 300 μL of TCEP solution and 100 μL of miRNA solution to a 2 mL tube. Vortex mixture and place it on an orbital shaker for 1 h.
2. Add 50 μL of NaAc solution and 1.5 mL of 200 proof ethyl alcohol to the reaction mixture. Vortex and divide solution in two aliquots of 975 μL each.
3. Place aliquots at −80 °C for 25 min.
4. Spin aliquots at 13,000 × *g* at 4 °C for 25 min.
5. Pour the supernatant from the aliquots into a 2 mL tube and save it.
6. Slowly add 200 μL of 95% ethanol to the aliquots. Wash the wall of the tubes, and discard the solvent (*see* **Note 8**).
7. Repeat previous step once and dry the aliquots under vacuum overnight.
8. Add 100 μL of water to each aliquot. Vortex and check sulfhydryl-activated miRNA concentration at $\lambda = 260$ nm in a NanoDrop 2000c Spectrophotometer (*see* **Note 9**).
9. Store solution at −80 °C (*see* **Note 10**).

3.4 Synthesis of GSH-Sensitive miRNA Conjugate (miRNA-S-S- DOPE)

1. Transfer 250 μL of DOPE solution and 250 μL of activated miRNA solution into an 8 mL glass vial. Pour N_2 into the system before closing the vial and stir for 24 h.
2. Dry the reaction mixture under N_2 (*see* **Note 11**).
3. Freeze the remaining sample at −80 °C for 30 min.

4. Freeze-dry the solution for 1 h (or until complete removal of water).
5. Suspend the powder in 1 mL of chloroform and centrifuge for 3 min at 10,000 × *g*. Transfer the supernatant to an 8 mL glass vial and save it. Repeat twice (*see* **Note 12**).
6. Dry the pellet under vacuum for 2 h.
7. Hydrate powder in 0.5 mL of water. Vortex until complete solubilization, and filter it through a 0.20 μm PES filter into a 2 mL tube.
8. Store the miRNA conjugate (miRNA-S-S-DOPE) solution at −80 °C.
9. Check miRNA concentration at $\lambda = 260$ nm in a NanoDrop 2000c Spectrophotometer.
10. Check efficiency of the conjugation by electrophoresis.
    (a) Dilute the miRNA conjugate stock solution up to 50 μg of miRNA/mL.
    (b) Add 20 μL of the dilution to each well of a 2% agarose E-gel.
    (c) Perform electrophoresis at 60 V for 30 min.
    (d) Check the gel under UV light in a Molecular Imager® ChemiDoc™ XRS (Fig. 3) (*see* **Note 13**).
11. Check the molecular weight of miRNA conjugate by MALDI-TOF mass spectrometry (Fig. 4).
    (a) Dissolve 20 mg of 2′,4′,6′-trihydroxyacetophenone monohydrate (THAP) in 1 mL of acetonitrile.
    (b) Dissolve 2 mg of ammonium citrate dibasic in 1 mL of water.

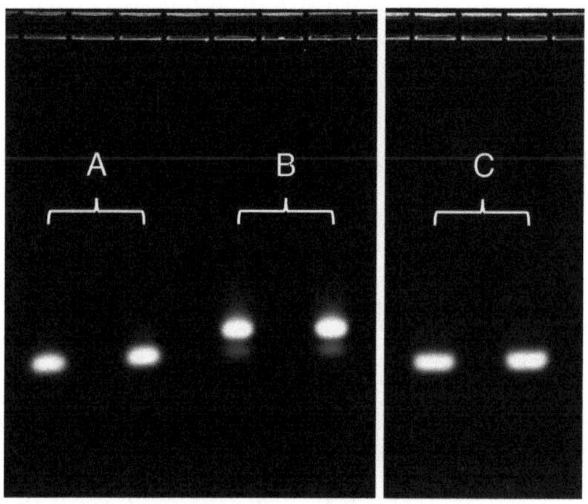

Fig. 3 UV signal of (**a**) miRNA, (**b**) miRNA-S-S-DOPE conjugate, and (**c**) reduced miRNA conjugate submitted to gel electrophoresis

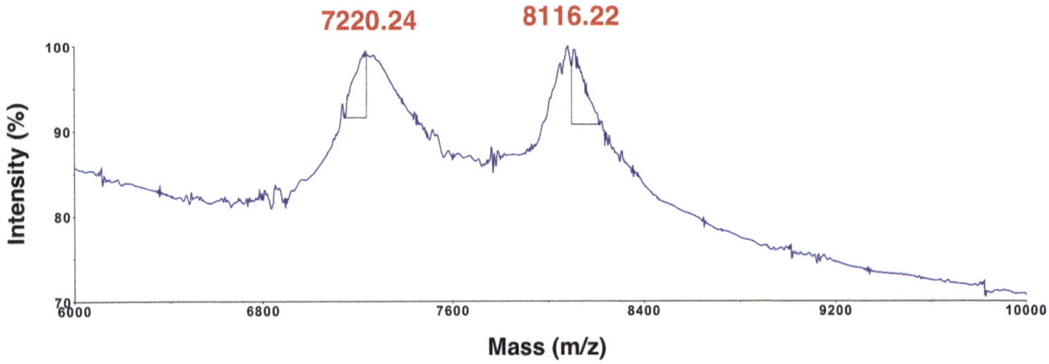

Fig. 4 MALDI-TOF mass spectrum of miRNA-S-S-DOPE conjugate showing unmodified (7220.2 *m/z*) and lipid-modified (8116.2 *m/z*) strands of the miRNA

(c) Prepare matrix solution by mixing 0.5 mL of THAP solution with 0.5 mL of ammonium citrate dibasic solution.

(d) Dilute the miRNA conjugate stock solution up to 200 μg of miRNA/mL.

(e) Mix 10 μL of the dilution with 90 μL of matrix solution.

(f) Deposit 0.7 μL of miRNA sample per spot of a MALDI plate.

(g) Perform mass spectrometry in a TOF/TOFTM 5800 system using positive ion, midrange linear mode (*see* **Note 14**).

12. Check the sensitivity of miRNA conjugate to reduction by GSH.

(a) Dilute the miRNA conjugate stock solution up to 200 μg of miRNA/mL.

(b) Mix 20 μL of the dilution with 60 μL of GSH solution (*see* **Note 15**).

(c) Incubate at 37 °C for 4 h, and analyze the samples by electrophoresis as described in **step 10** (Fig. 3). MALDI-TOF can also be used.

3.5 Preparation of Dual-Sensitive Mixed Micelles

1. Transfer 200 μL of PEG1k-PE solution to a glass vial.
2. Add 25 μL of PEG2k-CLV-Dox solution (0.5 mg of conjugate in total) to the vial and vortex.
3. Dry under N_2 and vacuum to form a thin layer film in the bottom of the glass vial.
4. Hydrate the film with 100 μL of miRNA-S-S-DOPE solution. Vortex for 5 min or until complete solubilization of the thin layer film (*see* **Note 16**).
5. Filter the formulation through a 0.20 μm PES filter into a 2 mL tube.
6. Check *Z*-average size and polydispersity index by dynamic light scattering in a Zetasizer Nano Z (*see* **Note 17**).

4 Notes

1. TEA must be added to convert the salt to free base form. Only the free base is able to react with PEG2k-CLV conjugate. Exposure to light can cause degradation of doxorubicin.
2. The concentration of miRNA in the solution of conjugate is given in ng/μL on NanoDrop software. The value of 0.0066% m/v is equivalent to 66 ng/μL.
3. Creating a N_2 atmosphere in the vial avoids degradation of the lipid by O_2.
4. DMF is being exchanged for water. Due to this solvent exchange, dialysis tube swells considerably. It is recommended to check the system each 30 min to verify any leakage of reaction mixture from the tube due to high pressure. If it happens, tubing must be changed. At this step, a red precipitate of byproducts and unreacted material is formed inside the tubing.
5. Water-MeOH mixture is required to extract any unreacted Dox (free base). Make sure the manufacturer guarantees the dialysis tubing can be exposed to methanol.
6. Drug loading varies from 25 to 35 μg Dox/mg conjugate.
7. The area of the peak at 12.9 min (Dox prodrug) will decrease. The peak at 11.4 min (free Dox) will reappear. Maximum cleavage ratio is around 60% depending on incubation time and potency of the enzyme.
8. After discard of the ethanol, a pellet will appear on the wall of the tubes. In case there is no pellet formation, the supernatant saved previously might be submitted to spin one more time.
9. Purity of the sample is confirmed by 260/280 ratio ≥ 2.0 and 260/230 ratio ≥ 2.2.
10. Ideally, the miRNA should be used immediately after reduction. Disulfides can reform over time.
11. The average time to dry 500 μL of the reaction mixture under N_2 is 20 min. A white precipitate is formed on the wall of the vial.
12. Chloroform removes unreacted lipid from the system. The miRNA conjugate is not soluble in chloroform.
13. Conjugation is confirmed by appearance of a new band of higher molecular weight in comparison to free miRNA used as a control. A less intense signal of free miRNA can still be detected in the conjugate sample, which indicates the reaction yield is not 100%.
14. The peaks seen in the mass spectrum might not be an exact match with the theoretical molecular weights expected for the

miRNA strands. Lipids and nucleic acids are very complex organic structures. Unpredicted reactions are likely to happen during the lipid modification of miRNA. Cytotoxicity and gene regulation tests in cell culture are encouraged to confirm the efficiency of the conjugate.

15. A 100 mM solution of DL-dithiothreitol solution (DTT) can also cleave the conjugate.
16. The components of the formulation self-assemble into mixed micelles. Ultrasound can be applied to the formulation if film remains attached to the walls of the glass vial after vortexing. Concentrations of each component can vary to fit desired treatment dosage. Very low amounts of PEG1k-PE can cause increase in average size of the particles.
17. Particle size can vary from 15 to 20 nm. Make sure to also check average size by "number of particles" and "volume."

Acknowledgments

This work was supported by the NIH Grant (1R21CA179286) to V. P. Torchilin and CAPES Foundation Scholarship (BEX 13275/13-5) to D. F. Costa. The authors thank Dr. W. Hartner for helpful comments on the preparation of the manuscript and Dr. R. W. Giese and Dr. P. Wang for assistance on the mass spectrometry analysis.

References

1. Wang Y, Kohane DS (2017) External triggering and triggered targeting strategies for drug delivery. Nat Rev Mat 2:17020
2. Torchilin VP (2014) Multifunctional, stimuli-sensitive nanoparticulate systems for drug delivery. Nat Rev Drug Discov 13(11):813–827
3. Insua-Rodriguez J, Oskarsson T (2016) The extracellular matrix in breast cancer. Adv Drug Deliv Rev 97:41–55
4. Heissig B, Eiamboonsert S, Salama Y et al (2016) Cancer therapy targeting the fibrinolytic system. Adv Drug Deliv Rev 99(Pt B):172–179
5. Hsieh CL, Liu CM, Chen HA et al (2017) Reactive oxygen species-mediated switching expression of MMP-3 in stromal fibroblasts and cancer cells during prostate cancer progression. Sci Rep 7(1):9065
6. Davaa E, Lee J, Jenjob R et al (2017) MT1-MMP responsive doxorubicin conjugated poly(lactic-co-glycolic acid)/poly(styrene-alt-maleic anhydride) core/shell microparticles for intrahepatic arterial chemotherapy of hepatic cancer. ACS Appl Mater Interfaces 9(1):71–79
7. Gerspach J, Müller D, Münkel S et al (2006) Restoration of membrane TNF-like activity by cell surface targeting and matrix metalloproteinase-mediated processing of a TNF prodrug. Cell Death Differ 13(2):273–284
8. Ansari C, Tikhomirov GA, Hong SH et al (2014) Development of novel tumor-targeted theranostic nanoparticles activated by membrane-type matrix metalloproteinases for combined cancer magnetic resonance imaging and therapy. Small 10(3):566–575
9. Xue HY, Guo P, Wen WC et al (2015) Lipid-based nanocarriers for RNA delivery. Curr Pharm Des 21(22):3140–3147
10. Campani V, Salzano G, Lusa S et al (2016) Lipid nanovectors to deliver RNA oligonucleotides in cancer. Nanomaterials (Basel) 6(7):131
11. Salzano G, Costa DF, Torchilin VP (2015) siRNA delivery by stimuli-sensitive nanocarriers. Curr Pharm Des 21(31):4566–4573

12. Musacchio T, Vaze O, D'Souza G et al (2010) Effective stabilization and delivery of siRNA: reversible siRNA-phospholipid conjugate in nanosized mixed polymeric micelles. Bioconjug Chem 21(8):1530–1536
13. Salzano G, Navarro G, Trivedi MS et al (2015) Multifunctional polymeric micelles co-loaded with anti-survivin siRNA and paclitaxel overcome drug resistance in an animal model of ovarian cancer. Mol Cancer Ther 14(4):1075–1084
14. Salzano G, Costa DF, Sarisozen C et al (2016) Mixed nanosized polymeric micelles as promoter of doxorubicin and miRNA-34a co-delivery triggered by dual stimuli in tumor tissue. Small 12(35):4837–4848

Chapter 9

Minicell-Based Targeted Delivery of shRNA to Cancer Cells: An Experimental Protocol

Mehul Jivrajani and Manish Nivsarkar

Abstract

Bacterial minicell has emerged as a novel targeted delivery system for RNAi-based therapeutics. In this chapter, we have described the detailed protocol for the preparation of minicell-based targeted delivery system for shRNA. Initially, minicell-producing parent bacterial cells were transformed with plasmid vector containing shRNA. Subsequently, shRNA-packaged minicells were purified from parent bacterial cells. Purified minicells were characterized by fluorescence microscopy and transmission electron microscopy. In the next step, targeting ligand was conjugated on the minicell surface for the active targeting of cancer cell surface receptors. Eventually, target-specific delivery of minicells was explored in vitro in selected cancer cell line and in vivo in mice bearing tumor xenograft.

Key words Bacterial minicells, shRNA delivery, Folate receptor targeted delivery, RNA interference, Cancer

1 Introduction

RNA interference (RNAi)-based therapeutic approaches are under vigorous exploration by scientists worldwide to search a cure for various diseases. RNAi-based therapeutics work by silencing targeted endogenous mRNA before translation and preventing cells from making a specific protein that contributes to disease. RNAi can be induced artificially by introducing synthetic si (short interfering), sh (short hairpin), or mi (micro) RNAs into cells [1]. Hence, RNAi has emerged as a novel therapy in many diseases where overexpression and/or faulty expression of genes are responsible for the occurrence of disease [2].

Cancer is an example of disease which arises from more than one genetic change where many genes are overexpressed or mutated. Downregulation of such overexpressed or mutated genes by RNAi-based therapeutics restrains proliferation of cancerous cell and its metastasis which eventually reduces the tumor growth.

Lekha Dinesh Kumar (ed.), *RNA Interference and Cancer Therapy: Methods and Protocols*, Methods in Molecular Biology, vol. 1974, https://doi.org/10.1007/978-1-4939-9220-1_9, © Springer Science+Business Media, LLC, part of Springer Nature 2019

Theoretically, RNAi can be exploited to silence virtually any disease-causing gene in the cells [2].

However, regardless of significant potential of RNAi therapeutics, most approaches to RNAi-mediated gene silencing for cancer therapy have been demonstrated in vitro in the laboratory, and key hurdle in the transition to clinical use is lack of appropriate delivery system [3]. To confer with drug-like properties which include stability, tissue bioavailability, and cellular delivery to RNAi therapeutics, several delivery strategies have been developed such as viral vectors [4], aptamers [5], lipid-based delivery [6], polymer-mediated delivery [7], and bacterial minicell-based delivery system [2, 8]. Recently, bacterial minicells have got much attention as targeted delivery systems after the pioneering demonstration by MacDiarmid et al. Bacterial minicells are nano-sized anucleated cells produced during the abnormal cell division by depressing cryptic polar sites of cell fission. This is achieved by inactivating bacterial genes, i.e., minCDE chromosomal deletion, which controls the normal bacterial cell division [9–12]. Minicells possess all of the molecular components of the parent cell, except the chromosome. Most importantly, minicells do not possess the capacity to divide, and their production does not hamper normal cell division which occurs simultaneously. Minicells have been exploited as an efficient delivery system for RNAi-based therapeutics due to several advantages like convenient and efficient packaging of sufficient amount of siRNA (~12,000 copies) and shRNA (~100 copies), stability of packaged si/shRNA, easy surface modification with antibodies/ligand for active targeting, etc. [13]. Further, si/shRNA-packaged minicells can be selectively targeted to tumor cell surface receptors by targeting moiety conjugated on the minicells. Consequently, receptor engagement results in minicell endocytosis and intracellular degradation that releases si/shRNA which exerts its effect by undergoing RNAi pathway.

In this chapter, we have illustrated stepwise protocol for the development of minicell-based targeted delivery system for shRNA. Protocol includes packaging of shRNA into minicells, minicell purification and its characterization, conjugation of targeting ligand (here, folic acid) on minicell surface for active targeting, and its in vitro and in vivo delivery. Key steps of protocol are depicted in the Scheme 1.

2 Materials

All media, reagents, and other chemicals used here are of either molecular biology or cell culture grade. Prepare all solutions using ultrapure water (prepared by purifying deionized water, to attain a sensitivity of 18 MΩ cm at 25 °C) and analytical grade reagents. All the microbiological media such as Luria Bertani (LB) broth, LB

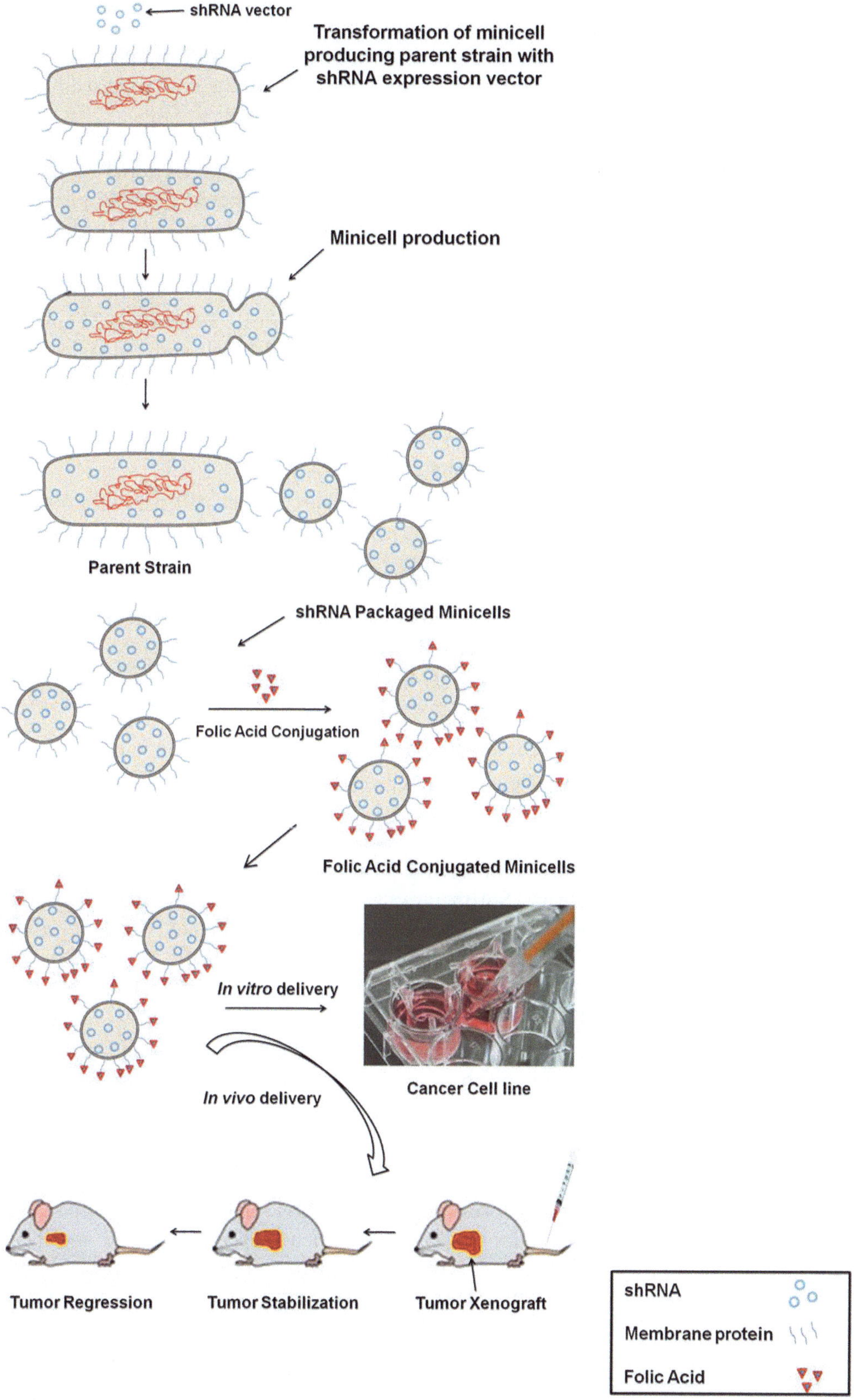

Scheme 1 Schematic representation of protocol. Schematic representation of shRNA packaging, folic acid conjugation on minicells' surface, and its in vitro and in vivo delivery

agar, thioglycollate medium, antibiotics, and gram's staining kit used here are of microbiology grade. All the solvents used here are of analytical grade.

2.1 Characterization of Minicells Producing Bacterial Strain

E. coli PB114, minicell producing mutant bacterial strain, was exploited to develop minicell-based targeted delivery system for shRNA. *E. coli* PB114 was kindly provided by Dr. Lawrence Rothfield, University of Connecticut, USA (*see* **Note 1**). Other requirements are given as follows:

1. Sterile LB broth: Weigh 25 g LB broth powder in a 2 l autoclavable bottle or flask. Add 1000 ml of ultrapure distilled water and dissolve powder thoroughly. Sterilize by autoclaving at 15 lbs pressure, 121 °C for 15 min. Cool down, aliquot aseptically in sterile flask, and use for inoculation. Store LB broth at room temperature (*see* **Note 2**).
2. Kanamycin (50 μg/ml): Prepare a stock of 1 mg/ml in distilled water, filter sterilize with 0.22 μm filter, and store at 2–8 °C. Add appropriate volume of kanamycin from stock to achieve final concentration of 50 μg/ml in LB broth or LB agar. For long-term storage, prepare several aliquots of antibiotic stock, and store at −20 °C. Avoid repeated freeze-thaw cycle.
3. Gram's staining kit having 2% crystal violet, gram's iodine, 95% ethanol, and 0.5% safranin.
4. Upright microscope with 100× oil immersion lens.
5. Transmission electron microscope (TEM).

2.2 Packaging of shRNA Expression Vector into Minicells

Plasmid based expression vectors such as pEZ43G-D, pSUPERneo, and vector having VEGFA shRNA were kindly provided by Dr. Neeta Shrivastava (Department of Pharmacognosy and Phytochemistry, PERD Centre, Ahmedabad, Gujarat, India). pSUPERneo is a backbone vector in which shRNA sequence specific for VEGFA and scramble sequence (control shRNA) has been cloned.

1. Prepare sterile LB broth and kanamycin (50 μg/ml) as mentioned in Subheading 2.1.
2. 0.1 M calcium chloride [$CaCl_2$]: Weigh 1.11 g of $CaCl_2$ and dissolve in 100 ml of ultrapure distilled water. Filter sterilize through a 0.22 μm filter and store at 2–8 °C.
3. Plasmid expression vectors: pEZ43G-D plasmid, and pSUPERneo backbone vector in which shRNA sequence specific for VEGFA and scramble sequence (non-specific shRNA) have been cloned. Here, we have used shRNA specific for VEGFA as a test shRNA (*see* **Note 3**).
4. Ampicillin, (100 μg/ml): Prepare a stock of 1 mg/ml, filter sterilize with 0.22 μm filter, and store at 2–8 °C. Add appropriate

volume of ampicillin from stock to achieve final concentration 100 μg/ml in LB broth or LB agar. For long-term storage, prepare several aliquots of stock, and store at −20 °C. Avoid repeated freeze-thaw cycle.

5. LB agar plates: Suspend 40 g of LB agar in 1000 ml of distilled water. Heat to boiling to dissolve the medium completely. Sterilize by autoclaving at 15 lbs pressure, 121 °C for 15 min. Cool down to ~45 °C and add desired concentration of antibiotics. Mix well and pour into sterile petri plates.

2.3 Minicell Purification

1. LB broth and LB agar.
2. Kanamycin (50 μg/ml).
3. Ampicillin (100 μg/ml).
4. Ceftriaxone (100 μg/ml): Prepare a stock of 1 mg/ml and store at 2–8 °C. Add appropriate volume of ceftriaxone from stock to achieve final concentration 100 μg/ml. For long-term storage, prepare several aliquots of stock, and store at −20 °C. Avoid repeated freeze-thaw cycle.
5. Thioglycollate medium: Suspend 29.75 g in 1000 ml of distilled water. Heat to boiling to dissolve the medium completely. Sterilize by autoclaving at 15 lbs pressure, 121 °C for 15 min. Cool to 25 °C and store in a cool dark place preferably below 25 °C.
6. Upright fluorescence microscope with 100× oil immersion lens.
7. Transmission electron microscope (TEM).

2.4 Analysis of Packaged shRNA Expression Vector

1. Alkaline lysis solution-I: 50 mM glucose, 25 mM Tris–Cl (pH 8.0), and 10 mM EDTA (pH 8.0). Prepare solution-I from standard stocks in batches of approx. 100 ml, sterilize by autoclaving, and store at 4 °C.
2. Alkaline lysis solution-II: 0.2 N NaOH, 1% (w/v) SDS. Prepare solution-II fresh and use at room temperature.
3. Alkaline lysis solution-III: 5 M potassium acetate, 60.0 ml; glacial acetic acid, 11.5 ml; and H_2O, 28.5 ml. Store the solution at 4 °C and transfer it to ice bucket just before use.
4. Isopropanol.
5. 70% ethanol in distilled water.
6. 10× Tris EDTA (TE) buffer: 100 mM Tris–Cl (pH 8.0), 10 mM EDTA (pH 8.0). Adjust the pH by adding concentrated HCl. Sterilize the buffer by autoclaving and store at room temperature. Use 1× TE buffer to dissolve plasmid DNA.

2.4.1 Agarose Gel Electrophoresis

1. Electrophoresis (Tris-acetate-EDTA (TAE) buffer: Prepare 50× stock solution of TAE buffer. For this weigh 242 g of Tris base. Transfer this to a 1000 ml beaker. Prepare EDTA solution (0.5 M, pH 8.0) by weighing 9.31 g of EDTA, and dissolve it in 40 ml distilled water. EDTA is insoluble, and it can be made soluble by adding sodium hydroxide pellets. Check the pH using pH meter. Make the solution 100 ml by adding distilled water. Pipette out 57.1 ml of glacial acetic acid. Mix the Tris base, EDTA solution, and glacial acetic acid, and add distilled water to make the volume to 1000 ml. Prepare 1× TAE buffer to fill the electrophoresis tank and to cast the gel.
2. Agarose solution: Weigh 0.9 g of agarose and transfer it to an Erlenmeyer flask. Add 100 ml of electrophoresis (1× TAE) buffer. Heat the slurry in a boiling water bath or a microwave oven until the agarose dissolves. Allow it to cool down to 45–50 °C and add 0.5 μg/ml of ethidium bromide. Mix the gel solution thoroughly by gentle swirling and cast the gel.
3. Ethidium bromide: Prepare 10 mg/ml stock solution of ethidium bromide, and transfer the solution to a dark bottle, or wrap the container in aluminum foil, and store at room temperature. Add 5 μl of ethidium bromide from stock to the 100 ml of agarose solution to get a final concentration 0.5 μg/ml (*see* **Note 4**).
4. Plasmid DNA samples isolated from minicells and purified plasmid DNA as a reference.
5. 6× gel loading buffer: Components of 6× gel loading buffer include 60% glycerol, 10 mM Tris–HCl (pH 7.6), 60 mM EDTA, 0.03% bromophenol blue, and 0.03% xylene cyanol FF. Mix DNA samples with 6× gel loading buffer in such a way that final concentration of gel loading buffer becomes 1×. Use these samples for agarose gel electrophoresis.
6. λ Hind III ladder.
7. Horizontal gel electrophoresis unit along with power pack.

2.5 Conjugation of Folic Acid on Minicells

1. Phosphate buffer saline (PBS): Weigh 8 g of NaCl, 0.2 g of KCl, 1.44 g of Na_2HPO_4, and 0.24 g of KH_2PO_4, and add in 800 ml of distilled water. Adjust the pH to 7.4 with HCl and make up volume to 1000 ml. For this experiment, prepare second set of PBS and adjust pH to 7.6.
2. shRNA-packaged minicells (1×10^{10}/ml).
3. Folic acid solution dissolved in PBS (1.0 ml, 1 mg/ml, total 5 mg, pH 7.6) in a 10 ml round-bottomed flask.
4. EDC (1-ethyl-3-(-3-dimethylaminopropyl) carbodiimide hydrochloride), (10.0 mg, 50 μmol).
5. Sulfo-NHS (*N*-hydroxysuccinimide), (22.0 mg, 100 μmol).

6. DMAP (4-dimethylaminopyridine) (1 mg for 5 ml of reaction mixture).

2.6 In Vitro Uptake of Folic Acid Conjugated Minicells

1. Cell lines: Cancer cell line, KB, has been procured from the National Centre for Cell Sciences (NCCS), Pune, India. KB has been selected as a choice of cell line due to high level of folate receptor expression [14] (*see* **Note 5**). Maintain KB cell line in DMEM medium supplemented with 100 U/ml penicillin, 100 μg/ml streptomycin, 0.25 μg/ml amphotericin B, and 10% FBS at 37 °C under a humidified atmosphere containing 5% CO2. Acclimatize KB cells in folate-free DMEM medium supplemented with 10% FBS a week before in vitro uptake study.
2. minicells$_{VEGFA}$ (minicells packaged with VEGFA shRNA).
3. ^{FA}minicells$_{VEGFA}$ (minicells packaged with VEGFA shRNA and conjugated with folic acid).
4. FITC: Prepare 1 mg/ml solution of FITC in PBS (pH 7.4) for staining of minicells.
5. 1% Triton X-100: Prepare 1% Triton X-100 (v/v) in PBS (pH 7.4).
6. BCA protein assay: Estimate protein concentration of cell lysate by BCA protein assay kit as per the manufacturer's protocol.
7. Inverted fluorescence microscope.
8. Spectrofluorometer.

2.7 In Vivo Delivery of Folic Acid Conjugated Minicells

Animal: Healthy C57 BL/6 mice are used to study in vivo delivery. Keep mice in individually ventilated cages (IVC), with a relative humidity of 60 ± 5% and a temperature of 25 °C ± 2 °C. A 12:12 h light/dark cycle should be regulated for these animals. Provide autoclaved balanced rodent food pellet and water ad libitum. All the experimental protocols pertaining to animal use should be reviewed and accepted by Institutional Animal Ethics Committee (IAEC) prior to initiation of experiments. Here, all our experimental protocols were reviewed and accepted (PERD/IAEC/2013/014) by the IAEC prior to initiation of the experiments.

2.7.1 Development of tumor xenograft in pharmacologically immunocompromised mice

1. Cyclosporine.
2. Ketoconazole.
3. Cyclophosphamide.
4. Ampoxin.
5. Hematology analyzer with veterinary mode.

2.7.2 Preparation and Implantation of Cancer Cells

1. Pharmacologically immunocompromised C57 BL/6 mice. Here, in this study we have exploited previously described immunosuppression protocol using cyclosporine, ketoconazole, and cyclophosphamide drug regimen to develop pharmacologically immunocompromised C57 BL/6 mice and subsequent implantation of cancer cells [15].
2. Semi-confluent KB cells.
3. Trypsin-EDTA solution (0.25%).
4. Trypan blue solution (0.4%) in PBS (pH 7.4).
5. Folate-free rodent diet.
6. Depilatory cream.
7. 70% alcohol in distilled water.
8. 1 ml syringe with 27 G needle.
9. Digital caliper.

2.7.3 In Vivo Delivery of ^{FA}minicells$_{VEGFA}$

1. Sterile physiological saline (0.9%).
2. ^{FA}minicells$_{scramble}$ (minicells packaged with scramble (non-specific) shRNA and conjugated with folic acid).
3. minicells$_{VEGFA}$ (minicells packaged with VEGFA shRNA).
4. ^{FA}minicells$_{VEGFA}$ (minicells packaged with VEGFA shRNA and conjugated with folic acid).
5. Digital caliper.

2.8 In Vivo Biodistribution of ^{FA}minicells$_{VEGFA}$

1. C57 BL/6 mice bearing KB tumor xenograft.
2. FITC-loaded ^{FA}minicells$_{VEGFA.}$
3. Liquid nitrogen.
4. Cryotome.
5. Hematoxylin and eosin (H&E) stain.
6. Fluorescence microscope.

3 Methods

Perform all the procedures at room temperature unless otherwise specified.

3.1 Characterization of Minicells Producing Bacterial Strain

1. Perform enrichment of *E. coli* PB114 by inoculating a single colony in Luria Bertani (LB) broth supplemented with 50 μg/ml kanamycin in 25 ml flask (*see* **Note 6**).
2. Incubate flask at 37 °C, 180 rpm overnight (~16 h) in shaking incubator, and observe the growth after incubation.
3. Perform gram's staining in order to observe the morphology, presence of gram-negative organism, i.e., *E. coli*, as well as presence of minicells (Fig. 1a).

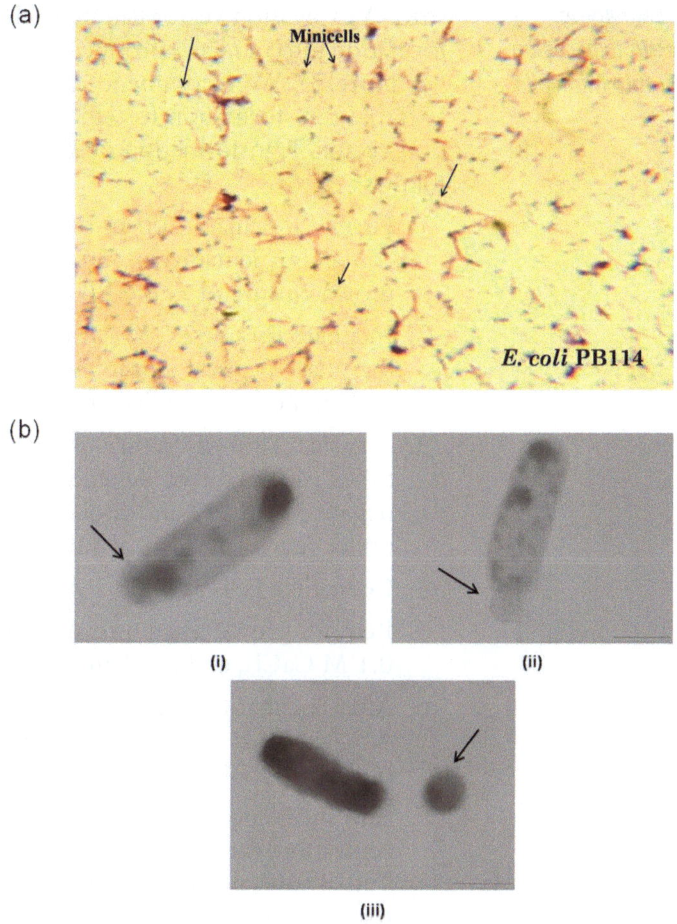

Fig. 1 Characterization of minicells producing bacterial strain (*E. coli* PB114). (**a**) Light microscopy image of *E. coli* PB114 after gram's staining (×1000 magnification). *E. coli* PB114 turned pink after gram's staining thus confirming the presence of gram-negative organism. It also showed the presence of large number of minicells which are indicated with arrows. (**b**) Transmission electron microscopy image of *E. coli* PB114. (×*20,000 magnification*). (**b**i–iii) Sequential minicell production from the parent *E. coli* PB114 which is indicated by arrows. Scale bars 500 nm, 1000 nm, and 1000 nm, respectively. Thus detailed morphology and minicell production can be observed more closely by TEM (*Reproduced from ref.* 2)

4. Subsequently, characterize cells by transmission electron microscopy (TEM) to observe morphology of cells in more detail. For TEM, apply 100 μl of overnight grown culture of *E. coli* PB114 to copper TEM grid for 5 min. During this time bacterial cells get absorbed onto the supporting film.
5. Absorb remaining culture from the edge of the grid by wedge of filter paper.
6. Here, we recommend the use of transmission electron microscope (Tecnai 20, 200 kV, Philips, Holland) at an accelerating voltage of 90 kV (Fig. 1b).

3.2 Packaging of shRNA Expression Vector into Minicells

shRNA expression vector can be packaged into minicells by transformation into minicells producing parent strain, i.e., *E. coli* PB114 (*see* **Note 7**). We recommend calcium chloride ($CaCl_2$)-mediated bacterial transformation for efficient packaging of shRNA expression vector. Transformation of *E. coli* PB114 can be performed as follows:

1. Freshly inoculate 200 μl of culture into 10 ml of LB broth containing kanamycin (50 μg/ml) from overnight (~16 h) grown culture of *E. coli* PB114.
2. Incubate flask at 37 °C with shaking till the cells reach to the log phase (optical density at 600 nm = 0.4). Generally, a 3-h shaking period is sufficient.
3. Transfer 2 ml of culture aseptically in 2 ml microcentrifuge tubes (no. of tubes depends upon no. of tests, positive and negative control), and cool down to 4 °C by storing the tubes on ice (perform all subsequent steps in aseptic condition).
4. Centrifuge cells at 6000 × *g* for 5 min at 4 °C.
5. Remove growth medium, and resuspend cell pellet in 1 ml of 0.1 M $CaCl_2$, and incubate on ice for 15 min.
6. After incubation, centrifuge cells at 6000 × *g* for 5 min at 4 °C. Discard supernatant and resuspend cell pellet in 200 μl of 0.1 M $CaCl_2$. Incubate on ice for 30 min (*see* **Note 8**).
7. Add shRNA containing plasmid vector in three different concentration, for instance, Test-1 (120 ng), Test-2 (240 ng), and Test-3 (600 ng) to competent cells. Do not add plasmid vector in positive and negative control tubes. Mix well and incubate on ice for 30 min. Here, along with shRNA expression vector, we have also taken pEZ43G-D plasmid for *E. coli* PB114 transformation which possesses GFP as a reporter gene along with ampicillin-resistant gene, to track minicells purification under fluorescence microscope.
8. After incubation, provide heat shock to the cells at 42 °C for 2 min in a water bath, and rapidly transfer to ice, and chill the cells for 2 min.
9. Transfer cells to tubes containing 2 ml LB broth, and incubate at 37 °C for 30 min at 200 rpm in shaking incubator.
10. Transfer cells to 2 ml microcentrifuge tube, and centrifuge it at 6000 × *g* for 5 min at 4 °C.
11. Remove medium and resuspend cells in 100 μl of LB broth.
12. Plate these cultures in different petri dishes such as negative control (ampicillin, 100 μg/ml; kanamycin, 50 μg/ml; no plasmid DNA), positive control (no ampicillin; kanamycin, 50 μg/ml; no plasmid DNA), Test-1 (ampicillin, 100 μg/ml; kanamycin, 50 μg/ml) (120 ng/1 μl plasmid DNA), Test-2

(ampicillin, 100 μg/ml; kanamycin, 50 μg/ml) (240 ng/2 μl plasmid DNA), and Test-3 (ampicillin, 100 μg/ml; kanamycin, 50 μg/ml) (600 ng/1 μl plasmid DNA) (*see* **Note 9**).

13. Incubate plates overnight at 37 °C in incubator.
14. Observe plates for the growth of transformed cells after the incubation (Fig. 2).

3.3 Minicell Purification

Minicells can be purified from the living parent bacteria by combination of antibiotic treatment (ceftriaxone) and filtration (*see* **Note 10**). This method is based on the principle of nondividing nature of minicells. Ceftriaxone acts on dividing cells by inhibiting its cell wall synthesis. Hence, it kills parent cells but does not harm the minicells. Further, residual parent bacteria and cell debris and endotoxins can be removed by filtration through 0.45 μm and 0.22 μm filter, respectively. Hence, combination of ceftriaxone and filtration method results in high yield of ultrapure minicells [17].

The detailed method is given below.

1. Take 200 ml of overnight (~16 h) grown culture of *E. coli* PB114 transformed with shRNA expression vector. Here, we have taken *E. coli* PB114 transformed with pEZ43G-D plasmid to track purification on Luria agar plate and fluorescence microscopy.
2. Centrifuge cells at 2000 × *g*, room temperature for 10 min to remove majority of parent bacterial cells.
3. Collect supernatant in a fresh tube and discard the pellet.
4. Centrifuge supernatant at 10,000 × *g* for 10 min at room temperature to pellet the minicells.
5. Resuspend minicell pellet in 50 ml of fresh LB broth.
6. Incubate suspension at 37 °C with vigorous aeration for 30–40 min to permit reinitiation of parent cell growth.
7. After incubation, add ceftriaxone at a dose of 100 μg/ml, and incubate flask for 30–45 min in shaking incubator at 37 °C.
8. In the next step, centrifuge broth at 2000 × *g* to remove any viable parent cells and cell debris.
9. Collect supernatant in a separate tube and discard the pellet.
10. Centrifuge supernatant at 10,000 × *g* for 10 min at room temperature to pellet the minicells.
11. Resuspend minicell pellet in fresh broth and filter through 0.45 μm filter.
12. Centrifuge filtrate at 10,000 × *g* for 10 min at room temperature to pellet minicells.
13. Subsequently, filter minicells through 0.22 μm filter to remove any cell debris.

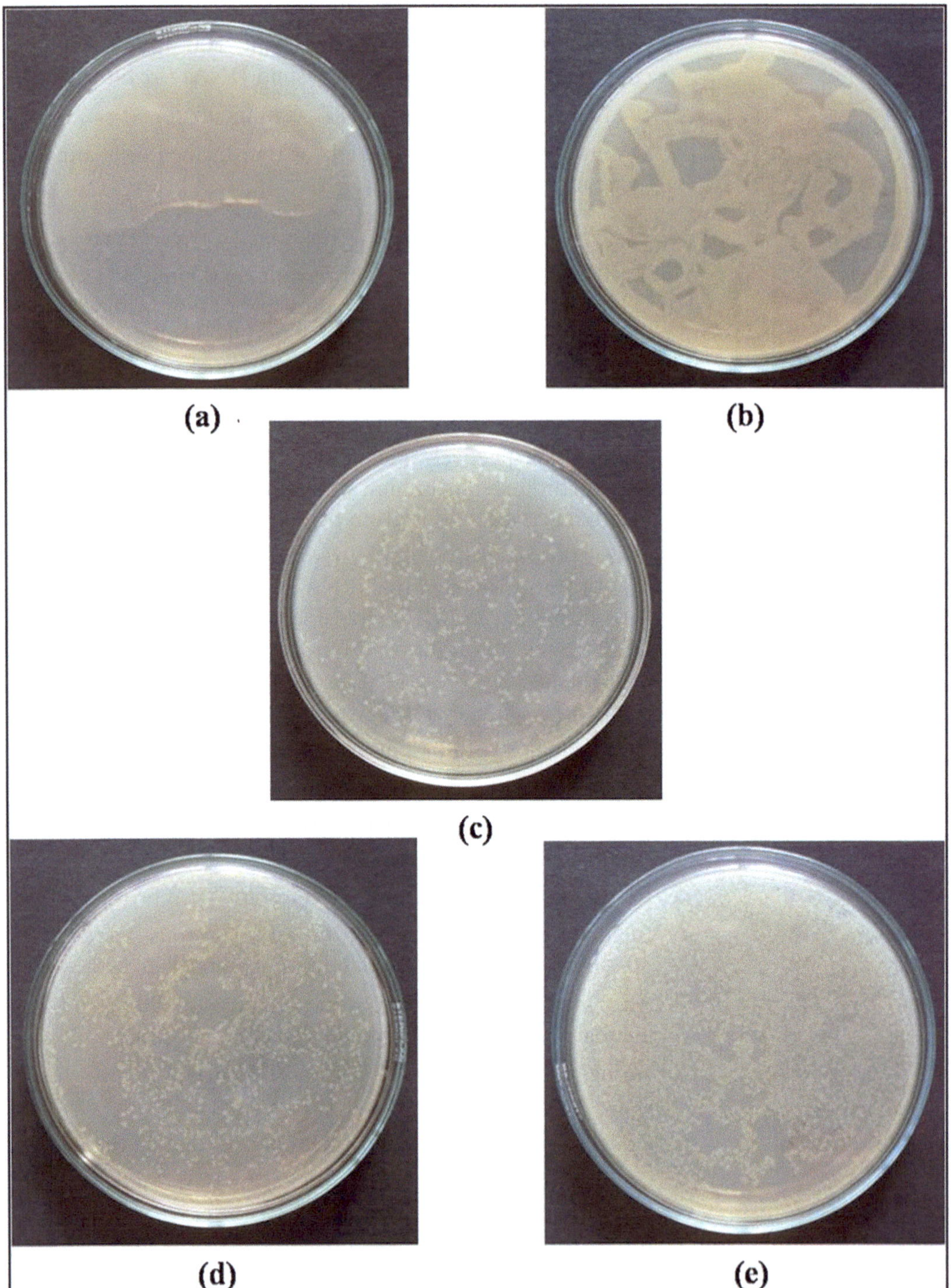

Fig. 2 Packaging of shRNA expression vector into minicells. LB agar plates showing (**a**) negative control, (**b**) positive control, (**c**) Test-1 (120 ng plasmid DNA), (**d**) Test-2 (240 ng plasmid DNA), and (**e**) Test-3 (600 ng plasmid DNA) plates. Test1–3 show *E. coli* PB114 transformed with pEZ 43G-D plasmid. In all the test plates, ampicillin was added at a concentration of 100 μg/ml. So only those cells can grow which have acquired pEZ43G-D. Further, it can be observed that the number of transformed colonies increased with the increasing concentration of plasmid DNA used for transformation. From these results, it can be inferred that transformation of *E. coli* PB114 with pEZ 43G-D plasmid has been successfully achieved (*Reproduced from ref.* 16)

3.3.1 Quality Check of Purified Minicells

1. Wash minicell pellet in fresh broth, and concentrate and plate on LB agar plate after each step of purification to check the presence of any viable parent cell.
2. Observe the same minicell sample under fluorescent microscope after each step of purification to assess the purity of purified minicells.
3. Additionally, purity of minicells can be confirmed further by measuring the size of cells at every step of purification through dynamic light scattering method (Zeta sizer ZS 90, Malvern) (*see* **Note 11**).
4. If required, observe morphology of purified minicells by TEM as describe above.
5. Eventually, inoculate purified minicells in fluid thioglycollate medium, and incubate for 14 days to check the presence of any slow-growing contaminant (Fig. 3) (Table 1).

It can be seen from the Table 1 that average size and polydispersity index (PDI) decreased gradually after each step of purification. Average size of cells decreased significantly from 929.4 nm (starting culture) to 522.3 nm (final fraction), suggesting the purity of minicells increases after each step of purification. Further, polydispersity index (PDI) value also decreases from 1.0 to 0.042 which suggest homogeneity of minicell population. Further, minicell number can be determined spectrophotometrically by taking optical density at 600 nm and employing the following equation [18]:

$$A_{600} \times 5.0 \times 10^{10} / \text{ml}$$

3.4 Analysis of Packaged shRNA Expression Vector

Packaging of shRNA expression vector into minicells can be confirmed by isolation of plasmid DNA from the purified minicells by alkaline lysis method, subsequent gel electrophoresis, and spectrophotometric analysis of isolated plasmid DNA. Detailed steps of method can be employed as follows:

1. Take purified minicells obtained from 200 ml of culture (here, we took 3.81×10^{10} minicells obtained from 200 ml of culture).
2. Centrifuge minicells at 10,000 × *g* for 10 min at 4 °C, and decant supernatant and dab tube on tissue paper.
3. Resuspend pellet in 100 μl of ice-cold alkaline lysis solution-I, and incubate on the ice bath for 5 min.
4. Add 200 μl of freshly prepared alkaline lysis solution-II, and mix the content of the tube by gently inverting the tube up and down three to four times. After that, incubate the tube for 10 min on ice bath.

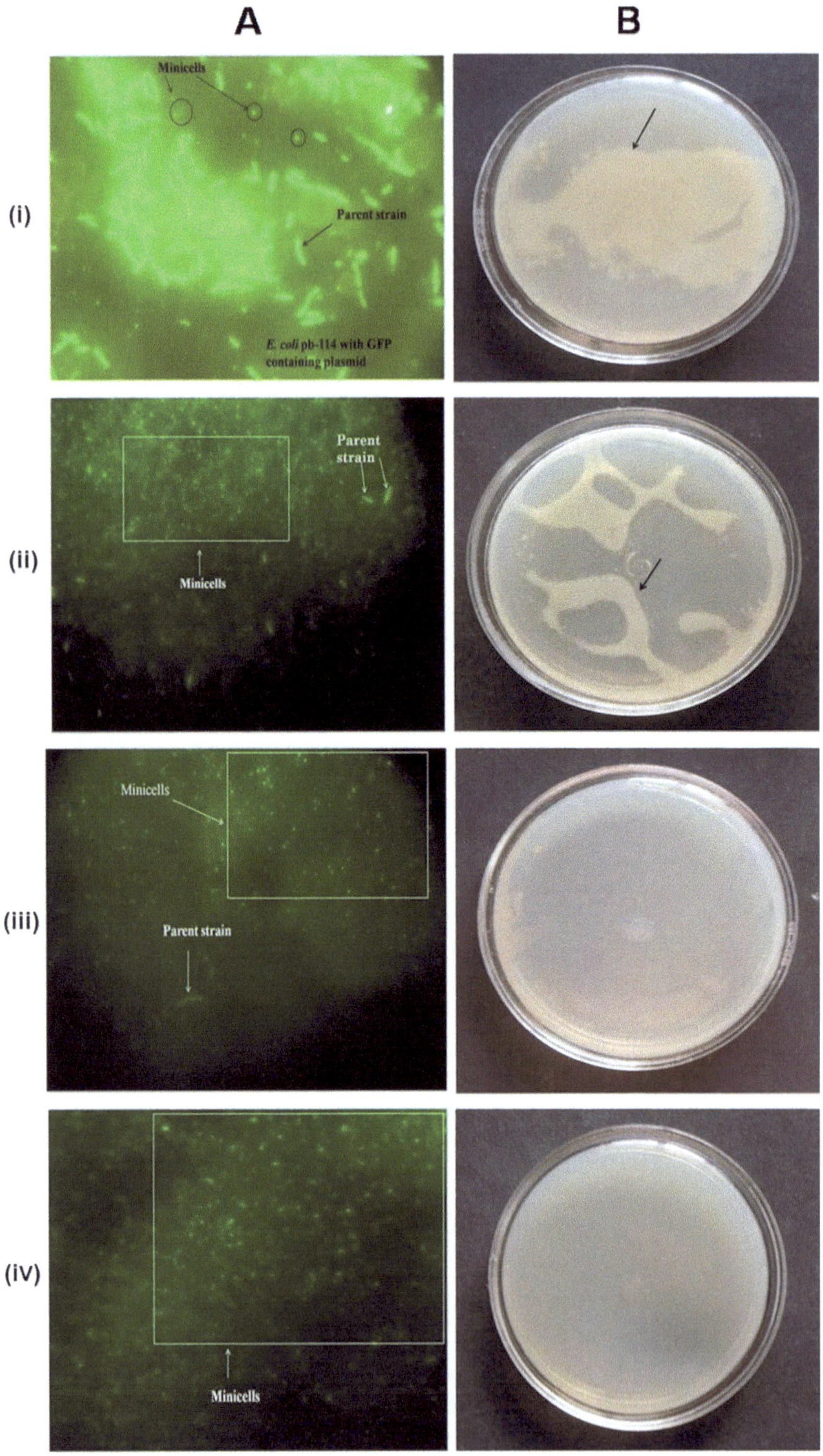

(a)

Fig. 3 (**a**) Minicell purification. Different stages of minicell purification by ceftriaxone lysis method: (A) fluorescence microscopy image and (B) growth of *E. coli* PB114 on LB agar plates; (i) starting culture of pEZ 43G-D transformed *E. coli* PB114, (ii) culture of *E. coli* PB114 pEZ 43G-D after differential centrifugation,

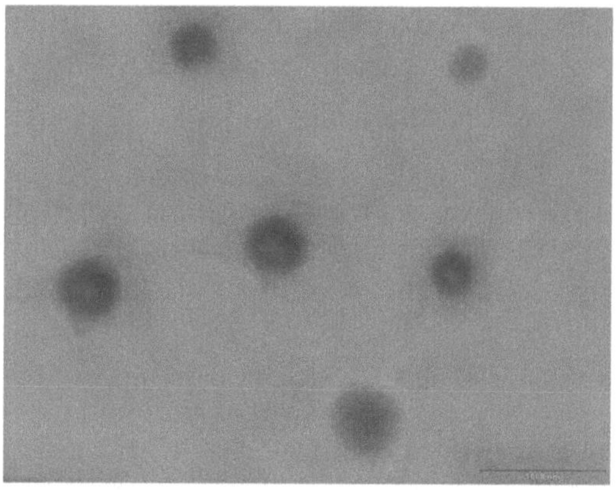

(b)

Fig. 3 (continued) (iii) minicells after ceftriaxone treatment, and (iv) minicells after successive filtration with 0.45 μm and 0.22 μm filter. It can be visualized from (**a**) that after every step of minicell purification, parent cells decreased which was confirmed by the fluorescence microscopy as well as corresponding growth of minicell culture on LB agar plates. Thus, ultrapure minicells can be obtained by combining two methods: ceftriaxone lysis and filtration, (**b**) transmission electron microscopy (×20,000 magnification). Scale bar 1000 nm. TEM image showed that minicells possess intact structure and of well-formed spherical shape [(**a**) *reproduced from ref.* 17; (**b**) *reproduced from ref.* 2]

Table 1
Average size and polydispersity index of different fractions of minicell purification

S. no.	Fraction	Size (nm)	PDI (polydispersity index)
1	Starting culture of pEZ 43G-D transformed *E. coli* PB114	929.4	1.0
2	Culture of *E. coli* PB114 pEZ 43G-D after differential centrifugation	639.9	0.156
3	Minicells after antibiotic treatment	584.9	0.081
4	Minicells after successive filtration with 0.45 μm and 0.22 μm filter	522.3	0.042

5. Add 150 μl of cold alkaline lysis solution-III, and mix the contents gently by inversion. Incubate tube for 10 min on ice bath.
6. Centrifuge tube at 12,000 × *g* for 10 min at 4 °C.
7. Next, separate supernatant carefully, and transfer to a new microcentrifuge tube.
8. Add 0.7th volume of isopropanol, and mix and keep at −20 °C temperature overnight (~16 h).
9. After the incubation, centrifuge tube at 12,000 × *g* for 20 min at 4 °C to pellet down plasmid DNA.
10. Decant supernatant and add 0.7 ml of 70% ethanol to wash the pellet.

11. After that, centrifuge tube at 10,000 × *g* for 10 min at 24 °C and decant supernatant.
12. Air-dry the pellet and add TE buffer to dissolve the DNA pellet.
13. Analyze sample by agarose gel electrophoresis (0.9%) and spectrophotometric method (A_{260} nm, data not shown) to confirm the presence of plasmid DNA (Fig. 4).

3.5 Conjugation of Folic Acid on Minicells

Active targeting of shRNA-packaged minicells can be accomplished by conjugation of ligands or antibodies specific for receptors which overexpress on cancer cells. In this protocol, we have described conjugation of folic acid to minicells' surface through peptide

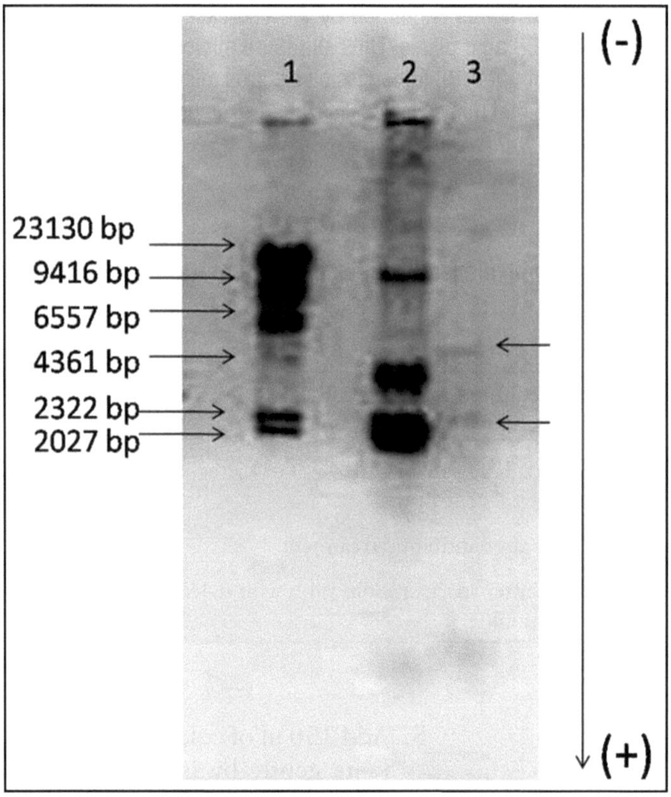

Fig. 4 Analysis of packaged shRNA expression vector. Gel electrophoresis of isolated plasmid DNA from minicells. (1) λ Hind III ladder, (2) standard pEZ43G-D plasmid, (3) plasmid isolated from minicells. Bands of isolated plasmid DNA are indicated by arrows. The plasmid band in Lane 3 corresponds to purified plasmid DNA in Lane 2. This suggests that isolated plasmid DNA is the same as purified plasmid DNA, i.e., pEZ43G-D, and confirmed the presence of plasmid DNA in the minicells. This finding was in agreement with the theory that when minicells produce from the parent cells, plasmid DNA segregates into the minicells (*Reproduced from ref.* 16)

coupling (*see* **Note 12**). Here, free amines of bacterial surface protein are conjugated with γ carboxyl group of folic acid.

Perform the following method to conjugate folic acid on the minicells.

1. Mix minicells (1×10^{10}/ml) with the folic acid solution dissolved in PBS (pH 7.6) in a 10 ml round-bottomed flask.
2. Then add EDC and sulfo-NHS to the mixture.
3. Finally, add DMAP and make up volume of final reaction mixture to 5 ml (*see* **Note 13**).
4. Perform reaction for about 16 h at room temperature with gentle agitation. Next, remove un-reactants fully by quick centrifugation at 1000 × *g* for 30 s at room temperature for three times.
5. Redisperse FA-modified minicells in appropriate volume of PBS (pH 7.4).
6. Obtained FA-modified minicells can be loaded with FITC and observed in fluorescent microscope for its integrity (*see* **Note 14**).
7. Moreover, optical density can also be measured at 600 nm to analyze integrity of minicells.
8. Finally, carry out transmission electron microscopy to check minicell morphology after folic acid conjugation.

3.5.1 Characterization of Folic Acid Conjugation

Success of folic acid conjugation can be determined by the following several methods.

(1) Quantification of the remaining amount of folic acid in the reaction mixture, (2) MALDI-TOF analysis of minicell surface protein, (3) in vitro uptake study of folic acid conjugated minicells, etc. Here, we have illustrated spectrophotometric quantification of folic acid and in vitro cellular uptake study (described in Sect. 3.6).

Spectrophotometric Method

Amount of folic acid conjugation can be quantified using spectrophotometric method by analyzing reaction mixtures before and after folic acid conjugation. Amount of folic acid can be quantified by employing the following steps.

1. Prepare standard curve of folic acid with a range of 4–28 μg/ml through serial dilution from the folic acid stock solution.
2. Centrifuge reaction mixtures before and after the conjugation reactions at 10,000 × *g* for 10 min at 24 °C to remove minicells.
3. Subsequently, analyze reaction mixtures at 363 nm, and plot the absorbance on standard curve to obtain amount of folic acid conjugation (Fig. 5).

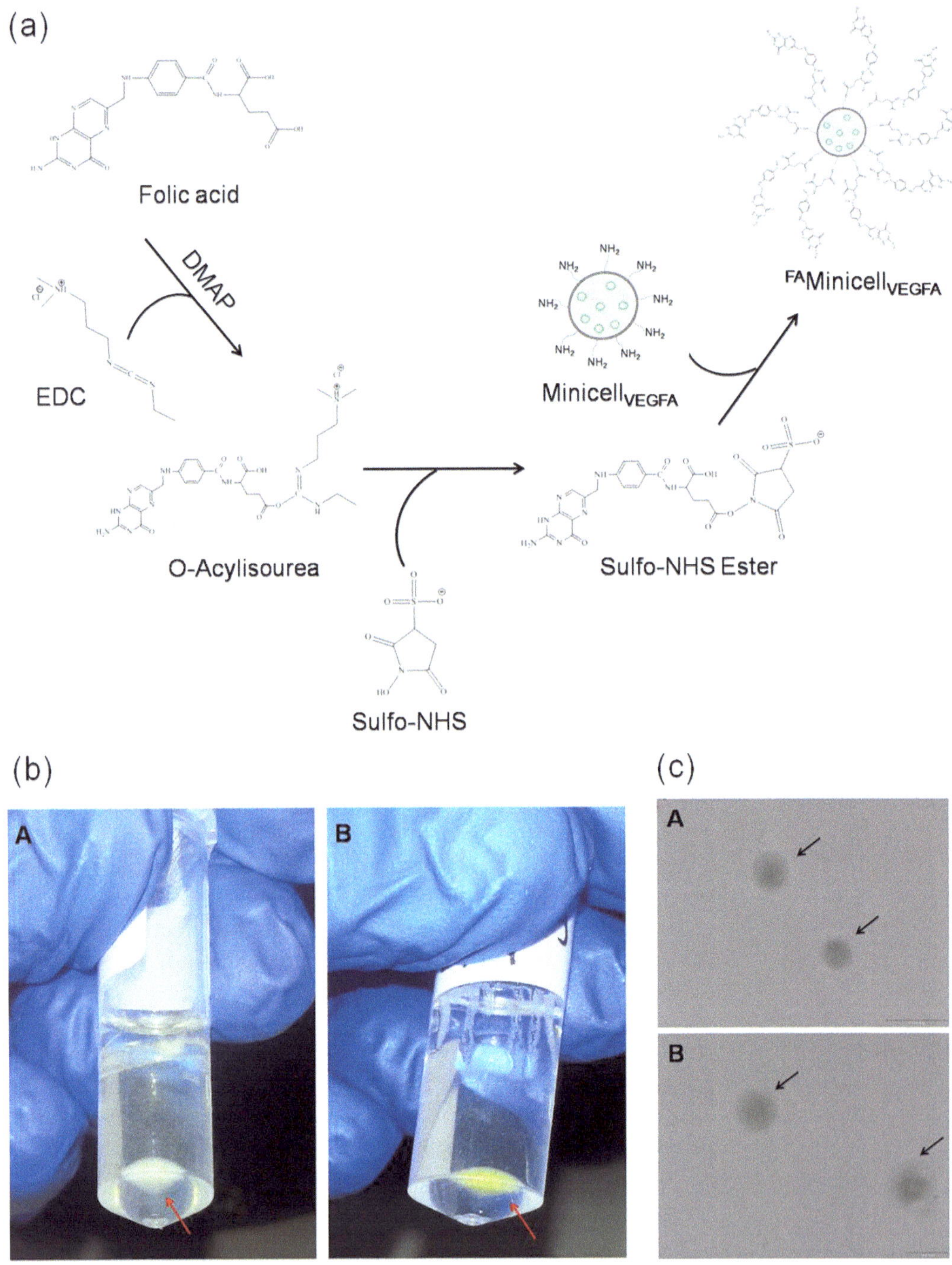

Fig. 5 Conjugation of folic acid on minicells. (**a**) Schematic representation of folic acid conjugation with shRNA-packaged minicells; EDC reacts with γ carboxyl group of folic acid to form *O*-acylisourea intermediate which reacts with free amine present on the minicell surface to form amide bond. However, *O*-acylisourea intermediate is highly unstable and quickly hydrolyze to form isourea. But the presence of sulfo-NHS in the reaction reacts with γ carboxyl group of folic acid to produce amine-reactive sulfo-NHS ester. This in turn reacts with free amine present on the minicell surface proteins to form amide bond and produce folic acid conjugated minicells. Addition of DMAP in the reaction increases the yield of folic acid conjugated minicells. (**b**) Minicell pellet.

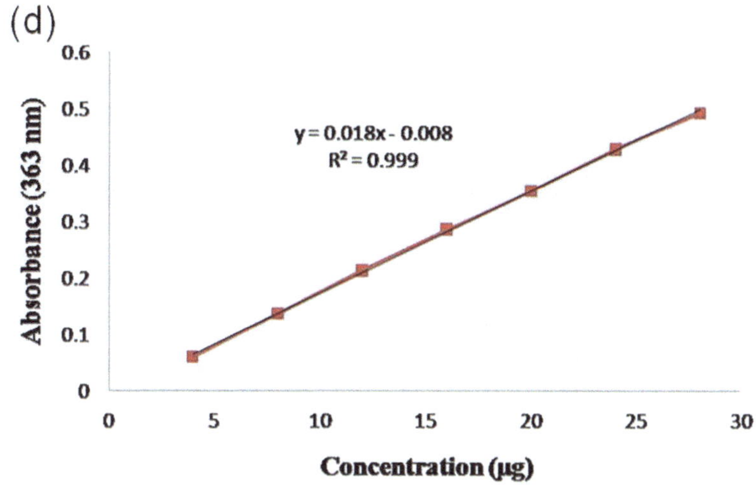

Fig. 5 (continued) (A) Before folic acid conjugation reaction and (B) after folic acid conjugation reaction. It can be observed from the image that minicell pellet turned yellow in color after the folic acid conjugation. Moreover this yellow color did not fade out even after repeated washing with PBS. This confirmed the folic acid conjugation on the minicells. Hence, yellow color of minicell pellet indicates successful folic acid conjugation. (Minicell pellet indicated with arrow.) (**c**) TEM images before and after folic acid conjugation. TEM images (A) before folic acid conjugation and (B) after folic acid conjugation (×20,000 magnification, minicells are indicated with arrows). Scale bar 1000 nm and 400 nm, respectively. TEM image showed that minicells were morphologically similar with unconjugated minicells with well-characterized cell wall and cell membrane. (**d**) Standard curve of folic acid. Amount of folic acid was calculated from the calibration curve. Amount of folic acid conjugated was found to be 1972.733 μg ± 112~2.0 mg/10^{10} minicells [(**a–c**) *reproduced from ref.* 2; (**d**) *reproduced from ref.* 16]

3.6 In Vitro Uptake of Folic Acid Conjugated Minicells

Success of folic acid conjugation on minicells can be further confirmed by in vitro uptake studies in folate receptor overexpressing cell lines (*see* **Note 15**). Execute the following steps to study the uptake mechanism of folic acid-modified minicells:

1. Grow KB cells in folate-free DMEM medium supplemented with 10% FBS.
2. Plate cells in 24-well plate at a density of 10^6 cells per plate, a day before study.
3. Treat cells with 10^9 FITC-loaded minicells as respective treatment (1) minicells$_{VEGFA}$ at 37 °C, (2) ^{FA}minicells$_{VEGFA}$ at 37 °C, (3) ^{FA}minicells$_{VEGFA}$ at 4 °C, and (4) ^{FA}minicells$_{VEGFA}$ with free folic acid at 37 °C for 1 h in humidified incubator with 5% carbon dioxide and 95% air.
4. After incubation, wash cells thrice with sterile PBS (pH 7.4), and observe under fluorescence microscope to visualize minicell uptake.
5. Moreover, uptake of minicells can also be evaluated quantitatively (in terms of FITC) by spectrofluorometer. Treat cells with 1% Triton X-100 for 20 min at 4 °C to release FITC. Determine total protein concentration of cells by micro BCA method.

Analyze cell lysate by spectrofluorometer at excitation and emission wavelengths of 490 and 515, respectively. Represent total fluorescence (mV) as per milligram of protein (Fig. 6) (*see* **Note 16**).

3.7 In Vivo Delivery of Folic Acid Conjugated Minicells

After deducing uptake mechanism of folic acid conjugated minicells in cell line, in vivo delivery provides concrete evidence of biodistribution and subsequent tumor-targeted delivery of minicells. Implement the following procedure for the development of tumor xenograft in mice and succeeding in vivo delivery of folic acid conjugated minicells.

3.7.1 Development of tumor xenograft in pharmacologically immunocompromised mice

In vivo efficacy study of anticancer agents or drug delivery systems majorly employs tumor xenograft model in mice. Tumor xenograft can be easily established in genetically immunodeficient mice such as athymic nude mice and scid mice. Along with these models, pharmacological immunosuppression protocols are also widely employed to create immunodeficient condition and develop tumor xenograft in mice [15]. In this protocol we have demonstrated in vivo delivery of ^{FA}minicells$_{VEGFA}$ in pharmacologically immunocompromised C57BL/6 mice as reported earlier [2]. Follow the given protocol to induce pharmacological immunosuppression in C57 BL/6 mice.

1. To induce pharmacological immunosuppression, take healthy male mice (C57 BL/6), 4–6 weeks old and administered with 10 mg/kg ketoconazole and 30 mg/kg cyclosporine by oral and intraperitoneal route, respectively, every day for 7 days.
2. Keep a group of naive (control) animal for comparison.
3. Provide autoclaved rodent food pellet and water ad libitum. Give Ampoxin (0.1 μg/ml) to all the animals through drinking water during the study.
4. After completion of 7-day ketoconazole and cyclosporine regimen, initiate cyclophosphamide treatment.
5. Inject cyclophosphamide subcutaneously at a dose of 60 mg/kg on days 3 and 1 before tumor cell injection.
6. Immunosuppression can be confirmed by analysis of total white blood cells (WBC), lymphocyte and neutrophil count (Fig. 7).

3.7.2 Preparation of Cancer Cells

1. Trypsinize semi-confluent cancer cells (KB) by using 0.25% trypsin to detach the cells.
2. Centrifuge cells at 300 × *g* for 7 min at 4 °C, and resuspend and wash in their respective growth medium, i.e., DMEM.
3. After washing, again resuspend cells in their respective growth medium, and count using a Neubauer chamber to determine viability by trypan blue exclusion test.
4. Store viable cells on ice and inject immediately.

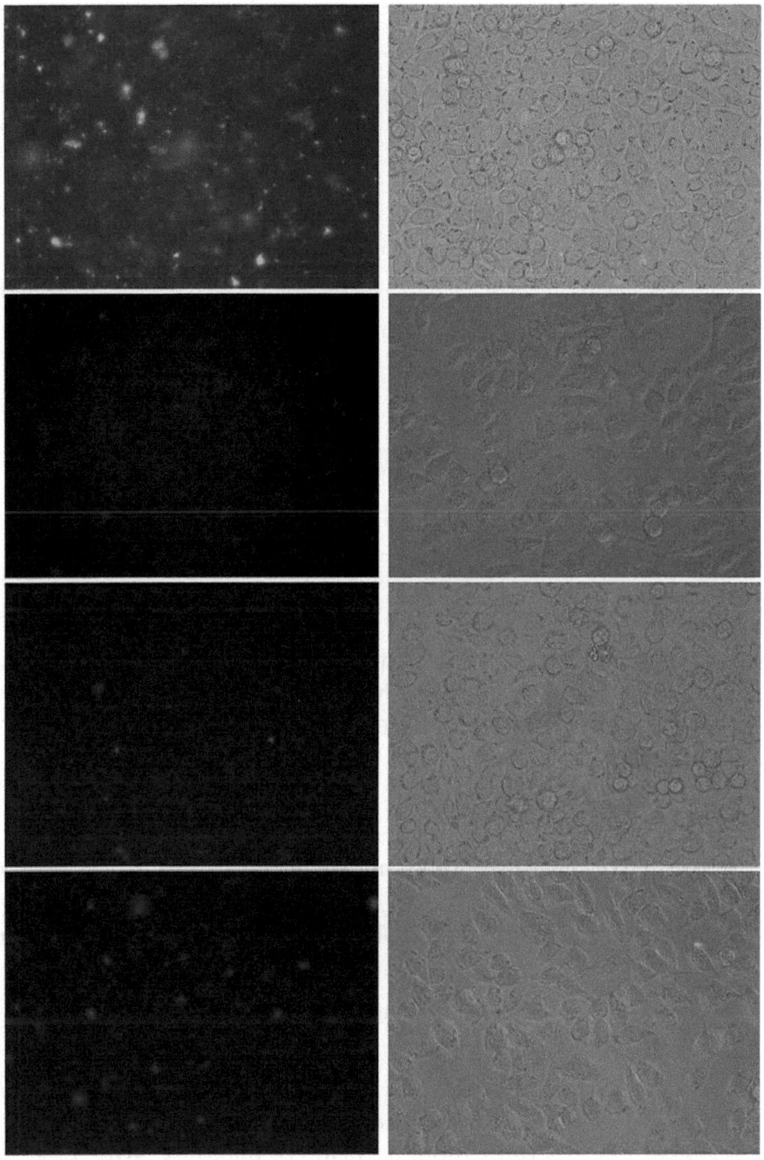

(a)

Fig. 6 In vitro uptake of folic acid conjugated minicells. (**a**) Microscopy image of KB cells after delivery of FITC-loaded minicells (i) ^{FA}minicells$_{VEGFA}$ at 37 °C, (ii) minicells$_{VEGFA}$ at 37 °C, (iii) ^{FA}minicells$_{VEGFA}$ at 4 °C, and (iv) ^{FA}minicells$_{VEGFA}$ with excess of free folic acid at 37 °C, (A) fluorescence image and (B) phase-contrast image (×200 magnification). (**b**) Graph showing quantitative minicells uptake in terms of mean fluorescence/mg of protein in different groups (as indicated above) of KB cells. Fluorescence was represented as mean ± SD values from at least three different experiments. Statistically significant when compared with *^{FA}minicells$_{VEGFA}$, $p < 0.05$. It can be observed from the (**a**) that there is a large amount of minicells uptake as evident by increased fluorescence intensity when incubated at 37 °C. In contrast, there is seldom uptake of minicells when incubated at 4 °C. Similarly, when minicells were delivered without targeting moiety, i.e. folic acid, there was no uptake. However, delivery of folic acid conjugated minicells in presence of free folic resulted in the

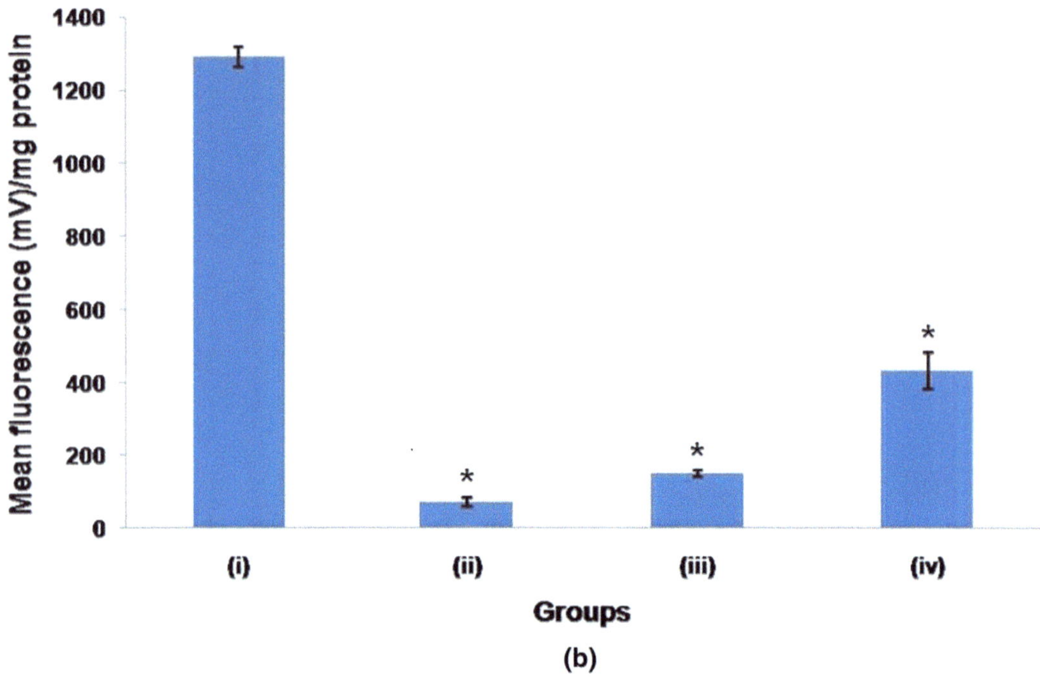

Fig. 6 (continued) decreased fluorescence intensity and hence significant reduction in the uptake of minicells. These results were further confirmed quantitatively in terms of total fluorescence and found to be in harmony with microscopy results. These results suggest the success of folic acid conjugation. Further, active targeting is essential for uptake of minicells by cancer cells (*Reproduced from ref.* 2)

3.7.3 Implantation of Cancer Cells

1. Depending upon the individual study protocol and type of cancer under investigation, use either sex of immunocompromised C57 BL/6 mice (4–6 weeks old) for the study.
2. Remove hair by waxing or applying depilatory cream from the shoulder blade of each animal 1 day before cancer cell injection.
3. Disinfect site of injection with 70% alcohol, and inject 0.1 ml of KB cells (approximately 5×10^6) subcutaneously into the right shoulder blade of mice (*see* **Note 17**).
4. Observe tumor growth at the site of injection. Measure tumor volume every week externally by digital caliper using the following formula [19]:

$$\text{Volume}\left(\text{mm}^3\right) = (\text{A}) \times \left(\text{B}^2\right) / 2$$

where A was the largest diameter (mm) and B the smallest (mm).

3.7.4 In Vivo Delivery of ^{FA}minicells$_{VEGFA}$

1. Randomized all the mice bearing tumor xenograft of KB cells on the basis of tumor volume (80–100 mm^3) and divided into four groups ($n = 6$) such as (1) saline, (2) ^{FA}minicells$_{scramble}$, (3) minicells$_{VEGFA}$, and (4) ^{FA}minicells$_{VEGFA}$.

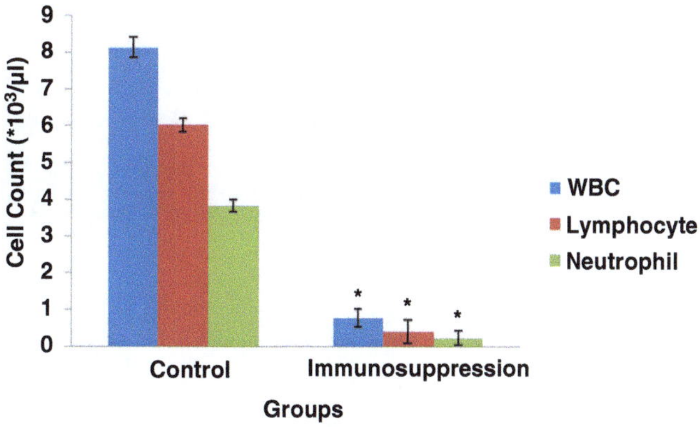

Fig. 7 Development of pharmacologically immunocompromised mice. Graph showing mean white blood cells (WBC), lymphocyte, and neutrophil count of control and of immunosuppressed mice after cyclophosphamide treatment ($n = 6$) ($*p < 0.05$). Significant reduction of WBC, lymphocyte, and neutrophil count after drug treatment confirmed the immunosuppression (*Reproduced from ref.* 15)

2. Provide autoclaved folate-free rodent food pellet and water ad libitum.
3. Administer 10^9 of respective minicells from each group intravenously through tail vein on first and third week (*see* **Note 18**).
4. Measure tumor volume every week externally by digital caliper as described above (Fig. 8) (*see* **Note 19**).

3.8 In Vivo Biodistribution of ^{FA}minicells$_{VEGFA}$

In vivo tumor-targeted delivery of shRNA can also be confirmed through biodistribution study of FITC-loaded ^{FA}minicells$_{VEGFA}$ after intravenous administration in mice bearing tumor xenograft.

Follow below steps to study biodistribution of ^{FA}minicells$_{VEGFA.}$

1. Inject 10^9 FITC-loaded ^{FA}minicells$_{VEGFA}$ intravenously through the tail vein in mice bearing KB tumor xenograft.
2. Three hours postinjection, sacrifice animals by CO_2 asphyxiation.
3. Excise vital organs such as the liver, heart, lung, kidney, brain, and spleen along with tumor, and snap frozen in liquid nitrogen.
4. Subsequently, take cryosection using cryotome cryostat.
5. Prepare section in duplicate, and keep (1) unstained for the fluorescence microscopy and (2) stained with H&E stain and observed under light microscope (Fig. 9).

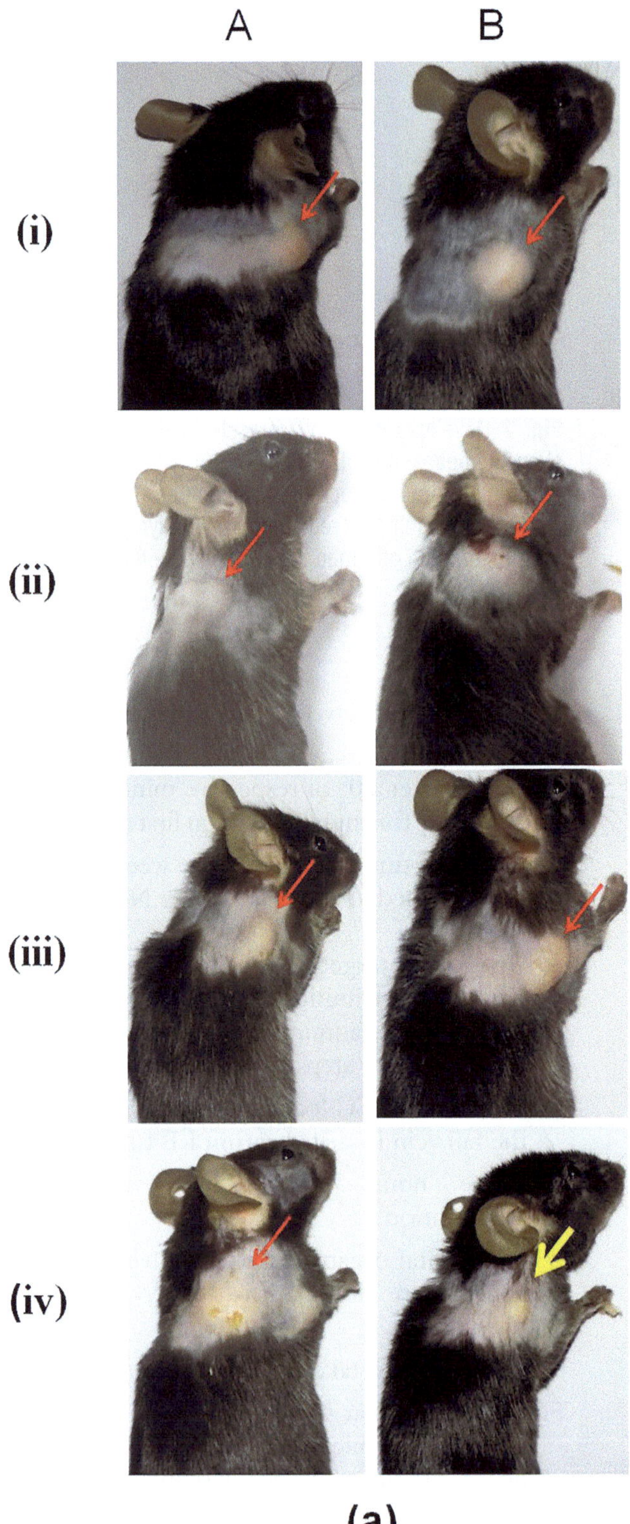

Fig. 8 In vivo delivery of folic acid conjugated minicells. Representative photograph of mice bearing tumor xenograft of (**a**) KB cells. (A) At start and (B) end of the treatment with (i) saline, (ii) ^{FA}minicells$_{Scramble}$, (iii) minicells$_{VEGFA}$, and (iv) ^{FA}minicells$_{VEGFA}$. Locations of tumors are indicated with arrows. (**b**) Graphs showing

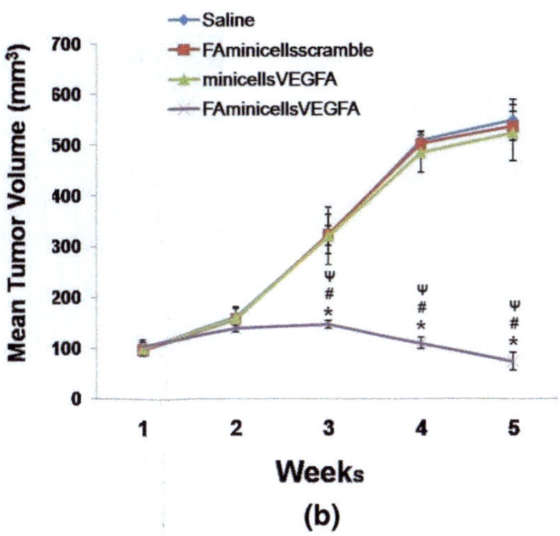

Fig. 8 (continued) mean tumor volume of KB xenograft from different groups at every week till the end of the treatment ($n = 6$). Statistically significant when compared with *saline, $^{\#FA}$minicells$_{scramble}$, and $^{\Psi}$minicells$_{VEGFA}$, $p < 0.05$. It can be observed that there is a significant decrease in tumor volume in ^{FA}minicells$_{VEGFA}$ treated group as compared to other groups. Reduction in tumor growth can be attributed to reduced angiogenesis because of folate receptor targeted delivery of VEGFA shRNA into cancer cells (*Reproduced from ref.* 2) (*see* **Note 20**)

4 Notes

1. Selection of minicells producing mutant strain is very decisive because a large number of minicells are required to develop minicell-based targeted delivery system. Hence, mutant strain which produces large number of minicells is ideal for such application. Here, we have selected *E. coli* PB114 among *E. coli* P678-54, *E. coli* X^{984}, and *E. coli* PB114 depending upon their minicell-producing potential.
2. Add antibiotics to LB broth freshly just before inoculation of culture. Similarly, add antibiotics to LB agar when it cools down to 45–50 °C just before inoculation of culture.
3. If possible, always incorporate GFP or other fluorescence gene as a selection marker in the expression vector along with the antibiotic-resistant gene to track plasmid transformation. Here, initially, we have transformed *E. coli* PB114 with pEZ43G-D plasmid which possesses GFP as a reporter gene to track minicell purification under fluorescence microscope as well as to confirm the segregation of expression vector into minicells. Subsequently, we have transformed *E. coli* PB114 with expression vectors having test and control shRNA to package these vectors into minicells (date not shown).

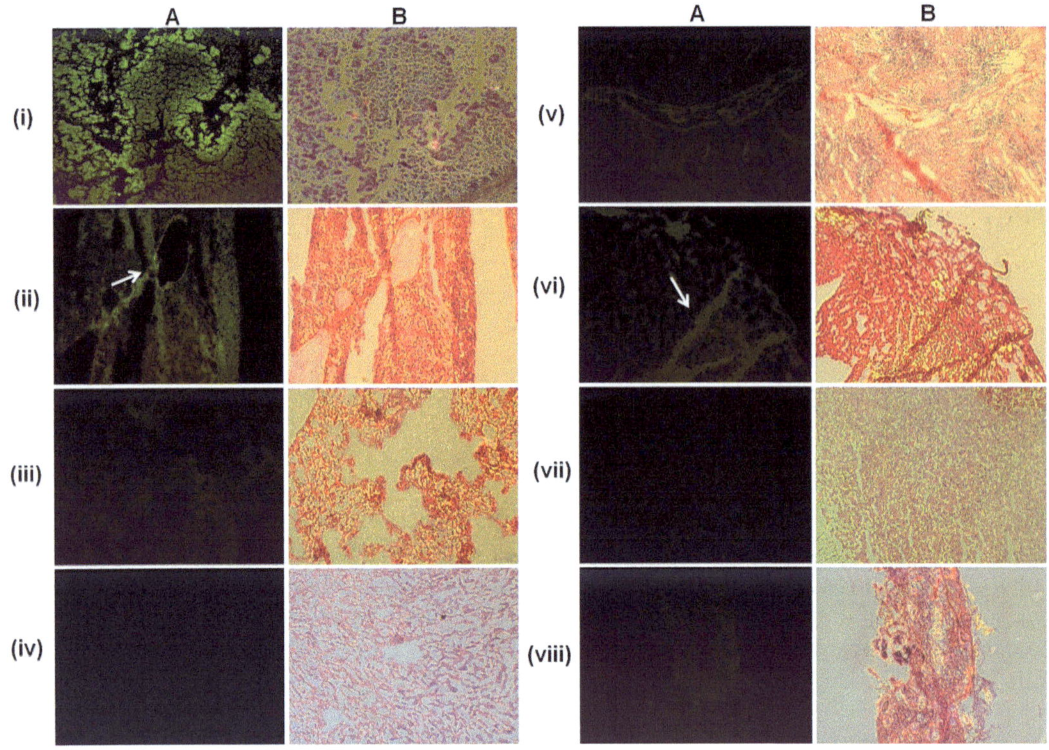

Fig. 9 In vivo biodistribution study of ^{FA}minicells$_{VEGFA}$. Microscopy image, (A) fluorescence image (B) light microscopy of H&E-stained cryosection of the (i) tumor, (ii) liver, (iii) lung, (iv) kidney, (v) spleen, (vi) heart, (vii) brain, and (viii) skin after in vivo delivery of FITC-loaded ^{FA}minicells$_{VEGFA}$ (×100 magnification). It can be visualized from the fluorescence image that majority of fluorescence is localized in the tumor. Cryosection of the liver and heart also showed substantial fluorescence; however it was majorly limited in blood vessels (indicated with arrow). Cryosection of the spleen also showed very weak fluorescence. Contrary, cryosection of the kidney, brain, lung, and skin did not show any fluorescence. This corroborates that majority of ^{FA}minicells$_{VEGFA}$ are distributed in tumor microenvironment which is due to both passive and active targeting (*Reproduced from ref.* 2)

4. Ethidium bromide is a mutagen and should be handled as a hazardous chemical. So wear gloves and suitable mask while handling ethidium bromide. Generally one should keep separate glassware as well as plasticware for ethidium bromide and for preparation of agarose gel solution and its subsequent use and disposal.
5. Here, we took KB cells as a model to explain in vitro delivery because of the highest level of folate receptor overexpression. Other cell lines such as A549 and CHO (folate receptor negative), Hela (folate receptor positive) can be included in the study depending upon the individual study protocol and objective.
6. *E. coli* PB114 possesses kanamycin-resistant gene. So kanamycin is supplemented in the LB broth for the selection of *E. coli* PB114.

7. As explained in the introduction section, minicells possess all of the molecular components of the parent cell except the chromosome. Hence, any extrachromosomal DNA such as plasmid DNA also segregates into minicells upon cell division. So, shRNA expression vector can be packaged into minicells by transformation into minicell-producing parent strain. Always preserve minicell-producing transformed parent bacterial culture on the solid media either in culture plate or slant because these transformed bacteria are highly unstable and cure shRNA expression vector very rapidly in liquid culture.
8. Disaggregate cells thoroughly after addition of $CaCl_2$ to enhance bacterial cell competency and subsequent transformation efficiency.
9. Always incorporate antibiotic-resistant gene in shRNA expression cassette to facilitate selection of transformed bacterial cells. Here, we have incorporated ampicillin-resistant gene in the shRNA expression vector. Further, *E. coli* PB114 possesses kanamycin-resistant gene. Hence, we have exploited dual antibiotic-based selection to pick transformed colonies.
10. It is essential to eliminate living parent bacteria and endotoxin from the minicell preparation before in vivo administration to reduce the risk of infection and any adverse reaction.
11. Zeta sizer measure size of nanoparticle by dynamic light scattering method. So average size and polydispersity index (PDI) of minicells at different stages of purification demonstrate purity of minicells. Average size and PDI should decrease with the decrease in the parent cell population.
12. Folic acid conjugation can be performed using EDC and sulfo-NHS coupling agents. EDC and sulfo-NHS are commonly used water-soluble coupling agents which activate carboxyl groups to react with primary amine and form covalent amide bond. EDC- and sulfo-NHS-based conjugation is very mild and can be performed in the aqueous medium. It is ideal for bacterial cell surface modification with ligand and/or antibodies because it does not harm bacterial cell membrane.
13. Concentration of folic acid, coupling agents (EDC and sulfo-NHS), no. of minicells, final reaction volume, and reaction time needs to be standardized by user of individual lab to obtain efficient conjugation. Further, conjugation efficiency may vary depending upon the use of minicells from gram-positive or gram-negative parent bacterial strain.
14. Obtained FA-modified minicells can be loaded with FITC and observed in fluorescent microscope for its integrity. Incubate minicells in FITC solution (1 mg/ml prepared in PBS, pH 7.4) for 2 h. After incubation, wash minicells three times

with PBS (pH 7.4) through centrifugation at 10,000 × *g*, 10 min at room temperature in order to remove excess FITC.

15. To study uptake mechanism, primarily minicells have been delivered with and without folic acid conjugation at 37 °C. In the third group, folic acid conjugated minicells have been delivered at 4 °C. Eventually, in the last group, folic acid conjugated minicells have been delivered in the presence of free folic acid to observe the delivery in the presence of free folic acid. Mechanism of minicell endocytosis can be studied in detail by blocking particular endocytosis pathway through specific inhibitor. For instance, phenylarsine oxide is an inhibitor of the clathrin-mediated endocytosis. Caveolae-mediated endocytosis is another major pathway for cellular uptake, which can be inhibited by filipin. Endocytosis through macropinocytosis can be inhibited by colchicine. Finally, sodium azide, which blocks cellular ATP synthesis, can be used to study the energy-dependent internalization process.
16. Success of in vitro shRNA delivery can be further scrutinized through mRNA or protein expression analysis of targeted gene by RT-PCR and/or western blotting, respectively.
17. Inject cell suspension prepared in 1:1 volume of media or PBS: Matrigel to accelerate growth of tumors. Matrigel is a gelatinous protein mixture which provides complex extracellular environment that resembles tumor microenvironment and boosts up tumor growth.
18. Optimum dose of minicells along with the treatment schedule can be determined on the basis of several factors such as specificity of shRNA, no. of shRNA carrying expression vector packaged in minicells, extent of tumor targeting, etc.
19. Tumor targeted delivery of ^{FA}minicells$_{VEGFA}$ can be tracked through biodistribution study of FITC-loaded ^{FA}minicells$_{VEGFA}$ in mice bearing tumor xenograft. Subsequently, observe fluorescence through whole-body in vivo imager. Alternatively, vital organs such as the liver, heart, lung, kidney, brain, and spleen along with tumor can be excised and snap frozen in liquid nitrogen. Subsequently, take cryosection using cryotome cryostat, and observe sections under the fluorescence microscope to visualize green fluorescence which corresponds to FITC-loaded minicells.
20. In vivo shRNA delivery can be further established by mRNA or protein expression analysis of targeted gene through RT-PCR or western blotting, respectively. Downregulation of targeted gene substantiates shRNA delivery.

Acknowledgments

The authors are thankful to B. V. Patel Pharmaceutical Education and Research Development (PERD) Centre, Ahmedabad, for providing all the facility for the successful completion of the work. Authors are also grateful to Dr. Lawrence Rothfield (University of Connecticut, USA) for providing *E. coli* PB114 as a generous gift. Additionally, authors would like to thank CSIR, India, for providing financial assistance to Mehul Jivrajani as Senior Research Fellow (# 113353/2K12/1) to carry out this work. Authors are also thankful to NIRMA University, Ahmedabad, Gujarat, India.

References

1. Elbashir SM, Harborth J, Lendeckel W, WK YA, Tuschl T (2001) Duplexes of 21-nucleotide RNAs mediate RNA interference in cultured mammalian cells. Nature 411:494–498
2. Jivrajani M, Nivsarkar M (2016) Ligand-targeted bacterial minicells: futuristic nano-sized drug delivery system for the efficient and cost effective delivery of shRNA to cancer cells. Nanomedicine: NBM 12:2485–2498
3. Oh YK, Park TG (2009) siRNA delivery systems for cancer treatment. Adv Drug Deliv Rev 61:850–862
4. Guo B, Zhang Y, Luo G, Li L, Zhang J (2009) Lentivirus-mediated small interfering RNA targeting VEGF-C inhibited tumor lymphangiogenesis and growth in breast carcinoma. Anat Rec 292:633–639
5. Zhao N, Bagaria HG, Wong MS, Zu Y (2011) A nanocomplex that is both tumor cell selective and cancer gene-specific for anaplastic large cell lymphoma. J Nanobiotechnol 9(2):1–12
6. Fenske DB, Cullis PR (2008) Liposomal nanomedicines. Expert Opin Drug Deliv 5(1):25–44
7. Jean M, Smaoui F, Lavertu M, Méthot S, Bouhdoud L, Buschmann MD et al (2009) Chitosan-plasmid nanoparticle formulations for IM and SC delivery of recombinant FGF-2 and PDGF-BB or generation of antibodies. Gene Ther 16(9):1097–1110
8. MacDiarmid JA, Mugridge NB, Weiss JM, Sedliarou I, Wetzel S, Kochar K et al (2009) Sequential treatment of drug-resistant tumors with targeted minicells containing siRNA or a cytotoxic drug. Nat Biotechnol 27(7):643–651
9. Adler HI, Fisher WD, Cohen A, Alice AH (1967) Miniature *Escherichia coli* cells deficient in DNA. Proc Natl Acad Sci U S A 57:321–326
10. Frazer AC, Curtiss R III (1975) Production, properties and utility of bacterial minicells. Curr Top Immunol Microbiol 69:1–84
11. De Boer PA, Crossley RE, Rothfield LI (1989) A division inhibitor and a topological specificity factor coded for by the minicell locus determine proper placement of the division septum in *E. coli*. Cell 56:641–649
12. MacDiarmid JA, Mugridge NB, Weiss JC, Phillips L, Bum AL, Paulin PR et al (2007) Bacterially derived 400 nm particles for encapsulation and cancer cell targeting of chemotherapeutics. Cancer Cell 11:431–445
13. Karagiannis ED, Anderson DG (2009) Minicells overcome tumor drug resistance. Nat Biotechnol 27(7):620–621
14. Parker N, Turk MJ, Westrick E, Lewis JD, Low PS, Leamon CP (2005) Folate receptor expression in carcinomas and normal tissues determined by a quantitative radioligand binding assay. Anal Biochem 338:284–293
15. Jivrajani M, Shaikh MV, Shrivastava N, Nivsarkar M (2014) An improved and versatile immunosuppression protocol for the development of tumor xenograft in mice. Anticancer Res 34:7177–7184
16. Jivrajani MN (2014) Targeted delivery of shRNA to cancer cell. Dissertation, Nirma University Ahmedabad
17. Jivrajani M, Shrivastava N, Nivsarkar M (2013) A combination approach for rapid and high yielding purification of bacterial minicells. J Microbiol Methods 92:340–343
18. Giacalone MJ, Zapata JC, Berkley NL, Sabbadini RA, Chu YL, Salvato MS, McGuire KL (2007) Immunization with non-replicating *E. coli* minicells delivering both protein antigen and DNA protects mice from lethal challenge with lymphocytic choriomeningitis virus. Vaccine 25:2279–2287
19. Carlsson G, Gullberg B, Hafstrom L (1983) Estimation of liver tumor volume using different formulas—an experimental study in rats. J Cancer Res Clin Oncol 105:20–23

Chapter 10

Targeted Delivery of miRNA Antagonists to Myeloid Cells In Vitro and In Vivo

Yu-Lin Su, Piotr Swiderski, Guido Marcucci, and Marcin Kortylewski

Abstract

Elevated levels of microRNAs in cancer cells are often associated with oncogenic effects and thus provide potential therapeutic targets. However, the lack of efficient delivery methods for synthetic miRNA inhibitors, antagomiR, or anti-miR oligonucleotides hindered clinical translation of such strategies. We recently developed an approach for targeted delivery of synthetic, 2′-O-methyl-modified antagomiR molecules to normal and malignant myeloid cells and B cells by tethering to the single-stranded, phosphorothioate oligodeoxynucleotides (PSO). The PSO-antagomiR are rapidly internalized through scavenger receptor-mediated endocytosis by human monocytes, dendritic cells, B cells, as well as myeloid leukemia and B-cell lymphoma cells, but not by T cells. Following internalization, the unformulated PSO-antagomiR potently reduces levels of target miRNA and modulates expression of downstream protein targets, both in vitro and in vivo. The simple design of PSO-antagomiR conjugates enable adaptation of this strategy for targeting oncogenic miRNAs in nonmalignant and malignant myeloid cells and B cells.

Key words MicroRNA, miRNA, Anti-miR, AntagomiR, Oligonucleotides, Myeloid cells, Leukemia

1 Introduction

MicroRNAs (miRNAs) are small noncoding RNAs (20–25 nucleotides in length) that regulate expression of multiple target genes [1]. The sequence-specific hybridization of the miRNA to the 3′ UTR of complementary mRNA results in inhibition of translation or degradation of target mRNA. Alteration of miRNA levels occurs in various diseases including cancer, where certain miRNAs play oncogenic role [1, 2]. The oncogenic miRNAs, or oncomiRs, provide an attractive therapeutic target, but the delivery of synthetic miRNA inhibitors remains a major challenge in clinical translation of miRNA-targeted therapeutics [3].

We previously developed an original method for cell-selective delivery of therapeutic oligonucleotides, such as siRNA molecules, by targeting scavenger receptors (SR) on their surface and the Toll-like receptor 9 (TLR9) located in endosomes [4–6]. Target

Lekha Dinesh Kumar (ed.), *RNA Interference and Cancer Therapy: Methods and Protocols*, Methods in Molecular Biology, vol. 1974, https://doi.org/10.1007/978-1-4939-9220-1_10,

cells, such as myeloid immune cells and B cells, recognize single-stranded and phosphorothioate oligodeoxynucleotide (PSO) by SRs, leading to rapid endocytic uptake of siRNA conjugates [7]. Shortly after uptake into TLR9$^+$ early endosomes, the unmethylated CpG motifs in the PSO sequence trigger TLR9 activation, thereby facilitating cytoplasmic release of the siRNA following the Dicer-mediated cleavage of the conjugate [5]. Recently, we modified this strategy for the delivery of antagomiRs as single-stranded PSO-conjugates to myeloid cells in vitro and in vivo [8]. Due to chemical modification of both parts of the conjugate through phosphorothioation and 2′-O-methylation, the PSO-antagomiRs show improved stability and high nuclease resistance. In contrast to CpG-siRNAs, PSO-antagomiRs do not require TLR9-mediated cytoplasmic release and showed comparable target miRNA knockdown with or without immunostimulatory CpG motifs in the PSO. The nuclease-resistance of PSO-antagomiRs facilitates systemic administration of these oligonucleotides in vivo. The efficacy studies in syngeneic models of chronic myelogenous leukemia (CML) demonstrated that intravenous administration of the specific PSO-antagomiR results in the significant reduction of the oncogenic miR-126 in leukemic stem cells, thereby resulting in improved CML cell sensitivity to thymidine kinase inhibitor and extended mice survival [8]. Hereby, we describe the design and application of PSO-antagomiR conjugates for the delivery of synthetic miRNA inhibitors into specific human and mouse target cells both in vitro and using systemic administration in vivo. We also discuss potential limitations and critical troubleshooting points for target cell selection and maximized efficacy of PSO-antagomiRs.

2 Materials

2.1 PSO-AntagomiR Design and Synthesis

1. The single-stranded PSO-antagomiRs can be synthesized using Akta OligoPilot100 (GE) and consists of the phophorothioated ODN (20mer) and antagomiR sequences (20–23mer). The design of conjugates is demonstrated as follows (*see* **Note 1**):

   5′ NNNNNNNNNNNNNNNNNNNN xxxxx RRRRRRR RRRRRRRRRRRRRR 3′.

   N = deoxyribonucleotide with partially or completely phosphothioated sugar backbone.

   x = a single C3 carbon spacer unit = $(CH_2)_3$.

   R = 2′O-methyl-modified ribonucleotide (*see* **Note 2**).

   Human versions (based on CpG-D19 sequence)

   5′ G_SG_STGCAT<u>CG</u>ATGCA$G_SG_SG_SG_SG_S$G xxxxx RRRRRRRR RRRRRRRRRRRR 3′.

5′ $\mathrm{G_SG_STGCAT\underline{GC}ATGCAG_SG_SG_SG_SG}$ xxxxx RRRRRR RRRRRRRRRRRRRR 3′

Mouse versions (based on CpG-1668 sequence)

5′- $\mathrm{T_SC_SC_SA_ST_SG_SA_S\underline{CSG}_ST_ST_SC_SC_ST_SG_SA_ST_SG_SC_ST}$ xxxxx RRRRRRRRRRRRRRRRRRRR 3′.

5′- $\mathrm{T_SC_SC_SA_ST_SG_SA_S\underline{GSC}_ST_ST_SC_SC_ST_SG_SA_ST_SG_SC_ST}$ xxxxx RRRRRRRRRRRRRRRRRRRR 3′

Underline indicates immunostimulatory CpG motif or mutated/non-immunogenic GpC motif (*see* **Note 3**).

Optionally, PSO-antagomiRs can be labeled using various fluorescent dyes conjugated to the 3′ end (e.g., using Cy3).

2.2 7 M Urea/Polyacrylamide (PAGE) Gel Reagents

1. 10× TBE: for 500 mL, dissolve 54 g Tris, 27.6 g boric acid, and 10 mL 0.5 M EDTA (pH 8.0) in water, adjust to pH 8.3, dilute to 1× solution for gel running and staining.
2. 8 M urea: for 200 mL, dissolve 96.1 g urea in water, and filter the solution.
3. 7 M urea/20% acrylamide solution: for 200 mL, dissolve 88.2 g urea in 100 mL of 40% acrylamide/bis-acrylamide solution (19:1); adjust volume to 200 mL using milli-Q water, and filter the solution; *see* Table 1.
4. Ethidium bromide: use 2 μL of ethidium bromide solution (10 mg/mL) in 25 mL of 1× TBE buffer for gel staining.
5. Double-stranded RNA marker.
6. Nuclease-free water: DNase/RNase free water.

2.3 Flow Cytometry

1. Staining buffer: 1× PBS, 2% FBS, and 0.1% NaN_3.
2. Antibodies: fluorescently labeled antibodies specific to cell surface markers including CD14, CD303, CD1c, CD19, and CD3.
3. Fc blocker: Anti-CD16/CD32 antibody was used to block unspecific binding by Fc receptors.
4. Viability dye: 7AAD.
5. Flow cytometry cell analyzer.

2.4 Confocal Microscopy

1. Wash buffer: 1× PBS, 0.01% Tween 20.
2. Fixative: 1× PBS, 2% paraformaldehyde.
3. Cytospin: Cytospin 4 centrifuge with Cytofunnels, Cytoslides, and Cytoclips.
4. Mounting solution: Vectashield hard-set medium.
5. Microscope: Zeiss LSM880 with Airyscan, LD C-Apo 40×/1.1 was used as objective.

Table 1
Composition of 7 M urea/15% PAGE

Reagent	For 10 mL
7 M urea/20% acrylamide solution	7.5 mL
8 M urea	1.5 mL
10× TBE	1 mL
10% APS	80 μL
TEMED	7 μL

6. Acquisition software: ZEN2.1 black; post-acquisition analysis software: ZEN2 blue.

2.5 Quantitative Real-Time PCR (qPCR)

1. Maxwell RSC Instrument for the automated isolation of miRNA using Maxwell® RSC miRNA Tissue Kit.
2. Reverse transcription: TaqMan microRNA Reverse Transcription Kit.
3. qPCR: TaqMan CFX96 real-time PCR detection system.

2.6 In Vivo Delivery of PSO-AntagomiRs

1. Syringes: 0.5 mL U-100 insulin syringe with 28G needle was used to inject PSO-AMO conjugates diluted in 1× PBS retroorbitally into mice (5 mg/kg).
2. Anesthetic agents: anesthesia induction chamber including isoflurane and oxygen supply.
3. Mice: C57BL/6 male mice aged between 6 and 8 weeks were purchased from National Cancer Institute (Frederick, MD).
4. c-Kit+ cell enrichment: EasySep Mouse CD117 (cKit) Positive Selection Kit.

3 Methods

3.1 Reconstitution of PSO-AntagomiRs and Quality Control

1. Reconstitute PSO-AMOs with DNase/RNase-free water to make 100 μM stock solution and mix well at 37 °C for 5 min. Make sure the oligonucleotides are completely dissolved, then aliquot and store at −80 °C.
2. Assemble the gel plates in the gel casting chamber, prepare 7 M urea/15% PAGE gel according to Table 1, and pour quickly between gel plates, insert comb, and allow polymerizing for 30 min.
3. Dismount the polymerized gel from casting chamber and assemble into gel apparatus according to manufacturer's instruction.
4. Fill the gel chamber with 1× TBE, carefully remove the comb, wash all wells, and pre-run for 30 min.

5. Dilute oligonucleotides to 10 μM, use 2 μL of diluted samples to mix with 0.5 μL of 0.5 M EDTA (pH 8.0) and 3 μL of 2× RNA loading dye, load into the wells together with the RNA marker, and run the gel at 250 V for 1 h.
6. Remove the gel and carefully transfer into a dish containing 1× TBE plus ethidium bromide, stain for 10–15 min, then examine the gel on the UV transilluminator (as in Fig. 1).

3.2 The Cytofluorometric Assessment of PSO-AntagomiR Uptake by Target Cells In Vitro

1. Culture 1–2 × 10^6 cells/mL in 24-well plate in an appropriate medium or, if using freshly isolated primary cells, allow cells to adjust to the in vitro culture conditions for 2–4 h.
2. Treat cells with various doses (10–500 nM) of fluorescently labeled PSO-antagomiRs for 4 h.
3. Collect cells, resuspend each sample in 100 μL of the staining buffer, add 1 μL of the Fc Block for 10 min, then stain cells with a combination of fluorescently labeled antibodies specific to immune cell markers for 20–30 min on ice, wash once with 2 mL of the staining buffer, and resuspend in the staining buffer plus 7AAD for exclusion of dead cells for the analysis using flow cytometry (as in Fig. 2).

3.3 Confocal Microscopy to Verify Internalization of PSO-AntagomiRs by Suspension Cells In Vitro

1. Plate and culture cells overnight at a desired density (usually 10^5 cells/well) in the appropriate medium.
2. Next day, add fluorescently labeled PSO-antagomiRs at various concentrations (10–500 nM) and incubate for desired times (1–24 h).
3. Wash cells twice with PBS and fix in 2% paraformaldehyde in PBS for 30 min.
4. Pipette 100 μL of cell suspensions into the prepared Cytofunnel with Cytoclip, centrifuge for 5 min using Cytospin centrifuge with slow acceleration mode.
5. Mount slides in Vectashield Hard-Set medium, cover with coverslips and seal, and store at 4 °C until the mounting medium is completely dry.
6. Acquire images using the inverted confocal microscope and analyze using the post-acquisition software (e.g., Zeiss Zen as in Fig. 3; *see* **Note 4**).

3.4 Verification of the *In Vitro* Target miRNA Knockdown Using qPCR

1. Plate and culture cells overnight at a desired density (usually 10^5 cells/well) in the appropriate medium.
2. Add various concentrations (10–500 nM) of PSO-antagomiRs or matched scrambled PSO-scrambled oligonucleotide as a negative control and incubate overnight.
3. Collect cells by centrifugation, then wash twice using PBS, and isolate RNA using the Maxwell RSC miRNA tissue kit.

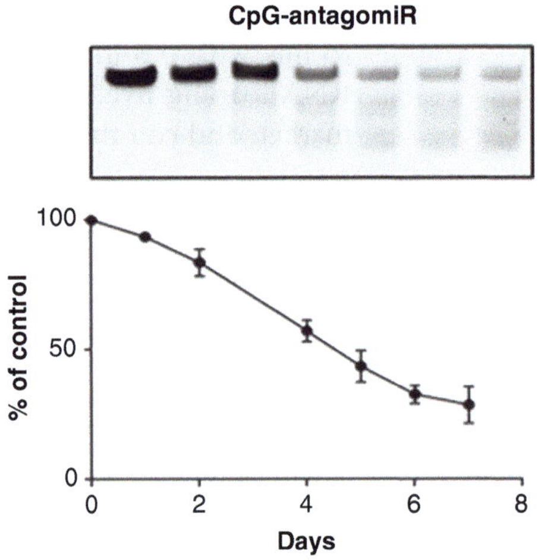

Fig. 1 Nuclease resistance of PSO$_{CpG}$-antagomiRs. PSO$_{CpG}$-antagomiR oligonucleotides were incubated in the presence of 50% human serum at 37 °C for up to 7 days. The oligonucleotides were then separated on the 7.5 M urea/15% PAGE gel and stained using ethidium bromide. Three independent experiments were performed; the representative images and quantified results are shown as means ± SEM

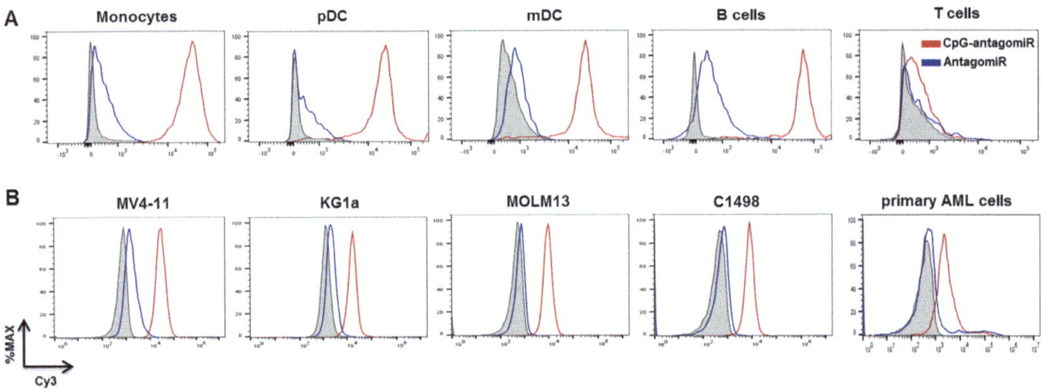

Fig. 2 Efficient internalization of PSO$_{CpG}$-antagomiR in immune cells and leukemia cells. (**a**) Human PBMCs were isolated from healthy donor and incubated with the fluorescently labeled PSO$_{CpG}$-antagomiR^{Cy3} or with the unconjugated antagomiR^{Cy3} at the concentration of 500 nM for 4 h without any transfection reagent. Intracellular uptake of fluorescently labeled constructs by different immune cells including monocytes (CD14^{+}), pDCs (CD303^{+}), mDCs (CD1c^{+}), B cells (CD19^{+}), and T cells (CD3^{+}) were detected using flow cytometry. (**b**) cultured human and mouse leukemia cells MV4-11, KG1a, MOLM13, C1498, and primary mouse AML cells were incubated with PSO$_{CpG}$-antagomiR^{Cy3} or with the unconjugated antagomiR^{Cy3} and analyzed using flow cytometry under the same conditions as described above

4. Quantify total RNA concentration and then reverse transcribe using TaqMan microRNA reverse transcription kit.
5. Perform qPCR using the appropriate TaqMan probe/primer sets with snoRNA234 as an internal control using CFX96 real-time PCR detection system (as in Fig. 4).

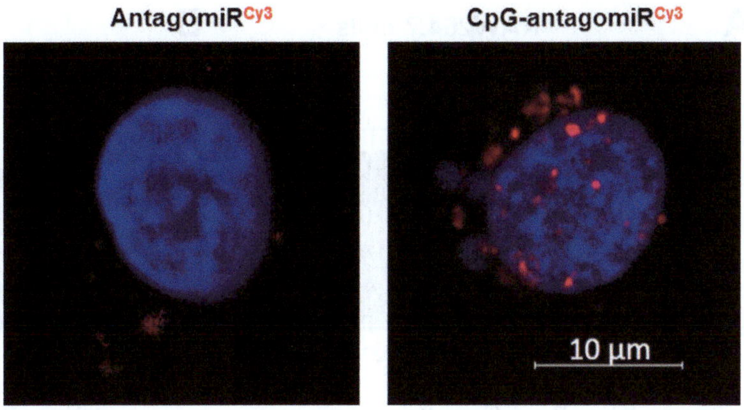

Fig. 3 Intracellular localization of PSO_{CpG}-antagomiR^{Cy3} in human MV4-11 cells. Human MV4-11 acute myeloid leukemia cells were incubated with 500 nM of PSO_{CpG}-antagomiR^{Cy3} or with the unconjugated antagomiR^{Cy3} for 1 h, and the intracellular localization of these oligonucleotides was confirmed using confocal microscopy

3.5 Intravenous Delivery of PSO-AntagomiRs to Target Myeloid Cell Populations In Vivo

1. Prepare 7–10 week old mice (C57BL/6 or NSG for syngeneic or xenotransplanted tumor models, respectively) and anesthetize using isoflurane inhalation.
2. Inject AML cells suspended in sterile PBS (usually 0.1–1 × 10^6 cells/100 μL) to engraft leukemia.
3. After the percentage of leukemic cells in blood exceeds 1%, as measured using flow cytometry (*see* Subheading 3.7), prepare PSO-antagomiRs for injections at 1–5 mg/kg.
4. Position the anesthetized mouse on its side, use index finger and thumb to draw back skin of above and below the eye, then carefully insert the 28G needle at ~45° angle at the corner of the eye, lateral to the medial canthus (*see* **Note 5**).
5. Slowly inject the PSO-antagomiR in 100 μL of sterile PBS into the retrobulbar sinus, then remove needle gently, and apply ointment to reduce strain (*see* **Note 5**).
6. After the desired time (usually 1–24 h after the last injection), euthanize mice, collect blood, and harvest various organs (e.g., spleen, bone marrow, etc.) to evaluate target miRNA knockdown (as in Fig. 4, *see* **Note 6**).

3.6 Isolation of Cells from Soft Tissues and Tumors

1. Transfer harvested tissues into Petri dishes and wash twice with HBSS.
2. Add 5–10 mL of 1× Collagenase D (400 U/mL) and 1× DNase I (1 mg/mL) and mince tissues into small cubes (~1–2 mm in diameter) using surgical scissors and incubate for 30 min/37 °C.

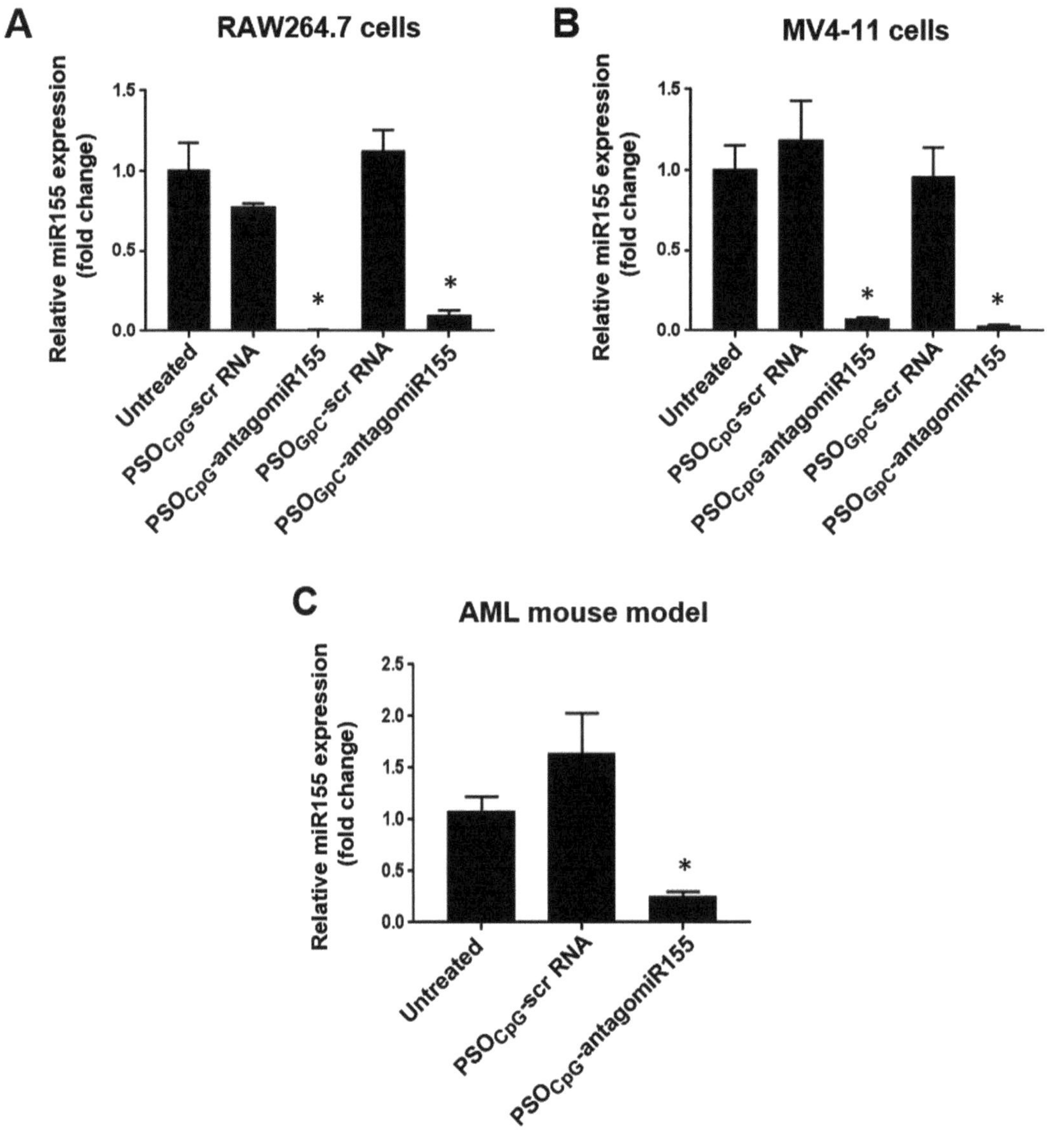

Fig. 4 Knockdown efficiency of PSO-antagomiR conjugates in vitro and in vivo. (**a**, **b**) miRNA silencing was assessed in vitro in mouse RAW264.7 macrophages (**a**) or in human MV4-11 AML cells (**b**) using qPCR after treatment with 500 nM of PSO_{CpG}-antimiR155, PSO_{GpC}-antimiR155, or the respective control RNA conjugates for 24 h. (c) Protein expression level of the miR-155 downstream target SHIP1 were assessed using Western blot after treatment with 500 nM of PSOCpG-antimiR155, PSOGpC-antimiR155 or the control RNA conjugates for 48 h. (**d**) For the in vivo experiment, C57BL/6 mice bearing *Cbfb-MYH11/Mpl* + AML [10, 11] were injected with PSO_{CpG}-antimiR155 or control PSO_{CpG}-scrRNA (5 mg/kg) every day for 6 days; miRNA expression levels in AML cells were assessed using qPCR. Results are shown as means ± SEM; *$p < 0.05$ compared to untreated group

3. Add 10 mM EDTA (tissue culture grade) to stop enzymatic reaction then pipette up and down to disperse tissues into single-cell suspension.
4. Filter cell suspensions through 70 μm cell strainers into 50 mL tubes, adjust volume to 20–30 mL with HBSS without calcium and magnesium, and spin down at 500 × *g*/5 min/18 °C.

5. To lyse erythrocytes, add 5 mL of 1× ACS buffer to the cell pellet, resuspend, and incubate for 3 min/RT.
6. Add 25 mL HBSS without calcium and magnesium to stop cell lysis, spin down, and resuspend cells into 10 mL HBSS without calcium and magnesium. To isolate viable cells, pipet 10 mL of cell suspension into 50 mL tube then carefully underlay with 5 mL of Histopaque®-1083 equilibrated to RT and centrifuge tubes at 400 × *g*/30 min/18 °C.
7. Collect viable cells from the interphase without touching the pellet, resuspend in HBSS without calcium and magnesium, spin down again, and then resuspend again for cell counting.

3.7 Processing of Blood Cells

1. Transfer collected blood into 15 mL tube containing 3–4 mL ACS buffer to lyse blood cells.
2. Incubate 5–10 min at RT and spin down 500 × *g* for 5 min.
3. Wash cell pellet twice with PBS.
4. Count cell number and proceed to next experiment.

4 Notes

1. PSO sequences used for targeting human or mouse myeloid cells are based on two classes of CpG oligodeoxynucleotides, A and B. Class A ODNs, with G-rich 3′ end, tend to form multimers and are optimally internalized by human myeloid cells. In contrast, the monomeric class B ODNs, lacking the G-rich region, are more effective in targeting human B cells or mouse target cells.
2. The seed sequence of the miRNA is a conserved sequence usually localized at positions 2–7 from the 5′-end of miRNA, which should be fully complementary to its target mRNA. For the PSO-antagomiR design, different lengths of antagomiR sequence (17–23 nt) can be tested to find the optimal design containing the seed region.
3. The presence of the CpG motif in the PSO sequence confers the immunostimulatory effect of the oligonucleotide. While TLR9 stimulation facilitates release of the oligonucleotide from endosomes to cytoplasm, which is critical for the effect of CpG-siRNA, it is not essential for PSO-antagomiR effect. Due to improved stability, PSO-antagomiRs show comparable target miRNA knockdown with and without CpG motif in their sequence. Also, the presence of CpG is not required for the ODN uptake by SR-positive target cells [5, 9].
4. Confocal microscopy is indispensable for confirming that oligonucleotides are localized intracellularly rather than associated with target cell surface to verify positive signal detected by flow cytometry.

5. In case of retroorbital injections, mice should not receive more than single injection per day. Additional injections should be done only into alternate eyes and no often than twice a week.
6. Even though PSO-antagomiRs are chemically modified for improved resistance to serum nucleases, their half-life in circulation can be limited by rapid kidney clearance due to relatively small size of these oligonucleotides (less than 20 kD). Nevertheless, it is sufficient for selective uptake by specialized immune cell types such as myeloid cells and B cells.

References

1. Garzon R, Marcucci G, Croce CM (2010) Targeting microRNAs in cancer: rationale, strategies and challenges. Nat Rev Drug Discov 9(10):775–789. https://doi.org/10.1038/nrd3179
2. Rupaimoole R, Slack FJ (2017) MicroRNA therapeutics: towards a new era for the management of cancer and other diseases. Nat Rev Drug Discov 16(3):203–222. https://doi.org/10.1038/nrd.2016.246
3. Zhang Y, Wang Z, Gemeinhart RA (2013) Progress in microRNA delivery. J Control Release 172(3):962–974. https://doi.org/10.1016/j.jconrel.2013.09.015
4. Kortylewski M, Swiderski P, Herrmann A, Wang L, Kowolik C, Kujawski M et al (2009) In vivo delivery of siRNA to immune cells by conjugation to a TLR9 agonist enhances antitumor immune responses. Nat Biotechnol 27(10):925–932. https://doi.org/10.1038/nbt.1564
5. Nechaev S, Gao C, Moreira D, Swiderski P, Jozwiak A, Kowolik CM et al (2013) Intracellular processing of immunostimulatory CpG-siRNA: Toll-like receptor 9 facilitates siRNA dicing and endosomal escape. J Control Release 170(3):307–315. https://doi.org/10.1016/j.jconrel.2013.06.007
6. Kortylewski M, Moreira D (2017) Myeloid cells as a target for oligonucleotide therapeutics: turning obstacles into opportunities. Cancer Immunol Immunother 66(8):979–988. https://doi.org/10.1007/s00262-017-1966-2
7. Zhang Q, Hossain DMS, Nechaev S, Kozlowska A, Zhang W, Liu Y et al (2013) TLR9-mediated siRNA delivery for targeting of normal and malignant human hematopoietic cells in vivo. Blood 121(8):1304–1315. https://doi.org/10.1182/blood-2012-07-442590
8. Zhang B, Nguyen LXT, Li L, Zhao D, Kumar B, Wu H et al (2018) Bone marrow niche trafficking of miR-126 controls the self-renewal of leukemia stem cells in chronic myelogenous leukemia. Nat Med 24(4):450–462. https://doi.org/10.1038/nm.4499
9. Latz E, Verma A, Visintin A, Gong M, Sirois CM, Klein DCG et al (2007) Ligand-induced conformational changes allosterically activate Toll-like receptor 9. Nat Immunol 8(7):772–779. https://doi.org/10.1038/ni1479
10. Kuo Y-H, Landrette SF, Heilman SA, Perrat PN, Garrett L, Liu PP et al (2006) Cbf beta-SMMHC induces distinct abnormal myeloid progenitors able to develop acute myeloid leukemia. Cancer Cell 9(1):57–68. https://doi.org/10.1016/j.ccr.2005.12.014
11. Hossain DMS, Santos CD, Zhang Q, Kozlowska A, Liu H, Gao C et al (2014) Leukemia cell-targeted STAT3 silencing and TLR9 triggering generate systemic antitumor immunity. Blood 123(1):15–25. https://doi.org/10.1182/blood-2013-07-517987

Chapter 11

Systemic Modulation of Lymphocyte Subsets Using siRNAs Delivered via Targeted Lipid Nanoparticles

Inbal Hazan-Halevy, Daniel Rosenblum, Srinivas Ramishetti, and Dan Peer

Abstract

Systemic delivery of RNA interference (RNAi) payloads for manipulation of gene expression in lymphocytes holds a great potential as a novel therapeutic modality for hematological malignancies and autoimmune disorders. However, lymphocytes are among the most difficult cells to transfect with RNAi, as they are resistant to conventional transfection reagents and are dispersed throughout the body, making it a challenge to successfully deliver these payloads via systemic administration route. We have developed a strategy to target lymphocytes and deliver RNAi payloads in a cell-specific manner to induce therapeutic gene silencing. This approach utilizes antibodies that decorate lipid nanoparticle surfaces to home into lymphocyte subsets. This approach opens new avenues for discovery of new drug targets and potentially for therapeutics.

Key words RNA interference (RNAi), Gene silencing, Small interfering RNAs (siRNAs), Lipid nanoparticles (LNPs)

1 Introduction

Lipid nanoparticles (LNPs) are the most advanced nanocarriers for siRNA delivery, due to their high siRNA encapsulation efficiency, low immunogenic properties, and potent gene knockdown in humans [1–3]. LNPs are composed of ionizable lipids and other helper lipids such as cholesterol, DSPC, and polyethylene glycol-conjugated (PEGylated) lipids [4]. LNPs can be produced using commercial microfluidic mixing technology [3, 5–7]. The amine head group of ionizable lipids maintains neutral charge at physiological pH 7.4, but at acidic pH, these lipids get protonated and electrostatically complexed with nucleic acids. After systemic administration, LNPs are associated with apolipoproteins in blood circulation, which mediates their binding and endocytosis primarily by hepatocytes and monocytes [8, 9]. Once LNPs are inside the endosomes, the acidic pH of endosomes makes the ionizable lipid

Lekha Dinesh Kumar (ed.), *RNA Interference and Cancer Therapy: Methods and Protocols*, Methods in Molecular Biology, vol. 1974, https://doi.org/10.1007/978-1-4939-9220-1_11,

positively charged, which facilitates the interaction with endosomal membrane followed by its destabilization and nucleic acid release. LNPs are currently in clinical trials for the treatment of liver diseases due to their accumulation mostly by liver cells after systemic administration [1–4, 10]. Exploiting this technology to target systemic lymphocytes and treat disseminated diseases such as leukemia and lymphoma is more challenging [7].

To this end, we described the use of targeted lipid nanoparticles (tLNPs) as a novel promising approach, for systemically targeting lymphocyte subsets and delivering RNAi payloads in a cell-specific manner. We synthesized LNPs composed of the ionizable lipid Dlin-MC3-DMA, cholesterol, DSPC, and PEGylated lipids. Functional PEGylated lipid such as PEG-maleimide was used to conjugate the antibodies chemically on the surface of LNPs. The antibodies were chosen according to their specificity and ability to internalize into the lymphocyte subpopulations [11].

We recently described a strategy to target specifically CD4+ T lymphocytes. LNPs were decorated with anti-CD4 monoclonal antibodies and entrapped siRNA against CD45, a pan leukocyte marker. Systemic administration of these CD4-targeted LNPs led to efficient binding and uptake into CD4+ T lymphocytes, followed by downregulation of CD45 specifically in CD4+ T lymphocytes, whereas the expression of CD45 on other lymphocyte populations was unaffected [11].

A similar strategy was described by Weinstein et al. [12], to deliver siRNAs against cyclin D1 to mantle cell lymphoma (MCL) cells in human MCL-xenografted mice, a prototypic blood cancer of B-cell malignancies. MCL is an aggressive B-cell malignancy characterized by an overexpression of cyclin D1, which is present in 85% of patients. We have previously shown that downregulation of cyclin D1 inhibits proliferation and induces cell cycle arrest in MCL cells [13]. We synthesized LNPs entrapped siRNAs against cyclin D1 that were further decorated with anti-CD38 monoclonal antibodies (mAbs). CD38 receptor is highly expressed on most MCL patients' cells and MCL cell lines. LNPs coated with anti-CD38 mAbs and loaded with siRNAs against cyclin D1 induced potent gene silencing in MCL cells and prolonged survival of tumor-bearing mice with no observed adverse effects. These results present a novel RNAi delivery system that opens new therapeutic opportunities for treating MCL and other B-cell malignancies.

2 Materials

2.1 tLNPs Production

1. Lipids: 1,2-distearoyl-sn-glycero-3-phosphocholine (DSPC), cholesterol (Chol). 1,2-Dimyristoyl-sn-glycerol, methoxypolyethylene glycol (DMG-PEG 2000), *N*-[(3-maleimide-1-oxopropyl) aminopropyl polyethyleneglycol-carbamyl] distear-

oylphosphatidyl-ethanolamine (DSPE-PEG-MAL) (NOF, American corporation). Dlin-MC3-DMA was synthesized according to previously described method [14].

2. NanoAssemblr microfluidic mixing device.
3. 1 M acetate buffer, pH 4.5: diluted to 25 mM with nuclease-free H_2O.
4. Ethanol, absolute.
5. 1× phosphate buffered saline (PBS) without calcium and magnesium, pH 7.4, at 20 °C.
6. GeBAflex dialysis tubes 14 kDa.
7. Monoclonal antibodies: antihuman CD38 clone THB7 and isotype control mouse IgG1 clone MOPC-21.
8. 100 mM 1,4-dithiothreitol (DTT).
9. 500 mM ethylenediaminetetraacetic acid (EDTA).
10. Desalting columns, Zeba™ spin, 7K MWCO, 0.5 mL size exclusion column with sepharose CL-4B beads.
11. Fraction collector.
12. Spectrophotometer (NanoDrop 1000).
13. 100k Amicon Ultra-4 microfuge.
14. 0.2 μm filter membrane.

2.2 Characterization of tLNPs: Particle and Zeta Potential Analysis

1. Malvern Zetasizer Nano ZS™.
2. 1× PBS, pH 7.4, at 20 °C.

2.3 Binding of tLNPs to Cells

1. Binding buffer: PBS, pH 7.4, 1% FBS, 0.01% sodium azide.
2. Granta519 (DSMZ): grown in Dulbecco's MEM (4.5 g/L glucose) supplemented with 1% antibiotics (penicillin and streptomycin), 2 mM L-glutamine, and 10% fetal calf serum (FCS).
3. Primary antibodies: antihuman CD38 clone THB7 and isotype control mouse IgG1 clone MOPC-21 (BioXcell).
4. Secondary antibodies: Alexa488-labeled anti-mouse IgG1.

2.4 siRNA Entrapment Efficiency in tLNPs

1. Chemically modified siRNAs against cyclin D1 were synthesized.
   Cyclin D1 sense: 5′-GUAGGACUCUCAUUCGGGATT-3′.
   Luciferase sense: 5′-CUUACGCUGAGUACUUCGATT-3′.
2. Nuclease-free water.
3. Quant-iT RiboGreen RNA assay kit for percent entrapment efficiency.

2.5 Silencing Efficiency of tLNPs

1. RNA purification reagent (we used EZ-RNA total RNA isolation kit).
2. Spectrophotometer (NanoDrop 1000).
3. cDNA synthesis kit.
4. qRT-PCR Kit5. Primers—expression of cyclin D1 was normalized to the two "housekeeping" genes, eIF3a and eIF3c, using the multiple endogenous controls option. The primers used for amplification are the following (5′–3′):

   CCND1 F—GAGGAGCCCCAACAACTTCC and R—GTCCGGGTCACACTTGATCAC;

   eIF3a F—TCCAGAGAGCCAGTCCATGC and R—CCTGCCACAATTCCATGCT;

   eIF3c F—ACCAAGAGAGTTGTCCGCAGTG and R—TCATGGCATTACGGATGGTCC.

3 Methods

3.1 tLNPs Production and Purification

tLNPs are nanometer-sized lipid nanoparticles possessing targeting moieties (antibodies to receptors on lymphocytes) on their surface. The preparation involves two critical steps: (1) preparation of lipid nanoparticles (LNPs) by microfluidic mixing and (2) introduction of targeting molecules (mAbs) on the surface of the LNPs (Fig. 1).

1. Prepare stock solutions of lipids in absolute ethanol (*see* **Note 1**).
2. Dissolve siRNA in nuclease-free water to final concentration of 20 mg/mL.

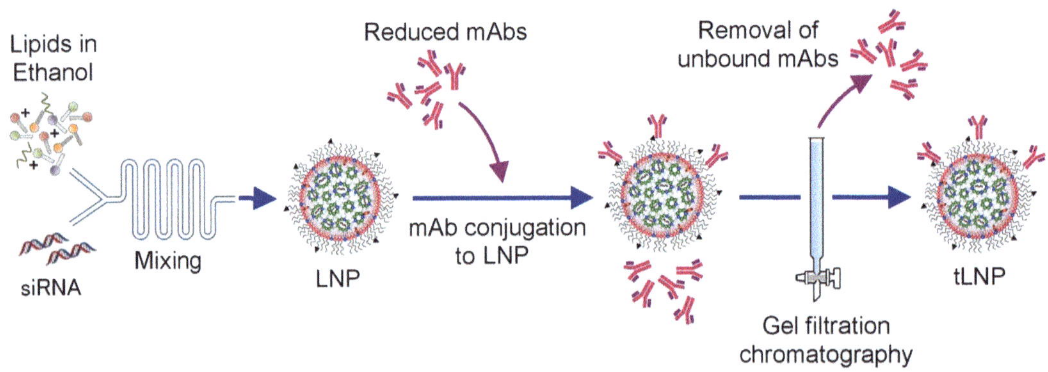

Fig. 1 Schematic diagram of the production process of targeted LNPs (tLNPs). Lipids including the ionizable lipid and PEG-maleimide dissolved in ethanol and siRNAs in acetate buffer are combined by the microfluidic mixer to produce maleimide-functionalized LNPs. Reduced immunoglobulins (IgGs) are conjugated with maleimide-functionalized LNPs to produce targeted LNPs (tLNPs). To remove unconjugated antibody, tLNPs are purified by gel filtration chromatography. Reproduced by permission of the American Society of Gene Therapy: Molecular Therapy 25(7):1491–1500, 2017

3. Prepare desired amount of siRNA duplexes in 25 mM sodium acetate buffer pH 4.0.
4. Prepare a lipid mixture solution at a molar ratio of 50:10:38:1.5:0.5 (Dlin-MC3-DMA, DSPC, Chol, DMG-PEG, and DSPE-PEG Mal, respectively).
5. Prepare LNPs with a final lipid concentration of 9.6 mM by mixing one volume of lipid mixture solution in ethanol with three volumes of an aqueous phase containing siRNA duplexes in 25 mM acetate buffer (N/P charge ratio of 1–5) using NanoAssemblr™ micromixer.
6. Transfer the assembled particles (LNPs) to 14 kDa GeBAflex dialysis tubes, and dialyze against PBS (pH 7.4) for 4 h, followed by overnight dialysis to remove the ethanol completely.
7. Collect the LNPs into a glass vial and store at 4 °C.

   Perform the binding reaction of LNPs to mAbs of choice to produce tLNPs.
8. Reduce the mAbs in order to react with maleimide groups present on the surface of LNPs. DTT was dissolved in 1× PBS containing 10 mM EDTA and added to mAbs (anti-CD38, THB7 clone or isotype mAbs, MOPC-21 clone) for 30 min at 37 °C (the final concentration of DTT depends on the ability to reduce the mAbs; *see* **Note 2**).
9. Remove DTT from reduced antibodies with 7k Zeba spin desalting columns according to the manufacturer's protocol.
10. Incubate 1 mL of LNPs with either 0.5 μg of reduced CD38 mAbs or isotype for 2 h with gentle shaking at room temperature followed by overnight incubation at 4 °C, to produce targeted LNPs (tLNPs).
11. Purify the tLNPs from unbound mAbs by using size exclusion column chromatography loaded with sepharose CL-4B beads equilibrated with PBS.
12. Collect the eluted fractions, and recognize the fraction that includes tLNPs or the free mAbs by taking sample of each fraction to a spectrophotometer measurement (*see* **Note 3**).
13. Concentrate the fractions that include tLNPs using 100k Amicon Ultra-4 according to the manufacturer's protocol.
14. Filter tLNPs through a 0.2 μm membrane and store at 4 °C until further use.

3.2 siRNA Entrapment Efficiency Determination

1. Determine siRNA entrapment efficiencies by a Quant-iT™ RiboGreen™ RNA assay (Molecular Probes, Invitrogen) according to the manufacturer's protocol in the presence and in the absence of 0.5% Triton X-100. In the absence of detergent, fluorescence can be measured from accessible (free or surface-

Table 1
Particle size and zeta potential measurements

Particle type	Hydrodynamic diameter (nm)	*z* potential (mV)
LNPs	59.7 ± 1.6	−16.7 ± 0.9 mV
CD38-tLNPs	89.2 + 1.06	−17.4 ± 0.5 mV

All measurements were done in PBS pH 6.7 (with 10 mM NaCl) at 20 °C in a Zetasizer Nano ZS, Malvern. Data presented as an average ± SD from n = 3 independent experiments

bound) siRNA only, whereas in the presence of detergent, fluorescence is measured from total siRNA; thus, the % encapsulation is described by the following equation:

$$\%\text{siRNA encapsulation} = \left[1 - \left(\text{free siRNA conc} / \text{total siRNA conc}\right)\right] \times 100$$

3.3 Particle and Zeta Potential Analysis

The diameter and surface charge (zeta potential) of nanoparticles are measured using a Malvern Zetasizer Nano ZS™ (Table 1).

3.4 In Vitro Binding Efficiency of tLNPs

Flow cytometry is used to ensure that conjugated antibodies are functional.

1. Dispense Granta519 cells (CD38 expressing cells) to suitable tubes at 0.5×10^6 cells per tube.
2. Add to each tube 1 mL of binding buffer; centrifuge at $250 \times g$, 5 min, 4 °C; and aspirate supernatant.
3. Add to each tube 100 μL of binding buffer.
4. For binding competition experiment, add 1 μg of free antibody, CD38 (THB7 clone) to the appropriate tubes.
5. Incubate samples on ice for 15 min.
6. Add 0.5 μg/mL of the appropriate LNPs (untargeted or tLNPs).
7. Incubate samples on ice for 30 min.
8. Add to each tube 1 mL of binding buffer; centrifuge at $250 \times g$, 5 min, 4 °C; and aspirate supernatant.
9. Add to each tube 100 μL of binding buffer containing 1 μg/mL secondary antibody Alexa488-anti-mouse Ab IgG1.
10. Incubate on ice for 20–30 min.
11. Add to each tube 1 mL of binding buffer; centrifuge at $250 \times g$, 5 min, 4 °C; and aspirate supernatant.
12. Resuspend the cell pellets in appropriate volume of binding buffer, and analyze samples by flow cytometer (Fig. 2).

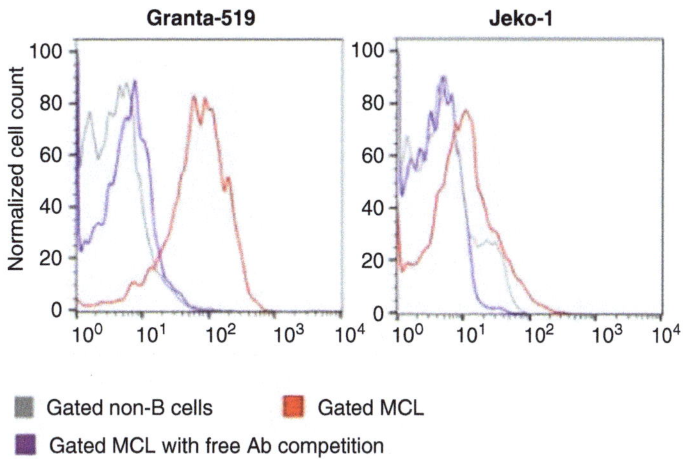

Fig. 2 αCD38-LNPs-siRNA mediate delivery of siRNA specifically into MCL cells. Granta-GFP (left) or Jeko-GFP (right) was cocultured with TK-1 (murine T lymphoma) cells and incubated with αCD38-LNPs-siRNA entrapping labeled siRNA. siRNA-LNPs binding to non-B cells (gray), MCL cells (red), or MCL cells in samples incubated with free competing αCD38 mAbs before αCD38-LNPs-siRNA incubation (purple) (Proc Natl Acad Sci U S A 113:E16–22)

3.5 In Vitro Silencing Efficiency of tLNPs

1. Seed Granta cells in a 6-well microtiter plate at 2.5×10^5 cells in 1 mL per well. Culture cells overnight at 37 °C, 5% CO_2.
2. Add to each well 0.2–5 μg/mL of tLNP entrapping cyclin D1 siRNA.
3. Appropriate controls should be included: cells with no treatment; cells treated with tLNPs entrapping negative control siRNA (e.g., silencer firefly luciferase siRNA or scrambled siRNA).
4. Culture cells for 48–72 h at 37 °C, 5% CO_2.
5. Collect the cells and add to each tube 1 mL of PBS; centrifuge at 250 × *g*, 5 min, 4 °C; and aspirate supernatant.
6. Purify total RNA from the cell pellets using total RNA isolation kit, and generate cDNA using cDNA synthesis kit according to the manufacturer's instructions.
7. Perform qRT-PCR using primers for the silence gene and endogenous control primers (Fig. 3).

4 Notes

1. Lipids need to be dissolved in ethanol completely. Cholesterol and DSPC solutions may need to be warmed up to 37 °C.
2. The concentration of DTT and the incubation time of DTT with mAbs need to be calibrated for each IgG clone. The IgG

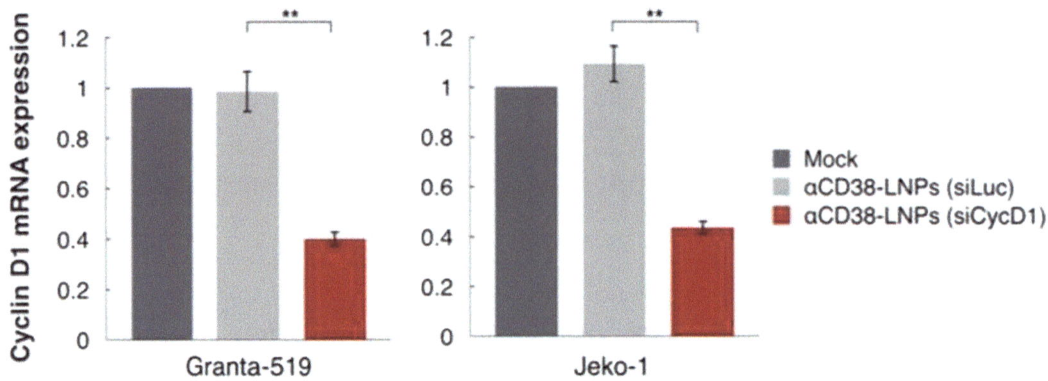

Fig. 3 αCD38-LNPs-siCycD1 mediate cycD1 mRNA knockdown in MCL cell lines. qRT-PCR quantification of cyclin D1 transcripts in MCL cell lines. RNA was extracted from cells 48 h following treatment with mock (black), αCD38-LNPs-siLuc (gray), or αCD38-LNPs-siCycD1 (red). Bar plots represent mean ± SEM of cycD1 expression relative to mock ($n = 3$ independent experiments per cell line; $**p < 0.01$; one-way ANOVA test with Bonferroni correction) (Proc Natl Acad Sci U S A 113:E16–22)

antibodies are optimally reduced only in the hinge region, resulting in two Fab fragments with Mw of 75 kDa. Try to avoid reduction of the disulfide bonds between the light and the heavy chains.

We used 1 mM DTT and 5 mM DTT to reduce anti-CD38 (THB7 clone) and isotype mAbs (MOPC-21), respectively.

3. Free antibodies emitted light with a peak around 280 nm, while tLNPs emitted light with a peak around 230 nm.

Acknowledgments

This work was supported in part by grants from the Dotan Hemato-oncology Center at Tel Aviv University, by the Lewis Trust for blood cancers and by the Leona M. and Harry B. Helmsley Nanotechnology Research Fund awarded to D.P.

References

1. Coelho T, Adams D, Silva A et al (2013) Safety and efficacy of RNAi therapy for transthyretin amyloidosis. N Engl J Med 369:819–829
2. Tabernero J, Shapiro GI, LoRusso PM et al (2013) First-in-humans trial of an RNA interference therapeutic targeting VEGF and KSP in cancer patients with liver involvement. Cancer Discov 3:406–417
3. Peer D, Karp JM, Hong S et al (2007) Nanocarriers as an emerging platform for cancer therapy. Nat Nanotechnol 2:751–760
4. Rietwyk S, Peer D (2017) Next-generation lipids in RNA interference therapeutics. ACS Nano 11:7572–7586
5. Leung AK, Tam YY, Chen S et al (2015) Microfluidic mixing: a general method for encapsulating macromolecules in lipid nanoparticle systems. J Phys Chem B 119:8698–8706
6. Mizrahy S, Hazan-Halevy I, Dammes N et al (2017) Current progress in non-viral RNAi-based delivery strategies to lymphocytes. Mol Ther 25:1491–1500

7. Ramishetti S, Landesman-Milo D, Peer D (2016) Advances in RNAi therapeutic delivery to leukocytes using lipid nanoparticles. J Drug Target 24:1–7
8. Akinc A, Querbes W, De S et al (2010) Targeted delivery of RNAi therapeutics with endogenous and exogenous ligand-based mechanisms. Mol Ther 18:1357–1364
9. Novobrantseva TI, Borodovsky A, Wong J et al (2012) Systemic RNAi-mediated gene silencing in nonhuman primate and rodent myeloid cells. Mol Ther Nucleic Acids 1:e4
10. Morrissey DV, Lockridge JA, Shaw L et al (2005) Potent and persistent in vivo anti-HBV activity of chemically modified siRNAs. Nat Biotechnol 23:1002–1007
11. Ramishetti S, Kedmi R, Goldsmith M et al (2015) Systemic gene silencing in primary T lymphocytes using targeted lipid nanoparticles. ACS Nano 9:6706–6016
12. Weinstein S, Toker IA, Emmanuel R et al (2016) Harnessing RNAi-based nanomedicines for therapeutic gene silencing in B-cell malignancies. Proc Natl Acad Sci U S A 113:E16–E22
13. Weinstein S, Emmanuel R, Jacobi AM et al (2012) RNA inhibition highlights cyclin D1 as a potential therapeutic target for mantle cell lymphoma. PLoS One 7:e43343
14. Jayaraman M, Ansell SM, Mui BL et al (2012) Maximizing the potency of siRNA lipid nanoparticles for hepatic gene silencing in vivo. Angewandte Chemie 51:8529–8533

Chapter 12

Targeting Cancer with Peptide RNAi Nanoplexes

A. James Mixson, Qixin Leng, Szu-Ting Chou, and Martin C. Woodle

Abstract

With the recent explosion of genomic information on the root causes of disease, there is an increased interest in nucleic acid therapeutics, including siRNA and gene therapy, all of which require delivery of highly charged nucleic acids from siRNA with a molecular weight of about 1.4×10^4 to plasmids with an approximate molecular weight of $2.0–3.0 \times 10^6$. This chapter describes the delivery of shRNA via plasmid or siRNA with a peptide-based carrier. We focus on the histidine-lysine peptide which serves as an example for other peptides and polymeric carrier systems. When the HK peptide and nucleic acids are mixed together and interact with one another through ionic and nonionic interactions, nanoplexes are formed. These nanoplexes, carrying either shRNA or siRNA that target oncogenes, provide promising options for the treatment of cancer. We describe methods of preparation and characterization of these nanoplexes using dynamic light scattering, zeta potential, and gel retardation assays. We also provide protocols for transfection in vitro and in vivo for these nanoplexes.

Key words Peptides, Polymers, RNAi, siRNA, shRNA, Tumor, Polyplexes, Nanoplexes

1 Introduction

RNAi silencing of oncogenes via siRNA or plasmid-based shRNA has significant potential for cancer treatment. Despite the promise of RNAi therapy, it has a similar delivery problem as gene and antisense therapies. That is, nucleic acids which are highly negatively charged cannot easily be transported through the bloodstream to their cytosolic targets in the tumor. Another significant and related issue, which limits the efficacy of RNAi, is its escape from the endosomes. As a result, continued advances in the carrier design are required to shift the intracellular trafficking of the RNAi away from lysosomal degradation to endosomal lysis, thereby enabling RNAi to reach its target in the cytosol. Several carriers for siRNA are now being tried in animal disease models including cell-penetrating peptide conjugates [1, 2], aptamers [3], synthetic non-peptide polymers [4], peptides and peptide-like polymers [5–8], cyclodextrin [9], atelocollagen [10], and neutral and cationic liposomes [11–13]. Whereas there have been no FDA-approved RNAi

Lekha Dinesh Kumar (ed.), *RNA Interference and Cancer Therapy: Methods and Protocols*, Methods in Molecular Biology, vol. 1974, https://doi.org/10.1007/978-1-4939-9220-1_12,

carriers for systemic treatment, several have advanced to clinical trials [14–17] (*see* Review of [18]).

Two approaches utilize RNAi to inhibit target genes: shRNA (short hairpin RNA) and siRNA (small interfering RNA). While the shRNA approach is usually promoter-dependent and can be delivered by both viral (e.g., lentivirus, adeno-associated virus) and nonviral (plasmid-based) methods, siRNA is a chemically synthesized RNA duplex and is generally delivered by nonviral delivery systems. Although groups including ours have shown efficient nonviral delivery to tumors in vivo with siRNA [19, 20], efforts with nonviral delivery of plasmids to tumor xenografts have been thwarted by inefficient delivery. Though the vast majority of nanoparticles are dependent on the enhanced permeability and retention mechanism for their accumulation in the tumor, other mechanisms of transport such as the neuropilin-1 system may yield more efficient transport of shRNA plasmid-based therapy [21, 22].

RNAi utilizes the microRNA machinery of the cell for its activity, and this provides the basis for efficacy and toxicity of RNAi. In contrast to chemically synthesized siRNA, the promoter-based shRNA approach requires multiple enzymatic and/or transport steps (e.g., nuclear entry, transcription, nuclear export, Drosha and Dicer processing) before interaction with the RNA-induced silencing complex (RISC), which results in cleavage of the targeted mRNA. Thus, in contrast to synthetic siRNA that initially targets RISC within the cytosol, plasmid-based RNAi therapy is significantly limited for in vivo applications by its need for nuclear transport and processing.

Regardless of whether RNAi is synthetically made (i.e., siRNA) or derived from plasmids (i.e., shRNA), 19–23 siRNA duplexes with 5′-phosphorylated ends and 2-nucleotide unpaired and unphosphorylated 3′-ends are incorporated into RISC [23, 24]. RISC is a cytosolic protein-RNA complex that separates the strands of the RNAi duplex and discards the sense (passenger) strand. The antisense (guide) RNA strand then guides the activated RISC to anneal and cleave the target mRNA [25]. The endonuclease argonaute-2 plays a key role in unwinding the duplex (sense and antisense siRNA strands) and cleaving the target mRNA [26]. By recycling the target mRNA, the activated RISC complex may show a therapeutic effect for up to 7 days in dividing cells and for several weeks in nondividing cells.

To avoid the cytokine induction and promote stability, siRNA is frequently modified. For siRNA, cytokine induction is primarily through its interaction with TLR7 in mice and TLR 7/8 in humans [27–29]. These TLR7(8) receptors are located in acidic endosomes of plasmacytoid dendritic cells and monocytes, and pH-buffering agents can reduce induction of cytokines [30–32]. A more common approach to reduce cytokine levels and enhance the stability of siRNA nanoplexes is by selective modification of the siRNA such as with 2-O methyl bases [28]. Furthermore, limited modification with phosphorothioate linkages of siRNA has frequently been used to increase the stability of the siRNA [33]. Several other recent modifications have demonstrated promise to augment the stability of siRNA [33, 34]. Cytokine induction may also occur with plasmids through their interaction with TLR9. Notably, investigators have determined that CpG-free plasmids can markedly reduce the induction of cytokines [35, 36].

This chapter will focus on peptide delivery of RNAi, although other delivery systems of RNAi such as cationic liposomes and other polymer systems will be discussed. Peptide carriers usually bind through ionic interactions to the negatively charged phosphate backbone of siRNA, although additional non-covalent bonds may stabilize the interaction between the polymer and nucleic acids [5]. The judicious use of disulfide bonds and hydrophobic groups may also further enhance the stability and efficacy of peptide-based nanoparticles [37, 38]. Alternatively, cationic peptides such as cell-penetrating peptides (CPP) may be covalently conjugated to siRNA to promote entry into the cell [39, 40]. To prevent electrostatic interactions between CPP and siRNA in these conjugates, reduced negatively charged nucleic acid analogues such as peptide nucleic acid (PNA) or phosphorodiamidate morpholino oligonucleotides (PMO) have been used [39, 40]. Regardless of the type of linkage between the peptide and nucleic acids, peptide-mediated delivery enters the cell primarily through the endosomal pathway [32, 41–43]. Even cell-penetrating peptide nanoplexes enter cells mainly through macropinocytosis [43], although a small percentage of these nanoplexes may directly enter through a non-endosomal mechanism [44].

2 Materials

Solutions are prepared using ultrapure water (using deionized water to attain a sensitivity of 18.2 MΩ-cm at 25 °C).

2.1 Cells, Media, and Reagents

1. MDA-MB-435 (a malignant cell line), MDA-MB-231 (human breast cancer), 4T1 (mouse breast cancer), HL460 (human lung cancer), and N2A neuroblastoma (ATCC, Manassas, VA).
2. Dulbecco's Modified Eagle's Medium (DMEM).
3. Dulbecco Modified Eagle Medium containing high glucose (25 mM), HEPES (25.03 mM), and no phenol.
4. Fetal calf serum (FCS).
5. Glutamine.
6. Opti-MEM.
7. DH5α bacteria.

8. Terrific Broth: To formulate 1 l of Terrific Broth (TB), prepare two solutions—A and B. For solution A, combine the ingredients (yeast extract, 24 g; tryptone, 12 g; glycerol, 4 g) in a final volume to 900 ml with ultrapure water, and then autoclave. For solution B, mix the KH_2PO_4 (2.3 g) and K_2HPO_4 (16.4 g) with 100 ml of ultrapure water, and then autoclave. To prepare TB, mix 900 ml of solution A with 100 ml of solution B.
9. Luciferase Assay System kit (containing the assay buffer and lyophilized assay (luciferin) substrate).
10. D-Luciferin.
11. Ion-exchange chromatography columns for DNA purification with buffers as part of the EndoFree Maxi Plasmid kits.
12. Coomassie Plus Protein Assay Reagent.
13. Hydrogen peroxide (3%) solution.
14. Agarose.
15. NuSieve GTG agarose.
16. Ethidium bromide solution (10 mg/ml).
17. Rabbit antihuman polyclonal Raf-1 antibody.
18. Goat horseradish peroxidase-labeled antirabbit antibody.
19. 3,3′-Diaminobenzidine (DAB) substrate (dissolve DAB in 50 mM Tris-HCl, pH 7.2 at 1 mg/ml. Immediately before use, add an equal volume of 0.02% hydrogen peroxide).
20. TRIzol reagent.
21. SuperScript III One-Step RT-PCR system kit.
22. Quantum RNA-β actin Internal Standards kit.

2.2 Buffers and Solutions

1. Formalin solution, neutral buffer (10%) (formaldehyde 37–40% 100 ml/l, deionized water 900 ml/l, NaH_2PO_4 4 g/l, Na_2HPO_4 (anhydrous) 6.5 g/l).
2. Citrate buffer (10 mM citric acid, 0.05% Tween 20, pH 6.0).
3. Antibody dilution buffer, phosphate-buffered saline (PBS), 10×, pH 7.4 (90 g/l NaCl, 1.44 g/l KH_2PO_4, 7.95 g/l Na_2HPO_4 (anhydrous); sterile filtered). Use as 1×.
4. Tris-buffer saline, 1× (TBS) 2.42 g Tris base/l, 8.0 g NaCl/l; adjust pH to 7.6 with HCl.
5. Wash buffer, TBS with Triton X-100 (TBS-Triton) (0.2%). Dilute 1.0 ml of Triton X-100 solution into 500 ml of 1× TBS. Heat in microwave for 20 s, and stir for 20 min before use.
6. Blocking solution: Normal goat serum (dilute to 5% with PBS).
7. Tris-borate-EDTA buffer, 10× (TBE). (Dissolve 108 g Tris and 55 g boric acid in 900 ml distilled water; add 40 ml 0.5 M

Na_2EDTA (pH 8.0); adjust volume to 1 l; store at RT.) Use as 1×.

8. Tris-borate-EDTA buffer containing ethidium bromide. Add 30 μg of ethidium bromide per 100 ml of TBE.
9. Reporter lysis solution for cells.
10. Tumor lysis buffer (0.1 M Tris, pH 7.5, 1.5 mM EDTA, 0.5% Triton X-100).

2.3 RNAi

2.3.1 siRNA Sequences

1. The siRNA duplex targeting luciferase (siLuc) is as follows: sense, 5′-CUGCACAAGGCCAUGAAGAdTdT-3′; antisense, 5′-UCUUCAUGGCCUUGUGCAGdTdT-3′, targeting 5′-CUGCACAAGGCCAUGAAGA-3′ (Dharmacon, Lafayette, CO).
2. The Raf-1 siRNA duplex (siRafl) is as follows: sense, 5′-UGUCCACAUGGUCAGCACCdTdT-3′; antisense, 5′ GGUGCUGACCAUGUGGACAdTdT-3′ (Dharmacon, Lafayette, CO).
3. Control siRNA (siControl) sense, 5′-CAG UUG CGC AGC CUG AAU GdTdT-3′; antisense, 5′-CAU UCA GGC UGC GCA ACU GdTdT-3′ (no target; Qiagen).
4. Fluorescent-labeled siRNA Alexa Fluor 555 Allstars siRNA (no target; Qiagen).
5. Reconstitute the siRNA with buffer supplied by the manufacturer and deionized ultrapure water.
6. Measure siRNA concentration and purity, respectively, at 260λ and the 260/280λ ratio with spectrophotometer.
7. Adjust concentration of siRNA with ultrapure water to 1–2 mg/ml.

2.3.2 shRNA Plasmids

1. Companies provide several shRNA sequences inserted into a plasmid (i.e., PSI-U6) to test their efficacy against a particular target and/or oncogene.
2. Effective target sequences for Raf-1 and luciferase shRNA are (1) Raf-1, UGUCCACAUGGUCAGCACC (h,m) [6], and (2) luciferase, CUGCACAAGGCCAUGAAGA [45] (*see* **Notes 1–4**).
3. The control shRNA sequence is CAGUUGCG CAGCCUGAAUG (no target) (h, human; m, mouse).
4. Isolate and purify PSI-U6-shLuc and PSI-U6-shRNA-1 from *E. coli* (DH5α) after growth in Terrific Broth using plasmid purification kits (*see* **Note 5**).
5. Measure plasmid concentration and purity at 260λ and the 260/280λ ratio, respectively, with a spectrophotometer.
6. Adjust concentration of RNAi plasmid with ultrapure water to about 2 mg/ml.

2.4 Peptides

1. H3K(+H)4b, a four-branched peptide, with each branch having a sequence of KHHHKHHHKHHHHKHHHK, was made on a solid-phase synthesizer (Fig. 1). The four branches emanate from a three-lysine core. HK peptides such as H3K(+H)4b peptides can be obtained from GenScript with 90% or greater purity (*see* **Notes 6** and **7**).
2. The linear H2K has a sequence of KHKHHKHHKHHK HHKHHKHK.
3. To validate the synthesis and purity, mass spectroscopy and HPLC of the final peptide should be provided (Fig. 2).
4. Reconstitute the peptides with ultrapure water to a concentration of 30 mg/ml.

2.5 Maintaining Cells and Culture Medium

MDA-MB-435, MDA-MB-231, C6, 4T1, N2A, and HL460 cells are maintained in DMEM containing 10% FCS and 2 mM glutamine.

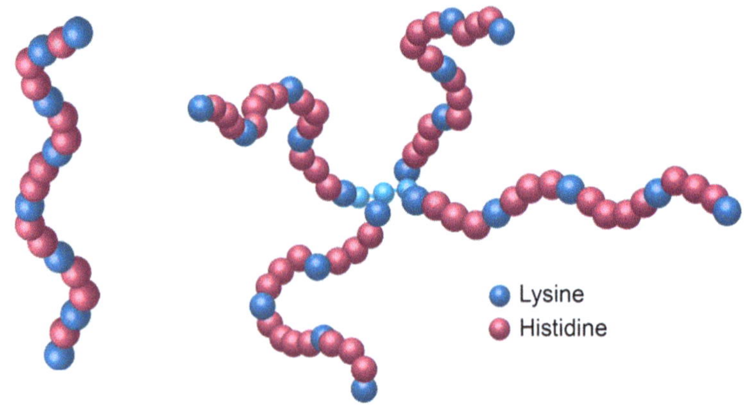

Fig. 1 Schematic model of the linear H2K and branched H3K(+H)4b peptides

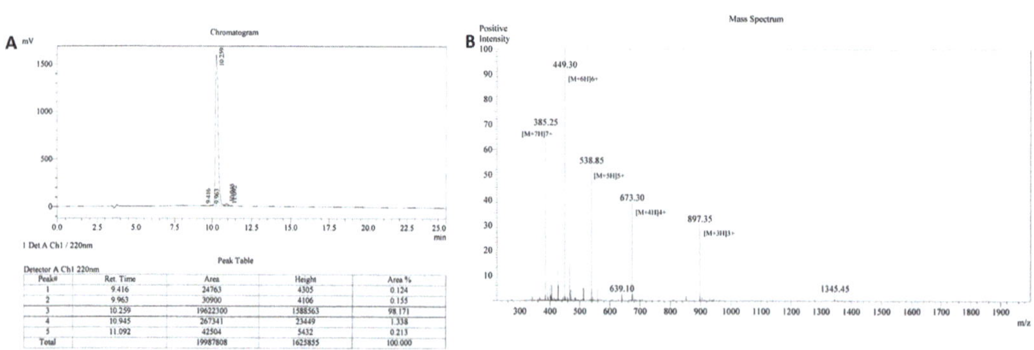

Peak#	Ret. Time	Area	Height	Area %
1	9.416	24763	4305	0.124
2	9.963	30900	4106	0.155
3	10.259	19622300	1588563	98.171
4	10.945	267341	23449	1.338
5	11.092	42504	5432	0.213
Total		19987808	1625855	100.000

Fig. 2 HPLC and mass spectroscopy of the H2K peptide

2.6 Animals

Female athymic mice (4–8 weeks old, body weight ~20 g) are purchased from the Jackson Laboratory (Bar Harbor, ME). All experiments were performed in accordance with regulations by the Institutional Animal Care and Use Committee.

2.7 Equipment (Fig. 3)

1. Zetasizer Nano.
2. IVIS-200 optical imaging system/software with XGI-8 anesthesia system.
3. Luminometer.
4. Gel imaging system with gel software.
5. Bench centrifuge.
6. UV-vis spectrometer.
7. Mini-beadbeater-1, Mini-beadbeater-8, 1 μm beads.
8. Power supply.
9. Horizontal gel electrophoresis system.
10. Analytical scale.
11. Tridek stepper with 1 ml disposable syringes.
12. Heating lamp with reflector infrared heating light bulb.
13. Mouse Tailveiner restrainer.
14. Thermocycler.

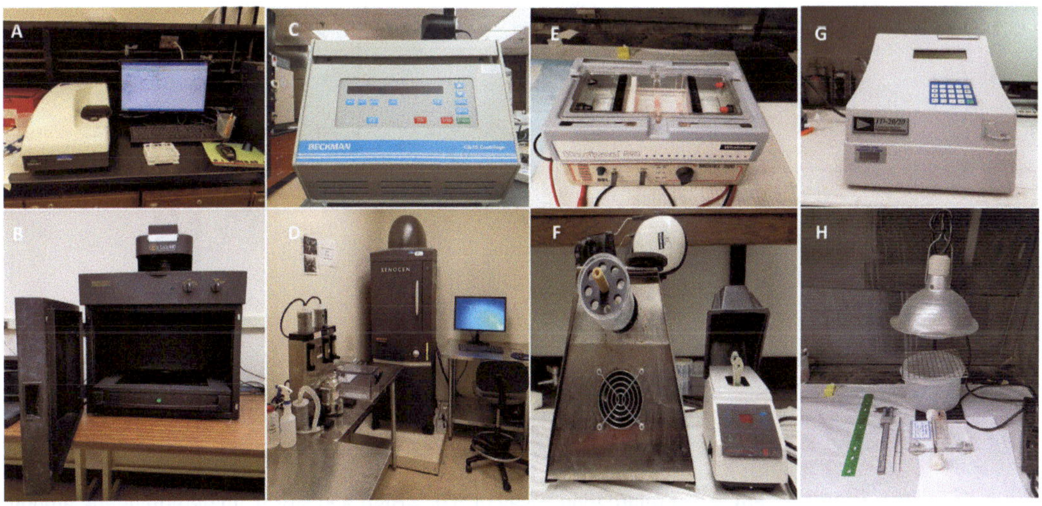

Fig. 3 Equipment to prepare and analyze peptide nanoplexes. (**a**) Zetasizer determines the size and surface charge of nanoplexes. (**b**) Gel image analyzer. (**c**) Bench centrifuge. (**d**) Xenogen imaging with anesthesia. (**e**) Horizontal gel electrophoresis with power supply. (**f**) Mini-beadbeater-8 and Mini-beadbeater-1. (**g**) Turner 20/20 luminometer. (**h**) Skin calipers, Tailveiner restrainer, warming lamp with container

3 Methods

3.1 HK:siRNA Nanoparticle Preparation

3.1.1 In Vitro Preparation

1. For in vitro experiments, HK siRNA nanoplexes are prepared by briefly mixing siRNA (1 μg) in 50 μl of Opti-MEM with various amounts of HK peptides as indicated in Subheading 3.1.2.
2. The siRNA nanoplexes are maintained at room temperature for 30 min before size measurements, gel retardation studies, and/or their incubation with cells.

3.1.2 Gel Retardation Assay

1. The amount of peptide to retain the siRNA within the well is determined by the gel retardation assay.
2. Varying amounts of HK peptides are mixed with 1 μg of siRNA and incubated for 30 min at room temperature. For example, the following peptide siRNA ratios (w/w) are a representative range for H3K(+H)4b that retains the nanoplex in the well (1:1; 1.5:1; 2:1; 2.2:1; 2.5:1) (*see* **Note 8**) (Fig. 4).
3. The zeta potential and size of all nanoplexes are determined as described below.
4. After the HK siRNA nanoplex is loaded onto the gel (3% agarose), electrophoresis is carried out at a constant voltage of

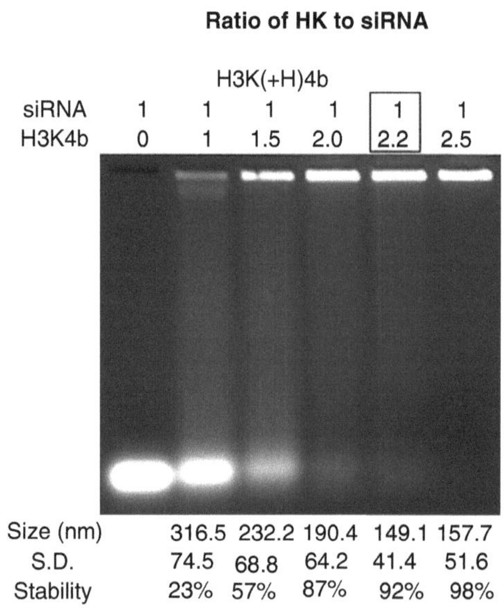

Fig. 4 Gel retardation assay of H3K(+H)4b and siRNA. To determine the amount of HK that retards siRNA migration, various ratios of unmodified H3K(þH)4b in complex with siRNA (1 μg) were prepared and then subjected to gel electrophoresis for 30 min (3% gel). Different ratios of HK polymers to siRNA are represented above the gel, and the particle size is shown below the gel. The ratio enclosed in the box represents the selected in vitro and in vivo silencing conditions based on the size of the particle (~150 nm) and siRNA retardation. Reproduced from ref. 20 with permission from Springer

75 V for 45 min in Tris-borate-EDTA buffer containing ethidium bromide.

5. The siRNA band densities are then visualized under an ultraviolet imaging system. On the basis of fluorescence of siRNA that has migrated into each lane, the stability of HK siRNA nanoplex is determined with gel imaging software.
6. The nanoplex that is retained near 100% in the well at the lowest peptide/siRNA ratio as well as the size of the nanoplex that is less than 150 nm is utilized for silencing studies in vivo (*see* **Notes 9–11**).

3.1.3 DLS and Zeta Potential

1. Turn the Zetasizer Nano on and start the software.
2. In a small tube, add 30 μl of the formulation of the nanoplex to 170 μl of water.
3. Pipette up and down three to four times and add 200 μl of the mix to the disposable cuvette.
4. Place the cuvette into the Zetasizer instrument.
5. Measure the size and PDI characterization of the nanoplex at least ten times under the automatic mode according to the equipment protocol.
6. Add 550 μl of water to the nanoplex (200 μl), mix gently, and transfer the nanoplex solution (750 μl) to a folded capillary cell.
7. Place folded cell into instrument and measure zeta (ζ) potential.

3.2 In Vitro Luciferase Assays

1. MDA-MB-435 cells expressing luciferase are plated in a 24-well plate (0.5 ml of Dulbecco's minimal essential medium, 10% serum) at a density of 3×10^4 cells per well.
2. Different amounts of the peptides in complex with siLuc (1 μg) are prepared as described above, and these nanoplexes are incubated with MDA-MB-435 cells for 48 h (*see* **Notes 12** and **13**).
3. The cells are then exposed to the reporter lysis buffer (200 μl) for 10 min followed by centrifugation at $11{,}500 \times g$ for 5 min.
4. The supernatant fraction (1 μl) is then added to the luciferase buffer (50 μl), and the luciferase activity is measured immediately with a luminometer.
5. To normalize relative light units, protein content of cells is measured with the Coomassie reagent.

3.3 In Vivo Studies

3.3.1 Establishment of Xenograft or Allograft Tumors in Mice

1. Using a stepper (which delivers a set volume) and a 27 gauge needle, the malignant MB-MDA-435 cells (5×10^5 cells) are injected subcutaneously (*see* **Notes 14–16**).
2. After the implanted tumors are established (*see* **Note 17**) (20–100 mm^3), mice are separated into various groups, which are dependent on the number of peptide carriers to be screened.

3. Tumors are measured with skin calipers (*see* **Note 18**).
4. Tumor volume is determined using the formula $1/2 \times \text{length} \times \text{width}^2$.

3.3.2 In Vivo siRNA Nanoplex Formation

1. For tumor xenografts experiments in vivo, mice are treated by injection in the tail vein with H3K(+H)4b nanoplexes containing 40 μg of Luc or control siRNA.
2. After the H3K(+H)4b peptide (120 μg in 150 μl) is mixed with siRNA (40 μg in 150 μl) in sterile water, the resulting nanoplex forms at room temperature for 40 min (*see* **Notes 19–22**).
3. After the tail is warmed with a heating lamp and mouse is restrained with a "Tailveiner restrainer" (*see* Fig. 3h), the mouse receives 0.3 ml of the HK carrier siRNA nanoplex (*see* **Notes 23** and **24**).

3.3.3 Bioluminescence Imaging

1. MDA-MB-435 xenografts that stably express luciferase are established by injecting 2×10^6 cells into midclavicular line of female nude mice (*see* **Note 25**).
2. After the tumors reach about 75–100 mm^3, the mice receive 150 mg/kg of D-luciferin IP (in PBS).
3. Twenty minutes after the injection, the mice are anesthetized via nasal cones with a 2.5% isoflurane/oxygen mixture.
4. The luciferase activity of the tumors is then measured with the IVIS-200 optical imaging system.
5. During imaging, the imaging chamber is maintained at 37 °C, and animal welfare is followed by monitoring breathing and skin color, and anesthesia is confirmed by the toe-pinch test.
6. One hour after imaging, the mice are injected with H3K(+H)4b nanoplexes containing 40 μg of siLuc (or siControl or saline) administered via the tail veins (*see* **Note 26**).
7. Forty-eight hours after inoculation of the nanoplexes, luciferase activity of the tumors is again measured with the IVIS-200 optical imaging system in real-time living mice (*see* **Notes 27** and **28**).
8. Photon emission is measured in the imaging chamber 30 min after the administration of luciferin.
9. To normalize the photon emission (that is, relative light units), the region of interest at times 0 and 48 h areas is of equal size (*see* **Note 29**).
10. Tumor volume is again determined.

3.3.4 In Vivo Luciferase Experiments

1. To validate the bioluminescent imaging study, luciferase activity can be measured in the lysates of MDA-MB-435 xenografts that stably express luciferase (*see* **Note 30**).

2. After the tumor reaches 75–100 mm^3, the mice are injected with nanoplexes comprised of various ratios of the peptide and siRNA (*see* **Note 31**).
3. After imaging, the mice are euthanized, and the tumors are removed and weighed.
4. Tumors are cut into pieces with a razor blade.
5. Place tumor in 2 ml screw top tube which is 2/3 filled with 1 μm beads and 500 μl of the tumor lysis buffer (kept at 4 °C).
6. Fill remainder of tube with ice-cold lysis buffer and place on ice.
7. Homogenize tumor sample with Mini-beadbeater-1 at setting 5 for 50 s.
8. Place sample on ice for at least 15 min.
9. Homogenize sample(s) (one to eight samples) a second time for 2 min with the Mini-beadbeater-8, and afterward place samples on ice for 15 min (*see* **Notes 32** and **33**).
10. Centrifuge for 5 min. at a setting of 11,500 × *g*.
11. Mix 10 μl (and 100 μl) of the supernatant with 100 μl of luciferase assay buffer.
12. Determine RLU levels of each sample with a luminometer (*see* **Note 23**).
13. Normalize RLU values per tumor weight (*see* **Note 34**).

3.3.5 Systemic Therapy of RNAi Therapy Reducing Tumors

1. Prior to treatment in vivo, it is essential to determine efficacy of siRNA or shRNA in vitro to reduce target (*see* **Note 35**).
2. After establishment of tumors of 20 mm^3 (~4 days), systemic therapy with the H3K(+H)4b: siRaf-1 nanoplex is initiated (*see* **Notes 36–38**).
3. As previously described, peptide nanoplexes containing Raf-1 siRNA (40 μg) are administered intravenously (total volume—300 μl) (*see* **Notes 39** and **40**).
4. Nanoplexes are administered IV every 3 days for three to six injections (*see* **Note 41**).
5. The sizes of the tumors are measured prior to each injection of nanoplexes (Fig. 5).
6. Three days after the last injection, the tumors are processed for RNA analysis and immunostaining.

3.3.6 Detection of Raf-1 mRNA by RT-PCR in Tumors

1. The expression of Raf-1 mRNA and β-actin mRNA as an internal standard can be determined by reverse transcriptase-polymerase chain reaction.
2. Total RNA is isolated from treated and untreated MDA-MB-435 cells by using TRIzol reagent.

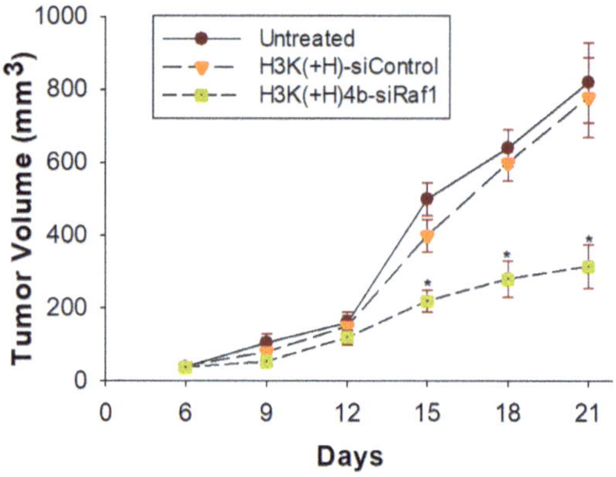

Fig. 5 Tumor inhibition targeting the Raf-1 oncogene. Systemic injection of HK Raf-1 siRNA nanoplexes into nude mice with MDA-MB-435 tumors. For these mice, tumor volume (mm^3) was measured in the groups (untreated, H3K(+H)4b-siControl, H3K(+H)4b-siRaf-1) before the five treatment times and three days after the last treatment. *$P < 0.05$, Raf-1 siRNA vs untreated or siRNA control. Initial tumor size prior to treatment ~20 mm^3

3. RNA integrity is then established by gel electrophoresis.
4. RNA concentration is determined spectrophotometrically in triplicate for each sample.
5. The primers used for RT-PCR of human Raf-1 are 5′-TCAGACTTCTCCACGAACAC-3′ (sense) and 5′-AACAGACTCTCGCATACGAT-3′ (antisense), resulting in a PCR fragment of 368 bp.
6. Primers for β-actin Internal Standards kit are used as a control, resulting in a PCR fragment of 294 bp.
7. RT-PCR is performed as follows: cDNA is synthesized at 61 °C for 30 min, followed by denaturation, 94 °C for 2 min; for a 30 cycle amplification, the conditions are denaturing, 94 °C, 15 s; annealing, 62 °C, 30 s; extension, 68 °C, 45 s; 30 cycles. The final extension is for 5 min at 68 °C.
8. RT-PCR is performed on a thermocycler, and the linearity of the amplified products of Raf-1 and β-actin can be verified by analyzing the products at various cycles.
9. RT-PCR products are then subjected to electrophoresis on a 3% agarose gel (2% agarose/1% NuSieve GTG agarose).
10. A digital image is then taken, and density of the bands is analyzed by the imaging system and gel software.

3.3.7 Immunostaining for Raf-1

1. Mice are euthanized and tumors are removed and cut in half.
2. Tumors are fixed in formalin buffer for 24 h.
3. Tumor sections are then deparaffinized in xylene for two times, 5 min each.

4. Transfer slides to 100% alcohol, for two times, 3 min each, and then transfer once through 95%, 70%, and 50% alcohols, respectively, for 3 min each.
5. Antigens are retrieved by maintaining tissue in 10 mM citrate buffer at boiling point for 40 min.
6. Block endogenous peroxidase activity by incubating sections in 3% H_2O_2 solution in methanol at room temperature for 10 min.
7. Rinse with TBS-Triton for two times, 5 min each.
8. Nonspecific binding is blocked with 5% goat serum (100 μl) for 30 min.
9. Blot off excess blocking buffer.
10. For detection of Raf-1, tissue sections are incubated with rabbit antihuman polyclonal antibody targeting Raf-1 (100 μl; 1:50 dilution with PBS) at 4 °C overnight.
11. Rinse with TBS-Triton for two times, 5 min each.
12. The secondary horseradish peroxidase-labeled antirabbit antibody (100 μl) is added for 30 min.
13. Rinse with TBS-Triton for two times, 5 min each.
14. The chromogen diaminobenzidine (DAB) (100 μl) is applied for 5 min to permit color development.
15. Rinse the slides in running tap water for 10 min.
16. The tissue is dehydrated and incubated four times with alcohol (95%, 95%, 100%, and 100%), 5 min each.
17. Immerse the tissue slides in xylene and fix coverslip using a mounting solution.
18. Four randomly picked bright-field images are converted to normalized blue images, allowing automatic classification (quantification) of positive DAB staining.

4 Notes

1. The order of the transcribed shRNA sequences is the sense sequence followed by the loop sequence which is underlined and then the antisense sequence: (1) shRafl, H1 or U6 promoter TGTCCACATGGTCAGCACC TCAAGAG GGTGCTGACCATGTGGACATT; (2) shLuc, H1 or U6 promoter CTGCACAAGGCCATGAAGA TCAAGAG TCTTCATGGCCTTGTGCAGTT; and (3) shControl, H1/U6 promoter CAGTTGCGCA-GCCUGAATG TCAAGAG CATTCAGGCTGCGCAACTGTT.
2. GeneCopoeia inserted these shRNA sequences into PSI-U6 in vitro, and we tested these sequences in vitro against targets.

In plasmids, shRNA expression can be driven by a U6 or H1 promoter cassettes. U6 is generally considered a stronger promoter but has been associated with toxicity [46]. In our case, we did not notice a significant difference between the silencing effects in vitro of shRNA driven by U6 or H1. For our studies in vivo, shRNA expression was driven by U6 promoter cassettes.

3. shRNA sequences can be digested and inserted to other plasmid vectors. Alternatively, DNA containing shRNA sequences with the promoter can be synthetically made and inserted into the desired plasmid. Although GeneCopoeia sent us four sequences, we found in limited testing that the siRNA sequence target was as or more effective than the other three sequences (reducing 80% or more of the target by PCR). To evaluate the efficacy of the shRNA, it is important to use a carrier that can effectively transport the shRNA plasmids into cells. This may not necessarily be the carrier utilized for in vivo studies.
4. shRNA and siRNA should be screened for homology to decrease off-target effects.
5. Assaying plasmids on an analytical agarose gel ensures that there is no contamination with other nucleic acids including bacterial chromosomal DNA.
6. Although many companies provide peptide synthesis services, very few have been able to synthesize branched peptides such as H3K(+H)4b successfully. In contrast, most peptide companies will be able to synthesize the linear peptides such as H2K.
7. H3K(+H)4b is an effective carrier of siRNA and plasmids in vitro and in vivo.
8. In addition to polyplexes, gel retardation has an important role in determining the optimal liposome/siRNA ratio for lipoplexes.
9. Ratios of the peptide/RNAi higher than the minimal ratio necessary to retain the nanoplex in the well should be used to determine the optimal transfection in vivo. For our studies, the lowest ratio that retains the nucleic acid (siRNA or the plasmid) within the well and results in size of the polyplex less than 150 nm was determined to be the optimal ratio to deliver intravenously.
10. These ratios may vary depending on the affinity that the cationic polymer or liposome has for the RNAi.
11. In tumors with small vascular pores, nanoparticles which are dependent on EPR may need to be less than 50 nm for their adequate transport into tumors [47].
12. The HK to siRNA ratios, tested for in vitro bioluminescent experiments, are determined by gel electrophoresis. The low-

est peptide to siRNA ratio that retained the nanoplex in the well during electrophoresis as well higher peptide/siRNA ratios is reasonable to utilize.

13. In vitro silencing results may give misleading results for in vivo studies. Notably, to evaluate the silencing efficacy of the RNAi, an effective carrier in vitro is required which is not necessarily the carrier you are interested in using in the in vivo studies. As a result, we think the most important aspect of the silencing assay is to determine the efficacy of the siRNA, not to screen for the carrier to be used in vivo.
14. Cells in Dulbecco's Modified Eagle Medium (high glucose, no phenol) are injected into mice.
15. Cells (0.1 ml; 5 X 10^5 cells/100 μl) are injected with the Tridek stepper.
16. All experiments involving animals will require approval by the institution of the researcher.
17. Although subcutaneous tumors are a reasonable model to test initially the efficacy of the therapy, other models such as the orthotopic and transgenic tumor models should be tested, particularly if the therapy is effective with the subcutaneous models [48].
18. Fluorescence activity (tumors expressing fluorescent proteins), bioluminescence activity, tumor weight, histochemistry, and advanced imaging methods such as microCT and MRI are alternative strategies to determine the efficacy of the therapy for localized and/or metastatic tumors [49, 50].
19. Although branched HK peptides are effective carriers of siRNA and shRNA in vivo, linear HK peptides are more effective carriers of shRNA-expressing plasmids in vivo. The optimal peptide/shRNA ratio is 1: 2 with 36 μg of the plasmid administered intravenously [21, 22]. The size of these plasmid polyplexes is about 130 nm.
20. The ratio of polymers to siRNA for in vivo bioluminescent studies is based on the gel retardation and particle size measurements.
21. The diluent is frequently 5% dextrose, which has the same osmolality as serum. With HK nanoplexes, we determined that water could be substituted for 5% dextrose.
22. At higher concentrations used for injections in vivo, solubility of polyplexes or lipoplexes may be an issue. For HK nanoplexes, adding HK to the nucleic acids at a lower pH (final pH~5) reduces aggregation markedly. Because as little as 100 μl of serum can neutralize the solution containing the HK nanoplex, it is unlikely that the nanoplex affects the pH of the blood of the mice in vivo.

23. The mice are warmed with a heating lamp for less than 5 min. Care should be taken not to overheat the mouse or the container.
24. Volumes in excess of 300 μl should be avoided except with the hydrodynamic approach, particularly when water is the primary diluent for the nanoplex.
25. These bioluminescence imaging studies can readily be adapted to other tumors such as MDA-MB-231 cells, which stably express luciferase.
26. Modifying H3K(+H)4b with cRGD and PEG markedly enhances the delivery of siRNA to tumors [20]. Recent advances have enabled addition of PEG and ligands to peptide-like oligoamines as part of the solid phase synthesis of the carrier [51] (*see* Review by [52]). Similar to peptide nanoplexes, targeting ligands and PEG augment delivery of lipoplexes to tumors [53].
27. The Xenogen imaging system can be used to determine tumor accumulation of fluorescently labeled peptide nanoplexes at various times [19].
28. Although somewhat controversial [54], multiple injections of D-luciferin may inhibit tumor growth and affect analysis [55], especially in tumors expressing high levels of luciferase. Two injections of D-luciferin did not affect the growth of MDA-MB-435 xenografts expressing luciferase compared to untreated controls.
29. Photon emission is measured within a region of interest using IVIS Image software. The background signal detected in the areas without xenografted cells was similar between the different mice and approximately two to three orders of magnitude lower than that of the tumors.
30. Measurement of luciferase in tumor lysates can substitute for Xenogen imaging system. The advantage of the Xenogen system is that the tumors can be reimaged at various time points in the same mice without euthanasia.
31. Enhanced permeability and retention (EPR) is the primary attribute of nanoparticles for delivery to solid tumors. Deep penetration of the nanoparticles into the tumor has been an issue with this mechanism, and other mechanisms of tumor entry should be sought. In contrast to H3K(+H)4b RNAi nanoplexes which utilize EPR to enter tumors, linear H2K nanoplexes are likely transported throughout the tumor by a non-EPR mechanism (the neuropilin-1 delivery system) [21, 22].
32. Prolonged shaking of samples with Mini-beadbeater-8 (greater than 3 min) warms samples above room temperature, result-

ing in a decrease in luciferase activity. After shaking, samples should be near room temperature.

33. Noise level from the Mini-beadbeater-8 can exceed 85 decibels, resulting in hearing loss. Consequently, the users should wear an earmuffs or ear plugs and/or leave the room during the operation of the Mini-beadbeater-8.
34. It is not clear if tumor weight, tumor protein, or per tumor gives appropriate normalization with systemic injection. This is unlike studies done in tissue culture where most cells are exposed to the transfection agent. In vivo, most nanoparticles administered systemically cannot penetrate deeply into the tumor parenchyma, so only a small percentage of tumor cells are exposed to the nanoparticle. However, if the tumors are of similar size without obvious signs of necrosis, the dilemma of which normalization method to use becomes less important.
35. Therapy should be carried out with a carrier that effectively delivers therapeutic nucleic acid in vitro. This step is important not to determine an effective carrier in vivo but to determine if the siRNA can potently silence its target.
36. Size of tumor, when therapy is initially begun, is dependent on efficacy of therapy. Initial size of tumor may vary between 20 mm^3 and 150 mm^3. The more effective the therapy, the larger the tumor should be initially to evaluate differences among treatment groups. In addition, the rate of growth of the tumor may significantly affect the efficacy of therapy.
37. The size of group is dependent on power analysis of treatment which should be done prior to initiation of the study.
38. Similar experiments can be used to establish subcutaneous tumors with C6 rat glioma xenografts (initial injection, 1×10^6 cells), human breast MDA-MB-231 xenografts (initial injection, 1×10^6 cells), mouse breast 4T1 allografts (initial injection, 0.5×10^6 cells), human lung HL460 xenografts (initial injection, 1×10^6), and mouse N2A neuroblastoma allografts (initial injection, 1×10^6 cells).
39. Peptide: siRaf1 ratios of 4:1 were utilized for H3K(+H)4b in published reports [20]. Lower peptide ratios were later determined to be more effective in targeting Raf-1.
40. siRaf-1 can directly target mitogenic activity of tumor endothelial cells, tumor cells alone, or both. Additionally, siRaf1 can induce apoptosis in many tumor cells such as MDA-MB-435 cells.
41. Alternative schedules can also be effective [13, 19], and frequency may be dependent on the half-life of the nanoplex and the siRNA within the target cell.

Acknowledgments

This study was supported by the National Cancer Institute CA70394.

References

1. Hayashi Y, Yamauchi J, Khalil IA, Kajimoto K, Akita H, Harashima H (2011) Cell penetrating peptide-mediated systemic siRNA delivery to the liver. Int J Pharm 419:308–313
2. Shukla RS, Qin B, Cheng K (2014) Peptides used in the delivery of small noncoding RNA. Mol Pharm 11:3395–3408
3. Subramanian N, Kanwar JR, Kanwar RK, Sreemanthula J, Biswas J, Khetan V, Krishnakumar S (2015) EpCAM aptamer-siRNA chimera targets and regress epithelial cancer. PLoS One 10:e0132407
4. Schiffelers RM, Ansari A, Xu J, Zhou Q, Tang Q, Storm G, Molema G, Lu PY, Scaria PV, Woodle MC (2004) Cancer siRNA therapy by tumor selective delivery with ligand-targeted sterically stabilized nanoparticle. Nucleic Acids Res 32:e149
5. Chou ST, Hom K, Zhang D, Leng Q, Tricoli LJ, Hustedt JM, Lee A, Shapiro MJ, Seog J, Kahn JD, Mixson AJ (2014) Enhanced silencing and stabilization of siRNA polyplexes by histidine-mediated hydrogen bonds. Biomaterials 35:846–855
6. Leng Q, Mixson AJ (2005) Small interfering RNA targeting Raf-1 inhibits tumor growth in vitro and in vivo. Cancer Gene Ther 12:682–690
7. Leng Q, Scaria P, Lu P, Woodle MC, Mixson AJ (2008) Systemic delivery of HK Raf-1 siRNA polyplexes inhibits MDA-MB-435 xenografts. Cancer Gene Ther 15:485–495
8. Lee DJ, Kessel E, Edinger D, He D, Klein PM, Voith von Voithenberg L, Lamb DC, Lachelt U, Lehto T, Wagner E (2016) Dual antitumoral potency of EG5 siRNA nanoplexes armed with cytotoxic bifunctional glutamyl-methotrexate targeting ligand. Biomaterials 77:98–110
9. Bartlett DW, Davis ME (2008) Impact of tumor-specific targeting and dosing schedule on tumor growth inhibition after intravenous administration of siRNA-containing nanoparticles. Biotechnol Bioeng 99:975–985
10. Takeshita F, Minakuchi Y, Nagahara S, Honma K, Sasaki H, Hirai K, Teratani T, Namatame N, Yamamoto Y, Hanai K, Kato T, Sano A, Ochiya T (2005) Efficient delivery of small interfering RNA to bone-metastatic tumors by using atelocollagen in vivo. Proc Natl Acad Sci U S A 102:12177–12182
11. Villares GJ, Zigler M, Wang H, Melnikova VO, Wu H, Friedman R, Leslie MC, Vivas-Mejia PE, Lopez-Berestein G, Sood AK, Bar-Eli M (2008) Targeting melanoma growth and metastasis with systemic delivery of liposome-incorporated protease-activated receptor-1 small interfering RNA. Cancer Res 68:9078–9086
12. Yano J, Hirabayashi K, Nakagawa S, Yamaguchi T, Nogawa M, Kashimori I, Naito H, Kitagawa H, Ishiyama K, Ohgi T, Irimura T (2004) Antitumor activity of small interfering RNA/cationic liposome complex in mouse models of cancer. Clin Cancer Res 10:7721–7726
13. Chen Y, Wu JJ, Huang L (2010) Nanoparticles targeted with NGR motif deliver c-myc siRNA and doxorubicin for anticancer therapy. Mol Ther 18:828–834
14. Davis ME, Zuckerman JE, Choi CH, Seligson D, Tolcher A, Alabi CA, Yen Y, Heidel JD, Ribas A (2010) Evidence of RNAi in humans from systemically administered siRNA via targeted nanoparticles. Nature 464:1067–1070
15. Strumberg D, Schultheis B, Traugott U, Vank C, Santel A, Keil O, Giese K, Kaufmann J, Drevs J (2012) Phase I clinical development of Atu027, a siRNA formulation targeting PKN3 in patients with advanced solid tumors. Int J Clin Pharmacol Ther 50:76–78
16. Alsina M, Tabernero J, Shapiro G, Burris H, Infante JR, Weiss GJ, Cervantes-Ruiperez C, Gounder MM, Paz-Ares L, Falzone R, Hill J, Cehelsky J, Vaishnaw A, Gollob J, LoRusso P (2012) Open-label extension study of the RNAi therapeutic ALN-VSP02 in cancer patients responding to therapy. ASCO annual meeting. American Society of Clinical Oncology, Chicago, IL, 2012
17. Brower V (2010) RNA interference advances to early-stage clinical trials. J Natl Cancer Inst 102:1459–1461
18. Zuckerman JE, Davis ME (2015) Clinical experiences with systemically administered siRNA-based therapeutics in cancer. Nat Rev Drug Discov 14:843–856
19. Chou ST, Leng Q, Scaria P, Kahn JD, Tricoli LJ, Woodle M, Mixson AJ (2013) Surface-

modified HK:siRNA nanoplexes with enhanced pharmacokinetics and tumor growth inhibition. Biomacromolecules 14:752–760

20. Chou ST, Leng Q, Scaria P, Woodle M, Mixson AJ (2011) Selective modification of HK peptides enhances siRNA silencing of tumor targets in vivo. Cancer Gene Ther 18:707–716
21. Leng Q, Mixson AJ (2016) The neuropilin-1 receptor mediates enhanced tumor delivery of H2K polyplexes. J Gene Med 18:134–144
22. Leng Q, Chou ST, Scaria PV, Woodle MC, Mixson AJ (2014) Increased tumor distribution and expression of histidine-rich plasmid polyplexes. J Gene Med 16:317–328
23. Elbashir SM, Harborth J, Lendeckel W, Yalcin A, Weber K, Tuschl T (2001) Duplexes of 21-nucleotide RNAs mediate RNA interference in cultured mammalian cells. Nature 411:494–498
24. Amarzguioui M, Rossi JJ, Kim D (2005) Approaches for chemically synthesized siRNA and vector-mediated RNAi. FEBS Lett 579:5974–5981
25. Hammond SM, Bernstein E, Beach D, Hannon GJ (2000) An RNA-directed nuclease mediates post-transcriptional gene silencing in Drosophila cells. Nature 404:293–296
26. Hammond SM, Boettcher S, Caudy AA, Kobayashi R, Hannon GJ (2001) Argonaute2, a link between genetic and biochemical analyses of RNAi. Science 293:1146–1150
27. Hornung V, Guenthner-Biller M, Bourquin C, Ablasser A, Schlee M, Uematsu S, Noronha A, Manoharan M, Akira S, de Fougerolles A, Endres S, Hartmann G (2005) Sequence-specific potent induction of IFN-alpha by short interfering RNA in plasmacytoid dendritic cells through TLR7. Nat Med 11:263–270
28. Robbins M, Judge A, Liang L, McClintock K, Yaworski E, MacLachlan I (2007) 2′-O-methyl-modified RNAs act as TLR7 antagonists. Mol Ther 15:1663–1669
29. Schlee M, Hornung V, Hartmann G (2006) siRNA and isRNA: two edges of one sword. Mol Ther 14:463–470
30. Hansmann L, Groeger S, von Wulffen W, Bein G, Hackstein H (2008) Human monocytes represent a competitive source of interferon-alpha in peripheral blood. Clin Immunol 127:252–264
31. Sioud M (2005) Induction of inflammatory cytokines and interferon responses by double-stranded and single-stranded siRNAs is sequence-dependent and requires endosomal localization. J Mol Biol 348:1079–1090
32. Leng Q, Chou ST, Scaria PV, Woodle MC, Mixson AJ (2012) Buffering capacity and size of siRNA polyplexes influence cytokine levels. Mol Ther 20:2282–2290
33. Kurreck J (2003) Antisense technologies. Improvement through novel chemical modifications. Eur J Biochem 270:1628–1644
34. Janas MM, Jiang Y, Schlegel MK, Waldron S, Kuchimanchi S, Barros SA (2017) Impact of oligonucleotide structure, chemistry, and delivery method on in vitro cytotoxicity. Nucleic Acid Ther 27:11–22
35. Scheule RK (2000) The role of CpG motifs in immunostimulation and gene therapy. Adv Drug Deliv Rev 44:119–134
36. Yew NS, Zhao H, Przybylska M, Wu IH, Tousignant JD, Scheule RK, Cheng SH (2002) CpG-depleted plasmid DNA vectors with enhanced safety and long-term gene expression in vivo. Mol Ther 5:731–738
37. Lachelt U, Wittmann V, Muller K, Edinger D, Kos P, Hohn M, Wagner E (2014) Synthetic polyglutamylation of dual-functional MTX ligands for enhanced combined cytotoxicity of poly(I:C) nanoplexes. Mol Pharm 11:2631–2639
38. Miyata K (2016) Smart polymeric nanocarriers for small nucleic acid delivery. Drug Discov Ther 10:236–247
39. Muratovska A, Eccles MR (2004) Conjugate for efficient delivery of short interfering RNA (siRNA) into mammalian cells. FEBS Lett 558:63–68
40. Jarver P, Coursindel T, Andaloussi SE, Godfrey C, Wood MJ, Gait MJ (2012) Peptide-mediated cell and in vivo delivery of antisense oligonucleotides and siRNA. Mol Ther Nucleic Acids 1:e27
41. Wagner E, Plank C, Zatloukal K, Cotten M, Birnstiel ML (1992) Influenza virus hemagglutinin HA-2 N-terminal fusogenic peptides augment gene transfer by transferrin-polylysine-DNA complexes: toward a synthetic virus-like gene-transfer vehicle. Proc Natl Acad Sci U S A 89:7934–7938
42. Chen QR, Zhang L, Luther PW, Mixson AJ (2002) Optimal transfection with the HK polymer depends on its degree of branching and the pH of endocytic vesicles. Nucleic Acids Res 30:1338–1345
43. Palm-Apergi C, Lonn P, Dowdy SF (2012) Do cell-penetrating peptides actually "penetrate" cellular membranes? Mol Ther 20:695–697
44. Hirose H, Takeuchi T, Osakada H, Pujals S, Katayama S, Nakase I, Kobayashi S, Haraguchi T, Futaki S (2012) Transient focal membrane deformation induced by arginine-rich peptides leads to their direct penetration into cells. Mol Ther 20:984–993

45. Kim B, Tang Q, Biswas PS, Xu J, Schiffelers RM, Xie FY, Ansari AM, Scaria PV, Woodle MC, Lu P, Rouse BT (2004) Inhibition of ocular angiogenesis by siRNA targeting vascular endothelial growth factor pathway genes: therapeutic strategy for herpetic stromal keratitis. Am J Pathol 165:2177–2185
46. An DS, Qin FX, Auyeung VC, Mao SH, Kung SK, Baltimore D, Chen IS (2006) Optimization and functional effects of stable short hairpin RNA expression in primary human lymphocytes via lentiviral vectors. Mol Ther 14:494–504
47. Cabral H, Matsumoto Y, Mizuno K, Chen Q, Murakami M, Kimura M, Terada Y, Kano MR, Miyazono K, Uesaka M, Nishiyama N, Kataoka K (2011) Accumulation of sub-100 nm polymeric micelles in poorly permeable tumours depends on size. Nat Nanotechnol 6:815–823
48. Talmadge JE, Singh RK, Fidler IJ, Raz A (2007) Murine models to evaluate novel and conventional therapeutic strategies for cancer. Am J Pathol 170:793–804
49. Caysa H, Hoffmann S, Luetzkendorf J, Mueller LP, Unverzagt S, Mader K, Mueller T (2012) Monitoring of xenograft tumor growth and response to chemotherapy by non-invasive in vivo multispectral fluorescence imaging. PLoS One 7:e47927
50. Baklaushev VP, Grinenko NF, Yusubalieva GM, Abakumov MA, Gubskii IL, Cherepanov SA, Kashparov IA, Burenkov MS, Rabinovich EZ, Ivanova NV, Antonova OM, Chekhonin VP (2015) Modeling and integral X-ray, optical, and MRI visualization of multiorgan metastases of orthotopic 4T1 breast carcinoma in BALB/c mice. Bull Exp Biol Med 158:581–588
51. Martin I, Dohmen C, Mas-Moruno C, Troiber C, Kos P, Schaffert D, Lachelt U, Teixido M, Gunther M, Kessler H, Giralt E, Wagner E (2012) Solid-phase-assisted synthesis of targeting peptide-PEG-oligo(ethane amino)amides for receptor-mediated gene delivery. Org Biomol Chem 10:3258–3268
52. Krhac Levacic A, Morys S, Wagner E (2017) Solid-phase supported design of carriers for therapeutic nucleic acid delivery. Biosci Rep 37:BSR20160617
53. Leng Q, Woodle MC, Mixson AJ (2017) Targeted delivery of siRNA therapeutics to malignant tumors. J Drug Deliv 2017:6971297
54. Tiffen JC, Bailey CG, Ng C, Rasko JE, Holst J (2010) Luciferase expression and bioluminescence does not affect tumor cell growth in vitro or in vivo. Mol Cancer 9:299
55. Brutkiewicz S, Mendonca M, Stantz K, Comerford K, Bigsby R, Hutchins G, Goebl M, Harrington M (2007) The expression level of luciferase within tumour cells can alter tumour growth upon in vivo bioluminescence imaging. Luminescence 22:221–228

Chapter 13

Preparation of a Carrier to Achieve In Vivo Delivery of siRNA: The Example of Chitosan-Coated Poly(isobutylcyanoacrylate) Nanoparticles

Christine Vauthier

Abstract

This chapter describes the preparation of chitosan-coated poly(isobutylcyanoacrylate) nanoparticles as a suitable carrier to deliver siRNAs to two types of xenograft tumor models of mice. The nanoparticles are prepared by a method of emulsion polymerization that includes steps of polymerization and purification. The polymerization method is carried out in a single pot in an aqueous medium. siRNAs are coupled with the nanoparticles at the end of the preparation by adsorption. The protocol also explains how to determine optimum yield/the titer of association of siRNA with the nanoparticles. It is described for a preparation scale at 4 mL of nanoparticle dispersion at a concentration of 42–46 mg nanoparticles/mL. Optimal loading capacity of the nanoparticles with the siRNA can be achieved by performing an association yield above 90% using a mass ratio of 1 mg siRNA/50 mg of nanoparticles (20 μg siRNA/mg nanoparticles, 1 nmol siRNA (M_w 14 kDa)/mg nanoparticles).

Key words Chitosan, Delivery system, Emulsion polymerization, Poly(alkylcyanoacrylate) nanoparticle, siRNA

1 Introduction

Many types of carriers have been developed to achieve in vivo delivery of siRNAs [1–11]. Several carriers have shown encouraging results delivering antitumoral siRNAs to xenograft tumor models where the formulation was administered by the intravenous route. One of those carriers consisted of poly(isobutylcyanoacrylate) (PIBCA) nanoparticles decorated with a chitosan-containing corona. It was applied to the delivery of two anticancer siRNAs targeted against junction oncogenes including the RET/PTC 1 found in papilloma thyroid carcinoma [12, 13] and the EWS/Fli 1 junction oncogenes found in Ewing sarcoma [14]. After in vivo delivery through an intravenous injection in mice bearing the corresponding xenografted tumor model, a reduction of the tumor growth volume was reported compared with controls in experiments

Lekha Dinesh Kumar (ed.), *RNA Interference and Cancer Therapy: Methods and Protocols*, Methods in Molecular Biology, vol. 1974, https://doi.org/10.1007/978-1-4939-9220-1_13,

considering papilloma thyroid sarcoma [13]. The antitumor activity of the siRNA directed towards the junction oncogene EWS/Fli 1 appeared as a reduction of the expression level of the mRNA and the protein coded by the corresponding junction oncogene [14]. The nanoparticle being the carrier of both siRNAs was prepared by emulsion polymerization of isobutylcyanoacrylate. The method includes a one-pot process for the synthesis of the nanoparticles which is performed in an aqueous environment. After synthesis, the nanoparticles are purified, and siRNAs are associated by surface adsorption on the preformed purified nanoparticles. The colloidal stability of the nanoparticles is ensured using a blend of poloxamer F68 and chitosan that are grafted on the nanoparticle surface during the step of the synthesis of the nanoparticles [15]. Chitosan is a key component of the formulation of the carrier. It is required to promote association of siRNA with the nanoparticles by surface adsorption. This component is very often used as an ingredient to promote association of nucleic acids with drug carriers [1, 5, 10, 16–26]. Association of siRNA to carriers using chitosan is achieved through multivalent interactions due to the polycationic character of the polysaccharide and the polyanionic character of the nucleic acid. The polyelectrolyte complexes that are formed between the nucleic acids and chitosan are generally very stable compared to those considered in the early developments based on interactions occurring on mono-cationic compounds [27–31].

The aim of the present protocol was to describe the preparation of siRNA-loaded PIBCA nanoparticles. The protocol was divided in four parts: (1) preparation of the nanoparticles, (2) purification of the nanoparticles, (3) association of siRNA with the nanoparticles, and (4) optimization of the yield of association of siRNA with the nanoparticles. The produced nanoparticles are expected to display a hydrodynamic diameter ranging from 60 nm to 75 nm and an apparent zeta potential evaluated in NaCl 1 mM around +26 mV [13–15]. Size was not affected by association of siRNA although zeta potential decreased to reach negative values (−17 mV [13]). A yield of association of siRNA above 90% is expected, while association is performed in experimental conditions considering a siRNA/nanoparticle mass ratio of 20 μg siRNA/mg nanoparticles (1 nmol siRNA (M_w 14 kDa)/mg nanoparticles).

2 Materials

2.1 Preparation and Purification of the Nanoparticles

1. Argon gas or nitrogen gas (*see* **Note 1**).
2. Cerium IV ammonium nitrate: 80 mM in nitric acid 0.2 M (*see* **Note 2**).
3. Chitosan 20 kDa (*see* **Note 3**).
4. Isobutylcyanoacrylate (*see* **Notes 4–6**).

5. Milli-Q® water (*see* **Note 7**).
6. NaCl: 1 mM (*see* **Note 8**).
7. Nitric acid: 0.2 M (*see* **Note 9**).
8. Poloxamer 188 (*see* **Note 10**).
9. Polymerization medium composed of 1.37% chitosan 20 kDa, 3% poloxamer 188 and 16 mM cerium (Ce IV) ammonium nitrate, and 5% isobutylcyanoacrylate.
10. Magnetic plate equipped with heat control.
11. Graduated glass pipettes (2 and 10 mL) (*see* **Note 11**).
12. Teflon®-covered magnetic bars (cylindrical shape, 15 × 4.5 mm) (*see* **Note 12**).
13. Flexible tubing (*see* **Note 13**).
14. Pasteur pipettes (*see* **Note 14**).
15. Vials 20 mL (*see* Fig. 1) (*see* **Note 15**).
16. Volumetric flasks (*see* **Note 16**).

2.2 Association of siRNA with the Nanoparticles

1. Nanoparticles prepared according to Subheading 3.1.
2. HEPES (4-(2-hydroxyethyl)-1-piperazineethanesulfonic acid) (10 mM) buffer pH 7.3, 100 mM NaCl (*see* **Note 17**).
3. RNase-free water.
4. Small interfering RNA (siRNA) (*see* **Note 18**).
5. Zeta sizer (*see* **Note 19**).
6. RNase-free microtubes.
7. RNase-free tips for the automatic pipettes.

2.3 Optimization of the Yield of Association of siRNA with the Nanoparticles

1. Nanoparticles prepared according to Subheading 3.1.
2. HEPES (4-(2-hydroxyethyl)-1-piperazineethanesulfonic acid) buffer (10 mM) pH 7.3, 100 mM NaCl (*see* **Note 17**).
3. RNase-free water.
4. Small interfering RNA (siRNA) (*see* **Note 18**).
5. Centrifuged for microtubes.
6. Laboratory for the manipulation of radiolabeled compounds (^{32}P- or ^{33}P-labelled siRNA).
7. Precision balance.
8. Radioactivity counter.
9. Ultracentrifuged + consumable (tubes).
10. UV-Vis spectrophotometer working on small size samples (drop size samples).
11. RNase-free microtubes.
12. RNase-free tips for the automatic pipettes.

3 Methods

3.1 Preparation of the Nanoparticles

The protocol describes the preparation of 5 mL of nanoparticle dispersion at a final concentration of 42–46 mg nanoparticles/mL (*see* **Note 20**).

1. Preparation of the setup for the polymerization including a water bath at 42 °C.

   The setup for the polymerization includes the magnetic plate equipped with a temperature-controlling device, a water bath, and a static bar to which the vial serving as very simple reactor to carry on the polymerization will be attached. This setting needs to be mounted near a source of argon or nitrogen gas. A scheme of the setting is shown on Fig. 1.

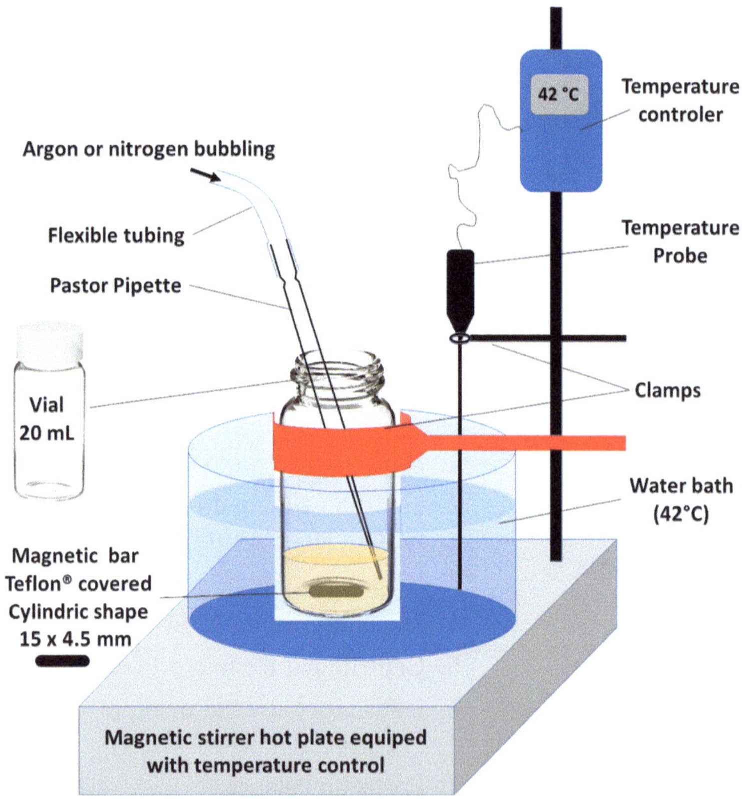

Fig. 1 Scheme of an example of simple setting that can be used to prepare chitosan-coated PIBCA nanoparticles by emulsion polymerization. The polymerization medium is introduced in the vial to equilibrate at 42 °C. It is then purged with nitrogen or argon gas for 10 min prior to the start of the polymerization by sequential addition of the initiator (cerium (IV) ammonium nitrate solution) and the monomer (IBCA)

2. Preparation of nitric acid 0.2 M from commercial solution: A solution of nitric acid at 0.2 M is prepared from the commercial solution of concentrated nitric acid by dilution in Milli-Q® water using a volumetric flask (50 mL or 200 mL). The minimum amount required to perform the protocol is 6 mL including the void volume.
3. Preparation of cerium (IV) ammonium nitrate solution: Cerium (IV) ammonium nitrate is dissolved in nitric acid 0.2 N to obtain a solution at 80 mM. The amount needed to perform the protocol is 1.5 mL including the void volume.
4. Prepare the polymerization medium: Weigh 0.0688 g chitosan 20 kDa and 0.15 g Poloxamer 188, and introduce these compounds in the 20 mL vial. Then, dissolve chitosan and poloxamer in 4 mL nitric acid 0.2 M, and place the polymerization medium in the water bath for equilibration at the temperature of 42 °C. Start of the polymerization.
   (a) When the polymerization medium has reached the temperature of 42 °C, turn on argon or nitrogen bubbling for 10 min to eliminate oxygen from the solution (*see* **Note 1**).
   (b) Prepare a graduated glass pipette of 1 mL; it will be used to add cerium (IV) ammonium nitrate (80 mM in 0.2 N nitric acid) in the polymerization medium.
   (c) Prepare an automatic pipette to sample 0.250 mL of the monomer from the stock container (*see* **Notes 4–6**).
   (d) After 10 min bubbling, adjust agitation of the polymerization medium to the maximum speed to obtain the formation of a vortex. Then add successively 1 mL of the cerium ammonium nitrate solution and 0.25 mL of IBCA in the polymerization medium. Continue bubbling for 5 min, and check that the agitation remains homogeneous (*see* **Note 21**). **Note 22** indicates how to label the nanoparticles with a fluorescent probe.
   (e) After 10 min, remove argon or nitrogen bubbling and close the vial. Let the polymerization to continue under vigorous agitation for an additional 50 min.
5. Recover nanoparticle dispersion at the end of the polymerization.
   (a) Prepare an ice bath.
   (b) Sixty minutes after addition of the monomer, stop the agitation, and pour the vial in an ice bath for a few minutes to reduce the temperature of the sample down to 12–15 °C.
   (c) At this step, the synthesis is competed. The sample occurred as a milky dispersion. It is generally purified right after the synthesis.

3.2 Purification of the Nanoparticles

1. Place the nanoparticle dispersion sample in the dialysis bag. Check for tight closure of the clips.
2. Immerse the dialysis bag containing the sample in a large volume (500 mL) of Milli-Q® water (*see* **Notes 23** and **24**).
3. Place the dialysis on a magnetic plate with a gentle agitation.
4. Change the counter dialysis media after 30 min, 1 h, 2 h, and 3 h, and allow the dialysis to proceed overnight (*see* **Note 25**).
5. The next day, collect the purified nanoparticle dispersion from the inside of the dialyzing bag.
6. Determine the concentration of nanoparticles in the dispersion (*see* **Note 26**).
7. Purified nanoparticle dispersions can be stored at +4 °C until use. In general, nanoparticles are prepared and purified a few days before association of the siRNA to be exposed to cells or to be used in animal studies.

3.3 Association of the siRNA

1. Adsorb siRNA on the nanoparticle surface (*see* **Note 27**).
   (a) Transfer nanoparticles into an HEPES buffer. This can be done by dialysis, dialyzing the nanoparticle dispersion against the HEPES buffer. Alternatively, this transfer can also be done by adding 1/10 volume of a ten times concentrated HEPES buffer in 9/10 volumes of the nanoparticle dispersion. Note that the initial dispersion of nanoparticles is diluted by 9/10 with the last method.
   (b) Prepare a solution of siRNA at 5 μM in HEPES buffer (10 mM HEPES pH 7.3, 100 mM NaCl).
   (c) Adjust nanoparticle dispersion to the required concentration by diluting the parent dispersion with HEPES buffer.
   (d) Add the required amount of the siRNA solution to nanoparticle dispersion to obtain the optimal nanoparticle/siRNA mass ratio, and complete the volume to 300 μL with the HEPES buffer (*see* **Note 28**).
   (e) The tubes are let to incubate for 10 min at ambient temperature before use.
2. Evaluate association of siRNA with the nanoparticles.
   (a) Dose the amount of the remaining free siRNA after separation of the nanoparticles by ultracentrifugation (average g-force 138,915 × *g*, 4 °C, 1 h). This determination can be done on the supernatant recovered after the centrifugation of the dispersion using a small sample size UV-visible spectrophotometer or by measuring the radioactivity in the supernatant in case of the use of radiolabeled siRNA (*see* **Note 18**).

(b) Evaluate zeta potential of the siRNA-loaded nanoparticles, and compare value with that of the parent nanoparticles (*see* **Notes 19** and **29**).

3.4 Optimization of the Yield of Association of siRNA with the Nanoparticle Dispersion

1. Prepare HEPES buffer.
2. Transfer nanoparticles into an HEPES buffer (10 mM HEPES pH 7.3, 100 mM NaCl). This can be done by dialysis; dialyze the nanoparticle dispersion against the HEPES buffer. Alternatively, this transfer can also be done by adding 1/10 volume of a ten times concentrated HEPES buffer in 9/10 volumes of the nanoparticle dispersion. Note that the initial dispersion of nanoparticles is diluted by 9/10 with the last method.
3. Prepare 5′-end radiolabeled siRNA (*see* **Note 18**).
4. Prepare a solution of radiolabeled siRNA at 5 μM in HEPES buffer.
5. Prepare a series of Eppendorf tubes (RNase-free and low bind) with dilutions of nanoparticle dispersions at different concentrations to reach nanoparticles/siRNA mass ratios ranging from 1/0 to 300/1 in a final volume of 300 μL. The dilution is performed with HEPES buffer. An example of series includes nanoparticle/siRNA mass ratios: 1/0, 10/1, 20/1, 25/1, 30/1, 40/1, 45/1, 50/1, 100/1, 200/1, and 300/1.
6. Add 30 μL of the siRNA solution to each Eppendorf tube, and complete volume to 300 μL with HEPES buffer.
7. Incubate samples for 10 min at room temperature.
8. After incubation, remove nanoparticles from the sample by centrifugation (average g-force 138,915 × *g*, 4 °C, 1 h).
9. Dose the amount of remaining free siRNA in the supernatant. This determination is done by measuring the radioactivity in the supernatant using radiolabeled siRNA (*see* **Note 18**).
10. Data from the dosage are used to draw the curve given the percentage of siRNA associated with the nanoparticles as the function of the nanoparticle/siRNA mass ratio. The percentage of siRNA associated with the nanoparticles is deduced from the difference between the amount of siRNA added in the sample and the amount of siRNA remaining free after incubation. Examples of curves are given in Fig. 2. Conditions to obtain optimal yield of siRNA loading on nanoparticles can be determined considering the minimal value of the nanoparticle/siRNA mass ratio reaching the plateau value giving the maximal yield of association of the siRNA (% associated siRNA). The plateau value is expected to occur for a percentage of associated siRNA above 90% (*see* Fig. 2).

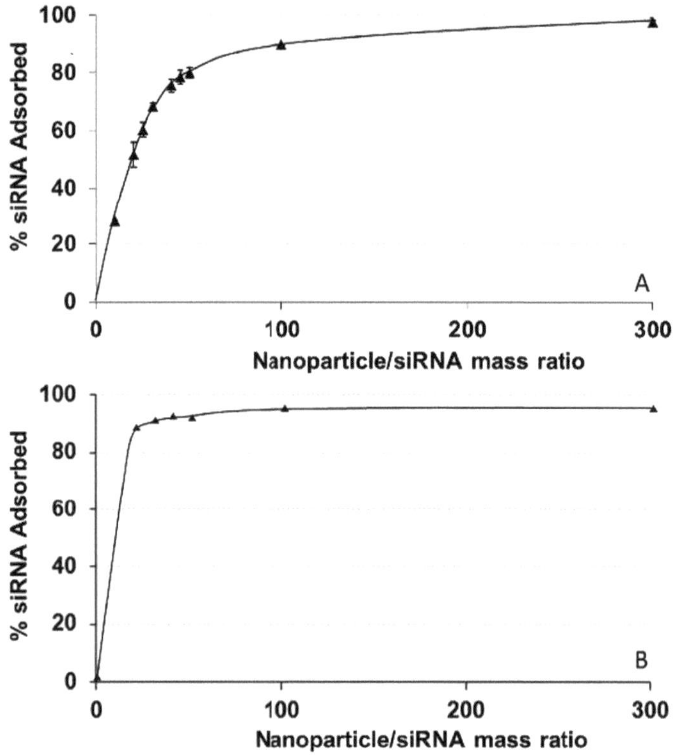

Fig. 2 Example of graphs given the percentage of association of the siRNA on nanoparticles evaluated at various nanoparticle/siRNA mass ratio. Conditions to obtain optimal yield of siRNA loading was determined at a mass ratio of 50 that corresponded to the minimal nanoparticle/siRNA mass ratio of the plateau indicating the maximal value of the percentage of siRNA that can associate with the nanoparticles. (**a**) Association of the RET/PTC 1 siRNA with the nanoparticles performed in water (adapted from [13], reproduced with permission of Springer), (**b**) association of the EWS/Fli 1 siRNA with the nanoparticles performed in HEPES/NaCl buffer (adapted from [14], reproduced with permission of Wiley), as described in the protocol. Association performed in water (**a**) was slightly less efficient than performed in HEPES buffer, 100 mM NaCl (**b**)

4 Notes

1. Argon gas or nitrogen gas is used to purge polymerization medium from oxygen that is a radical scavenger preventing initiation of radical polymerization.
2. Cerium IV ammonium nitrate is the initiator of the polymerization.
3. Chitosan 20 kDa is needed to associate siRNA to the nanoparticles and to ensure colloidal stability of the nanoparticle dispersion. The choice for the molar mass of chitosan is a critical parameter. Using chitosan with higher molar mass will produce larger nanoparticles [32]. Chitosan 20 kDa can be

purchased from few providers, but it is a good idea to check the molar mass (M_w) before use. This can be done by measuring the intrinsic viscosity, $[\eta]$, of the polysaccharide in acetic acid 0.1 M, NaCl 0.2 M at 25 °C. Using these measuring conditions, coefficients of the Mark-Houwink-Sakurada relationship ($[\eta] = K \cdot M_w^{\alpha}$) for chitosan are $K = 1.81 \times 10^{-3}$ and $\alpha = 0.93$ [33]. Alternatively, chitosan 20 kDa can be obtained from chitosan of higher molar mass through a chemical hydrolysis performed with sodium nitrite ($NaNO_2$). Starting from chitosan with medium viscosity, experimental conditions that can be used to perform the hydrolysis are 2 g of chitosan and 0.09–0.11 g $NaNO_2$ in 100 mL 6% acetic acid for 1 h [14, 15, 32]. The $NaNO_2$ concentration needs be tuned depending on the batch of chitosan. After hydrolysis, chitosan is purified by precipitation raising the pH to 9 with sodium hydroxide and filtration or centrifugation to collect the solid part and washing of the precipitate. Chitosan may be further purified by dialysis after dissolution in slightly acid solution of acetic acid. It can be stored at +4 °C until use after lyophilization.

4. Isobutylcyanoacrylate serves as monomer. It is a colorless liquid with a low viscosity. As far as it becomes viscous, it means that it has started polymerizing. In this case it cannot be used further.
5. This monomer as all cyanoacrylate monomers is highly reactive. They are stored at +4 °C until use in a tightly closed container. The stock bottle should be open a minimum of time as water can initiate the polymerization of this monomer. Sampling the monomer should be made very quickly. Take out of the stock a small excess of monomer (i.e., 300 μL), and transfer in an Eppendorf® tube. Close rapidly the stock monomer container and put it again in the refrigerator. This sampling should be done right before the start of the polymerization.
6. Isobutylcyanoacrylate can be replaced by other alkylcyanoacrylate monomers including isohexylcyanoacrylate or *n*-butylcyanoacrylate. It can also be replaced by methyl methacrylate. Note that nanoparticles prepared with methyl methacrylate are not biodegradable; hence they are unsuitable for in vivo delivery of siRNA and more generally of drugs.
7. Milli-Q® water is used to prepare all solutions.
8. NaCl solution at 1 mM is needed for zeta potential evaluation.
9. Nitric acid 0.2 M is prepared from concentrated commercial solution and used to prepare the polymerization medium and the polymerization initiator solution.
10. Poloxamer 188 can be found under various commercial names, Pluronic® F68 and Lutrol® F68. This nonionic surfactant

achieves the colloidal stability of the nanoparticle dispersion together with chitosan.

11. Graduated glass pipettes (1 or 2 mL) are used for sampling 0.2 M nitric acid.
12. Teflon®-covered magnetic bars are used to homogenize the polymerization medium during polymerization.
13. Flexible tubing is needed to carry the gas from the stock container to the polymerization medium.
14. Pasteur pipettes: placed at the free end of the flexible tubing to deliver the gas in the polymerization medium.
15. Vials 20 mL serve as very simple polymerization reactor.
16. Volumetric flasks: preparation of various solutions, nitric acid, buffers, and cerium ammonium nitrate.
17. HEPES buffer is used for the association of siRNA with nanoparticles and needed for in vivo administration of the dispersion.
18. Small interfering RNA (siRNA): optimization of conditions for the association of siRNA with the nanoparticles is easier to perform using radiolabeled siRNA. Obtaining radiolabeled siRNA is a very well-described method using T4 polynucleotide kinase and [γ-^{33}P or ^{32}P] ATP. During labelling, the radioactive [γ-^{33}P or ^{32}P] of ATP is transferred on the 5′-end of one strand of the siRNA. The labelled siRNA is then purified by gel filtration through a desalting column to remove all non-attached radioactive phosphate containing molecules and chemicals needed for the labelling. The protocol for the labelling is generally provided with T4-polynucleotide kinase. An example of protocol is summarized here. Prepare a stock solution of siRNA at 20 μM. Take 5 μL of this solution, and place in a RNase-free Eppendorf tube. Add 30 μL of RNase-free water, 5 μL of a ten times kinase buffer, and 5 μL of T4 polynucleotide kinase at a concentration of 10 U/μL. Then add 5 μL of gamma ATP at a specific radioactivity above 5000 Ci/mmol. The reaction media is then incubated at 37 °C for 1 h and 20 min at 65 °C. The labelled siRNA is purified using MicroCon YM3 (MWCO 3000, from Millipore). The sample to purify is placed in the upper chamber of the MicroCon YM3. Add 200 μL of RNase-free water and centrifuge at 4000 × *g*. The centrifugation is repeated ten times after adding 200 μL RNase-free water between each centrifugation. At the end of the procedure, the ^{33}P or ^{32}P 5′-end labelled siRNA can be recovered in the upper chamber of the purification device to achieve the dosage by determination of the radioactivity.
19. Zeta sizer is an instrument that can be used to evaluate the zeta potential of the nanoparticles. After association of siRNA,

the zeta potential of the nanoparticles is expected to decrease compared to zeta potential shown by the plain nanoparticles. This instrument can be used to verify the modification of the nanoparticle surface charge attributed to the adsorption of siRNA. Prior to the use of this approach, characterization of association of siRNA with nanoparticles must be performed by a more direct method of dosage based on the use of radioactive labelled siRNA (*see* **Note 20**).

20. Nanoparticle dispersions can be prepared up to a volume of 50 mL using the described protocol which can be scaled up by multiplying the amount of chemicals and adjusting the volume of the vessels. To produce a larger volume of nanoparticle dispersion, it would be required to change the setup.
21. This step is critical for the success of the synthesis. It is very important that the agitation remains homogenous, with the vortex formed and continuously for the duration of the polymerization once the monomer is added. It is also important to be fast adding successively the cerium IV ammonium nitrate solution and the monomer (IBCA). The order of addition of these two ingredients into the polymerization medium matters for the success of the synthesis, so the cerium IV ammonium nitrate solution must be added first and the monomer second.
22. Fluorescently labelled nanoparticles can be obtained by adding 1 mL of a solution of PolyFluor® 570 (4 mg/mL in acetonitrile) 2 min after the start of the polymerization. This compound that is methacryloxyethyl thiocarbamoyl rhodamine B copolymerizes with the IBCA labelling the core of the nanoparticles with rhodamine [14, 34].
23. The purification is generally performed right after preparation by dialysis against Milli-Q® water using a Spectra Pro® dialysis membrane at a MWCO 100,000. The dialysis has two roles: to raise the pH of the dispersion close to neutrality and to remove cerium ions and excess of reagents that may have not reacted during polymerization. You may use large Becher or graduated cylinders.
24. Milli-Q® water will serve here as counter dialyzing medium. In case aggregates form during dialysis, you can replace Milli-Q® water by slightly acid Milli-Q® water prepared by adding 1 mL of acetic acid 16 mM in 1 L of Milli-Q® water.
25. Renewal of the first and second dialysis needs to be performed after a short duration (30 min for each). This is to preserve the dialyzing membrane from damage due to the acid sample (0.2 M nitric acid). During dialysis, the pH is raised to neutral pH.
26. Concentration of nanoparticles in the dispersion can be determined by gravimetry. A sample of the dispersion (for instance,

100 μL) is weighted prior and after lyophilization. The concentration is then calculated from the ratio between the weight of the solid content deduced from the difference of the weighted mass of the sample and the weight of sample prior lyophilization. The determination is generally performed in triplicate.

27. Association of siRNA with the nanoparticles is performed by adsorption on the nanoparticle surface thanks to interactions with chitosan. The protocol is based on a simple incubation of the nanoparticles under gentle agitation with siRNA. A protocol to optimize the yield of association of siRNA with nanoparticles is described in subheading 3.4.
28. Optimal nanoparticle/siRNA mass ratio was 1/50 in the work of [13, 14]. With the two siRNAs tested, the balance was the same. However, it may depend on the sequence of the siRNA to be associated with the nanoparticles, and it is recommended to optimize this balance for each siRNA.
29. Association of siRNA with chitosan-coated nanoparticles induced a reduction of the zeta potential of the nanoparticles. Zeta potential of the nanoparticles before adsorption of siRNA should be frankly positive (+25 mV to +30 mV evaluated in 0.1 mM NaCl), while after adsorption of siRNA, it can even become negative (−17 mV evaluated in 0.1 mM NaCl) [13].

References

1. Mao S, Sun W, Kissel T (2010) Chitosan-based formulations for delivery of DNA and siRNA. Adv Drug Deliv Rev 62:12–27
2. Müller LK, Landfester K (2015) Natural liposomes and synthetic polymeric structures for biomedical applications. Biochem Biophys Res Commun 468:411–418. https://doi.org/10.1016/j.bbrc.2015.08.088
3. Navarro G, Pan J, Torchilin VP (2015) Micelle-like nanoparticles as carriers for DNA and siRNA. Mol Pharm 12:301–113. https://doi.org/10.1021/mp5007213
4. Park J, Park J, Pei Y, Xu J, Yeo Y (2016) Pharmacokinetics and biodistribution of recently-developed siRNA nanomedicines. Adv Drug Deliv Rev 104:93–109. https://doi.org/10.1016/j.addr.2015.12.004
5. Ragelle H, Vandermeulen G, Preat V (2013) Chitosan-based siRNA delivery systems. J Control Release 172:207–218
6. Reinhard S, Wagner E (2017) How to tackle the challenge of siRNA delivery with sequence-defined oligoamino amides. Macromol Biosci. https://doi.org/10.1002/mabi.201600152
7. Scomparin A, Polyak D, Krivitsky A, Satchi-Fainaro R (2015) Achieving successful delivery of oligonucleotides—from physico-chemical characterization to in vivo evaluation. Biotechnol Adv 33:1294–1309. https://doi.org/10.1016/j.biotechadv.2015.04.008
8. Shajari N, Mansoori B, Davudian S, Mohammadi A, Baradaran B (2017) Overcoming the challenges of sirna delivery: nanoparticle strategies. Curr Drug Deliv 14:36–46. https://doi.org/10.2174/1567201813666160816105408
9. Singh Y, Tomar S, Khan S, Meher JG, Pawar VK, Raval K, Sharma K, Singh PK, Chaurasia M, Surendar Reddy B, Chourasia MK (2015) Bridging small interfering RNA with giant therapeutic outcomes using nanometric liposomes. J Control Release 220:368–387. https://doi.org/10.1016/j.jconrel.2015.10.050
10. Vauthier C, Zandanel C, Ramon AL (2013) Chitosan-based nanoparticles for in vivo delivery of interfering agents including siRNA. Curr Opin Colloid Interface Sci 18:406–418. https://doi.org/10.1016/j.cocis.2013.06.005
11. Xu X, Li Z, Zhao X, Keen L, Kong X (2016) Calcium phosphate nanoparticles-based systems for siRNA delivery. Regen Biomater 3:187–195. https://doi.org/10.1093/rb/rbw010

12. De Martimprey H, Bertrand JR, Fusco A, Santoro M, Couvreur P, Vauthier C, Malvy C (2008) siRNA nanomedicine against the Ret/PTC1 junction oncogene is efficient in papillary thyroid carcinoma. Nucleic Acid Res 36:e2. https://doi.org/10.1093/nar/gkm1094. Erratum: Nucleic Acids Res (2008) 36: 6944
13. De Martimprey H, Bertrand JR, Malvy C, Couvreur P, Vauthier C (2010) New core-shell nanoparticles for the intravenous delivery of siRNA to experimental thyroid papillary carcinoma. Pharm Res 27:498–509. https://doi.org/10.1007/s11095-009-0043-8
14. Ramon AL, Bertrand JR, De Martimprey H, Bernard G, Ponchel G, Malvy C, Vauthier C (2013) SiRNA associated with immunonanoparticles directed against cd99 antigen improve gene expression inhibition in vivo in Ewing's sarcoma. J Mol Recogn 26:318–329. https://doi.org/10.1002/jmr.2276
15. Zandanel C, Vauthier C (2012) Poly(isobutylcyanoacrylate) nanoparticles decorated with chitosan: effect of conformation of chitosan chains at the surface on complement activation properties. J Coll Sci Biotechnol 1:68–81. https://doi.org/10.1166/jcsb.2012.1004
16. Karunaratne DN, Jafari M, Ranatunga RJ, Siriwardhana A (2015) Natural carriers for siRNA delivery. Curr Pharm Des 21: 4529–4540
17. Kastas H, Ghafoor Raja MA, Lam KL (2013) Development of chitosan nanoparticles as a stable drug delivery system for protein/siRNA. Int J Biomater:Article ID:146320. https://doi.org/10.1155/2013/146320
18. Kaur M, Sethi S, Bhatia A (2016) Chitosan nanoparticles: a therapeutic carrier for delivery of DNA, siRNA and CpG-ODNs. Nanosci Nanotechnol Asia 6:92–104. https://doi.org/10.2174/2210681206666160402004454
19. Lallana E, Rios de la Rosa JM, Tirella A, Pelliccia M, Gennari A, Stratford IJ, Puri S, Ashford M, Tirelli N (2017) Chitosan/hyaluronic acid nanoparticles: rational design revisited for RNA delivery. Mol Pharm 14:2422–2436. https://doi.org/10.1021/acs.molpharmaceut.7b00320
20. Martirosyan A, Olesen MJ, Howard KA (2014) Chitosan-based nanoparticles for mucosal delivery of RNAi therapeutics. Adv Genet 88:325–352. https://doi.org/10.1016/B978-0-12-800148-6.00011-0
21. Mokhtarzadeh A, Alibakhshi A, Hashemi M, Hejazi M, Hosseini V, de la Guardia M, Ramezani M (2017) Biodegradable nanopolymers as delivery vehicles for therapeutic small non-coding ribonucleic acids. J Control Release 245:116–126. https://doi.org/10.1016/j.jconrel.2016.11.017
22. Nascimento AV, Gattacceca F, Singh A, Bousbaa H, Ferreira D, Sarmento B, Amiji MM (2016) Biodistribution and pharmacokinetics of Mad2 siRNA-loaded EGFR-targeted chitosan nanoparticles in cisplatin sensitive and resistant lung cancer models. Nanomedicine. https://doi.org/10.2217/nnm.16.14
23. Raemdonck K, Martens TF, Braeckmans K, Demeester J, De Smedt SC (2013) Polysaccharide-based nucleic acid nanoformulations. Adv Drug Deliv Rev 65:1123–1147
24. Ragelle H, Riva R, Vandermeulen G, Naeye B, Pourcelle V, Le Duff CS, D'Haese C, Nysten B, Braeckmans K, De Smedt SC, Jerome C, Preat V (2014) Chitosan nanoparticles for siRNA delivery: optimizing formulation to increase stability and efficiency. J Control Release 176:54–63
25. Tavakoli Naeini A, Soliman OY, Alameh MG, Lavertu M, Buschmann MD (2017) Automated in-line mixing system for large scale production of chitosan-based polyplexes. J Colloid Interface Sci 500:253–263. https://doi.org/10.1016/j.jcis.2017.04.013
26. Yu S, Chen Y, Li X, Gao Z, Liu G (2017) Chitosan nanoparticle-delivered siRNA reduces CXCR4 expression and sensitizes breast cancer cells to cisplatin. Biosci Rep 37:BSR20170122. https://doi.org/10.1042/BSR20170122
27. Chavany C, Le Doan T, Couvreur P, Puisieux F, Hélène C (1992) Poly-alkylcyanoacrylate nanoparticles as polymeric carriers for antisense oligonucleotides. Pharm Res 9:441–449
28. Chavany C, Saison-Behmoaras T, Le Doan T, Couvreur P, Hélène C (1994) Adsorption of oligonucleotides onto polyisohexylcyanoacrylate nanoparticles protects them against nucleases and increases their cellular uptake. Pharm Res 11:1370–1378
29. Nakada Y, Fattal E, Foulquier M, Couvreur P (1996) Pharmacokinetics and biodistribution of oligonucleotide adsorbed onto poly(isobutylcyanoacrylate) nanoparticles after intravenous administration in mice. Pharm Res 13:38–43
30. Zimmer A (1999) Antisense oligonucleotide delivery with polyhexylcyanoacrylate nanoparticles as carriers. Methods 18(286–295):322
31. Zobel HP, Kreuter J, Werner D, Noe CR, Kümel G, Zimmer A (1997) Cationic polyhexylcyanoacrylate nanoparticles as carriers for antisense oligonucleotides. Antisense Nucleic Acid Drug Dev 7:483–493
32. Bertholon I, Vauthier C, Labarre D (2006) Complement activation by core-shell

poly(isobutylcyanoacrylate)-polysaccharide nanoparticles: influences of surface morphology, length, and type of polysaccharide. Pharm Res 23:1313–1323
33. Peh K, Khan T, Ch'ng H (2000) Mechanical, bioadhesive strength and biological evaluations of chitosan films for wound dressing. J Pharm Pharm Sci 3:303–311
34. Zandanel C, Vauthier C (2012) Characterization of fluorescent poly(isobutylcyanoacrylate) nanoparticles obtained by copolymerization of a fluorescent probe during redox radical emulsion polymerization (RREP). Eur J Pharm Biopharm 82:66–75. https://doi.org/10.1016/j.ejpb.2012.05.002

Chapter 14

Monoclonal Antibody-Conjugated Dendritic Nanostructures for siRNA Delivery

Hitesh Kulhari, Ashok K. Jangid, and David J. Adams

Abstract

Small interfering RNA (siRNA) is a promising tool for gene therapy-based disease treatments. However, delivery of siRNA to the target cells requires a specific and reliable carrier system. Herein we describe a targeted carrier system that can deliver siRNA to cancer cells overexpressing the human epidermal growth factor 2 (HER2) receptor. Trastuzumab-conjugated poly(amido)amine dendrimers can be synthesized using the protocols described here.

Key words siRNA delivery, Dendrimers, Trastuzumab, Bioconjugation, HER2 receptors

1 Introduction

In recent years, RNA interference (RNAi) has emerged as a powerful technique for treating diseases by interfering with the activity of disease-causing or promoting genes. RNAi is attracting increased attention due to several intrinsic properties that include effective gene silencing, high specificity, minor side effects, as well as easy and low cost of synthesis [1–3]. ShRNAs/siRNAs are a class of short double-stranded RNA molecules which can effectively silence a specified target gene in vitro/in vivo. However, to date, shRNA-/siRNA-based gene therapy suffers from several drawbacks [4, 5]. Firstly, naked or unmodified siRNAs are highly unstable in the bloodstream and are rapidly removed enzymatically by endo- and exonucleases. Secondly, given its anionic nature, siRNA cannot cross the cytoplasmic membrane by simple diffusion [6]. Thirdly, siRNAs are rapidly cleared from the body due to high aqueous solubility [7]. Fourthly, the potential for siRNA off-target effects. Therefore, the use of siRNAs/shRNAs in the human body is a challenging task that necessitates a carrier system to deliver the therapeutic siRNA to its target site [8].

Although various gene carrier systems have been investigated in the past decade, dendrimers are among the most attractive

Lekha Dinesh Kumar (ed.), *RNA Interference and Cancer Therapy: Methods and Protocols*, Methods in Molecular Biology, vol. 1974, https://doi.org/10.1007/978-1-4939-9220-1_14,

carrier system due to several inherent gene-carrying properties [9–12]. Dendrimers (Dend) are molecules consisting of radially symmetric tree-like arms or branches that compose into a well-defined, homogeneous, and monodisperse nanostructure [13]. Due to their structure, there are several void spaces within the dendrimers where siRNAs/shRNAs can be encapsulated providing high chemical and physical stability to the cargo within biological systems. Furthermore, cationic dendrimers can be easily complexed with anionic nucleic acid molecules by electrostatic interactions into "ready to go" nanoparticles; which also have a plethora of surface functional groups that facilitates specific targeting of dendrimers in order to avoid off-target effects of the therapeutic siRNA.

This chapter illustrates a dendrimer-based strategy for site-specific delivery of a siRNA to human epidermal growth factor 2 (HER2) receptor-positive breast cancer cells. About 20% of breast cancers express a high level of HER2 receptors [14, 15]. Trastuzumab (TZ) is a humanized monoclonal antibody that binds specifically to HER2 receptors and is currently used for the treatment of HER2-positive breast cancer [15, 16]. Therefore, TZ-conjugated dendrimers represent an efficient delivery system for the targeted delivery of drugs and genes to HER2-positive breast cancer cells.

2 Materials

1. G4 PAMAM dendrimers (MW: 14215 Da) with a diaminobutane core.
2. Trastuzumab (TZ), Nava Sanjivani drugs (Hyderabad, India).
3. MAL-PEG-NHS (MW: 865.92 Da).
4. Nylon membrane filters (pore size of 0.22 μm).
5. Fluorescein isothiocyanate (FITC).
6. Dialysis tubing (molecular weight cutoff 2000).
7. Dulbecco's Modified Eagle's Medium (DMEM).
8. 0.25% Trypsin–EDTA (1×) phenol red.
9. Antibiotic/antimycotic solution.
10. PBS (Ca^{2+}, Mg^{2+} free).
11. 3-(4,5-Dimethylthiazol-2-yl)-2,5-diphenyl tetrazolium bromide (MTT).
12. Dimethyl sulfoxide (DMSO).
13. Annexin V-FITC apoptosis detection kit.
14. Acridine orange (AO).
15. Ethidium bromide (EB).

16. Fetal bovine serum.
17. Human breast cancer cell lines MDA-MB-453 (HER2-positive) and MDA-MB-231 (HER2-negative).

3 Methods

3.1 Synthesis of FITC-Labelled Dendrimers

1. To synthesize FITC-labelled dendrimers (FITC-Dend), in a round-bottom flask, resuspend G4 PAMAM dendrimers in PBS (pH 7.4) to a final dendrimer concentration of 5 mg/mL (*see* **Note 1**).
2. Dissolve enough FITC in acetone to achieve a 1:6 molar ratio of FITC to dendrimers.
3. Add FITC solution to dendrimer solution, and stir at room temperature for 12 h.
4. Remove unconjugated FITC by dialysis against PBS for 24 h at room temperature.
5. Measure the absorbance at 495 nm using a UV-VIS spectrophotometer, and calculate the number of FITC molecules conjugated to per molecule of dendrimers. The amount of FITC conjugated to the dendrimers is measured by standard calibration of FITC.

3.2 Conjugation of Hetero-Cross-Linker (NHS-PEG-MAL) to FITC-Dend

1. For the generation of Dend-PEG-MAL molecules, NHS-PEG-MAL is mixed with FITC-Dend conjugates in a molar ratio of 3:1 in PBS (pH 8.0) and allowed to react by stirring for 30 min at room temperature.
2. Purify the FITC-Dend-PEG-MAL conjugate by dialysis against PBS for 12 h at room temperature. Dialysis will remove the unconjugated or free NHS-PEG-MAL.
3. Lyophilize the purified FITC-Dend-PEG-MAL to obtain in powder form for ^{1}H-NMR spectroscopy (*see* **Note 2**).
4. Dissolve about 5 mg of conjugate in D_2O, and run a ^{1}H-NMR to confirm the conjugation of FITC-Dend-PEG-MAL.

3.3 Bioconjugation of TZ with Dendrimers

1. For thiolation of TZ, dissolve the TZ in PBS, and add 2-iminothiolane hydrochloride in 1:10 molar ratio. Stir the solution for 2 h at room temperature.
2. Place the solution in a dialysis bag, and dialyze against PBS to remove free 2-iminothiolane hydrochloride.
3. For the bioconjugation of thiolated TZ and dendrimer, thiolated TZ is coupled to FITC-Dend-PEG-MAL (10:1 ratio).
4. The FITC-Dend-PEG-MAL dendrimer solution is mixed with the thiolated TZ solution and allowed to react overnight by stirring at room temperature.

5. FITC-Dend-PEG-TZ conjugates are purified by PBS (pH 7.4) elution through a size exclusion chromatography Sephadex column (G-25 M).
6. The chemical reactions for Subheadings 3.1–3.3 are presented in Fig. 1.

3.4 Measurement of the Zeta Potential of FITC-Dend-PEG-MAL-TZ Conjugates

1. Determine zeta potential of different formulations using Zetasizer Nano ZS. Dilute the samples in deionized water, and analyze at 25 °C (*see* **Note 3**).

3.5 Preparation of Dendrimer-siRNA Complex

1. Resuspend dendrimer conjugates (FITC-Dend, FITC-Dend-PEG-TZ) in water, and mix with siRNA at different N/P ratios (*see* **Note 4**). The N/P ratio is defined by the number of amine groups present in the dendrimers and the number of anionic phosphate groups from siRNA.
2. Vortex dendrimer-siRNA mixture for 1 min, and incubate at room temperature for 30 min in order to ensure the formation of the complexes (Fig. 1).
3. Store the prepared dendrimer-siRNA (siRNA/FITC-Dend, siRNA/FITC-Dend-PEG-TZ) complexes in freezer (−4 °C) until use.

3.6 Cell Culture and Antiproliferation Assay

1. Grow MDA-MB-453 (HER2-positive) and MDA-MB-231 (HER2-negative) human breast cell lines in DMEM medium supplemented with 10% fetal bovine serum 100 U/mL penicillin, 100 mg/mL streptomycin, and 2 mM L-glutamine. Cells should be maintained at 37 °C and 5% CO_2 environment in a CO_2 incubator.
2. Seed the cells in 96-well plates at a density of 5×10^3 cells per well in 100 μL of medium, and allow them to adhere overnight before incubation with nonconjugated siRNA and conjugated siRNA/FITC-Dend-PEG-MAL or siRNA/FITC-Dend-PEG-MAL-TZ at equivalent siRNA concentrations (0–100 nM) for 48 h.
3. Replace the media with serum-free DMEM containing MTT (0.5 mg/mL), and incubate for 4 h.
4. Gently remove the media and add 150 μL DMSO.
5. Measure the absorbance at 570 nm using a microplate reader.

Fig. 1 Schematic diagram illustrating the synthesis of trastuzumab (TZ)-grafted, fluorescein isothiocyanate (FITC)-labelled G4 poly(amido) amine (PAMAM) dendrimers siRNA/FITC-Dend-PEG-TZ. *Step 1*, FITC was conjugated to G4 PAMAM dendrimers FITC-Dend. *Step 2*, FITC-labelled dendrimers were cross-linked with MAL-PEG-NHS to give FITC-Dend-PEG-MAL (Product A). *Step 3*, TZ was thiolated using Traut's reagent (2-iminothiolane) to give thiolated TZ. *Step 4*, FITC-Dend (Product A) was reacted to thiolated TZ (Product B) to synthesize TZ-grafted FITC-labelled G4 PAMAM dendrimers (FITC-Dend-PEG-TZ). *Step 5*, siRNA is reacted to generate siRNA/FITC-Dend-PEG-TZ

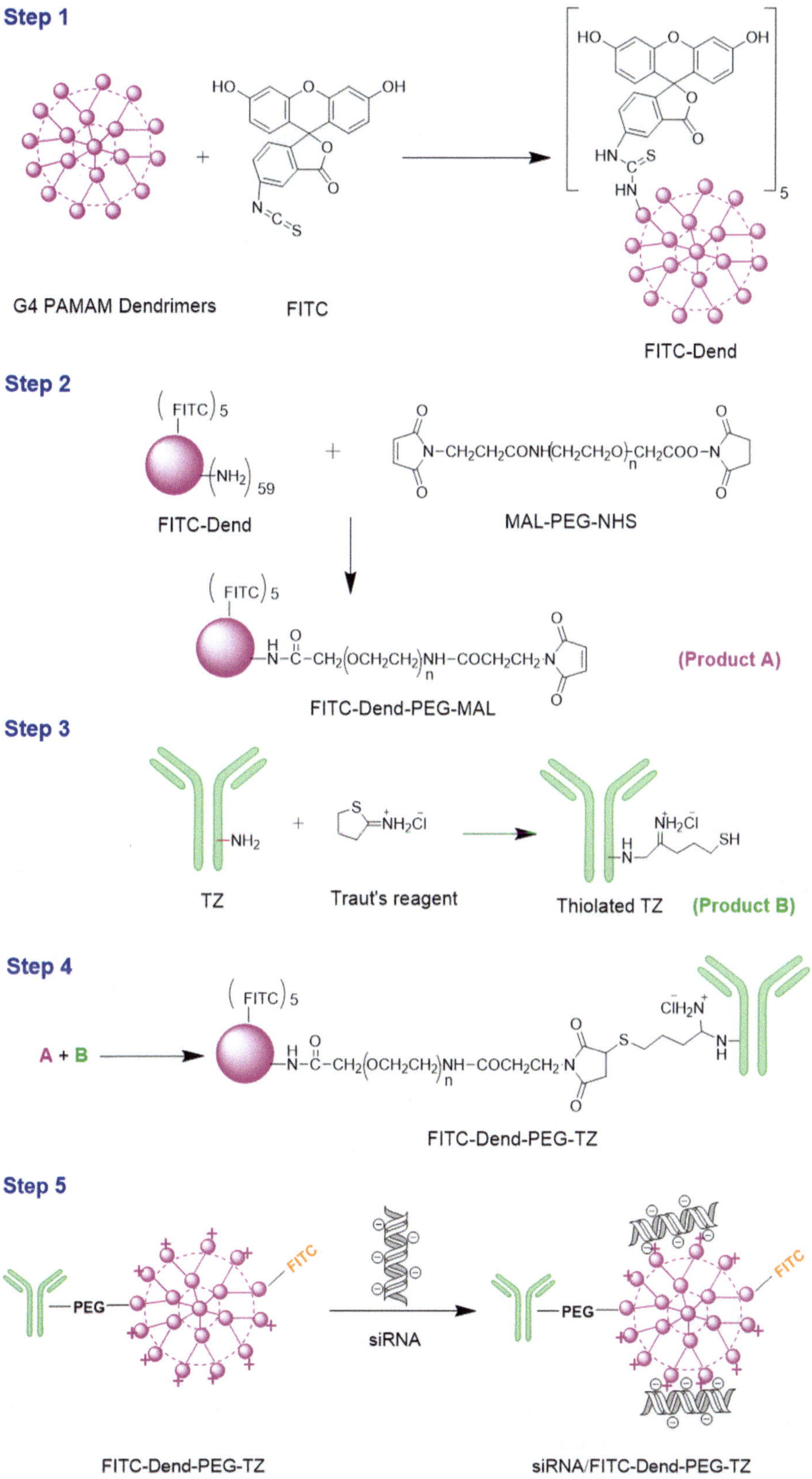
Step 1
HO
O
OH
N=C=S
G4 PAMAM Dendrimers
FITC
HN
C
S
HN
5
FITC-Dend
Step 2
FITC 5
NH2 59
FITC-Dend
N-CH2CH2CONH(CH2CH2O)n-CH2COO-N
MAL-PEG-NHS
N-C-CH2(OCH2CH2)n NH-COCH2CH2-N
(Product A)
FITC-Dend-PEG-MAL
Step 3
NH2
S
NH2Cl
TZ
Traut's reagent
SH
Thiolated TZ
(Product B)
Step 4
A + B
ClH2N
FITC-Dend-PEG-TZ
Step 5
PEG
FITC
siRNA
FITC-Dend-PEG-TZ
siRNA/FITC-Dend-PEG-TZ

6. Untreated cells are used as a negative control (100% viability), and IC_{50} is calculated by fitting the curve of cell viability against the drug concentration.

3.7 Cellular Uptake Studies

1. Seed cells in 12-well plates and allow them to adhere for 24 h.
2. Incubate cells with siRNA/FITC-Dend-PEG-MAL or siRNA/FITC-Dend-PEG-MAL-TZ conjugates, at an equivalent FITC concentration (100 μg/mL).
3. For competitive binding studies, preincubate the cells with an excess (20 mM) of TZ for 2 h, and then treat with siRNA/FITC-Dend-PEG-MAL or siRNA/FITC-Dend-PEG-MAL-TZ.
4. Remove the culture media, wash the cells twice with cold PBS, and observe the cells using a fluorescence microscope.
5. For quantitative studies, add 0.1% Triton X-100 in 0.2 M NaOH, and then add into each well to lyse cells.
6. Measure the fluorescence intensity using a microplate reader at an excitation wavelength of 495 nm and emission wavelength of 520 nm.

3.8 Stability Study

1. To check the stability of final formulation (siRNA/FITC-Dend-PEG-MAL-TZ), store the samples at 4 °C.
2. Monitor the formulations for any changes in consistency and sign of precipitation.
3. Measure the zeta potential using Zetasizer Nano ZS at predetermined time intervals.

4 Notes

1. The chemical reactions and purification must be performed in the dark. FITC is a light-sensitive compound.
2. Before lyophilization, sample should be pre-freeze using liquid nitrogen.
3. During the measurements of zeta potential using Zetasizer Nano ZS, the samples should be diluted appropriately to achieve count rate of between 50 and 300.
4. Dendrimer-siRNA complex can also be prepared in HEPES buffer pH 7.4.

Acknowledgments

H.K. acknowledges the Department of Science and Technology, New Delhi, for INSPIRE Faculty Award. A.K.J. acknowledges the University Grants Commission, New Delhi, for Ph.D. scholarship.

References

1. Xin Y, Huang M, Guo MM et al (2017) Nano-based delivery of RNAi in cancer therapy. Mol Cancer 16:134
2. Fellmann C, Lowe SW (2014) Stable RNA interference rules for silencing. Nat Cell Biol 16(1):10–18
3. Scherman D, Rousseau A, Bigey P et al (2017) Genetic pharmacology: progresses in siRNA delivery and therapeutic applications. Gene Ther 24:151–156
4. Haussecker D (2014) Current issues of RNAi therapeutics delivery and development. J Control Release 195:49–54
5. Mizrahy S, Hazan-Halevy I, Dammes N et al (2017) Current progress in non-viral RNAi-based delivery strategies to lymphocytes. Mol Ther 25:1–10
6. Wang J, Lu Z, Wientjes MG et al (2010) Delivery of siRNA therapeutics: barriers and carriers. AAPS J 12:492–503
7. Ozpolat B, Sood AK, Lopez-Berestein G (2014) Liposomal siRNA nanocarriers for cancer therapy. Adv Drug Deliv Rev 66:110–116
8. Wu SY, Lopez-Berestein G, Calin GA et al (2014) RNAi therapies: drugging the undruggable. Sci Transl Med 6(240):240ps7
9. Yang J, Zhang Q, Chang H, Cheng Y (2015) Surface-engineered dendrimers in gene delivery. Chem Rev 115:5274–5300
10. Biswas S, Torchilin VP (2013) Dendrimers for siRNA delivery. Pharmaceuticals 6:161–183
11. Zhou J, Wu J, Hafdi N et al (2006) PAMAM dendrimers for efficient siRNA delivery and potent gene silencing. Chem Commun (Camb) 22:2362–2364
12. Senkusa E, Cardosob F, Pagani O (2014) Time for more optimism in metastatic breast cancer? Cancer Treat Rev 40:220–228
13. Sampathkumar SG, Yarema KJ (2007) Dendrimers in cancer treatment and diagnosis. In: Kumar CSSR (ed) Nanotechnologies for the life sciences, vol 7. Wiley-VCH, Weinheim
14. Figueroa-Magalhães MC, Jelovac D, Connolly RM, Wolff AC (2014) Treatment of HER2-positive breast cancer. Breast 23:128–136
15. Kulhari H, Pooja D, Rompicharla SVK et al (2015) Biomedical applications of Trastuzumab: as a therapeutic agent and a targeting ligand. Med Res Rev 35:849–876
16. Kulhari H, Pooja D, Shrivastava S et al (2015) Trastuzumab-grafted PAMAM dendrimers for the selective delivery of anticancer drugs to HER2-positive breast cancer. Sci Rep 6:23179

Chapter 15

Synthesis of Gold Nanoparticles for Gene Silencing

Claudia Tortiglione and Jesús M de la Fuente

Abstract

Over the last decade, the capability of double-stranded RNAs to interfere with gene expression has driven new therapeutic approaches. Since small interfering RNAs (siRNAs, 21-base-pair double-stranded RNA) were shown able to elicit RNA interference (RNAi), efforts were directed toward the development of efficient delivery systems to preserve siRNA bioactivity throughout the delivery route, from administration site to the target cell. Starting from the synthesis of gold nanoparticle, here we describe comprehensive methodologies for functionalization with specific moieties (charged groups, peptides) and chemico-physical characterization. Moreover, we report two different strategies that can be used to bind siRNA molecules on the gold nanoparticle: a covalent approach, based on thiolated siRNA bound via a thiol bond to nanoparticle surface, and an ionic approach based on the electrostatic interaction between the negatively charged siRNA backbone and azide and quaternary ammonium groups positively charged anchored on the nanoparticle surface. Both methodologies were shown highly efficient for siRNA delivery, achieving specific gene silencing in in vitro and in vivo biological systems.

Key words Gold nanoparticles, RNA interference, Functionalization, Small interfering RNA

1 Introduction

Nanoparticles exhibit excellent physicochemical properties for application in drug delivery. Their reduced sizes and the high surface to volume ratio allow for surface functionalization with a plethora of molecules, for specific targeting and drug payloads. These structural features allow interfacing with biomolecules in a one-to-one scale. Among the numerous nanoparticle formulations designed to improve siRNA delivery and efficiency, those based on gold nanoparticles (AuNPs) have been extensively investigated without undesirable immune response or off-target effects [1, 2]. We describe here the effective conjugation strategies to combine specific biomolecules to the surface of AuNPs in a highly controlled way. These biomolecules include (a) biofunctional spacers; (b) cell-penetrating peptides (CPPs) to overcome the lipophilic barrier of the cellular membranes; and (c) siRNA complementary to a master regulator gene. Polyethylene glycol (PEG) was used as

Lekha Dinesh Kumar (ed.), *RNA Interference and Cancer Therapy: Methods and Protocols*, Methods in Molecular Biology, vol. 1974, https://doi.org/10.1007/978-1-4939-9220-1_15,

stabilizer to increase solubility and biocompatibility. PEG was modified at one end with a thiol group to bond covalently to the gold nanoparticle and at the other end with carboxylic acid (thiol-PEG-COOH) or azide (thiol-PEG-N_3) functional groups. Depending on the nature of the interaction between siRNA and inorganic nanoparticles, we describe two main approaches, namely, ionic and covalent binding, producing Au. Overall these methodologies offer the possibility of fine-tuning siRNA concentrations and bonding and the combination of different chemical groups on the nanoparticle surface thus personalizing gene silencing for specific targets.

2 Materials

2.1 Gold Nanoparticle Synthesis

1. Hydrogen tetrachloroaurate (III) hydrate (0.2 g; 0.6 mmol).
2. Sodium citrate dihydrate (0.6 g; 5 mmol).

2.2 Gold Nanoparticle Derivatization

1. Sodium dodecyl sulfate (SDS) (0.028%).
2. PEG-COOH (1): SH-EG(8)-$(CH_2)_2$-COOH (39 μM).
3. PEG-N_3 (1): SH-$(CH_2)_3$-CONH-EG(6)-$(CH_2)_2$-N_3 (13 μM).
4. PEG-COOH (2): SH-EG(8)-$(CH_2)_2$-COOH (7 μM).
5. PEG-N_3 (2): SH-$(CH_2)_3$-CONH-EG(6)-$(CH_2)_2$-N_3 (7 μM).
6. Sodium hydroxide (0.1 M).
7. Citrate-gold nanoparticles (10 nM).

2.3 Gold Nanoparticle Functionalization

1. Buffer A: 13 mM sulfo-*N*-hydroxy-succinimide (sulfo-NHS) and 50 mM 2-(-morpholino)-ethanesulfonic acid (MES) pH 6.1.
2. (2-Aminoethyl)trimethylammonium chloride hydrochloride (R_4N^+) (38 mM).
3. (1-Ethyl-3-(3-dimethylaminopropyl)-carbodiimide) (EDC) (0.8 mM).
4. TAT peptide (H-KKK-GREKKRRQRRR-OH).
5. Buffer B: 6 mM sulfo-NHS, 25 mM MES buffer pH 6.1.
6. siRNA (small interfering RNA) sequences were designed using online tools from manufacturers (5 nmol/mL).
7. RGD peptide (G-RGD-S).
8. α-Amine-ω-azido hepta(ethylene glycol) (0.6 mmol).
9. Dimethyl sulfoxide (12.5 mL).
10. γ-Tiobutirolactone (0.63 mmol).
11. Triethylamine (5.7 mmol).
12. 1,4-Dithiothreitol (DTT) (1 mL; 0.1 M).

13. DEPC water.
14. AuNPs-ion (ionic interaction between siRNA and AuNP): AuNPs-ion (26 nM).
15. AuNPs-cov (covalent bonding between siRNA and AuNP): AuNPs-cov (20 nM).

2.4 Gold Nanoparticle Characterization

1. 5,5′-Dithio-bis(2-nitrobenzoic) acid (DTNB) (50 μL, 2 mM).
2. Buffer C: Tris buffer (100 μL, 1 M, pH 8).
3. Buffer D: Sodium acetate (NaAc) (50 mM).
4. Sodium borohydride ($NaBH_4$).
5. Buffer E: Phosphate buffer (0.05 M, pH 8).
6. Buffer F: Phosphate buffer (0.2 M, pH 7).
7. Bradford assay.
8. BCA Protein Assay Kit (2 mL).
9. GelRed.
10. NAP-5 column.
11. Buffer G: 6 mM sulfo-NHS, 0.8 nM EDC, 25 mM MES buffer pH 6.1.
12. Coomassie Brilliant Blue G-250 (0.15 mL).
13. Buffer F: TBE (0.5×).

3 Methods

3.1 Synthesis of Citrate-Gold Nanoparticles

We describe the synthesis of gold nanoparticles with an average diameter of ~14 nm, according to the method described by Lee and Meisel [3].

1. Hydrogen tetrachloroaurate (III) hydrate is dissolved in 500 mL of distilled water, heated, and stirred under reflux (*see* **Note 1**).
2. When the solution boils, sodium citrate dihydrate is added obtaining a red solution. The solution is kept under reflux for extra 30 min.
3. After that, the solution is kept at room temperature for cooling down (*see* **Note 2**).

3.2 Synthesis of PEG Heterofunctional Surfaces in Citrate-Gold Nanoparticles

To increase colloidal stability of AuNP, PEG coating is an excellent option (*see* **Note 3**). Thiol (SH) group is required for its bonding to AuNP. Additional functional groups such as carboxyl (COOH), amine (NH_2), or azide (N_3) could provide further chemical functionality and enhance colloidal stability due to electrostatic repulsion. We describe a protocol for tuning the percentage of PEG molecules covering the gold surface, as this can impact on the stability and on the interaction with a cell membrane. Two different PEG molecules have been used for such study:

1. 100% saturated PEG layer (AuNP-ion) functionalized with 75% carboxylic groups (SH-EG(8)-$(CH_2)_2$-COOH) and 25% of azide groups (SH-$(CH_2)_3$-CONH-EG(6)-$(CH_2)_2$-N_3)
2. 25% saturated PEG layer (AuNP-cov) functionalized with 50% carboxylic groups (SH-EG(8)-$(CH_2)_2$-COOH) and 50% of azide groups (SH-$(CH_2)_3$-CONH-EG(6)-$(CH_2)_2$-N_3).

Once 100% saturated PEG layer guarantees a perfect coating of the AuNPs and high colloidal stability in different media, lower coating, such as 25%, allows the incorporation of further thiolated components, such as the SH-siRNA.

1. Citrate-gold NPs, SDS, and PEG-COOH (1) and 1PEG-N_3 (1) for NP-ion and PEG-COOH (2) and 1PEG-N_3 (2) for AuNP-cov were mixed.
2. NaOH was added (final concentration of 25 mM) in order to deprotonate thiol groups and improve Au–S bonding.
3. The mixture was incubated for 16 h at room temperature.
4. The excess PEG chains were removed by centrifugation (16,080 × *g*, 30 min, 4 °C).
5. This washing process was repeated three times.
6. Finally the obtained pellet with AuNPs was resuspended in pure water. The suspension should be kept at room temperature and protected from light.

3.3 AuNP-Ion Functionalization with TAT Peptides and Quaternary Ammonium Groups by EDC/NHS Coupling Reactions

1. AuNP-ion was incubated in Buffer A [4, 5].
2. R_4N^+ was added to the mixture; EDC was added (3 mM final concentration).
3. The mixture was incubated for 20 min.
4. TAT peptide was added at a final concentration of 1 μM.
5. The suspension was kept at room temperature for 16 h.
6. Removal of peptide excess was done by centrifugation (3×, 16,080 × *g*, 30 min, 4 °C).
7. Supernatants were tested for peptide concentration by Bradford assay.

3.4 AuNP-Cov Functionalization with TAT Peptides by EDC/NHS Coupling Reactions

1. AuNP-cov was incubated in Buffer B.
2. TAT peptide was added to a final concentration of 0.5 μM followed by EDC.
3. The suspension was kept at room temperature for 16 h.

4. Removal of peptide excess was done by centrifugation (3×, 16,080 × *g* 30 min, 4 °C).
5. Supernatants were tested for peptide concentration by Bradford assay.

3.5 AuNP-Ion Functionalization with RGD Peptides and Quaternary Ammonium Groups by EDC/NHS Coupling Reactions

1. AuNP-ion was incubated in Buffer A [4, 5].
2. R_4N^+ was added to the mixture; EDC was added (3 mM final concentration).
3. The mixture was incubated for 20 min.
4. RGD peptide was added at a final concentration of 2 μM.
5. The suspension was kept at room temperature for 16 h.
6. Removal of peptide excess was done by centrifugation (3×, 16,080 × *g*, 30 min, 4 °C).

3.6 AuNP-Cov Functionalization with RGD Peptides by EDC/NHS Coupling Reactions

1. AuNP-cov was incubated in Buffer B.
2. RGD peptide was added to a final concentration of 0.5 μM followed by EDC.
3. The suspension was kept at room temperature for 16 h.
4. Removal of peptide excess was done by centrifugation (3×, 16,080 × *g*, 30 min, 4 °C).

3.7 Characterization of Citrate-Gold Nanoparticles

AuNPs were characterized by different techniques (Fig. 1).

1. Transmission electron microscopy (TEM, T20 200 keV; FEI) showing an average size of 14 nm (*see* **Note 4**).
2. Infrared spectra (FTIR, JASCO FT/IR 4100) confirmed the presence of the citrate groups covering the gold cores.
3. The UV-Vis molecular absorption spectra of the AuNPs show the characteristic surface plasmon resonant peak at 519 nm. They present at pH 7 with a negative *Z*-potential value of −12.32 mV.

3.8 Synthesis of N-(20-Azido-3,6,9,12,15,18-Hexaoxoicosane)-4-Mercaptobutanamide ($HS-(CH_2)_3-CONH-EG(6)-(CH_2)_2-N_3$)

1. α-Amine-ω-azido hepta(ethylene glycol) were added to dimethyl sulfoxide under argon atmosphere in a poly(ethylene glycol) bath at 50 °C and under continuous stirring.
2. γ-Tiobutirolactone and triethylamine were added.
3. After 5 h, the mixture is cooled down and lyophilized obtaining a yellow oil. The yield of the reaction was 93%, without any further purification step.
4. Characterization: The product was characterized by FTIR, 3301 cm^{-1} (NH st), 2870 cm^{-1} (–CH st), 2530 cm^{-1} (SH st), 2106 cm^{-1} (N_3 st), and 1663 cm^{-1} (C=O st); ^{1}H-RMN 300 MHz (δ), 2.00 (m, 2H), 2.25 (t, 2H, J = 7.30 Hz), 2.65 (t, 2H, J = 7.00 Hz), 3.01 (s, 1H, SH), 3.32 (t, 4H,

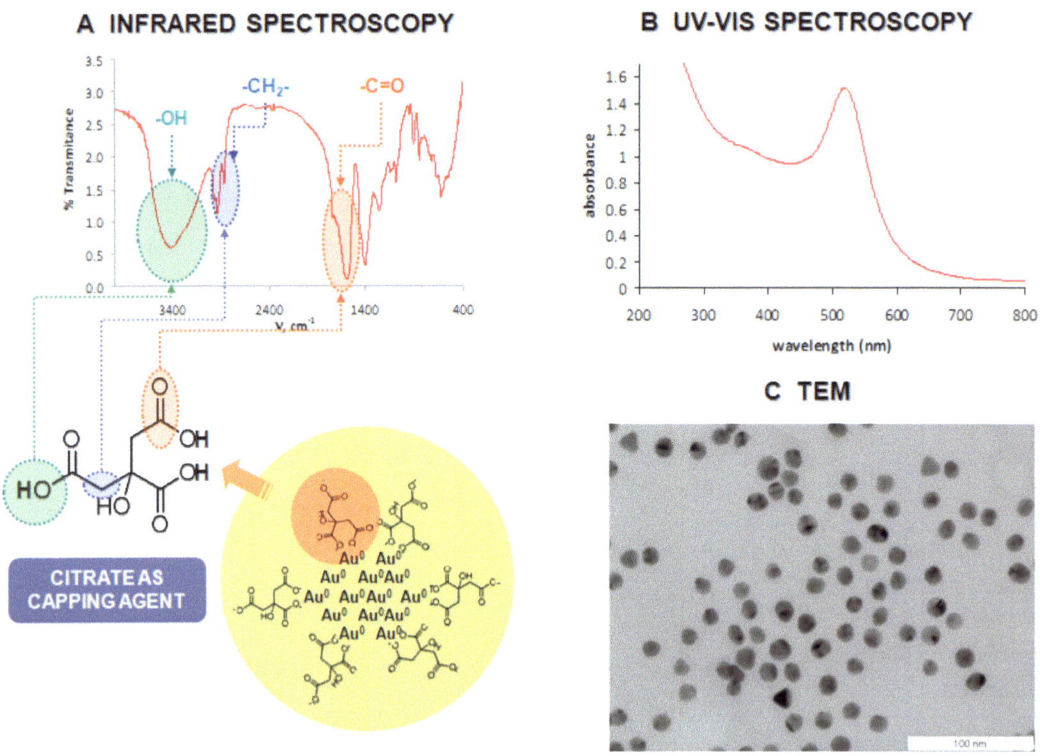

Fig. 1 Characterization of citrate-AuNPs (**a**) Infrared spectra, peaks appear at 3422.06 cm^{-1}, 2957.30 cm^{-1}, 2925.48 cm^{-1}, 2854.13 cm^{-1}, 1584.23 cm^{-1}, 1400.07 cm^{-1}, 1253.50 cm^{-1}, 1080.90 cm^{-1}; (**b**) UV-Vis molecular absorption spectra (absorbance peak at 520 nm); (**c**) transmission electron microscopy, scale bar 100 nm. Reprinted with permission from (ACS Nano 6, 9, 8316–8324). Copyright (2012) American Chemical Society

J = 5.00 Hz), 3.37 (t, 2H, *J* = 5.22 Hz), 3.59 (s, 22H), and 6.60 (t, 1H, *J* = 4.76 Hz); and mass spectroscopy [M–H$^+$] 451.1.

3.9 Quantification of SH-EG(8)-$(CH_2)_2$-COOH Bound to AuNPs

In order to quantify the level of functionalization of AuNP-thiolated PEG chains (SH-EG(8)-$(CH_2)_2$-COOH), a modification of Ellman's assay [6] was optimized. In this assay, the supernatant, containing the spacer with thiol groups on its structure, reacts with 5,5′-dithio-bis(2-nitrobenzoic) acid (DTNB) providing a colored product that can be measured by spectrometry.

1. Supernatants obtained from functionalization steps of the AuNP with thiolated PEG molecules after centrifugation (16,080 × *g*, 30 min, 4 °C) were used for such quantification.
2. 400 μL of either the stock solution of PEG-COOH (1) or the supernatant is mixed with Buffer C and DTNB in Buffer D.
3. After 10 min of incubation, the absorbance at 412 nm is measured. The linear range for the (SH-EG(8)-$(CH_2)_2$-COOH) chain obtained by this method is 1–100 μM

($Abs_{412} = 0.006 \cdot [(SH\text{-}EG(8)\text{-}(CH_2)_2\text{-}COOH), \mu M] + 0.1764$). The number of exchanged chains is given by the difference between the amount determined by this assay and the initial amount of chains in the incubation mixture with the NPs.

3.10 Quantification of SH-$(CH_2)_3$-CONH-EG (6)-$(CH_2)_2$-N_3 Bound to AuNPs

3.10.1 Direct Method

1. Firstly, the reduction of disulfide bonds should be carried out by sodium borohydride ($NaBH_4$) under inert conditions. The sample is dissolved in Buffer E.
2. $NaBH_4$ is added at final concentration of 0.12 M and incubated at room temperature for 90 min.
3. Acidification of the solution using HCl 3 M/Acetone (1:1) (v/v) for 10 min guarantees the complete elimination of excess borohydride.
4. Neutralization of the reaction was done by adding NaOH 3 M. The neutralized solution was buffered at pH 7 with Buffer F (*see* **Note 5**).
5. 400 μL of either the stock solution of PEG-N_3 (1) or the supernatant are mixed with Buffer C and DTNB 2 mM in Buffer D.
6. After 10 min of incubation, the absorbance at 412 nm is measured. The linear range for the SH-$(CH_2)_3$-CONH-EG (6)-$(CH_2)_2$-N_3 chain obtained by this method is 0.022–2.2 mM ($Abs_{412} = 0.5$ [SH-$(CH_2)_3$-CONH-EG(6)-$(CH_2)_2$-N_3, nM] + 0.1215). The number of exchanged chains is given by the difference between the amount determined by this assay and the initial amount of chains in the incubation mixture with the NPs.

3.10.2 Indirect Method

An indirect method of quantification can be used by adding a second addition of SH-EG(8)-$(CH_2)_2$-COOH (*see* **Note 6**). SH-EG (8)-$(CH_2)_2$-COOH was added to AuNPs until a complete saturation (0.1 mM of thiolated chain per 10 nM of gold nanoparticles).

1. AuNPs were functionalized with a controlled amount of thiolated chains in order to obtain NPs with different degrees of saturation—Fig. 2a (increasing amount of SH-$(CH_2)_3$-CONH-EG(6)-$(CH_2)_2$-N_3 at a constant SH-EG(8)-$(CH_2)_2$-COOH chains bound to AuNPs) and Fig. 2b (increasing amount of SH-EG(8)-$(CH_2)_2$-COOH at a constant SH-$(CH_2)_3$-CONH-EG(6)-$(CH_2)_2$-N_3 chains bound to AuNPs).
2. A fixed amount of a PEG-COOH (1) was added and incubated with the AuNPs (from 37 μM to 11 μM of total PEG chains per 10 nM of gold (Fig. 2c).
3. After obtaining supernatant from functionalization steps of the AuNP with thiolated PEG molecules after centrifugation

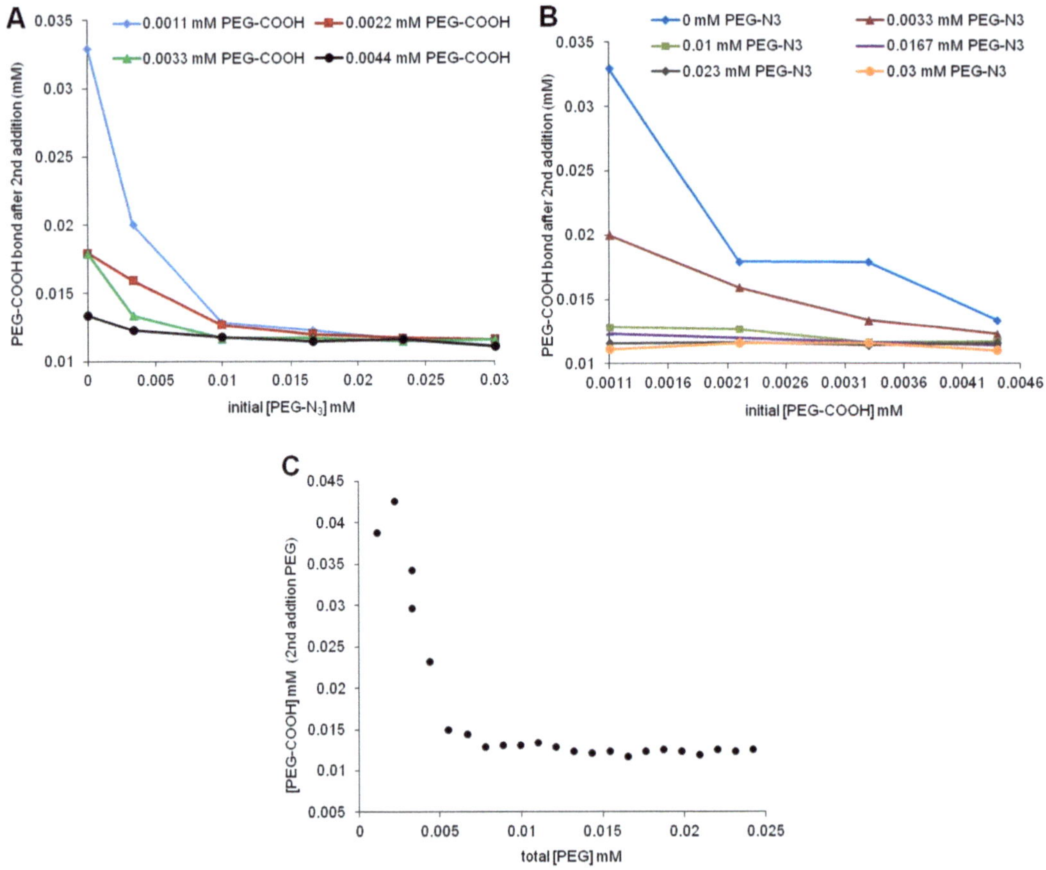

Fig. 2 Quantification of degree of saturation for the PEG chains (**a**) Increasing amount of SH-$(CH_2)_3$-CONH-EG(6)-$(CH_2)_2$-N_3 (PEG-N_3) at a constant SH-EG(8)-$(CH_2)_2$-COOH (PEG-COOH) chains bound to AuNPs. (**b**) Increasing amount of PEG-COOH at constant PEG-N_3 chains bound to AuNPs. (**c**) Optimum degree of saturation of AuNPs for the entrance of a third thiolated chain. Reprinted with permission from (ACS Nano 6, 9, 8316–8324). Copyright (2012) American Chemical Society

(16,080 × *g*, 30 min, 4 °C), supernatants were used for PEG-COOH (1) quantification.

4. 400 μL of either the stock solution of SH-EG (8)-$(CH_2)_2$-COOH or the supernatant is mixed with Buffer C and DTNB in Buffer D.
5. After 10 min of incubation, the absorbance at 412 nm is measured. The linear range for the (SH-EG (8)-$(CH_2)_2$-COOH) chain obtained by this method is 1–100 μM ($Abs_{412} = 0.006 \cdot [(\text{SH-EG(8)-}(CH_2)_2\text{-COOH}), \mu M] + 0.1764$). The number of exchanged chains is given by the difference between the amount determined by this assay and the initial amount of chains in the incubation mixture with the NPs.

3.11 AuNP Functionalization with Small Interfering RNAs (siRNA)

1. Thiolated siRNA (for AuNP-cov) was dissolved in DTT (*see* **Note 7**).
2. In order to remove remaining DTT and by-products, the solution was extracted three times with ethyl acetate, and further purification of the aqueous solution was carried out by a desalting NAP-5 column following manufacturer's instructions.
3. The siRNA (for AuNP-ion) was resuspended in DEPC water and incubated with RNase-free solution of the AuNP-ion@ R_4N^+@TAT (10 nM) for 16 h at 4 °C.
4. The purified thiolated siRNA (for AuNP-cov) was resuspended in DEPC water and incubated with RNase-free solution of the AuNP-cov@TAT, containing SDS for 16 h at 4 °C.
5. Subsequently, the salt concentration is increased to 0.1 M NaCl with brief ultrasonication following each addition to increase the coverage of oligonucleotides on the nanoparticle surface.
6. Particles were purified by centrifugation (16,080 × *g*, 20 min, 4 °C) and resuspended in DEPC water. This process was repeated three times (*see* **Note 8**).

3.12 Determination of the Bound TAT Peptide

In order to quantify the amount of TAT peptide linked to the AuNPs, the peptide in the supernatant after the coupling was quantified by a modification of Bradford method. Due to the interferences obtained by EDC and sulfo-NHS, the calibration curve was obtained under reaction conditions (Buffer C).

1. Supernatant were mixed with Coomassie Brilliant Blue G-250 (1:1) in a 96-well plate.
2. After 10 min of incubation at room temperature, the absorbance of the solution was measured at 570 nm.
3. Correlation of the obtained results using the calibration curve (Fig. 3) provides the non-bonded peptide amount.

3.13 Quantification of the RGD Peptide

The RGD quantification was achieved using the BCA Protein Assay.

1. Each standard and unknown sample (the supernatants) is mixed with BCA™ Working Reagent (1:20) to each tube and incubated at 60 °C for 2 h.
2. The tubes are cooled down to room temperature and the absorbance measured at 562 nm.
3. The standard curve is used to determine the RGD concentration of each unknown sample (supernatant). The calibration curve for a working range (0–18 μg/mL) is given by the following equation $Abs_{562\ nm} = -0.0001x^2 + 0.0081x + 0.0017$.

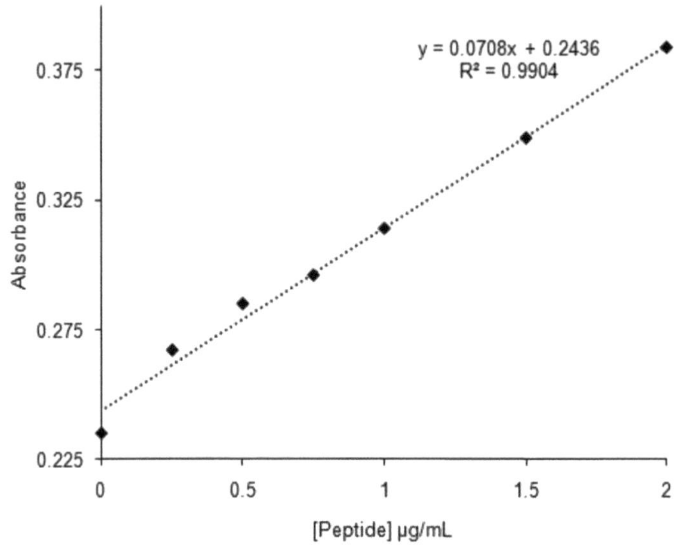

Fig. 3 Standard addition calibration curve for the TAT peptide in the conditions of the EDC coupling reaction. In the linear range, the calibration curve is given by the following equation: Abs_{570} = 0.071[peptide, μg/mL] + 0.24. Reprinted from [7]

3.14 Quantification of the siRNA Strands Loaded onto AuNPs

The method to quantify the amount of siRNA (as (dsRNA) in excess during the functionalization is performed using GelRed™ dye as a nucleic acid intercalator [7].

1. The supernatants are incubated with GelRed 100× and TBE 0.5×, and fluorescence is measured with emission at 602 nm. A correlation to a standard curve of known siRNA concentrations allows accurate measurement of siRNA in unknown samples (Emission at 602 nm = 439.96·[siRNA, nmol] + 32.697) (Fig. 4).

4 Notes

1. Glass material used for the synthesis of nanoparticles should be previously cleaned using piranha solution (3 parts of concentrated sulfuric acid and 1 part of 30% hydrogen peroxide).
2. Gold nanoparticle suspension should be stored at room temperature and at darkness.
3. It is very important to do a right selection of the length and functional groups on the PEG molecule in order to obtain a good colloidal stability. Bigger particles require longer PEG molecules.

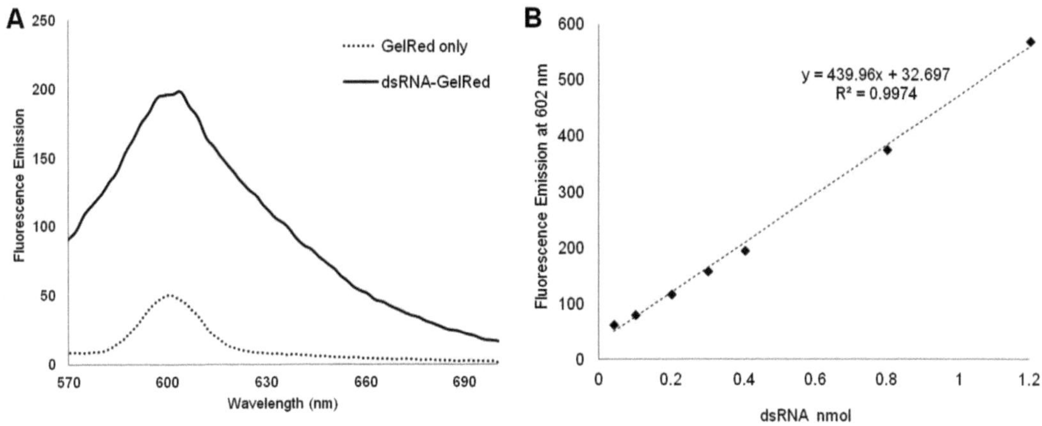

Fig. 4 (**a**) Fluorescence spectra of the GelRed only and dsRNA-GelRed. (**b**) Calibration curve for the thiolated dsRNA. In the linear range, the calibration curve is given by the following equation: emission at 602 nm = 439.96[dsRNA, nmol] + 32.697. Reprinted from [7]

4. Examination of the samples by transmission electron microscopy must be carried out using nanoparticle concentration around 0.1–1 mg·mL^{-1}.
5. To prevent re-oxidation of the thiols to disulfides, it is crucial to perform immediately the Ellman's assay.
6. Due to the lack of sensitivity in the case of SH-$(CH_2)_3$-CONH-EG(6)-$(CH_2)_2$-N_3 quantification, an indirect method of quantification can be used by adding a second addition of SH-EG(8)-$(CH_2)_2$-COOH. It is worthy to mention that the presence of azide-modified PEG linker has been confirmed by FTIR (2106 cm^{-1}, N_3 st).
7. The sequence design needs to be performed with the aid of online tools developed by oligonucleotide manufacturers. We experienced that different siRNAs targeting the same mRNA sequence may not induce RNAi with the same efficiency, and some sequence may induce off-target effects or may be completely ineffective. This is strongly depending on the gene targeted, the mRNA secondary structure, and the biological system. A selected region may work for a given organism, and not in another one. We suggest designing at least three different siRNAs and to produce different siRNA-AuNPs targeting the same gene, testing and comparing all of them. An alternative may be to use a mix of all of them as siRNA source for the synthesis. This procedure is faster; however the final AuNP conjugate may not show best performance, due to possible reduction of good siRNA molecules on the AuNP.
8. Functionalized gold nanoparticles cannot be sonicated in order to keep the stability of the ligands linked to the nanoparticle surface.

References

1. Conde J, Ambrosone A, Hernandez Y, Tian F, McCully M, Berry CC, Baptista PV, Tortiglione C, de la Fuente JM (2015) 15 years on siRNA delivery: beyond the state-of-the-art on inorganic nanoparticles for RNAi therapeutics. Nano Today 10:421–450
2. Conde J, Ambrosone A, Sanz V, Hernandez Y, Marchesano V, Tian F, Child H, Berry CC, Ibarra MR, Baptista PV, Tortiglione C, de la Fuente JM (2012) Design of multifunctional gold nanoparticles for in vitro and in vivo gene silencing. ACS Nano 6:8316–8324
3. Lee PC, Meisel D (1982) Adsorption and surface-enhanced Raman of dyes on silver and gold sols. J Phys Chem 86:3391–3395
4. Grabarek Z, Gergely J (1990) Zero-length crosslinking procedure with the use of active esters. Anal Biochem 185:131–135
5. Staros JV, Wright RW, Swingle DM (1986) Enhancement by N-hydroxysulfosuccinimide of water-soluble carbodiimide-mediated coupling reactions. Anal Biochem 156:220–222
6. Ellman GL (1959) Tissue sulfhydryl groups. Arch Biochem Biophys 82:70–77
7. Sanz V, Conde J, Hernández Y, Baptista PV, Ibarra MR, de la Fuente JM (2012) Effect of PEG biofunctional spacers and TAT peptide on dsRNA loading on gold nanoparticles. J Nanopart Res 14:917

Chapter 16

Design of Eco-Friendly Gold Nanoparticles for Cancer Treatment

Deep Pooja and Ramakrishna Sistla

Abstract

Gold nanoparticles (GNPs) have emerged as a potential scaffold for a wide range of biomedical applications such as biosensing, drug delivery, and imaging. However, the toxicity of nanoparticles remains a challenge for using them in biological system. The morphology and surface chemistry of GNP can be manipulated by their method of preparation. GNP can be synthesized and functionalized by various methods. This chapter illustrates the synthesis of highly biocompatible GNP using a natural gum, i.e., xanthan gum (XG). Moreover, due to the presence of mannose moiety in XG, these XG-stabilized GNP may also act as self-targeted drug carriers for the delivery of chemotherapeutic agents/siRNA/shRNA to mannose receptor overexpressing cancer cells.

Key words Xanthan gum, Gold nanoparticles, Biocompatibility, Chemotherapy, Drug delivery

1 Introduction

Nanoparticles are solid structures with a size below 200 nm and have their applications in sensing, in imaging, and in drug/gene delivery. Gold nanoparticles (GNPs) are one of the most commonly explored and used nanoparticles in drug delivery because of controlled size, improved efficacy, and targeting potential [1, 2]. As drug delivery carrier, GNPs have been used for delivering both hydrophilic and hydrophobic drugs [3–7]. The conventional methods for synthesis of GNP involve the use of chemicals like sodium borohydride, trisodium citrate, etc. as reducing agents [8–12]. These GNPs have been found to be very unstable and form aggregates with the slight change in pH and electrolyte concentration [13, 14]. The GNP prepared with gellan gum showed the stability between pH 4 and 8 and addition of NaCl up to 0.1 M [15]. Natural gums stabilize the inorganic nanoparticles by two mechanisms: firstly, by adsorbing to the surface of nanoparticles which creates steric repulsion among the particles. Secondly, they

Lekha Dinesh Kumar (ed.), *RNA Interference and Cancer Therapy: Methods and Protocols*, Methods in Molecular Biology, vol. 1974, https://doi.org/10.1007/978-1-4939-9220-1_16,

increase the viscosity of nanoparticle suspension and therefore slow down the aggregation processes [16–18].

Xanthan gum (XG) is an anionic, high molecular weight exopolysaccharide produced by aerobic fermentation of sugars by *Xanthomonas campestris*. Its basic chain consists of a β-(1 → 4) linked glucose backbone with the substitution of a charged trisaccharide side chains of [β-(1 → 3)-mannose-α-(1 → 2)-glucuronic acid-β-(1 → 4)-mannose] on alternate glucose residues [19]. It is nontoxic, hydrophilic, and biodegradable biopolymer. XG is being used in food, cosmetic, and pharmaceutical industries. The industrial applications of XG are based upon its exceptional rheological properties. XG is soluble in both cold and hot water, hydrates quickly, and produces high viscosity at low concentration [20, 21]. In pharmaceutical industries, XG has been reported for formulation of both solid and liquid dosage forms. In solid dosage formulation, it is used as controlled-release agent [19, 22–26], whereas in liquid formulations, it is used as thickening agent, suspending agent, and emulsion stabilizer [25].

Recently, XG has also been used in the preparation and stabilization of inorganic iron and palladium nanoparticles [18, 26–28]. The aim of this investigation was to synthesize the environment-friendly gold nanoparticles using XG as reducing agent. The nanoparticle preparation was optimized for gum concentration and volume, gold concentration, temperature, and reaction time. Gum-stabilized GNPs were studied for their utility as drug delivery carriers using doxorubicin hydrochloride as model drug. Doxorubicin hydrochloride is hydrophilic in nature and has been clinically used for the treatment of various cancers, hematological malignancies, soft tissue sarcomas, and solid tumors [29, 30]. Two major limitations of DOX are non-specific cytotoxicity and multidrug resistance. Multidrug resistance to DOX is due to drug efflux by P-glycoproteins. Doxorubicin is a substrate for P-glycoproteins which effluxes DOX and decreases intracellular drug level. DOX-loaded GNP has shown to overcome both the problems of non-specific toxicity and drug resistance [31]. The prepared nanoparticles were also investigated for biocompatibility, stability, and cytotoxicity study in lung cancer cells.

2 Materials

1. Sodium acetate buffer pH 5.0: 2.99 g sodium acetate, 1.66 mL glacial acetic acid in 1000 mL distilled water. Adjust the pH at 4.5 with glacial acetic acid or NaOH.
2. Phosphate-buffered saline pH 7.4: 8 g NaCl, 0.2 g KCl, 1.44 g Na_2HPO_4, 0.24 g KH_2PO_4 in 1000 mL distilled water. pH adjusted with 0.1 N HCl or NaOH.

3. DMEM medium: 10% fetal bovine serum, 100 mg/mL streptomycin, 2 mM L-glutamine, 100 U/mL penicillin.
4. Xanthan gum solution: 150 mg of xanthan gum in 100 mL of distilled water.
5. Normal saline solution: 9 g NaCl in 1 L of distilled water.

3 Methods

3.1 Preparation of Xanthan Gum Solution

1. Prepare a stock solution of xanthan gum (5 mg/mL), and stir it overnight at room temperature.
2. Remove the insoluble materials by centrifuging the gum solution at 25000 × *g*, and lyophilize the supernatant.
3. Dissolve the lyophilized dry powder in water to get desired concentration of gum solution.

3.2 Synthesis of Gold Nanoparticles

1. Prepare an aqueous solution of $HAuCl_4$ (10 mM), and heat at 80 °C in the presence of XG solution for 2 h for reduction of Au^{3+} to Au^0 (*see* **Notes 1** and **2**).
2. Cool the colloidal solution at room temperature after observing change in color from colorless to ruby red, and store in dark at 4 °C.

3.3 Particle Size Determination of Gum-Stabilized Gold Nanoparticles

1. Dilute the colloidal solution with distilled water appropriately to get particle count rate in the range 50–250 kcps (kilo count per second), and measure the hydrodynamic diameter, polydispersity index and surface charge, and size using particle size analyzer (*see* **Note 3**).
2. For determination of size, shape, and surface morphology using transmission electron microscopy (TEM), place a drop of diluted sample on carbon-coated grid. Dry at room temperature and stain with 2% uranyl acetate. Observe under the microscope.

3.4 Fourier Transform Infrared (FTIR) Analysis

1. Mix lyophilized GNP with potassium bromide in a ratio of 1:100 (w/w), and compress to form a pellet.
2. Place the pellet in sample holder, and scan for the measurement of % transmittance in the wave number ranging from 4000 to 450 cm^{-1}.

3.5 Drug Loading

1. Disperse the blank nanoparticles in PBS, and incubate with drug/siRNA/shRNA solution at room temperature overnight.
2. Collect the drug-loaded nanoparticles by centrifuging at 25,000 × *g* for 30 min.

3. Measure % drug loading (DL) by estimating free drug content in supernatant.

$$\%\text{Drug loading} = \{1 - (D_S / D_T)\} \times 100$$

where D_S is amount of drug present in supernatant and D_T total amount of drug loaded initially.

3.6 In Vitro Drug Release Studies

1. Disperse the drug-loaded GNP (equivalent to 1 mg of drug) in 1 mL of distilled water, and place in dialysis tubing (MWCO 12–14 kDa) (*see* **Note 4**). Place the dialysis bag in 100 mL of release medium (PBS or SAB) maintained at 37 ± 0.5 °C, and stir magnetically at 100 rpm.
2. Withdraw 1 mL of sample at predetermined time intervals, and replace with the same volume of fresh medium to maintain sink conditions.
3. Analyze the samples for drug content.

3.7 Hemolytic Toxicity Studies

1. Biocompatibility of GNP can be determined by estimating the % hemolysis, caused by GNP. Disperse the blank GNP in normal physiological saline.
2. Add different concentrations of dispersion to red blood cell (RBC) suspension (2% v/v), mix properly with gentle shaking, and incubate for 1 h at 37 °C. Incubate RBC suspension separately with normal saline and distilled water as control and standard, respectively.
3. After 1 h, centrifuge the samples at 1000 × *g* for 5 min.
4. Measure the absorbance of supernatant at 540 nm spectrophotometrically, and calculate the % hemolysis as follows:

$$\%\text{Hemolysis} = (A_{\text{sample}} - A_{\text{control}}) / (A_{\text{standard}} - A_{\text{control}}) \times 100$$

3.8 Cell Culture and Anticancer Activity

1. Grow A549 human lung cell lines in cell culture medium, and maintain at 37 °C and 5% CO_2 environment in a cell culture incubator.
2. Seed the cells at a density of 1×10^4 cells per/well in 96-well plates containing 100 μL of DMEM medium, and allow to attach overnight in incubator.
3. Incubate the seeded cells with varying concentration (0.1–10 μg/mL) of plain drug, blank gold nanoparticles (positive control), and drug-loaded GNP for 48 h.
4. Replace the media with serum-free DMEM, add 10 μL of MTT reagent (5 mg/mL), and incubate for 4 h.
5. Remove the media, add 200 μL of DMSO, and measure the absorbance at 570 nm.

6. The half maximum inhibitory concentration (IC_{50}) is calculated by plotting the curve cell viability vs. drug concentration. The absorbance of untreated cells is used as a negative control.

3.9 Stability Study

1. Store the drug-loaded GNP at 4 °C for 1 month.
2. Determine the particle size, zeta potential, and drug content at predetermined time intervals.

3.10 Effect of pH and Ionic Concentration

1. Incubate the drug-loaded GNP with solution of different pH (pH 3–11).
2. Incubate nanoparticles with PBS as a control.
3. Measure the absorbance after 24 h, and compare with absorbance of control.
4. For studying the effect of ionic strength on the stability, incubate the nanoparticles with varying concentrations (0.1–2 M) of sodium chloride (NaCl) in deionizing water.
5. Measure the absorbance after 24 h, and compare with absorbance of control.

3.11 Serum Stability

1. Disperse the nanoparticles in 10% v/v serum.
2. Measure the particles size, zeta potential, and absorbance at different time intervals up to 72 h.

4 Notes

1. The synthesis of nanoparticles depends on the various formulation and process variables such as gum concentration, gold concentration, gold to gum volume ratio, reaction time, and temperature. These variables need to be optimized for synthesis of GNP with reproducible size of particles. While optimizing, change one parameter at a time, and keep rest parameters constant.
2. During synthesis of GNP, temperature should be maintained properly. A slight change in temperature may lead to polydispersity in particle size. Prepared nanoparticles should be stored in dark and cool atmosphere.
3. Samples should be diluted appropriately for determination of particle size.
4. During drug loading, the therapeutic agent should be added as aqueous solution to avoid its precipitation and to achieve high drug loading.
5. Care should be taken in the selection of dialysis membrane. The membrane should have a pore size which allows free drug diffusion, but the nanoparticles should not pass through it.

Acknowledgments

The work described in this book chapter has been published as Deep Pooja et al., Xanthan gum stabilized gold nanoparticles: Characterization, biocompatibility, stability and cytotoxicity. Carbohydrate Polymers 2014;110:1–9. The work is reused after permission from Elsevier under license number 4236430677285.

References

1. Almeida JP, Figueroa ER, Drezek RA (2014) Gold nanoparticle mediated cancer immunotherapy. Nanomedicine 10(3):503–514
2. Pissuwan D, Niidome T, Cortie MB (2011) The forthcoming applications of gold nanoparticles in drug and gene delivery systems. J Control Release 149(1):65–71
3. Oliveira R, Zhao P, Li N et al (2013) Synthesis and in vitro studies of gold nanoparticles loaded with docetaxel. Adv Colloid Interf Sci 199(200):44–58
4. Chen YH, Tsai CY, Huang PY et al (2007) Methotrexate conjugated to gold nanoparticles inhibits tumor growth in a syngeneic lung tumor model. Mol Pharm 4(5):713–722
5. Gibson JD, Khanal BP, Zubarev ER (2007) Paclitaxel-functionalized gold nanoparticles. J Am Chem Soc 129(37):11653–11661
6. Aryal S, Grailer JJ, Pilla S et al (2009) Doxorubicin conjugated gold nanoparticles as water-soluble and pH-responsive anticancer drug nanocarriers. J Mater Chem 19: 7879–7884
7. Prabaharan M, Grailer JJ, Pilla S et al (2009) Gold nanoparticles with a monolayer of doxorubicin-conjugated amphiphilic block copolymer for tumor-targeted drug delivery. Biomaterials 30(30):6065–6075
8. Burygin GL, Khlebtsov BN, Shantrokha AN et al (2009) On the enhanced antibacterial activity of antibiotics mixed with gold nanoparticles. Nanoscale Res Lett 4(8):794–801
9. Duncan B, Kim C, Rotello VM (2010) Gold nanoparticle platforms as drug and biomacromolecule delivery systems. J Control Release 148:122–127
10. Jung J, Park S, Hong S et al (2014) Synthesis of gold nanoparticles with glycosides: synthetic trends based on the structures of glycones and aglycones. Carbohydr Res 386:57–61
11. Khan MS, Vishakantea D, Siddaramaiah H (2013) Gold nanoparticles: a paradigm shift in biomedical applications. Adv Colloid Interf Sci 199–200:44–58
12. Saha B, Bhattacharya J, Mukherjee A et al (2007) In vitro structural and functional evaluation of gold nanoparticles conjugated antibiotics. Nanoscale Res Lett 2(12): 614–622
13. Mirza AZ, Shamshad H (2011) Preparation and characterization of doxorubicin functionalized gold nanoparticles. Eur J Med Chem 46(5):1857–1860
14. Rouhana LL, Jaber JA, Schlenoff JB (2007) Aggregation-resistant water-soluble gold nanoparticles. Langmuir 23:12799–12801
15. Dhar S, Reddy EM, Shiras A et al (2008) Natural gum reduced/stabilized gold nanoparticles for drug delivery formulations. Chem Eur J 14(33):10244–10250
16. Tiraferri A, Chen KL, Sethi R et al (2008) Reduced aggregation and sedimentation of zero-valent iron nanoparticles in the presence of guar gum. J Colloid Interface Sci 324(1–2):71–79
17. Comba S, Sethi R (2009) Stabilization of highly concentrated suspensions of iron nanoparticles using shear-thinning gels of xanthan gum. Water Res 43(15):3717–3726
18. Xue D, Sethi R (2012) Viscoelastic gels of guar and xanthan gum mixtures provide long-term stabilization of iron micro- and nanoparticles. J Nanopart Res 14:1239–1253
19. Jian H, Zhu L, Zhang W et al (2012) Galactomannan (from *Gleditsia sinensis* Lam.) and xanthan gum matrix tablets for controlled delivery of theophylline: in vitro drug release and swelling behaviour. Carbohydr Polym 87:2176–2182
20. Sereno NM, Hill SE, Mitchell JR (2007) Impact of the extrusion process on xanthan gum behaviour. Carbohydr Res 342(10): 1333–1342
21. Sharma BR, Naresh L, Dhuldhoya NC et al (2006) Xanthan gum—a boon to food industry. Food promotion chronicle 1(5):27–30
22. Phaechamud T, Ritthidej GC (2007) Sustained-release from layered matrix system

comprising chitosan and xanthan gum. Drug Dev Ind Pharm 33:595–605

23. Santos H, Veiga F, Pina ME et al (2005) Compaction compression and drug release properties of diclofenac sodium and ibuprofen pellets comprising xanthan gum as a sustained release agent. Int J Pharm 295(1–2):15–27
24. Sinha VR, Mittal BR, Bhutani KK et al (2004) Colonic drug delivery of 5-fluorouracil: an in vitro evaluation. Int J Pharm 269(1): 101–108
25. Desplanques S, Renou F, Grisel M et al (2012) Impact of chemical composition of xanthan and acacia gums on the emulsification and stability of oil-in-water emulsions. Food Hydrocoll 27:401–410
26. Fan G, Cang L, Qin W et al (2013) Surfactants-enhanced electrokinetic transport of xanthan gum stabilized nanoPd/Fe for the remediation of PCBs contaminated soils. Sep Purif Technol 114:64–72
27. Comba S, Dalmazzo D, Santagata E et al (2011) Rheological characterization of xanthan suspensions of nanoscale iron for injection in porous media. J Hazard Mater 185(2–3): 598–605
28. Vecchia ED, Luna M, Sethi R (2009) Transport in porous media of highly concentrated iron micro- and nanoparticles in the presence of xanthan gum. Environ Sci Technol 43(23): 8942–8947
29. Carvalho C, Santos RX, Cardoso S et al (2009) Doxorubicin: the good, the bad and the ugly effect. Curr Med Chem 16(25):3267–3285
30. Laginha KM, Verwoert S, Charrois GJR et al (2005) Determination of doxorubicin levels in whole tumor and tumor nuclei in murine breast cancer tumors. Clin Cancer Res 11:6944–6949
31. Gu YJ, Cheng J, Man CW et al (2012) Gold-doxorubicin nanoconjugates for overcoming multidrug resistance. Nanomedicine 8(2): 204–211

Chapter 17

^{64}Cu-Labeled Aptamers for Tumor-Targeted Radionuclide Delivery

Lei Kang, Zachary T. Rosenkrans, and Weibo Cai

Abstract

Aptamers are a class of oligonucleotides with high binding affinity and specificity with their targets. Additionally, aptamers are nontoxic, very thermally stable, and able to reversibly undergo denaturation and have a small size. Cancer-related aptamers can be used for tumor-targeted drug delivery, such as to deliver diagnostic and therapeutic radionuclides to target cancers. We describe the process for preparing a ^{64}Cu-labeled modified A10 aptamer to target prostate cancer by conjugating and radiolabeling. The modified A10 aptamer was conjugated with p-SCN-Bn-NOTA as the chelator. Following this, the aptamer can be radiolabeled with the ^{64}Cu radioisotope. NOTA was selected as the chelator of choice due to its commercial availability and widely demonstrated in vivo stability with the ^{64}Cu radioisotope. Using this system, prostate cancer could potentially be targeted for noninvasive imaging using positron emission tomography (PET). Closely following this protocol allows many aptamers to be successfully radiolabeled to accurately and quantitatively trace their distribution in vivo for a wide range of medical applications.

Key words Aptamer, A10, Radionuclide, Copper-64 (^{64}Cu), Radiolabel, Tumor, Delivery

1 Introduction

Aptamers, single-stranded DNA (ssDNA) or RNA molecules, are a class of oligonucleotides with many biomedical applications. The technique used to generate aptamers was first reported in 1990 and is known as systematic evolution of ligands by exponential enrichment (SELEX) [1, 2]. Briefly, large oligonucleotide libraries are screened against the desired target, and then nonbinding molecules are separated by washing. Bound oligonucleotides are then separated and amplified using a polymerase chain reaction (PCR). These nucleic acids can then be used as ligands for a wide range of targets that include small molecules [3, 4], proteins [5, 6], carbohydrates [7, 8], and entire cells [9, 10].

Researchers use aptamers for a variety of reasons. Perhaps the most important is the high binding affinity (typically nanomolar levels) and specificity of aptamers with their targets. In this respect,

Lekha Dinesh Kumar (ed.), *RNA Interference and Cancer Therapy: Methods and Protocols*, Methods in Molecular Biology, vol. 1974, https://doi.org/10.1007/978-1-4939-9220-1_17, © Springer Science+Business Media, LLC, part of Springer Nature 2019

aptamers are similar to antibodies but with several additional benefits. Another major advantage is that aptamers are chemically synthesized in vitro. Through this process, as previously described, large-scale amounts required for clinical use can be produced with little variability in a comparatively shorter time. Moreover, the SELEX process makes it possible to generate aptamers that target toxic or not immunogenic molecules, which is not possible using the immune system [11]. Additionally, aptamers have found to be neither toxic nor immunogenic themselves [12]. Aptamers have high thermal stability and can be quickly restored to their original conformation after denaturation occurs without significant loss [11]. Finally, the small size of aptamers allows them to penetrate tissues that may not be accessible to antibodies [13].

Due to these advantageous characteristics, aptamers have been utilized in a variety of applications in medicine. One of the most successful areas has been in therapeutics. In 2004, the first FDA-approved aptamer product, Mucagen®, was authorized for the treatment of age-related muscular degeneration [14]. Several other aptamers are currently undergoing clinical trials for pancreatic and colorectal cancer [15], diabetes [16], and others [17]. For cancer-related targeting, several aptamers were also developed against nucleolin, prostate-specific membrane antigen (PSMA), tenascin C, mucin 1 (MUC1), Ras-binding domain, and others [18]. These cancer-related aptamers can be used for tumor-targeted drug delivery, which can be categorized into several classes: aptamer–drug complexes [19], hydrogels [20], and nanoparticles [21, 22].

In the field of nuclear medicine, radiopharmaceuticals are helpful for the diagnosis and treatment of disease on the molecular level. Aptamers can also be used to deliver diagnostic and therapeutic radionuclides to target cancers [23]. A RNA aptamer A10 can bind to prostate-specific membrane antigen (PSMA), which is commonly over-expressed on prostate cancer cells [24]. Herein, we describe the process how to prepare a ^{64}Cu-radiolabeled A10-modified aptamer conjugated with p-SCN-Bn-NOTA. This chelator was chosen because of its commercial availability and widely demonstrated in vivo stability with the ^{64}Cu radioisotope [25, 26]. Utilizing this aptamer, it is possible to deliver ^{64}Cu to target tumors for the potential noninvasive imaging of prostate cancer using positron emission tomography (PET).

2 Materials

All solutions were prepared with diethylpyrocarbonate (DEPC)-treated ultrapure water and analytical grade reagents. DEPC-treated water is prepared by incubating ultrapure water with 0.1% DEPC and then autoclaving to remove DEPC (*see* **Note 1**).

Prepare and store all reagents at room temperature (unless indicated otherwise). Diligently follow all waste disposal regulations when disposing waste materials.

2.1 Aptamer

1. Aptamer: A10 aptamer purchased from Integrated DNA Technologies. This aptamer was truncated and modified to include a PEG and amine groups on either end (*see* **Note 2**). The final sequence is:

   5′-PEG-GGGAGGACGAUGCGGAUCAGCCAUGUUU ACGUCACUCCUAA-NH_2-3′.

   Additional considerations are required for handling RNA materials, and we recommend following any standard protocols (*see* **Note 3**).

2.2 Materials for Conjugation

1. Chelator: p-SCN-Bn-NOTA purchased from Macrocyclics (*see* **Note 4**).
2. Dimethyl sulfoxide (DMSO) (*see* **Note 5**).
3. Carbonate-bicarbonate buffer (pH 9.2): 0.2 M sodium carbonate buffer and 0.2 M sodium bicarbonate buffer.
4. Purifying column: PD-10 desalting column.

2.3 Materials for Radiolabeling

1. Sodium acetate buffer (0.1 M, pH 5.0) with Chelex 100 resin (*see* **Note 6**).
2. Chelex 100 chelating ion exchange resin: Analytical grade, sodium form, 50–100 dry mesh size.

3 Methods

Carry out all procedures at room temperature unless otherwise specified. All activities involving radioactive materials should be performed in accordance with the standard operating procedures set forth by the appropriate governing body.

3.1 Aptamer Preparation

1. Dissolve the lyophilized aptamer into DEPC-treated ultrapure water to the concentration of 1–10 mg/mL. For each conjugation reaction, no more than 500 μL of aptamer is used. For long-term storage, suspended aptamer should be stored at −80 °C or lower without repeated thawing.

3.2 Conjugation Calculations

The amount of p-SCN-Bn-NOTA required for each reaction depends on the amount of the aptamer to be labeled. When conjugating aptamer with p-SCN-Bn-NOTA, a five- to tenfold molar excess is optimal (*see* **Note 7**).

3.3 Aptamer Conjugation with NOTA

1. Remove p-SCN-Bn-NOTA from −20 °C freezer, and place at room temperature for at least 5 min prior to opening to avoid hydration (*see* **Note 8**).
2. Put an Eppendorf tube directly on the scale, and add the required amount of NOTA. Then add 10 μL of DMSO (*see* **Note 9**).
3. Vortex the tube vigorously, or put the tube into an ultrasonic machine until the NOTA is completely dissolved (*see* **Note 10**).
4. Prepare a 0.2 M carbonate buffer by weighing 10.6 g of anhydrous sodium carbonate, and add water to volume of 500 mL. Prepare a 0.2 M bicarbonate buffer by weighing 8.4 g of anhydrous sodium bicarbonate, and add water to a volume of 500 mL. Combine 10 mL of carbonate buffer, 125 mL of bicarbonate buffer, and water to volume of 500 mL to attain a carbonate-bicarbonate buffer with a pH of 9.2. Store at room temperature for no more than 6 months.
5. Add no more than 500 μL of aptamer solution into an Eppendorf tube with an equal volume of the carbonate-bicarbonate buffer.
6. Add calculated amount of NOTA (DMSO solvent) to the prepared aptamer solution, and vigorously vortex immediately for several seconds (*see* **Note 11**).
7. Adjust the pH of the solution to 8.5–9.0 using the carbonate-bicarbonate buffer (*see* **Note 12**).
8. Incubate and rotate up and down for 2 h at room temperature (*see* **Note 13**).
9. Prepare the PD-10 desalting column for use while the reaction is incubating. Wash the PD-10 column four times using excess PBS. Approximately 25 mL of PBS is required for all washing steps.
10. Purify product using a PD-10 column. Add the conjugation product to the PD-10 column. When the sample completely goes into the PD-10 column, add more PBS so that the total volume reaches 2500 μL.
11. Add PBS to the PD-10 column in 500 μL increments for five times. Collect the eluate in a separate vial after each time PBS is added to the column.
12. Measure the concentration of aptamer at 260 nm absorbance to obtain peak product, and then store at −20 °C. Avoid repeated thawing as much as possible.

3.4 ^{64}Cu Labeling

1. Prepare sodium acetate (NaOAc) solution by dissolving 0.82 g NaOAc in 100 mL of water. Adjust pH to 5.0 using 0.1 M HCl, and add a tablespoon of Chelex 100 (*see* **Note 14**).

2. Add 300 μL of NaOAc into an Eppendorf tube (*see* **Note 15**).
3. Add appropriate activity of $^{64}CuCl_2$ to the NaOAc solution (*see* **Note 16**).
4. Adjust the pH of the radioisotope solution to 4.5–5.5 by adding NaOAc buffer (*see* **Note 17**).
5. Allow the conjugated aptamer warm to room temperature, and then add no more than 500 μL to the radioisotope solution (*see* **Note 18**).
6. Allow radiolabeling reaction to occur for 60 min at 37 °C on a table shaker (*see* **Note 19**).
7. Purify radiolabeled aptamer using a PD-10 column as described previously for conjugation. However, for purifying this reaction, add PBS to the PD-10 column in 500 μL increments for ten times. Collect the eluate in a separate vial after each time PBS is added to the column.
8. Measure the radioactivity and 260 nm absorbance of each tube. The peak of radioactivity can be found at the peak tube on 260 nm absorbance (Fig. 1).

4 Notes

1. DEPC treatment is the most commonly used method for eliminating RNase from water, buffers, and other solutions. Some reagents that contain secondary or tertiary amines groups (e.g., HEPES) can react with DEPC and result in the failure to inactivate RNases. Since the DEPC-treated water is autoclaved, using this water is not an issue when preparing solutions with these reagents.
2. This modified A10 aptamer was truncated from original 71 nucleotides to 39 nucleotides. Modification of 2′-fluoropyrimidines is used to improve the stability to resist RNase. NH_2 is linked at 3′ end for conjugation with NOTA. A PEG group is added to 5′ end to increase the in vivo circulation half-life and serum stability of the aptamer chimera, without affecting its targeting and gene silencing capability [27].
3. RNA products can be degraded by RNases which exist in human body fluids such as tears, saliva, mucus, and perspiration. Therefore, when handling a RNA product, some basic precautions are needed such as gloves and a mask, using RNase-free chemicals, and equipment devoted solely to working with RNA.
4. This chelator is usually stored at −20 °C with desiccant to protect from light and moisture.
5. Purity of DMSO used should be greater than 99.7%.

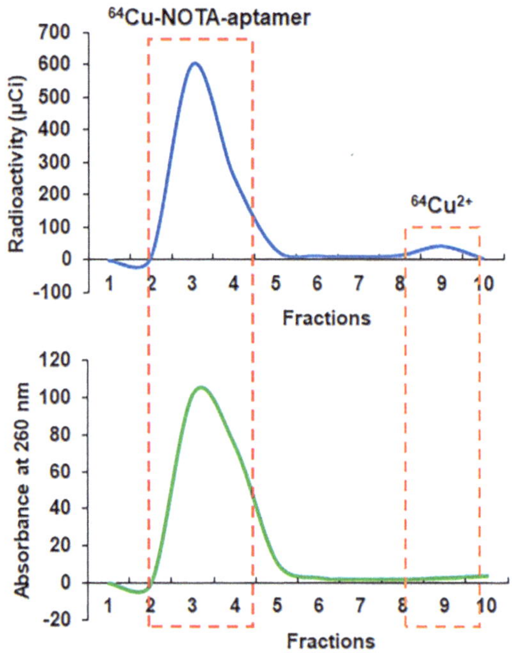

Fig. 1 An example of the radioactivity and 260 nm absorbance of an aptamer radiolabeled with 1 mCi of $^{64}CuCl_2$ at different portions after purification using PD-10 column. The peak of radioactivity can be found at the peak tube on 260 nm absorbance

6. Chelex 100 chelating ion exchange resin is added to the NaOAc solution to bind and remove metal impurities in the solution without altering nonmetal ions. It has a stronger selectivity for copper, iron, and other heavy metal ions over monovalent cations such as sodium and potassium.
7. Amount of NOTA should never be more than a 10:1 molar ratio. Excess NOTA is easily removed after purifying with a PD-10 desalting column.
8. It is important to wait until the vial completely reaches room temperature. Otherwise, warm air will rush into the vial once it is open and bring in moisture that induces the hydrolysis of the SCN-ester group in the chelator. The SCN-ester moiety readily hydrolyzes and becomes nonreactive. Therefore, do not prepare stock solutions for storage.
9. Since the calculated amount of p-SCN-Bn-NOTA is too little to accurately weigh, excess p-SCN-Bn-NOTA is typically measured initially. This often requires microgram amounts. Dissolve the excess amount in DMSO, and then use the corresponding volume required for the conjugation reaction.
10. Prepare this reagent immediately before use. Larger amounts of NOTA can be difficult to dissolve in 10 μL of DMSO. An

additional 5–10 μL of DMSO can be added in these situations, but the volume should never exceed 20 μL. Ultrasonication may help to dissolve NOTA.

11. Make sure the DMSO can be dispersed well in the solution as soon as possible.
12. pH test paper works better than a pH meter in this case due to the small amount of solution. In pH 7–9, SCN reacts efficiently with primary amino groups (–NH_2) to form stable amide bonds. The reaction results in the release of *N*-hydroxysuccinimide.
13. Mixing end over end is better preferred over horizontal shaking. Conjugation can be performed overnight if necessary.
14. The amount of Chelex required may vary, and an exact amount is not crucial, but enough should be used to cover the entire bottom of the container.
15. Be careful not to draw in resin on the bottom.
16. Higher concentrations of ^{64}Cu are preferred because the higher specific activity of the isotope will generate better radiolabeling yields. Avoid contamination whenever possible.
17. The pH value is important for successful radiolabeling in this reaction.
18. Large amounts of aptamer solution (greater than 500 μL) may dilute the reaction buffer and decrease the labeling yield. To achieve a high labeling yield, aptamer should be added in a ratio of more than 25 μg aptamer: 1 mCi ^{64}Cu. Under this condition, labeling yield more than 90% can be usually achieved.
19. Higher reaction temperatures can contribute higher labeling yield.

Acknowledgments

This work was supported by the University of Wisconsin–Madison, the National Institutes of Health (P30CA014520), the Beijing Nova Program (Z171100001117024), and the Beijing Capital Special Development Application Program (Z141107002514159).

References

1. Tuerk C, Gold L (1990) Systematic evolution of ligands by exponential enrichment: RNA ligands to bacteriophage T4 DNA polymerase. Science 249(4968):505–510. https://doi.org/10.1126/science.2200121
2. Ellington AD, Szostak JW (1990) In vitro selection of RNA molecules that bind specific ligands. Nature 346:818. https://doi.org/10.1038/346818a0
3. Yang J, Bowser MT (2013) Capillary electrophoresis–SELEX selection of catalytic DNA Aptamers for a small-molecule porphyrin target. Anal Chem 85(3):1525–1530. https://doi.org/10.1021/ac302721j

4. He X, Zhao Y, He D, Wang K, Xu F, Tang J (2012) ATP-responsive controlled release system using aptamer-functionalized mesoporous silica nanoparticles. Langmuir 28(35):12909–12915. https://doi.org/10.1021/la302767b
5. Bock LC, Griffin LC, Latham JA, Vermaas EH, Toole JJ (1992) Selection of single-stranded DNA molecules that bind and inhibit human thrombin. Nature 355:564. https://doi.org/10.1038/355564a0
6. Li N, Nguyen HH, Byrom M, Ellington AD (2011) Inhibition of cell proliferation by an anti-EGFR Aptamer. PLoS One 6(6):e20299. https://doi.org/10.1371/journal.pone.0020299
7. Srisawat C, Goldstein IJ, Engelke DR (2001) Sephadex-binding RNA ligands: rapid affinity purification of RNA from complex RNA mixtures. Nucleic Acids Res 29(2):e4–e4. https://doi.org/10.1093/nar/29.2.e4
8. Jeong S, Eom T-Y, Kim S-J, Lee S-W, Yu J (2001) In vitro selection of the RNA aptamer against the Sialyl Lewis X and its inhibition of the cell adhesion. Biochem Biophys Res Commun 281(1):237–243. https://doi.org/10.1006/bbrc.2001.4327
9. Esposito CL, Cerchia L, Catuogno S, De Vita G, Dassie JP, Santamaria G, Swiderski P, Condorelli G, Giangrande PH, de Franciscis V (2014) Multifunctional aptamer-miRNA conjugates for targeted cancer therapy. Mol Ther 22(6):1151–1163. https://doi.org/10.1038/mt.2014.5
10. Daniels DA, Chen H, Hicke BJ, Swiderek KM, Gold L (2003) A tenascin-C aptamer identified by tumor cell SELEX: systematic evolution of ligands by exponential enrichment. Proc Natl Acad Sci U S A 100(26):15416–15421. https://doi.org/10.1073/pnas.2136683100
11. Jayasena SD (1999) Aptamers: an emerging class of molecules that rival antibodies in diagnostics. Clin Chem 45(9):1628–1650
12. Bouchard PR, Hutabarat RM, Thompson KM (2010) Discovery and development of therapeutic aptamers. Annu Rev Pharmacol Toxicol 50(1):237–257. https://doi.org/10.1146/annurev.pharmtox.010909.105547
13. Lee JF, Stovall GM, Ellington AD (2006) Aptamer therapeutics advance. Curr Opin Chem Biol 10(3):282–289. https://doi.org/10.1016/j.cbpa.2006.03.015
14. Fine SL, Martin DF, Kirkpatrick P (2005) Pegaptanib sodium. Nat Rev Drug Discov 4:187. https://doi.org/10.1038/nrd1677
15. US National Library of Medicine (2017) clinicaltrials.gov. https://www.clinicaltrials.gov/ct2/show/NCT03168139?term=NCT03168139&rank=1. Accessed 2 Dec 2017
16. US National Library of Medicine (2014) clinicaltrials.gov. https://www.clinicaltrials.gov/ct2/show/NCT01547897?term=NCT01547897&rank=1. Accessed 2 Dec 2017
17. US National Library of Medicine (2017) clinicaltrials.gov. https://www.clinicaltrials.gov/ct2/show/NCT03364153?term=NCT03364153&rank=1. Accessed 2 Dec 2017
18. Pei X, Zhang J, Liu J (2014) Clinical applications of nucleic acid aptamers in cancer. Mol Clin Oncol 2(3):341–348. https://doi.org/10.3892/mco.2014.255
19. Yazdian-Robati R, Ramezani M, Jalalian SH, Abnous K, Taghdisi SM (2016) Targeted delivery of Epirubicin to cancer cells by polyvalent aptamer system in vitro and in vivo. Pharm Res 33(9):2289–2297. https://doi.org/10.1007/s11095-016-1967-4
20. Li J, Zheng C, Cansiz S, Wu C, Xu J, Cui C, Liu Y, Hou W, Wang Y, Zhang L, Teng I, Yang H-H, Tan W (2015) Self-assembly of DNA nanohydrogels with controllable size and stimuli-responsive property for targeted gene regulation therapy. J Am Chem Soc 137(4):1412–1415. https://doi.org/10.1021/ja512293f
21. Leach J, Wang A, Ye K, Jin S (2016) A RNA-DNA hybrid aptamer for nanoparticle-based prostate tumor targeted drug delivery. Int J Mol Sci 17(3):380
22. Taghavi S, HashemNia A, Mosaffa F, Askarian S, Abnous K, Ramezani M (2016) Preparation and evaluation of polyethylenimine-functionalized carbon nanotubes tagged with 5TR1 aptamer for targeted delivery of Bcl-xL shRNA into breast cancer cells. Colloids Surf B Biointerfaces 140(Supplement C):28–39. https://doi.org/10.1016/j.colsurfb.2015.12.021
23. Gijs M, Aerts A, Impens N, Baatout S, Luxen A (2016) Aptamers as radiopharmaceuticals for nuclear imaging and therapy. Nucl Med Biol 43(4):253–271. https://doi.org/10.1016/j.nucmedbio.2015.09.005
24. McNamara JO 2nd, Andrechek ER, Wang Y, Viles KD, Rempel RE, Gilboa E, Sullenger BA, Giangrande PH (2006) Cell type-specific delivery of siRNAs with aptamer-siRNA chimeras. Nat Biotechnol 24(8):1005–1015. https://doi.org/10.1038/nbt1223
25. Jiang D, Im H-J, Sun H, Valdovinos HF, England CG, Ehlerding EB, Nickles RJ, Lee DS, Cho SY, Huang P, Cai W (2017) Radiolabeled pertuzumab for imaging of human epidermal growth factor receptor 2

expression in ovarian cancer. Eur J Nucl Med Mol Imaging 44(8):1296–1305. https://doi.org/10.1007/s00259-017-3663-y

26. Yang Y, Hernandez R, Rao J, Yin L, Qu Y, Wu J, England CG, Graves SA, Lewis CM, Wang P, Meyerand ME, Nickles RJ, X-w B, Cai W (2015) Targeting CD146 with a ^{64}Cu-labeled antibody enables in vivo immunoPET imaging of high-grade gliomas. Proc Natl Acad Sci U S A 112(47):E6525–E6534. https://doi.org/10.1073/pnas.1502648112

27. Dassie JP, Liu XY, Thomas GS, Whitaker RM, Thiel KW, Stockdale KR, Meyerholz DK, McCaffrey AP, McNamara JO 2nd, Giangrande PH (2009) Systemic administration of optimized aptamer-siRNA chimeras promotes regression of PSMA-expressing tumors. Nat Biotechnol 27(9):839–849. https://doi.org/10.1038/nbt.1560

Chapter 18

Silencing PRDM14 via Oligonucleotide Therapeutics Suppresses Tumorigenicity and Metastasis of Breast Cancer

Hiroaki Taniguchi and Kohzoh Imai

Abstract

The PRDI-BF1 and RIZ (PR) domain zinc finger protein 14 (PRDM14) is upregulated in approximately 60% of breast cancers, some of which exhibit gene amplification. In contrast, PRDM14 is not expressed in normal, and differentiated tissues. PRDM14$^+$ breast cancer cells are resistant to chemotherapy drugs, are tumorigenic, and metastasize to the lungs. It is commonly assumed that genes that are overexpressed in cancers, such as PRDM14, are effective targets for new therapies that specifically abrogate the expression of these genes. RNA interference of *PRDM14*, a gene expressed by breast cancer cells, reduced the size of tumors and lung metastases in nude mice. In this chapter, we introduce the concept and methods to develop and apply systematically injected small interfering RNA therapy for breast cancer models in vivo.

Key words PRDM14, Breast cancer, Metastasis, siRNA therapy, Chimera siRNA, Drug delivery systems

1 Introduction

The PRDI-BF1 and RIZ (PR) domain zinc finger protein 14 (PRDM14) has a simple structure, which comprises a PR domain that shows homology to a Su(var)3–9, Enhancer-of-zeste and Trithorax (SET) methyltransferase domain, and six zinc finger motifs [1]. PRDM14 is specifically expressed in embryonic stem cells (ESCs) and primordial germ cells, and is required for the maintenance of ESC pluripotency via epigenetic mechanisms [2, 3]. On the other hand, PRDM14 is not expressed in normal, and differentiated tissues [2]. We and others have reported that PRDM14 expression is markedly increased in many different cancer types, including breast cancer, lung cancer, pancreatic cancer, ovarian cancer, testicular cancer, and others [4–7]. PRDM14 confers malignant phenotypes on breast cancer cells, such as resistance to chemotherapy and distant metastasis [6]. It is difficult to achieve sufficient treatment via antibodies or small molecules against

Lekha Dinesh Kumar (ed.), *RNA Interference and Cancer Therapy: Methods and Protocols*, Methods in Molecular Biology, vol. 1974, https://doi.org/10.1007/978-1-4939-9220-1_18,

PRDM14, because PRDM14 is mainly localized in the nucleus. Thus, we decided to develop a small interfering RNA (siRNA) therapy against *PRDM14*.

There are significant barriers in applying siRNA therapies to clinical treatment. These barriers can arise from the siRNA sequences (referred to as off-target effects), the stability of the siRNA in the blood, immunogenicity via the innate immune system, and delivery into target organs or lesions (Fig. 1). Ui-Tei et al. [8] have already reported guidelines for the selection of highly effective siRNA sequences for RNA interference (RNAi). Functional siRNAs simultaneously satisfy the following four conditions: adenine/uracil (A/U) at the 5′ end of the guide strand; guanine/cytosine (G/C) at the 5′ end of the passenger strand; at least four A/U residues in the 7 bp at the 5′ terminal of the guide strand; and G/C stretches longer than 9 bp should be avoided. To avoid off-target silencing, the selected sequences of siRNAs should display low thermodynamic stability (T_m < 21.5 °C) in the seed region of both strands and should not be similar to any other non-targeted transcripts, via at least two mismatches [8, 9]. This algorithm is used to design siRNA sequences against genes of interest using the siDirect 2.0 online software (http://siDirect2.RNAi.jp/) [9].

When cancer has spread outside of the primary organ (e.g., cancer invasion or metastasis), patients have to be treated with antitumor drugs via systemic administration and not via local injection. In general, when siRNAs are administered by systemic injection, they are easily degraded by ribonuclease in the blood, are filtered by the kidney owing to their size, accumulate poorly in target sites, and activate the innate immune system via Toll-like receptors. Furthermore, water-soluble siRNAs cannot readily diffuse across lipophilic cell membranes, and siRNAs must escape from endosomes to reach their target messenger RNAs.

We overcame some of the significant barriers by using a double-stranded RNA/DNA chimera (dsRDC, or a so-called chimera siRNA). dsRDC is a kind of siRNA with an RNA–DNA chimera modification that avoids off-target effects owing to the formation of an RNA-induced silencing complex by the sense strand and exhibits excellent stability in the bloodstream and low immunogenicity via Toll-like receptors in vivo [10].

We also need innovative drug delivery systems (DDSs) with high retention in the blood, such that siRNAs can accumulate in the target cancer tissues, instead of in the liver and spleen, and demonstrate advantages over typical DDSs, like lipid nanoparticles (LNP). The combination of a chimera siRNA with calcium phosphate (CaP) hybrid micelles or with poly(ethylene glycol)-poly(L-lysine) (PEG-PLL), invented and developed by Prof. Kataoka, enables the more efficient delivery of siRNAs to tumors through enhanced permeability and retention (EPR) effects, as a result of

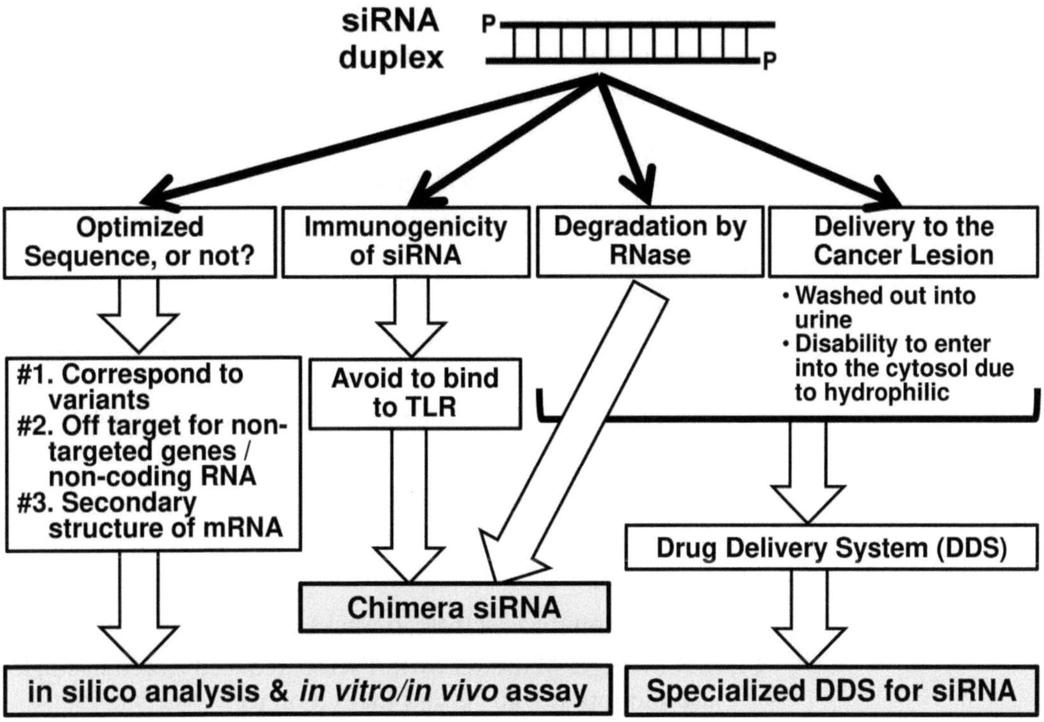

Fig. 1 Technical difficulty of siRNA therapy. Roughly speaking, there are four technical difficulties: (1) siRNA sequence; (2) immunogenicity of siRNA; (3) degradation by RNase; and (4) delivery to the cancer lesion. We can overcome these difficulties using a multidisciplinary approach

their narrow diameter distribution (30–40 nm for CaP hybrid micelles and 10–20 nm for PEG-PLL) [11, 12]. The PEG segment of CaP hybrid micelles and PEG-PLL reduces non-specific interactions in the bloodstream. As a consequence, CaP hybrid micelles and PEG-PLL protect the siRNA from degradation in the blood for a prolonged time as well from kidney filtration, following standard systemic administration [11, 12]. Furthermore, the polyanion segment of CaP hybrid micelles confers sensitivity to acidic pH, thereby permitting endosomal escape after endocytic internalization [11].

The present chapter reviews methods to analyze functions of genes that are overexpressed in cancers, such as PRDM14, and using innovative siRNA therapy that targets them for orthograft or metastatic breast tumor models by systemic injection in vivo.

2 Materials

2.1 Orthotopic Inoculation Model of Breast Cancer Cells into the Mammary Fat Pad of Mice

1. Tumor cell lines (e.g., MDA-MB-231 and HCC1937) obtained from the American Type Culture Collection (ATCC, Manassas, VA, USA) or other sources.

2. Lentiviral construct for expression or knockdown (shRNA) of the target gene and lentiviral packaging plasmids.
3. HEK293T/17 cell line (ATCC).
4. 0.45 μm filter.
5. Opti-MEM® I Reduced-Serum Medium.
6. Polybrene®: 10 mg/mL solution dissolved in 150 mM NaCl and sterile-filtered.
7. Penicillin–streptomycin.
8. Antibiotics for the selection of stably transduced cells: puromycin, hygromycin B, or neomycin (G-418).
9. An adequate culture medium containing 10% fetal bovine serum (FBS).
10. Phosphate-buffered saline (PBS).
11. 0.25% trypsin in 1 mM EDTA.
12. Isoflurane.
13. BALB/c-nu mice (female, 6 weeks of age; CLEA Japan).
14. BD Matrigel Matrix Growth Factor Reduced stored in single-use aliquots at −20 °C and thawed on ice.

2.2 Orthotopic Lung Metastatic Model of Breast Cancer Cells

1. Tumor cell lines (e.g. MDA-MB-231) obtained from ATCC or other sources.
2. Lentiviral construct for expression or knockdown of the target gene and lentiviral packaging plasmids.
3. HEK293T/17 cells.
4. 0.45 μm filter.
5. Opti-MEM® I Reduced-Serum Medium.
6. Polybrene®: 10 mg/mL solution dissolved in 150 mM NaCl and sterile-filtered.
7. Penicillin–streptomycin.
8. Antibiotics for selection of stably transduced cells: puromycin, hygromycin B, or neomycin (G-418).
9. An adequate culture medium containing 10% FBS.
10. PBS.
11. 0.25% trypsin in 1 mM EDTA.
12. BALB/c-nu mice (female, 6 weeks of age; CLEA Japan).
13. Retaining appliance for intravenous injection of mice.
14. Tuberculin syringe with a 27G needle.

2.3 In Vivo Treatment Model for an Orthograft Model of Breast Cancer Cells

1. Tumor cell lines (wild type) expressing the target gene.
2. An adequate culture medium containing 10% FBS.
3. PBS.
4. 0.25% trypsin in 1 mM EDTA.
5. Isoflurane.
6. BALB/c-nu mice (female, 6 weeks of age; CLEA Japan).
7. BD Matrigel Matrix Growth Factor Reduced stored in single-use aliquots at −20 °C and thawed on ice.
8. Custom siRNA or small interfering RNA/DNA chimera (so-called chimera siRNA). One of the drug delivery systems indicated below:
   (a) In vivo-jetPEI (polyethyleneimine; Polyplus-transfection, New York, NY, USA) in 5% glucose solution.
   (b) Calcium phosphate (CaP) hybrid micelles (a gift from Prof. Kataoka) [11].
   (c) Poly(ethylene glycol)-poly(L-lysine) (PEG-PLL: a gift from Prof. Kataoka) [12].
9. Doxorubicin hydrochloride, dissolved in sterile 0.9% saline and sterilized by filtration.
10. Docetaxel, diluted in PBS/dimethyl sulfoxide (ratio 1:1) and sterilized by filtration.

2.4 In Vivo Treatment Model for an Orthotopic Metastatic Model of Breast Cancer Cells

1. Tumor cell lines (wild type) expressing the target gene.
2. An adequate culture medium containing 10% FBS.
3. PBS.
4. 0.25% trypsin in 1 mM EDTA.
5. BALB/c-nu mice (female, 6 weeks of age; CLEA Japan).
6. Retaining appliance for intravenous injection of mice.
7. Tuberculin syringe with a 27G (for injection of tumor cells) and 29G needle (for drug injection).
8. Custom siRNA or small interfering RNA/DNA chimera (so-called chimera siRNA; RNAi Inc., Tokyo, Japan).
9. One of the drug delivery systems (DDSs) indicated below:
   (a) In vivo-jetPEI in 5% glucose solution.
   (b) CaP hybrid micelles.
   (c) PEG-PLL.

3 Methods

3.1 Orthotopic Inoculation Model of Breast Cancer Cells into the Mammary Fat Pad of Mice

To analyze gene functions contributing to tumorigenesis, breast cancer cells can be engineered to stably express or knockdown target genes (e.g., PRDM14), via lentiviral-mediated gene delivery. The stepwise process is summarized below:

1. Produce the lentivirus by co-transfection of HEK293T cells with a lentiviral construct and lentiviral packaging plasmids (*see* **Note 1**).
2. Collect lentiviral particles 48 h after transfection, and filter through a 0.45 μm filter, before the addition of Polybrene (10 μg/mL).
3. The day before transduction, plate the target cancer cells (5×10^4 cells) in a 24-well tissue culture plate, and culture them in a medium containing lentiviral particles for 48 h, before the addition of antibiotics for the selection of stably transduced cells.
4. After selection, confirm the transfection by mRNA and protein expression analyses.
5. Culture the engineered breast tumor cells with an adequate culture medium containing 10% FBS and penicillin–streptomycin.
6. Suspend the breast tumor cells (1×10^6 cells/mouse) in 100 μL of PBS, and mix with 100 μL of Matrigel on ice (*see* **Note 2**).
7. Orthotopically inoculate the mixture into the mammary fat pad of anesthetized 6-week-old female nude mice using a tuberculin syringe with a 27G needle under (Fig. 2a).
8. Measure the length and width of the tumor using calipers until the end point. Calculate the tumor volume using the formula: tumor volume = length × (width)^2 × 3.14/6.
9. At the end point, sacrifice the animals by CO_2 inhalation, and immediately resect the tumors.
10. After fixing the tumor tissues in formalin and embedding them in paraffin, stain the slices with hematoxylin and eosin (H&E), and incubate them overnight at 4 °C with antibodies for immunohistological analyses.

3.2 Orthotopic Lung Metastatic Model of Breast Cancer Cells

To analyze gene functions contributing to lung metastasis, breast cancer cells can be engineered to stably express or knockdown target genes using the same method as in the previous section.

1–5. Perform the same methods as in the previous section.

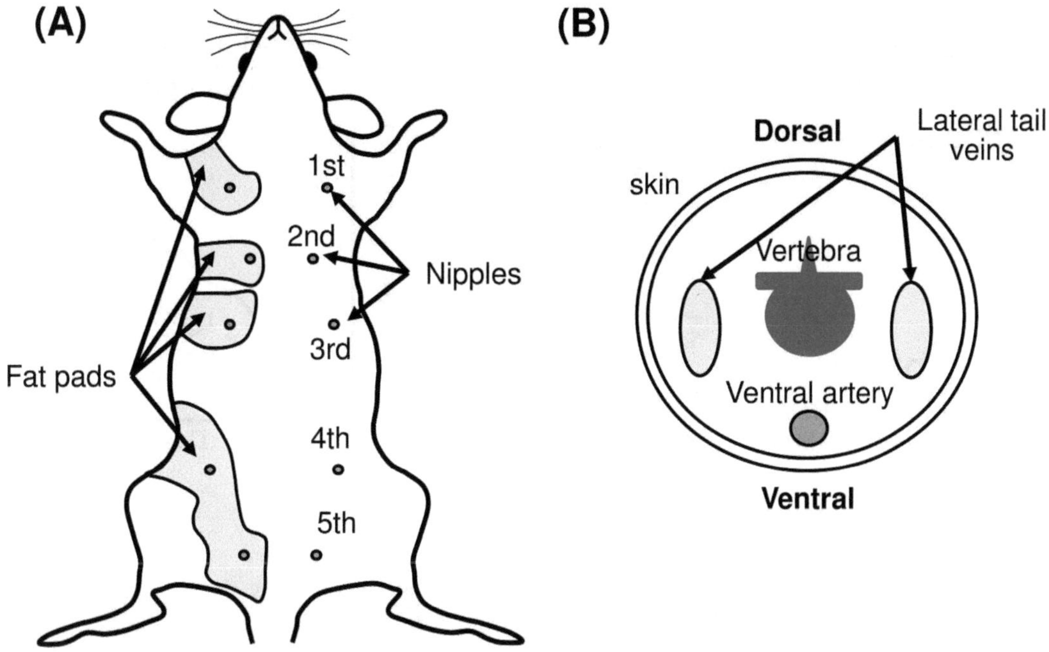

Fig. 2 Injection of tumor cells into mammary fat pad or tail vein of mice. Using an alcohol pad, disinfect the nipple or tail. (**a**) Inject breast cancer cells into the middle mammary fat pad from the fourth nipple of the nude mouse. (**b**) For tail vein injection, insert the needle (beveled edge facing up) horizontally into the lateral tail veins

6. Inject the engineered breast tumor cells (1×10^6 cells/mouse) prepared in PBS into nude mice via the tail vein using a tuberculin syringe with a 27G needle (Fig. 2b).
7. Several weeks later (4–6 weeks), sacrifice the animals by CO_2 inhalation, and assess them for lung weight and pulmonary metastases.
8. After fixing the tumor tissues in formalin and embedding them in paraffin, stain the slices with H&E, and incubate overnight at 4 °C with antibodies for immunohistological analyses.

3.3 In Vivo Treatment Model for Orthograft Model of Breast Cancer Cells (Fig. 3)

1. Select adequate siRNA sequences using the "siDirect" program (http://sidirect2.rnai.jp/), and prepare chimera siRNA against the target genes in advance (*see* **Note 3**).
2. Culture the wild-type breast tumor cell lines that express the target gene in an adequate culture medium containing 10% FBS and penicillin–streptomycin.
3. Suspend the breast tumor cells (1×10^6 cells/mouse) in 100 μL of PBS, and mix with 100 μL of Matrigel on ice (*see* **Note 2**).
4. Orthotopically inoculate the mixture into the mammary fat pad of 6-week-old female nude mice using a tuberculin syringe with a 27G needle under anesthesia (Fig. 2a).

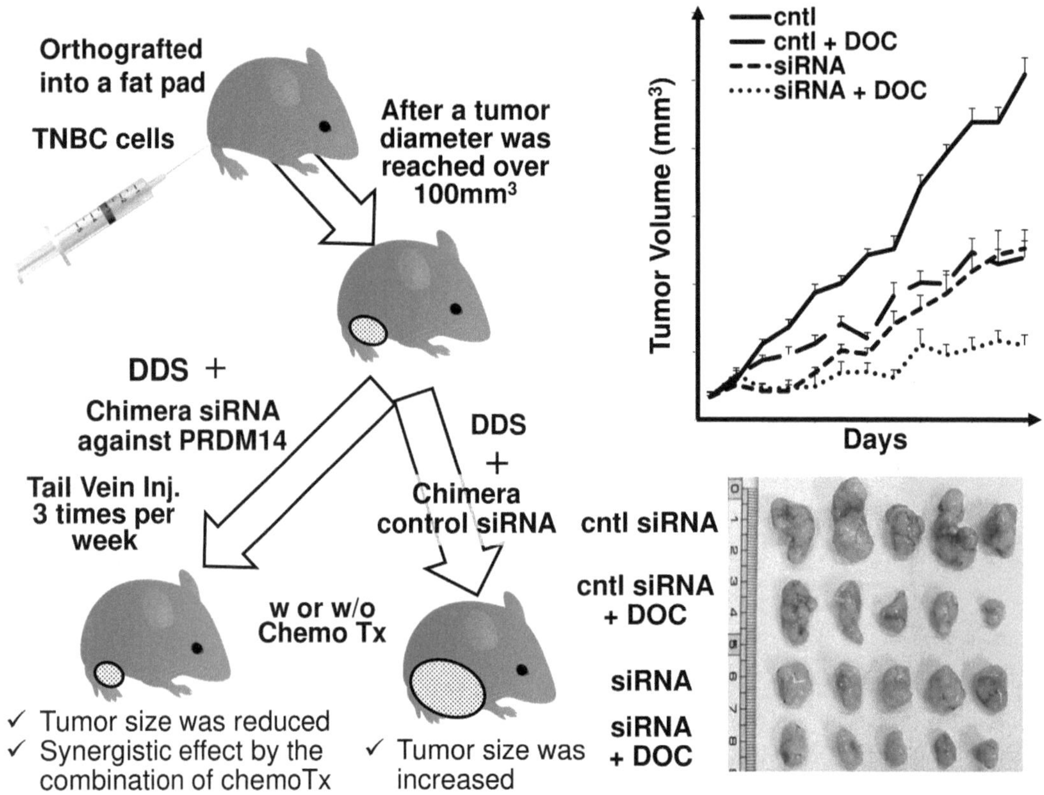

Fig. 3 PRDM14 siRNA reduces the tumorigenicity of breast cancer cells in vivo. Intravenous injection of PRDM14 siRNA/CaP hybrid micelles with or without DOC reduces the tumorigenicity of PRDM14+ breast cancer cells [6]

5. After tumor formation, measure the length and width of the tumor using calipers.
6. When the tumor volume exceeds 100 mm^3, initiate treatment with chimera siRNA and one of the DDSs indicated in Subheading 3.5 (*see* **Notes 3** and **4**).
7. Measure the length and width of the tumor using calipers until the end point.
8. Resect tumors immediately after the animals are sacrificed by CO_2 inhalation. After fixing the tumor tissues in formalin and embedding them in paraffin, stain the slices with H&E, and incubate them overnight at 4 °C with antibodies for immunohistological analyses.

3.4 Therapeutic Model for Lung Metastasis of Breast Cancer Cells (Fig. 4)

1. Select adequate siRNA sequences using the "siDirect" program, and prepare chimera siRNA against the target genes in advance.

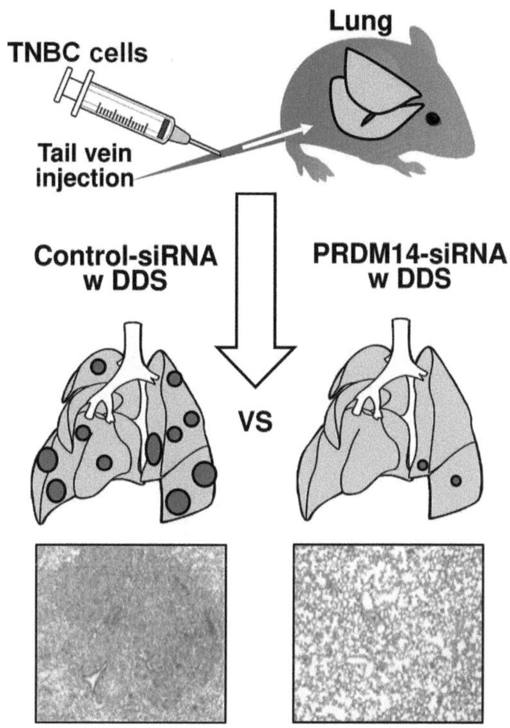

Fig. 4 Treatment with chimera siRNA and DDS for lung metastasis of PRDM14+ triple negative breast cancer (TNBC) cells. PRDM14+ TNBC cells were injected into mice via the tail vein. Three days following the injection, treatment with chimera siRNA against PRDM14 with DDS was begun. After approximately 45 days, liver or pulmonary metastases formed in the controls but not in PRDM14 siRNA-treated mice, clearly [6]

2. Culture wild-type breast tumor cells expressing the target genes with the capability to metastasize to the lung (e.g., MDA-MB-231 cells) in an adequate culture medium.
3. To create a mouse model for the treatment of pulmonary metastasis, inject breast tumor cells (1×10^6 cells/mouse), prepared in PBS, into nude mice via the tail vein (Fig. 2b).
4. 48 to 72 h after tumor cell challenge, initiate treatment with chimera siRNA and one of the DDSs indicated in Subheading 3.5.
5. 4 to 6 weeks later, euthanize the mice by CO_2 inhalation, and assess them for lung weight and pulmonary metastases.
6. After fixing the tumor tissues in formalin and embedding them in paraffin, stain the sections with H&E, and incubate overnight at 4 °C with antibodies for immunohistological analyses.

3.5 Preparation of Mixture Chimera siRNA with DDS for In Vivo Model

Mixture ratio of siRNA and DDS is different to apply in vivo-jetPEI to intratumoral or intravenous approach; on the other hand, it is unchanged to apply CaP hybrid micelles or PEG-PLL to them.

3.5.1 In Vivo-jetPEI

PEI is not so adequate for intravenous administration, due to its toxicity. However, we can use PEI as a DDS for siRNA, because PEI has the ability to accumulate in the lung via the bloodstream from the tail vein or orbital venous plexus.

1. For intratumoral injection, mix siRNA (20 μg/mouse) with in vivo-jetPEI (2.8 μL/mouse) at a nitrogen to phosphate (N/P) ratio of 7 (*see* **Note 5**), mixed with 40 μL of 5% glucose. For intravenous approach, mix siRNA (40 μg/mouse) with in vivo-jetPEI (6.4 μL/mouse) at an N/P ratio of 8, mixed with 200 μL of 5% glucose.
2. For intratumoral injection, inject the mixture directly into the tumors using a tuberculin syringe with a 29G needle three times per week. For intravenous approach, inject the mixture into the tail vein same as intratumoral injection.

3.5.2 CaP Hybrid Micelles

1. Mix 1 mg/kg chimera siRNA (RNAi, Inc., Tokyo, Japan) with calcium phosphate (CaP) hybrid micelles (kindly provided by Dr. Kazunori Kataoka) [11].
2. Inject the mixture into the tail vein using a tuberculin syringe with a 29G needle three times per week.

3.5.3 PEG-PLL

1. Mix chimera siRNA (1 mg/kg) with PEG-PLL at an N/P ratio of 5 to 10.
2. Inject the mixture into the tail vein using a tuberculin syringe with a 29G needle three times per week.

4 Notes

1. Use the lentiviral construct (GeneCopoeia, Rockville, Maryland, USA) and lentiviral packaging plasmids according to the Lenti-Pac HIV Expression Packaging Kit instructions (GeneCopoeia).
2. Dilute Matrigel with the PBS suspension of cancer cells (1:1), and keep this suspension on ice until injection into the mouse.
3. It is important to select target genes with limited expression in cancer cells and select the sequence of the siRNA while considering off-target effects and DDSs that promote accumulation at tumor sites through the enhanced permeability and retention effect.

4. If siRNA treatment is combined with chemotherapy drugs, administer doxorubicin (1 mg/kg) or docetaxel (10 mg/kg) intraperitoneally once per week.
5. The N/P ratio is the ratio of nitrogen to phosphate, which reflects the positive-to-negative charge ratio.

Acknowledgments

This work was supported by the Department of Research Promotion, Practical Research for Innovative Cancer Control, Ministry of Health, Labour and Welfare, and the Japan Agency for Medical Research and Development (AMED). We wish to thank Prof. Kazunori Kataoka from the Innovation Center of NanoMedicine for the DDSs and Prof. Yukikazu Natori and Associate professor Kumiko Ui-Tei from the University of Tokyo for chimera siRNA and selection of highly effective siRNA sequences for RNAi. We also would like to thank Editage (www.editage.jp) for English language editing.

References

1. Mzoughi S, Tan YX, Low D et al (2016) The role of PRDMs in cancer: one family, two sides. Curr Opin Genet Dev 36:83–91
2. Yamaji M, Seki Y, Kurimoto K et al (2008) Critical function of Prdm14 for the establishment of the germ cell lineage in mice. Nat Genet 40:1016–1022
3. Yamaji M, Ueda J, Hayashi K et al (2013) PRDM14 ensures naive pluripotency through dual regulation of signaling and epigenetic pathways in mouse embryonic stem cells. Cell Stem Cell 12:368–382
4. Nishikawa N, Toyota M, Suzuki H et al (2007) Gene amplification and overexpression of PRDM14 in breast cancers. Cancer Res 67:9649–9657
5. Dettman EJ, Justice MJ (2008) The zinc finger SET domain gene Prdm14 is overexpressed in lymphoblastic lymphomas with retroviral insertions at Evi32. PLoS One 3:e3823
6. Taniguchi H, Hoshino D, Moriya C et al (2017) Silencing PRDM14 expression by an innovative RNAi therapy inhibits stemness, tumorigenicity, and metastasis of breast cancer. Oncotarget 8:46856–46874
7. Moriya C, Taniguchi H, Miyata K et al (2017) Inhibition of PRDM14 expression in pancreatic cancer suppresses cancer stem-like properties and liver metastasis in mice. Carcinogenesis 38:638–648
8. Ui-Tei K, Naito Y, Saigo K (2006) Essential notes regarding the design of functional siRNAs for efficient mammalian RNAi. J Biomed Biotechnol 2006:1–8
9. Naito Y, Yoshimura J, Morishita S et al (2009) siDirect 2.0: updated software for designing functional siRNA with reduced seed-dependent off-target effect. BMC Bioinformatics 10:392
10. Ui-Tei K, Naito Y, Zenno S et al (2008) Functional dissection of siRNA sequence by systematic DNA substitution: modified siRNA with a DNA seed arm is a powerful tool for mammalian gene silencing with significantly reduced off-target effect. Nucleic Acids Res 36:2136–2151
11. Pittella F, Cabral H, Maeda Y et al (2014) Systemic siRNA delivery to a spontaneous pancreatic tumor model in transgenic mice by PEGylated calcium phosphate hybrid micelles. J Control Release 178:18–24
12. Shimizu H, Hori Y, Kaname S et al (2010) siRNA-based therapy ameliorates glomerulonephritis. J Am Soc Nephrol 21:622–633

Chapter 19

Biodrug Suppresses Breast and Colorectal Cancer in Murine Models

Syed Sultan Beevi, Naveen Kumar Tangudu, Vinod Kumar Verma, and Lekha Dinesh Kumar

Abstract

RNA interference (RNAi) remains one of the most promising and emerging strategies for the effective cancer treatment due to its high target specificity and greater potency. However, it is hindered due to lack of appropriate targeting technologies. Therefore, there is an imminent need to develop specific and robust delivery systems for successful gene silencing. Nanotechnology-based strategies have been in place to combat the shortcomings associated with viral-based delivery systems. Herein we describe protocols for successful in vitro and in vivo delivery of gene-specific nucleic acids such as siRNAs and shRNAs using PEI-PGMA nanoparticles for efficient cancer therapy.

Key words RNA interference, Gene silencing, Nanoparticles, Cancer therapy

1 Introduction

RNA interference (RNAi) is an important endogenous defense mechanism, through which eukaryotic cells can degrade exogenous genes like that of viruses [1, 2]. RNAi is a versatile research tool to develop new class of molecular therapeutics against disease-causing genes especially those that come under "non-druggable targets" which are not amenable to conventional therapies [3, 4]. Of late, RNAi technology has been proposed for cancer therapy to combat disease onset and progression [5, 6]. However, the crucial challenge to the use of RNAi technology in cancer therapy lies in the successful delivery of the desired molecule to the diseased site as well as accomplishing efficient intracellular delivery of these specific siRNAs or shRNAs for its integration into host DNA [7]. Although, viral vectors allow long-term genomic insertion of RNAi, there is a potential risk of genetic deregulation and oncogene activation by insertional mutagenesis [8]. Various studies involving non-viral nanoparticles have received immense attention

Lekha Dinesh Kumar (ed.), *RNA Interference and Cancer Therapy: Methods and Protocols*, Methods in Molecular Biology, vol. 1974, https://doi.org/10.1007/978-1-4939-9220-1_19,

in cancer therapy mainly due to its unique benefits compared to other carriers in reaching the target sites. Polyethylenimine (PEI) is a stable cationic polymer widely used as non-viral gene delivery systems both in vitro and in vivo [9]. Nevertheless, the major drawback of PEI as efficient gene delivery systems is its cytotoxicity when endeavoring for best transfection efficiency through increasing the molecular weight of the polymer [10–14]. Here, we demonstrate the methods for successful transfection of different types of nucleic acids such as siRNAs and shRNAs designed to target cancer-related genes using modified PEI-PGMA nanoparticles (NPs) [13].

2 Materials

Prepare all solutions and buffers with autoclaved ultrapure water. Sterilize culture media and other components for cell culture using 0.2 μm filters. Use surface-treated cell culture tested flasks or dishes for adherent culture and low attachment dishes for suspension culture. Prepare and store all reagents at 4 °C (unless otherwise indicated). Follow appropriate waste disposal SOPs when disposing infectious and other waste materials.

2.1 Biodrug Formulation

1. Gene-specific shRNAs in pCMV vector and siRNAs.
2. PEI-PGMA nanoparticles (NPs).
3. 20 mM HEPES buffer: To prepare 20 mM HEPES buffer containing 150 mM NaCl, dissolve 447 mg of HEPES free acid into 75 mL deionized water. Add 877 mg NaCl, and stir to dissolve the solutes completely. Adjust the pH to 7.0 and make up to 100 mL with deionized water (*see* **Note 1**).

2.2 Gel Retardation Assay

1. PEI-PGMA nanoparticles (NPs).
2. DNAs (shRNAs, siRNAs, and miRNAs).
3. 0.5 M EDTA: To prepare a 50 mL stock solution of 0.5 M EDTA, weigh 93.05 g of EDTA disodium. Dissolve it in 400 mL deionized water, and adjust the pH with sodium hydroxide (NaOH) to 8.0. Make up the solution to a final volume of 500 mL (*see* **Note 2**).
4. 50× TAE buffer: Make a stock solution (50×) of TAE buffer by weighing 242 g of Tris base and dissolving it in approximately 750 mL of deionized water. Carefully add 57.1 mL of glacial acetic acid and 100 mL of 0.5 M EDTA (pH 8.0). Adjust the solution to a final volume of 1 L (*see* **Note 3**).

5. 1× TAE buffer: Prepare working solution of 1× TAE buffer by diluting the stock solution by 50× in deionized water (*see* **Note 4**).
6. Ethidium bromide dye: Prepare ethidium bromide dye (1 mg/mL) by dissolving 10 mg of EtBr in 10 mL of deionized water (*see* **Note 5**).
7. 1.0% agarose gel: Prepare 1.0% agarose gel by weighing 0.5 g and dispensing into 50 mL of 1× TAE buffer (*see* **Note 6**).

2.3 Cytotoxicity of Biodrug in Mammalian Cells

1. PEI-PGMA nanoparticles (NPs).
2. Different cancer cell lines such as HEK293, MDA-MB231, MCF7, COLO205, and JURKAT (*see* **Note** 7).
3. Specific culture media for cell lines such as DMEM, RPMI-1640, etc. Prepare respective complete growth media consisting of incomplete culture medium along with supplements such as 10% fetal bovine serum (FBS) and 1% penicillin-streptomycin solution (*see* **Note 8**).
4. Fetal bovine serum.
5. Penicillin-streptomycin solution.
6. Phosphate-buffered saline: Make a stock solution (10×) of phosphate-buffered saline (PBS) without Ca^{2+} or Mg^{2+} by weighing 80 g of NaCl, 25.6 g of $Na_2HPO_4{\cdot}7H_2O$, 2.0 g of KCl, and 2.0 g of KH_2PO_4 and dissolving it in approximately 750 mL of deionized water. Adjust the pH to 7.2 and bring to 1 L with deionized water (*see* **Note 9**). Prepare working solution of 1× PBS by diluting the stock solution of 10× in deionized water.
7. Trypsin-EDTA: Prepare a stock of 10× trypsin-EDTA solution for passaging cancer cells. Dissolve 500 mg of porcine trypsin (0.5%) and 200 mg of EDTA disodium (0.2%) in 75 mL of phosphate-buffered saline (PBS) without Ca^{2+} or Mg^{2+}, and make up to 100 mL with PBS. Prepare working solution of 1× trypsin-EDTA by diluting the stock solution of 10× in 1× PBS (*see* **Note 10**).
8. MTT solution: Prepare MTT solution by dissolving 5 mg in 1 mL of PBS and filter sterilize (*see* **Note 11**).

2.4 In Vitro Transfection of Biodrug

1. Different cancer cell lines such as HEK293, MDA-MB231, MCF7, COLO205, and JURKAT (*see* **Note** 7).
2. Gene-specific GIPZ shRNA and GIPZ non-silencing shRNA complexed with PEI-PGMA nanoparticles (NPs)-biodrug (*see* **Note 12**).
3. Specific culture media for cell lines such as HDMEM, RPMI-1640, etc. Prepare respective complete growth media consisting of incomplete culture medium along with supplements

such as 10% fetal bovine serum (FBS) and 1% penicillin-streptomycin solution (*see* **Note 8**).

4. Opti-MEM.

2.5 In Vivo Cytotoxicity of Biodrug

1. Wild-type C57BL6/J mice from Jackson laboratory (*see* **Note 13**).
2. PEI-PGMA nanoparticles (NPs).
3. Gene-specific GIPZ shRNA and GIPZ non-silencing shRNA (*see* **Note 12**).
4. 2 mL syringe with $26_{1/2}$ gauge needle and oral gauge for mice.
5. Laminar airflow unit (*see* **Note 14**).

2.6 Biodrug Delivery to Mammary Tumors (BRCA2/p53 Knockout Models)

1. PEI-PGMA nanoparticles (NPs).
2. Gene-specific GIPZ shRNA and GIPZ non-silencing shRNA (*see* **Note 12**).
3. 2 mL syringe with $26_{1/2}$ gauge needle and oral gauge for mice.
4. Laminar airflow unit (*see* **Note 14**).

2.6.1 Divide BRCA2/p53 Knockout Mice into Different Cohort as Follows

1. Treatment with vehicle alone ($n = 5$).
2. Treatment with PEI-PGMA (NPs) alone ($n = 5$).
3. Treatment with gene-specific shRNA alone ($n = 5$).
4. Treatment with gene-specific biodrug complex ($n = 5$).
5. Treatment with scrambled shRNA + NPs complex ($n = 5$).

2.7 Biodrug Delivery to Colon Cancer Models (Apc fl/fl Models)

1. β-Naphthoflavone (10 mg/mL).
2. Tamoxifen (10 mg/mL).
3. PEI-PGMA nanoparticles (NPs).
4. Gene-specific GIPZ shRNA and GIPZ non-silencing shRNA (*see* **Note 12**).
5. 2 mL syringe with $26_{1/2}$ gauge needle and oral gauge for mice.
6. Laminar airflow unit (*see* **Note 14**).

2.7.1 Divide Apc fl/fl Mice into Different Cohort as Follows

1. Treatment with vehicle alone ($n = 5$).
2. Treatment with PEI-PGMA (NPs) alone ($n = 5$).
3. Treatment with gene-specific shRNA alone ($n = 5$).
4. Treatment with gene-specific shRNA + NPs (biodrug) complex ($n = 5$).
5. Treatment with scrambled shRNA + NPs complex ($n = 5$).

2.8 Quantitative PCR to Evaluate Gene-Specific Knockdown

2.8.1 RNA Isolation from Transfected Cancer Cell Lines and Tumor Tissues

1. TRIzol® Reagent (*see* **Note 15**).
2. Phosphate-buffered saline.
3. Ethanol.
4. Isopropanol.
5. RNase-free water (DEPC treated) (*see* **Note 16**).
6. RNase ZAP™.
7. 1.5 mL microcentrifuge tubes (DEPC treated) (*see* **Note 16**).
8. Disposable cell scraper.
9. 10CC syringes.
10. 18G syringe needles.
11. Autoclaved plastic mortar (to fit in 1.5 or 2 mL Eppendorf tube).
12. Pipette tips with filter plugs.

2.8.2 Synthesis of cDNA from RNA

1. High-capacity cDNA reverse transcriptase kit (*see* **Note 17**).
2. The kit consists of 10× RT buffer, 10× RT random primer, 25× dNTP mix (100 mM), MultiScribe™ reverse transcriptase (50 Unit/μL), and RNase inhibitor.
3. RNA at the concentration of at least 1–2 μg.
4. PCR instrument.

2.8.3 Quantitative PCR Using SYBR Green as Fluorescent Probe

1. SYBR green master mix.
2. cDNA.
3. Gene-specific forward and reverse primers.
4. 96-well PCR plate.
5. Pipette tips with filter plugs.
6. ViiA7 qPCR instrument with 96-well or 324-well block.

2.9 Immunoblotting to Evaluate Gene-Specific Knockdown

1. Frozen fresh tumor tissue.
2. Transfected cancer cells in 6-well plates.
3. Autoclaved plastic mortar (to fit in 1.5 or 2 mL Eppendorf tube).
4. RNase and DNase-free water.
5. Microcentrifuge tubes.
6. RIPA buffer: Mix together 5 mL of 1 M Tris–HCl (pH 7.4), 30 mL of 5 M NaCl, 5 mL of 20% Triton X-100, 5 mL of 10% sodium deoxycholate, 0.5 mL of 20% SDS, and 50 mL double-distilled water (*see* **Note 18**).
7. Pierce™ BCA assay kit (Cat No. 23227) for the measurement of protein concentration (*see* **Note 19**).

8. Precast SDS-PAGE gel (*see* **Note 20**).
9. Western blotting instrument for wet transfer.
10. Blocking solution [5% bovine serum albumin (BSA) or fat-free milk powder] (*see* **Note 21**).
11. Phosphate-buffered saline containing Tween 20 (*see* **Note 22**).
12. Appropriate primary and HRP-conjugated secondary antibody diluted at the ratio of 1:1000 for primary and 1:5000 for secondary antibody in 1% BSA solution.
13. Enhanced chemiluminescence solution.
14. ChemiDoc system.
15. Image J software.

2.10 Immunohistochemistry to Evaluate Gene-Specific Knockdown

1. Wax-embedded tumor tissue blocks.
2. Microtome.
3. Frosted and positively charged slides.
4. Glass cover slips.
5. Cryo-freezing media and plastic molding cassettes.
6. Dry ice and −80 °C freezer.
7. 4–5 μm thick wax tissue sections and 10 μm cryosections.
8. 100% xylene.
9. Graded series of 70%, 80%, 95%, and 100% ethanol.
10. Distilled water.
11. Antigen retrieval buffer: To prepare citrate buffer (10 mM citric acid, 0.05% Tween 20, pH 6.0), mix 1.92 g of anhydrous citric acid in 1000 mL of distilled water. Adjust the pH to 6.0 with 1 N sodium hydroxide, and then add 0.5 mL of Tween 20 (*see* **Note 23**).
12. Peroxidase blocking solution: Add one part 30% hydrogen peroxide to nine parts methanol.
13. Blocking solution [5% bovine serum albumin (BSA) or fat-free milk powder in 1× PBS] (*see* **Note 22**).
14. Appropriate primary and HRP-conjugated secondary antibody diluted at the ratio of 1:100 for primary and 1:500 for secondary antibody in 1% BSA solution in 1× PBS.
15. DAKO DAB kit.
16. Hematoxylin counterstaining.
17. Nonfluorescent mounting media.
18. Bright-field and confocal microscope.

3 Methods

3.1 Biodrug Formulation

1. Centrifuge 2 mL of PEI-PGMA NPs (2 μg/mL) at 16,000 × *g* for 30 min at room temperature.
2. Remove supernatant carefully without disturbing the NP pellet.
3. Suspend the NP pellet in 0.5 mL of HEPES buffer with gentle tapping.
4. Sterilize the NP solution under UV irradiation for 20 min.
5. Add DNAs such as shRNAs and siRNAs/miRNAs to the NPs at the different ratios such as 1:1, 4:1, 10:1, and 25:1. DNA without NPs, scrambled DNA, and NPs alone will serve as controls.
6. Incubate at room temperature for 45 min to induce nano-DNA complex formation called biodrug.
7. Check by gel retardation assay the most precise binding ratio by running the biodrug in a classical agarose gel electrophoresis.

3.2 Gel Retardation Assay

1. Prepare 1.0% agarose in TAE buffer (*see* **Note 6**).
2. Assemble gel tray and comb of appropriate size.
3. After the gel cools down to the temperature of 60 °C, add 1 μL of ethidium bromide solution per 10 mL of the gel, and swirl slowly (*see* **Note 5**).
4. Pour gently the gel into gel tray and let it cool down for 25–30 min.
5. Remove the comb, and submerge the gel in 1× TAE buffer in electrophoresis buffer tank (*see* **Note 24**).
6. Load 8–10 μL of nano-DNA complex and DNA size marker in separate wells.
7. Close the lid of the tank by connecting the appropriate electrodes.
8. Connect the electrical wires to the power supply.
9. Run the gel at 100 V for 30 min.
10. Power off the gel apparatus, and carefully lift the gel tray from the tank and visualize the DNA band under the UV light within gel documentation system.

3.3 Cytotoxicity of Biodrug in Mammalian Cells

1. Culture and maintain cancer cell lines such as HEK293 and MDA-MB231 in HDMEM media and MCF7, COLO205, and JURKAT in RPMI-1640 media.

2. Seed cancer cells in a 96-well plate at a seed rate of 2×10^4 cells/well, in 100 μL of respective complete media (*see* **Note 8**).
3. Incubate cells overnight at 37 °C in a humidified atmosphere containing 5% CO_2.
4. Treat the cultured cells with predetermined concentrations (0, 10, 50, 75, 150, 300 μg/mL) of NPs diluted appropriately with culture media in triplicates.
5. Mark series of wells as control, and treat the cells with respective culture medium without NPs.
6. Grow the cells for 48 h at 37 °C in a humidified atmosphere containing 5% CO_2.
7. After 48 h, carefully remove the media containing NPs by aspiration.
8. Add 100 μL of MTT solution to each well, and incubate further in the dark for 4 h.
9. After the incubation period, aspirate the MTT solution, and replace it with 100 μL of DMSO to solubilize the formazan crystals.
10. Incubate the 96-well plates at 37 °C for 45 min.
11. Determine the amount of formazan in each well by measuring the absorbance in multiskan microplate reader at 540 nm.
12. Present the data as percentage posttreatment recovery (% live cells), and define the absorbance from non-treated control cells as 100% live cells.

3.4 In Vitro Transfection of Biodrug

1. Culture and maintain cancer cell lines such as HEK293 and MDA-MB231 in HDMEM media and MCF7, COLO205, and JURKAT in RPMI-1640 media.
2. Seed cancer cells in a 24-well plate at a seed rate of 1×10^5 cells/well, in 500 μL of complete media.
3. Incubate cells overnight at 37 °C in a humidified atmosphere containing 5% CO_2.
4. Transfect the cultured cells with NPs complexed with shRNAs or siRNAs at the ratio of 25:1 for shRNA and 4:1 for siRNAs when it reaches 50–60% confluency.
5. Incubate the cells at 37 °C in a humidified atmosphere containing 5% CO_2 for 4 h for siRNAs and 6 h for shRNAs.
6. Replace the NP-DNA/RNA complex with fresh culture media, and incubate further at 37 °C in a humidified atmosphere containing 5% CO_2 for 48 or 72 h.

7. Evaluate the transfection efficacy of biodrug through downstreaming processing like immunoblot, immunocytochemistry, and qPCR (Fig. 1).

3.5 In Vivo Cytotoxicity of Biodrug

1. Divide wild-type mice into five cohorts (control, NPs, DNAs, and biodrug complex at two different ratios of 25:1 and 4:1) each consisting of five animals.
2. Each cohort receives a daily dose of 0, 250, 500, 1000, and 1500 μg of NPs or DNA or as a biodrug complex in a maximum volume of 250 μL by oral gavage for a period of 21 days.
3. Sacrifice the animals after the stipulated period of 21 days as per the ethics guidelines.
4. Remove the liver, kidney, lung, and spleen, and snap freeze immediately in liquid nitrogen.
5. Freeze dry the tissue using a Labconco system, and crush under aseptic conditions.
6. Analyze the freeze-dried pellets for the presence of characteristic peaks using a Bruker Vertex-70 FTIR spectrometer, with the ATR mode in the range of 400–4000 per cm.
7. Characteristic peaks represent intactness of macromolecular organization of the organs.

3.6 Biodrug Delivery to Mammary Tumors (BRCA2/p53 Knockout Models)

1. Breast cancer model (conditional Brca2/p53 knockout under control of Blg-cre transgene) that develops autochthonous tumors on any of the five pairs of mammary glands in a 6–15-month time window.
2. Divide animals into five groups (one group each for untreated control, shRNA + NPs, scrambled + NPs, c-Myc shRNA only, and NPs only) and each consisting of 5–6 animals.
3. Measure mammary tumor at 0.05 cm^3 before starting the treatment.
4. Complex 10 μg of c-Myc shRNA with 250 μg (1:25) of NPs in a maximum volume of 50 μL as described previously.
5. Treat all tumors in each cohort identically by giving intratumoral injection of 50 μL shRNA-NP complex along with control cohorts of scrambled shRNA + NPs (25:1), c-Myc shRNA only (10 μg), and NPs alone (250 μg) with concentrations only as described in **step 4** (*see* **Note 25**).
6. Test the penetrability and distribution of NPs in the tumor and its surrounding periphery by magnetic resonance imaging (MRI) and fluorescent confocal imaging.
7. Mice were humanely culled, and tumors were dissected and processed for downstream experiments.

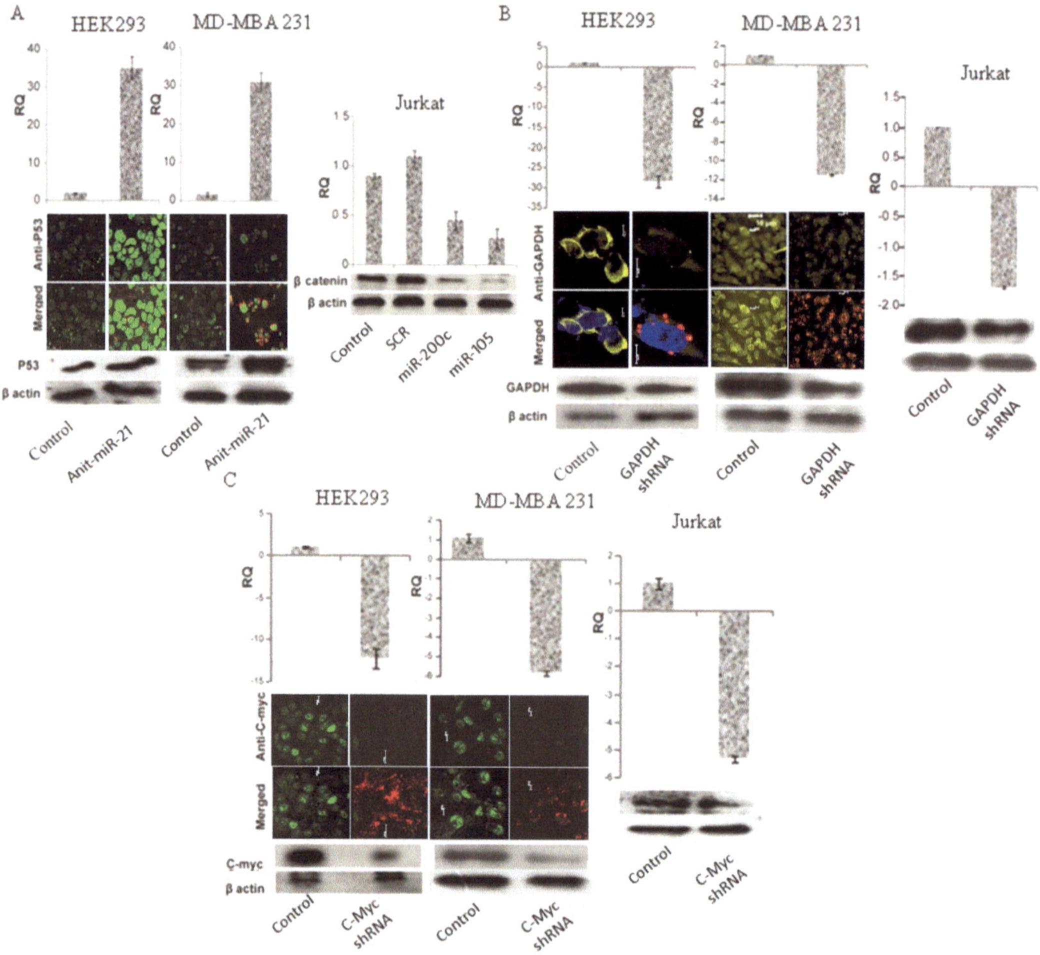

Fig. 1 Silencing of target genes using RNAi. (**a**) Upregulation of p53 using anti-miR-21 oligos linked to the rhodamine-labeled NPs. RNAi effect of miR-21 is demonstrated by upregulation of p53 mRNA by qPCR and protein by immunofluorescence and immunoblotting, respectively, in HEK293 and MDA-MB231 cell lines, as compared with untreated control cell lines or scrambled shRNA-treated controls. Similarly, knockdown of b-catenin by mimics of miR-200c and miR-105 is demonstrated by immunoblotting in Jurkat cell lines. (**b**) Downregulation of GAPDH using GAPDH shRNA linked to the nanoparticles. RNAi effect of delivered shRNA on GAPDH expression is demonstrated by qPCR, immunofluorescence, and immunoblotting in HEK293, MDA-MB231, and Jurkat cell lines as compared with controls. (**c**) Knockdown of the functional oncogene c-Myc using c-Myc shRNA linked to the NPs. RNAi effect of delivered shRNA on c-Myc is demonstrated by qPCR, immunofluorescence, and immunoblotting in HEK293, MDA-MB231, and Jurkat cell lines as compared with controls (untreated or scrambled shRNA-treated). b-Actin was used as a loading control in all experiments. All qPCR experiments were conducted in triplicate, each experiment having three biologic replicates and two technical replicates, as assessed by fluorescence. The relative quantitation (RQ) of expression of the respective genes in the treated cells in relation to their respective controls is presented here. All controls were without treatment of NPs or by treating with scrambled (scr) shRNA. Representative confocal images show the rhodamine B-labeled NPs (red), nuclei labeled with DAPI (blue), p53, c-Myc, and GAPDH labeled with anti-p53, anti-c-Myc (green, Alexa Fluor 488), and anti-GAPDH (yellow), respectively. All images were captured at 10 mm scale

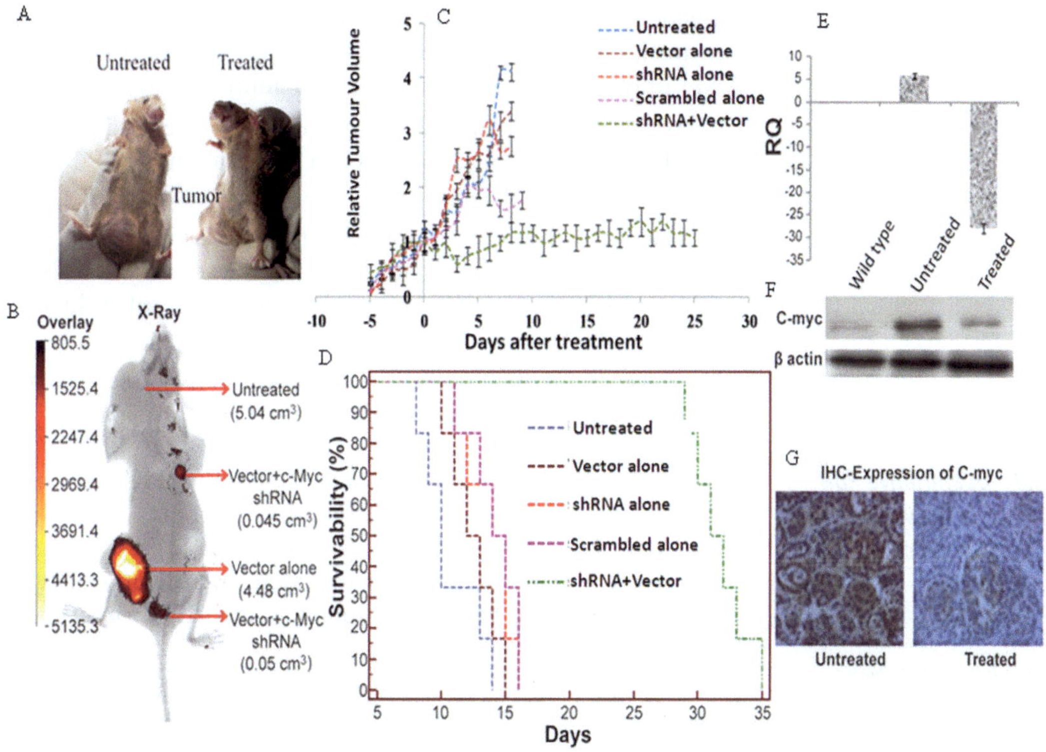

Fig. 2 Suppression of autochthonous mammary tumors by on-site delivery of nanoparticles to a mouse model of breast cancer (Brca2/p53 knockout). (**a**) Example of mammary tumor regression in conditional Brca2/p53 knockout mouse model upon treatment with c-Myc shRNA complexed with nanoparticles (NP). (**b**) For the purpose of live multispectral imaging, four similarly sized mammary tumors that appeared within a span of 10 days on the same mouse were treated individually for a period of 25 days, resulting in suppression of tumors treated with c-Myc shRNA-linked nanoparticles (third and fifth mammary) compared with continued growth of untreated tumor (first mammary) and tumor treated with scrambled shRNA-linked nanoparticles (fourth mammary). The red/yellow color in the tumor core versus periphery shows the intensity of the rhodamine expression as shown in the intensity scale. Suppression of tumor growth (**c**) and increased survival (**d**) in mice treated with c-Myc shRNA-NPs, as compared with control cohorts. (**e–g**) Reduction in c-Myc transcript and protein expression levels in tumors from the treated cohorts compared with wild-type tissue

8. Determine the change in the expression level of c-Myc through q-PCR, immunoblotting, and immunohistochemistry (Fig. 2).

3.7 Delivery of Biodrug to Colon Cancer Knockout Models (Apc fl/fl)

1. For colorectal cancer, use Apc knockout model (Apc fl/fl), an inducible colorectal cancer model which develops "crypt progenitor" phenotype in the intestine upon injection with β-naphthoflavone and tamoxifen.
2. Inject 6–8-week-old mice through intraperitoneal route with 80 mg/kg β-naphthoflavone and tamoxifen (dissolved together in corn oil at 10 mg/mL) once daily for 5 days (*see* **Note 26**).

3. Divide animals into five groups (one group each for untreated control, biodrug, scrambled shRNA + NPs, shRNA only, and Nps only) and each consisting of 5–6 animals.
4. Complex 50 μg of shRNA (in pGIPZ vector)-encoding plasmid DNA with 1.25 mg of NPs in a total volume of 250 μL as described previously.
5. Treat all tumors in each cohort identically by giving oral gavage of biodrug along with control cohorts of scrambled + NPs, c-Myc shRNA only, and NPs only from day 1 of induction until the end time point.
6. Mice were humanely culled, and tumors were dissected and processed for downstream experiments (Fig. 3).

3.8 Quantitative PCR Analysis to Evaluate Gene-Specific Knockdown

3.8.1 RNA Isolation from Transfected Cancer Cell Lines and Tumor Tissues

1. Cut a piece of frozen tissue on dry ice-cooled glass plate into an autoclaved tube from different mice cohorts [including wild-type C57BL6/J, Brca2/p53 knockout (breast cancer), and Apc fl/fl (colon cancer) mice treated with or without NPs, DNA alone, and biodrug complex], and thaw on ice for 30 min.
2. Gently add 1 mL of TRIzol under the laminar hood to avoid the phenol odor.
3. Homogenize the tissue in tissue homogenizer at a speed of 30 Hz, 1.5 min for three times.
4. Incubate homogenized samples at room temperature (around 25 °C) for 5 min.
5. Gently mix with 200 μL of clean chloroform, and mix the tube by turning up and down for ten times.
6. Centrifuge the samples at a speed of 10200 × *g* for 10 min at 4 °C.
7. Collect clear aqueous phase into a fresh autoclaved tube, and gently mix it with 0.7 volumes of isopropanol, and mix tube gently by turning up and down for ten times.
8. Centrifuge the samples at maximum speed (13,500 rpm) for 15 min at 4 °C.
9. Discard the liquid phase, and wash the pellet with 70% ethanol (add around 500 μL in each tube) by centrifuging the samples at maximum speed 10200 × *g* for 5 min at 4 °C.
10. Discard the liquid and dry the pellet for 5–10 min at RT.
11. Dissolve the pellet in an appropriate volume of DEPC water.
12. Check the RNA integrity by running a 2% agarose gel.
13. Measure the RNA concentration using NanoDrop 200 to get in ng/μL units.

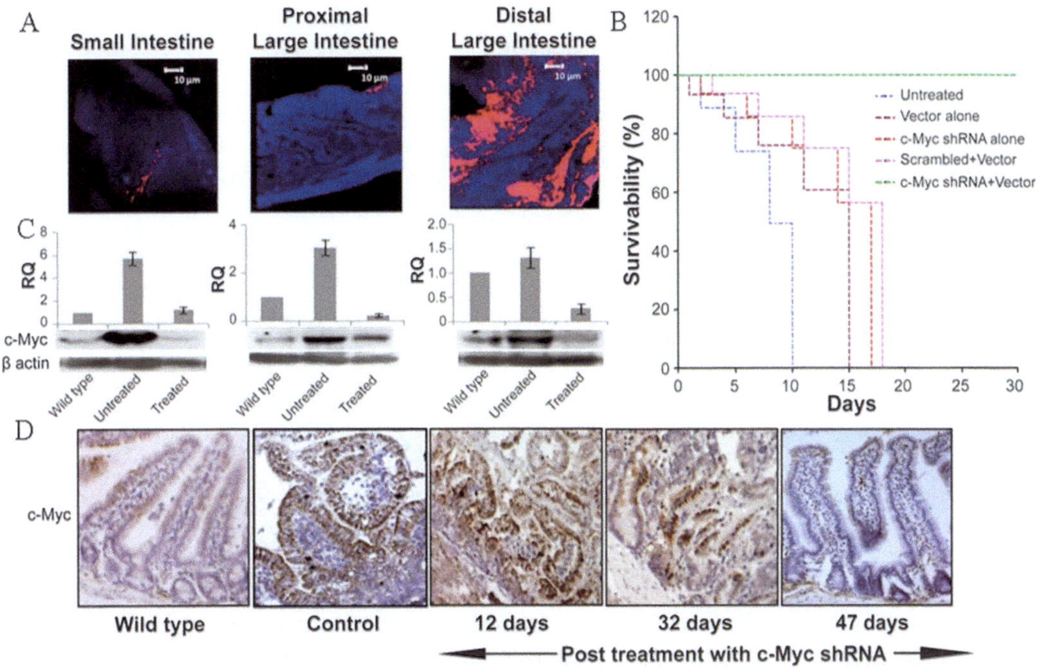

Fig. 3 Non-viral nanoparticle is suitable for oral delivery of shRNA to a mouse model of colorectal cancer (Apc knockout). (**a**) Confocal imaging of the fluorescent nanoparticles in Apc-deficient gut, following oral administration of 50 μg c-Myc shRNA-encoding plasmid DNA complexed with 1250μg nanoparticles, showing maximum retention of complex in the distal end of the large intestine. (**b**) Prolonged survival (up to 47 days) of mice treated with c-Myc shRNA-NP complex as compared with other treated cohorts [scrambled shRNA + nanoparticles, shRNA alone, nanoparticles alone, and induced but untreated mice (control)]. Wild type was taken as control tissue for all analyses. (**c**) Decreased c-Myc transcripts and protein levels on 12, 32, and 47 days after treatment, as demonstrated by RT-PCR using TaqMan gene expression assays and Western blotting, respectively. Mice were healthy but culled at these time points specifically to analyze expression. (**d**) Immunohistochemistry for c-Myc protein in treated cohort compared with other treated groups and control cohort (as described in **b**), again demonstrating gradual reduction in c-Myc protein levels during treatment

3.8.2 Synthesis of cDNA from RNA

1. Convert 1–2 μg of RNA in to cDNA using reverse transcriptase cDNA kit.
2. Gene-specific primers could be designed (Table 1).
3. Follow the manufacturer's instructions for synthesis of cDNA.

3.8.3 Quantitative PCR Using SYBR Green as Fluorescent Probe

1. Prepare a fresh 96- or 384-well qPCR plate, and add 5–25 ng of cDNA in each well from different samples.
2. Add appropriate amount of forward and reverse primers (10 pm/μl), distilled water, SYBR green master mix, and passive reference dye (ROX) in each well.
3. For each target gene, keep a negative control and water control.
4. Measure the gene expression by running the plate in ViiA7 qPCR instrument.

Table 1
Primers for qPCR analysis

PCR	Primer sequences (5′–3′)
c-Myc	c-Myc F = GCCTCTTTTCCACAGAAACAAC c-Myc R = CTACCCTCTCAACGACAGCAG
p53	p53 F = CCTCAGCATCTTATCCGAGTGG p53 R = TGGATGGTGGTACAGTCAGAGC
GAPDH	GAPDH F = AAGGTGAAGGTCGGAGTCAAC GAPDH R = ATGGGTGGAATCATATTGGAAC
U6	U6 F = CGCTTCGGCAGCACATATAC U6 R = AAAATATGGAACGCTTCACGA
NONO	NONO F = CATCAAGGAGGCTCGTGAGAAG NONO R = TGGTTGTGCAGCTCTTCCATCC

3.9 Protein Expression Analysis to Evaluate Gene-Specific Knockdown

1. Cut a piece of frozen tissue on dry ice-cooled glass plate into an autoclaved tube from different mice cohorts [including wild-type C57BL6/J, Brca2/p53 knockout (breast cancer), and Apc fl/fl (colon cancer) mice treated with or without Nps, DNA alone, and biodrug complex], and thaw on ice for 30 min.
2. Firstly prepare incomplete RIPA buffer with 150 mM NaCl, 1% NP-40, 0.5% sodium deoxycholate, 0.1% SDS, 50 mM Tris–HCl pH 8.0, and dH_2O.
3. Then prepare complete RIPA buffer freshly with the following composition: 818 μL incomplete RIPA buffer, 143 μL 7× protease inhibitor cocktail, 5 μL 200 mM sodium orthovanadate, 25 μL 1 M sodium fluoride, and 10 μL 100 mM PMSF.
4. Homogenize the tissue with autoclaved plastic pestle on top of ice till the tissue lyses.
5. Sonicate the tissue lysate for 1 min on ice for three times with 1 min interval each time.
6. Incubate on ice for 30 min.
7. Centrifuge the samples at maximum speed (11,000 × *g*) for 20 min at 4 °C.
8. Collect the lysate into a new autoclaved tube.
9. Freeze the samples immediately, or measure the protein concentration using BCA kit by following the manufacturer's instructions.
10. Load 30–50 μg of denatured protein (protein along with 1× SDS loading dye denatured by heating at 95 °C for 5 min).

11. Prepare 10%/12% SDS-PAGE according to the protocol and composition mentioned in *Molecular Cloning* textbook, written by Sambrook and Russell.
12. Transfer proteins from PAGE to 0.2 μm nitrocellulose membrane using wet transfer method (standard protocol mentioned in *Molecular Cloning* textbook, written by Sambrook and Russell).
13. Check the protein transfer from gel to membrane by incubating the membrane with Ponceau stain (a non-covalent protein binding stain) for 5 min, and immediately wash with water till the removal of extra stain.
14. The membranes were blocked with 5% BSA followed by overnight incubation of primary antibodies (rabbit anti-c-Myc, rabbit anti-β-catenin, mouse anti-GAPDH from Merck Millipore, USA, or Cell Signaling Technology, USA) at 4 °C with mild shaking in a rocker (500 rotations per minute).
15. Wash the membranes with 1× TBST for 10 min, three times.
16. Further incubate with rabbit and mouse secondary antibodies conjugated with HRP.
17. Wash the membranes with 1× TBST for 10 min, three times.
18. Add appropriate amount of chemiluminescence HRP substrate solution on membrane, and develop or detect the signal at ChemiDoc instrument.

3.10 Immunohistochemical Analysis to Evaluate Gene-Specific Knockdown

1. Cut 4–5 μm thickness of wax sections and 10 μm thickness of cryosections from tumor tissues of different cohorts including wild-type C57BL6/J, Brca2/p53 knockout (breast cancer), and Apc-CreERT (colon cancer) mice treated with or without Nps DNA alone and biodrug complex.
2. Incubate the slides at 65 °C for 30 min on hot plate.
3. De-wax the slides by incubating in xylene for 10 min, two times.
4. Rehydrate the slides by incubating the sections in decreasing concentration of alcohol (100%, 95%, 70% for 5 min each), and finally incubate in distilled water for 10 min.
5. Boil the slides (water bath at 95 °C) in antigen retrieval solution for 20–30 min (depends on the antibody), and later slowly cool down the slides to room temperature.
6. Incubate the slides with peroxidase blocking solution (0.3% or 3% H_2O_2 in methanol).
7. Wash the slides with PBS for 10 min, three times.
8. Block the sections with blocking solution (5% goat serum in 1% BSA in 1× PBS).

9. Round the sections with PAP pen (hydrophobic to avoid the wastage of primary antibody).
10. Incubate slides with primary antibody [rabbit anti-c-Myc (1:200), rabbit anti-β-catenin (1:200), Cell Signaling Technology, USA] at 4 °C for overnight in a humidified chamber.
11. Wash the slides with 1× PBST for 10 min, three times.
12. Incubate sections with HRP conjugated secondary antibody (follow the instructions mentioned in the kit).
13. Wash the slides with 1× PBST for 10 min, three times.
14. Incubate the slide with DAB substrate.
15. Wash the slides with water for 10 min.
16. Counterstain slides with hematoxylin for 1 min, and immediately wash under the running tap water.
17. Mount the sections with mounting media and appropriate cover slips.
18. Observe the staining under bright-field microscope.
19. Cryosections: Mount the sections with fluorescent mounting media (DAPI), and observe the signal under the confocal microscope.

4 Notes

1. Filter sterilize and store it at 4 °C. Remains stable for at least 1 month if stored at 4 °C.
2. Prepare a stock solution of 0.5 M EDTA well ahead of time. EDTA will not go completely into a solution until the pH is adjusted to about 8.0. This stock solution can be stored at room temperature.
3. This stock solution can be stored at room temperature. The pH of this buffer should not be adjusted and should be about 8.5.
4. Final solute concentrations in 1× TAE buffer are 40 mM/L Tris acetate and 1 mM/L EDTA.
5. Aliquot into 1 mL stock and store it at 4 °C for up to 1 year. Handle EtBr with appropriate safety procedure and must be treated as hazardous waste.
6. Dissolve agarose by boiling in microwave oven. Place flask with agarose solution in 60 °C in a water bath to cool.
7. Procure established cancer cell lines from ATCC culture collection.

8. Prepare incomplete culture media such as DMEM, RPMI-1640, etc., as per cancer cell types by following the instructions on the reagent label.
9. Autoclave for 40 min at 121 °C. This stock solution can be stored at room temperature.
10. Aliquot stock solution in 5 mL volume, and store it at −20 °C for up to 6 months. Final solute concentrations in 1× trypsin-EDTA are 0.05% porcine trypsin and 0.02% EDTA. Solution can be stored at 4 °C for up to 2 weeks without loss of enzyme activity.
11. Solution can be stored at 4 °C for up to 1 week without loss of activity.
12. Use GIPZ *c-Myc* shRNA as DNA constructs and GIPZ non-silencing shRNA as negative control to rule out off-target effects. Use the constructs as per the manufacturer's instruction.
13. Maintain all mice under constant 12 h light-dark cycle, and allow mice to food and water ad libitum in special pathogen-free facility.
14. Prepare a laminar hood by sterilizing it with UV light for 30 min, and later decontaminate with 70% ethanol before use it for experiment.
15. TRIzol® Reagent is very toxic and should only be used with caution. Work only in a fume hood with proper personal protective equipment (gloves, eye protection, etc.). TRIzol® waste is hazardous and should be handled and disposed of according to institutional hazardous waste protocols.
16. Water and other solutions should be treated with 0.1% DEPC. DEPC is a strong, but not absolute, inhibitor of RNases. It is used at a concentration of 0.1% to inactivate RNases on glass or plastic ware or to create RNase-free solutions and water. DEPC inactivates RNases by covalent modification.
17. Store all kit components at −15 to −25 °C.
18. RIPA buffer does not contain protease or phosphatase inhibitors. Protease and phosphatase inhibitors cocktail can be added to the RIPA buffer just before use to prevent proteolysis and maintain phosphorylation of proteins.
19. The BCA Protein Assay Kit is a two-component, high-precision, detergent-compatible assay reagent set to measure (A562 nm) total protein concentration compared to a protein standard.
20. Use 10–12% gel for proteins having molecular weight between 20 and 150 kDa.

21. Add BSA or fat-free milk powder in 1× PBS, and swirl gently to dissolve without forming any froth. Remain stable for a week if stored at 4 °C.
22. Add 0.1% Tween 20 to 1× PBS to prepare PBST. Cut the tip end for easy transfer of viscous Tween 20 to PBS solution. Remain stable at room temperature for at least 1 month.
23. Store at room temperature for up to 3 months; for extended storage, store at 4 °C.
24. Switch off the electrophoresis apparatus before placing the gel into the tank.
25. A volume more than 50 μL can cause leakage during intratumoral injections and not more than 250 μL in case of oral gavaging.
26. The drug has to be administered from day 1 of induction. The mice will be extremely sick, and you can see piloerection and blood discharge from the anus, hence providing mashed food.

Acknowledgments

We acknowledge all authors of original publication for their respective contribution to the original manuscript from where this protocol is derived.

References

1. Agrawal N, Dasaradhi PVN, Mohamed A, Malhotra P, Bhatnagar RK, Mukherjee SK (2003) RNA interference: biology, mechanism, and applications. Microbiol Mol Biol Rev 67(4):657–685
2. Sledz CA, Williams BRG (2005) RNA interference in biology and disease. Blood 106(3):787–794
3. Rao M, Sockanathan S (2005) Molecular mechanisms of RNAi: implications for development and disease. Birth Defects Res C Embryo Today 75(1):28–42
4. Caplen NJ (2004) Gene therapy progress and prospects. Downregulating gene expression: the impact of RNA interference. Gene Ther 11(16):1241–1248
5. Ren YJ, Zhang Y (2014) An update on RNA interference-mediated gene silencing in cancer therapy. Expert Opin Biol Ther 14(11):1581–1592
6. Phalon C, Rao DD, Nemunaitis J (2010) Potential use of RNA interference in cancer therapy. Expert Rev Mol Med 12:e26
7. Pecot CV, Calin GA, Coleman RL, Lopez-Berestein G, Sood AK (2011) RNA interference in the clinic: challenges and future directions. Nat Rev Cancer 11(1):59–67
8. Thomas CE, Ehrhardt A, Kay MA (2003) Progress and problems with the use of viral vectors for gene therapy. Nat Rev Genet 4:346–358
9. Höbel S, Aigner A (2010) Polyethylenimine (PEI)/siRNA-mediated gene knockdown in vitro and in vivo. Methods Mol Biol 623:283–297
10. Moghimi SM, Symonds P, Murray JC, Hunter AC, Debska G, Szewczyk A (2005) A two-stage poly(ethylenimine)-mediated cytotoxicity: implications for gene transfer/therapy. MolTher 11:990–995
11. Grzelinski M, Urban-Klein B, Martens T, Lamszus K, Bakowsky U, Hobel S et al (2006) RNA interference-mediated gene silencing of pleiotrophin through polyethylenimine-complexed small interfering RNAs in vivo exerts antitumoral effects in glioblastoma xenografts. Hum Gene Ther 17:751–766

12. Breunig M, Lungwitz U, Liebl R, Goepferich A (2007) Breaking up the correlation between efficacy and toxicity for nonviral gene delivery. Proc Natl Acad Sci U S A 104:14454–14459
13. Evans CW, Fitzgerald M, Clemons TD, House MJ, Padman BS, Shaw JA et al (2011) Multimodal analysis of PEI-mediated endocytosis of nanoparticles in neural Cells. ACS Nano 5:8640–8648
14. Iyer KS, Zdyrko B, Malz H, Pionteck J, Luzinov I (2003) Polystyrene layers grafted to macromolecular anchoring layer. Macromolecules 36:6519–6526

Chapter 20

Combinatorial Nanoparticle Delivery of siRNA and Antineoplastics for Lung Cancer Treatment

Rebaz Ahmed, Narsireddy Amreddy, Anish Babu, Anupama Munshi, and Rajagopal Ramesh

Abstract

Recent developments in nanotechnology, especially in drug delivery systems, are advanced by featuring novel multifunctional nanoparticles that promise safe, specific, and efficient therapeutic delivery for cancer treatment. Multifunctional nanoparticle-based drug delivery systems enable simultaneous delivery of multiple therapeutic agents for effective combination therapy for cancer. In this chapter, we provide detailed protocols for development and application of a multifunctional nanoparticle system for combinatorial delivery of a chemotherapeutic (cisplatin) and small interfering RNA (siRNA) for human antigen R (HuR) mRNA in cancer cells using a polyamidoamine (PAMAM) dendrimer platform. Protocols for nanoparticle functionalization with folic acid (FA) for targeted delivery of therapeutics toward folate receptor (FR)-overexpressing cancer cells are also described. Further, methods employed for physiochemical and functional characterization of the multifunctional nanoparticle system are discussed in detail. Using the methods described in this chapter, researchers would be able to develop PAMAM dendrimer-based multifunctional nanoparticles for targeted delivery of chemotherapeutics and siRNA combinations. We also provide an example showing the dendrimer-polyethyleneimine-*cis*-diamminedichloroplatinum-siRNA-folic acid (Den-PEI-CDDP-siRNA-FA) nanoparticle system developed was therapeutically effective toward non-small cell lung cancer (NSCLC) cell lines (H1299 and A549) while exhibiting reduced toxicity to normal lung fibroblast (MRC9) cells.

Key words PAMAM dendrimers, CDDP, HuR siRNA, Lung cancer, Folic acid

1 Introduction

The treatment of cancer often requires the combination of multiple agents with distinct properties and therapeutic efficiencies. To achieve enhanced therapeutic efficiency and to reduce the cytotoxicity of individual agents, such as chemotherapeutics, nucleic acid therapeutics, or small molecule inhibitors, combinatorial delivery methods have been developed. However, there is a continuing need for efficient and reliable drug delivery systems that can be tuned to accommodate combined delivery of two or more therapeutic agents with different characteristics. Recent developments

Lekha Dinesh Kumar (ed.), *RNA Interference and Cancer Therapy: Methods and Protocols*, Methods in Molecular Biology, vol. 1974, https://doi.org/10.1007/978-1-4939-9220-1_20,

in nanoparticle-based drug delivery systems have opened up new avenues for the combinatorial delivery of anticancer agents [1].

Nanoparticles based on liposomes, polymers, gold, and silica are commonly employed for the fabrication of combinatorial drug delivery systems [2–4]. Many have been successfully tested in in vitro and in vivo cancer models and have shown promise in combinatorial drug delivery for cancer treatment. Among these nanoparticle systems, dendrimers (Den) form a unique platform for the development of a biocompatible combinatorial drug delivery system due to their structural and functional characteristics, such as the nanoscale size, availability of functional groups for easy modification with therapeutic agents and/or targeting ligands, and the flexibility to accommodate multiple therapeutic molecules within the cavity-like structures [5–7].

Dendrimers (Den) are immensely branched, three-dimensional nanoparticle structures that are synthesized commonly by employing either (1) divergent or (2) convergent methods. In the divergent method, monomeric molecules that possess reactive groups are coupled in a stepwise fashion to a multifunctional core moiety. Each stepwise addition of reactive monomers from the multifunctional core, starting from generation "0" (G0), results in an upward increase in the generation of the dendrimer. In the convergent method, the dendrimers are synthesized by gradual assembly of surface units with reactive monomers that grow inward, resulting in more homogenous and precise molecular weight dendrimers. By following the above mentioned methods, we can control the size and customize the functionalities in a Den [8, 9].

Polyamidoamine (PAMAM) Den is known for its biocompatibility and for its highly branched polymeric network that can be utilized for the fabrication of a suitable nanoplatform for the combinatorial delivery of anticancer agents [10, 11]. In this chapter, we describe the methods for the preparation of a PAMAM Den-based targeted system for delivery of a combination of *cis*-diamminedichloroplatinum (CDDP) and small interfering RNA (siRNA) and also discuss its targeting ability and combinatorial therapeutic efficiency in cancer cells [12].

PAMAM Den has several primary and secondary amine groups. Those amine groups were utilized for coordinating with platinum (pt) through coordinate bonds via hydrolysis. SiRNA possesses a negative surface charge due to its phosphate backbones, and amine groups of Den exhibit positive surface charge in protonated buffer Tris–HCl (pH 7.4). When we incubate siRNA with PAMAM Den in Tris–HCl buffer, the siRNA binds to the Den by electrostatic interaction and condensation. Since Den have globular structures that each contains a cavity, Den hold therapeutic molecules with stronger interactions. For active targeted delivery, surface amine groups of Den were covalently conjugated with folic acid (FA)-polyethylene glycol (PEG)-*N*-hydroxysuccinimide (NHS), known as FA-PEG-NHS, through amide bonds. Each of the synthesis steps was confirmed by determining the changes in the surface charge, as measured by zeta potential. The size of the nanoparticles and dispersion were confirmed by transmission electron microscopy (TEM), as shown in Fig. 1a. The stability of the siRNA encapsulation and Den-siRNA complex in serum was confirmed by agarose gel electrophoresis. The siRNA and CDDP release kinetics in physiological pH 7.4 buffer were measured in the presence and absence of serum. The FA and CDDP concentrations were optimized using fluorescently labeled siRNA (siGLO) and cell viability, respectively, in non-small cell lung cancer (NSCLC) cell lines. The interaction of FA-functionalized nanoparticles with folate receptor alpha (FRA) on the tumor cell surface was confirmed by temperature-dependent endocytosis and receptor blocking studies.

Detailed methods are described for the combined delivery of CDDP and HuR (human antigen R) siRNA in NSCLC cell lines. Techniques used to evaluate the efficiency of Den-PEI-CDDP-siRNA or Den-PEI-CDDP-siRNA-FA, such as cell uptake, cell viability, and reverse transcriptase-polymerase chain reaction (RT-PCR) analysis, are described in detail. The protocols used in the comparative study between NSCLC cell lines and a normal cell line are also presented. The results presented demonstrate that normal lung fibroblast cells showed negligible effect with combined delivery, compared with delivery of individual therapeutics. The FA-targeted delivery was more therapeutically effective than was non-targeted delivery in tumor cells, as confirmed by cell viability, Western blot analysis (Fig. 1b), and apoptosis analysis by annexin V/PI staining.

2 Materials

2.1 Maintaining Adherent Cell Culture

1. Falcon® 6-well cell culture plate.
2. CELLSTAR® cell culture flasks 75 cm^2.
3. Bright-field inverted microscope.
4. 1× phosphate-buffered saline (PBS), pH 7.4.
5. Sterile disposable pipette tips (1–200 μL and 1 mL).
6. Sterile disposable serological pipettes (1, 5, and 10 mL).
7. Sterile centrifuge tubes (15 and 50 mL).
8. Laminar flow cabinet.
9. A 6 L water bath, 37 °C.
10. Hausser Scientific Bright-Line™ and Hy-Lite™ Counting Chamber.
11. Trypan blue 0.4% solution.
12. Disposable microcentrifuge tubes (1.5 mL).

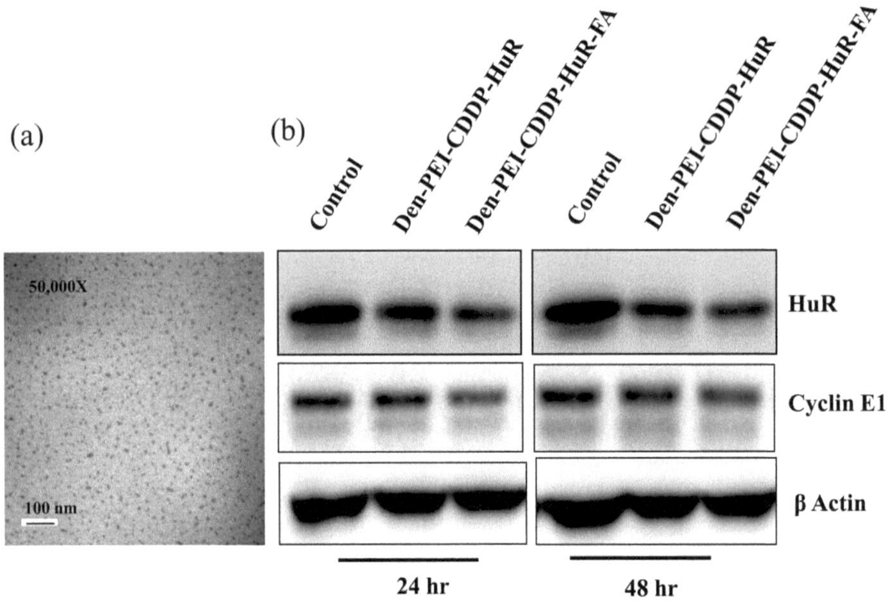

Fig. 1 (**a**) Transmission electron microscopy (TEM) image shows the shape and size of the dendrimer particles; (**b**) folate-targeted dendrimer-mediated co-delivery of HuR siRNA and CDDP in human H1299 lung cancer cells showed greater reduction in HuR and HuR-regulated cyclin E1 protein expression compared to non-targeted dendrimer-mediated delivery of HuR siRNA and CDDP

13. Coulter cell counter.
14. Benchtop refrigerated microcentrifuge and centrifuge with A-4-44 rotor.
15. A CO_2 incubator, 37 °C.
16. Ethyl alcohol (70%).
17. Normal lung fibroblast cell line (MRC-9) and lung cancer cell lines (H1299 and A549).
18. Tissue culture medium: RPMI-1640 and Minimum Essential Media.
19. Heat-inactivated fetal bovine serum.
20. Trypsin containing EDTA.
21. Pipettes (P-1000, P-200, and P-20).

2.2 Preparation of Den-PEI-CDDP-siRNA-FA Nanoparticles

1. Polyamidoamine (PAMAM) dendrimer fourth generation, 14,214 MW.
2. Tris (base).
3. Hydrochloric acid (HCl; 11.65 N).
4. Polyethyleneimine (PEI) 800 MW.
5. *N*-Hydroxysuccinimide (NHS).
6. Dialysis tubing (pore size: 8-kDa MW cutoff).

7. *Cis*-diamminedichloroplatinum (CDDP).
8. Ultrapure water (Milli-Q H_2O).
9. Sterile glass cylinder and beaker (1000 mL).
10. Magnetic stirrer.
11. pH meter.
12. Folic acid-PEG-NHS (FA).
13. UV-Vis spectrophotometer.
14. HuR-siRNA. (5′ UCA AAG ACG CCA ACU UGU A 3′) and control siRNA (5′ UAA GGC UAU GAA GAG AUA C 3′) were purchased from Dharmacon.
15. Sterile glass vials and bottles with caps.

2.3 Zeta Potential Analysis

1. Each step of synthesized nanoparticles for Den-PEI-CDDP-siRNA-FA.
2. Tris (base).
3. Hydrochloric acid (HCl; 11.65 N).
4. Dynamic light scattering instrument.
5. Pipettes (P-1000, P-200, and P-20).
6. Sterile disposable pipette tips (1–200 μL and 1 mL).
7. Deionized distilled water (dH_2O).

2.4 Transmission Electron Microscopy (TEM) Imaging

1. Den-PEI and Den-PEI-CDDP-siRNA-FA nanoparticle samples.
2. Deionized distilled water (dH_2O).
3. Millex® LCR syringe filters (pore size: 400 nm).
4. Syringes (5 mL).
5. Copper grids.
6. Whatman filter papers.
7. 2% phosphotungstic acid (PTA).
8. Transmission electron microscopy (TEM) (HITACHI, 80 kV).

2.5 CDDP Release Kinetics

1. Den-PEI-CDDP-siRNA and Den-PEI-CDDP-siRNA-FA samples.
2. Slide-A-Lyzer mini dialysis tubes with 200 μL capacity (3500 MW cutoff).
3. Disposable microcentrifuge tubes (1.5 mL).
4. Shaker incubator (37 °C).
5. UV-Vis spectrophotometer (DeNovix DS11).
6. Pipettes (P-1000, P-200, and P-20).
7. Sterile disposable pipette tips (1–200 μL and 1 mL).

2.6 siRNA Release Kinetics

1. Den-PEI-CDDP-siRNA and Den-PEI-CDDP-siRNA-FA nanoparticle samples.
2. Disposable microcentrifuge tubes (1.5 mL).
3. Shaker incubator (37 °C).
4. Benchtop refrigerated microcentrifuge and centrifuge with A-4-44 rotor.
5. PicoGreen assay reagent to measure siRNA.
6. Multiplate reader.

2.7 Agarose Gel Electrophoresis

1. HuR siRNA, Den-PEI-CDDP-siRNA, and Den-PEI-CDDP-siRNA-FA samples.
2. 2× sample loading buffer (1 mL sterile H_2O, 1 mL glycerol, and 0.05 mg bromophenol blue).
3. 1.2% agarose gel (0.6 g agarose in 50 mL TAE buffer, pH 8.0).
4. DNA ladder standard (100 kbp).
5. DNA electrophoresis supplies (gel casting tray, gel combs, electrophoresis tank, power supply).
6. Ethidium bromide solution.
7. TAE buffer (4.84 g Tris base, 1.14 mL glacial acetic acid, 2 mL 0.5 M EDTA (pH 8.0) in 1 L water).
8. A chemiluminescence imaging system.
9. Pipettes (P-1000, P-200, and P-20).
10. Sterile disposable pipette tips (1–200 μL and 1 mL).

2.8 Ortho-phenylenediamine (OPDA) assay

1. Orthophenylenediamine (OPDA).
2. Disposable microcentrifuge tubes (0.5 and 1.5 mL).
3. UV-Vis spectrophotometer (DeNovix DS11).
4. Pipettes (P-1000, P-200, and P-20).
5. Sterile disposable pipette tips (1–200 μL and 1 mL).
6. Heat block (temperature at 95 °C).
7. Sterile water.

2.9 Optimization of Folic Acid Concentration

1. Non-small cell lung cancer cell line (H1299).
2. Den-PEI-CDDP nanoparticles complex prepared.
3. Fluorescently labeled siRNA (siGLO red) at a concentration of 50 nM per well.
4. Different concentrations of folic acid-PEG-NHS linker (3.75, 7.5, 11, 18, 37, 56, or 75 μg/mL).
5. Multiplate reader.

6. Disposable microcentrifuge tubes (1.5 mL).
7. Pipettes (P-1000, P-200, and P-20).
8. Sterile disposable pipette tips (1–200 μL and 1 mL).
9. Sterile centrifuge tubes (15 mL).

2.10 Optimization of CDDP Concentration for Therapeutic Study

1. Non-small cell lung cancer cell line (H1299).
2. Den-PEI-CDDP-siHuR-FA nanoparticle complex prepared.
3. Different concentrations of CDDP (10, 25, 40, 50, 60, 80, or 100 μM).
4. Disposable microcentrifuge tubes (1.5 mL).
5. Pipettes (P-1000, P-200, and P-20).
6. Sterile disposable pipette tips (1–200 μL and 1 mL).
7. Sterile centrifuge tubes (15 mL).

2.11 Cell Uptake Study

1. Non-small cell lung cancer cell line (H1299).
2. Glass coverslips and slides.
3. Falcon® 6-well cell culture plate.
4. 1× phosphate-buffered saline (PBS), pH 7.4.
5. Fluorescent-siRNA (siGLO) containing Den-PEI-CDDP.
6. Paraformaldehyde (4%).
7. DNA fluorescent stain, DAPI (4′,6-diamidino-2-phenylindole).
8. Mounting medium.
9. Nail polish (for sealing coverslips).
10. A fluorescence microscope.
11. ImageJ analysis software.
12. LysoTracker green endosomal staining solution.

2.12 Receptor-Specific Cell Uptake Assay (Receptor Blocking by Varying Incubation Temperature)

1. Non-small cell lung cancer cell line (H1299).
2. RPMI-1640 culture media.
3. Falcon® 6-well cell culture plate.
4. Den-PEI-CDDP-siGLO or Den-PEI-CDDP-siGLO-FA nanoparticle prepared.
5. CO_2 incubator (37 °C).
6. Refrigerator (4 °C).
7. Multiplate reader.

2.13 Receptor-Specific Cell Uptake Assay (Receptor Blocking by Exogenous FA in the Incubation Medium)

1. Non-small cell lung cancer cell line (H1299).
2. Minimal, regular, and 1 mM of exogenous FA-containing RPMI-1640 media.
3. Den-PEI-siGLO-FA nanoparticles containing 50 nM siGLO concentrations.
4. Multiplate reader.

2.14 Trypan Blue Cell Viability Assay

1. Harvested lung cancer cells (H1299).
2. Hausser Scientific Bright-Line™ and Hy-Lite™ Counting Chamber.
3. Trypan blue 0.4% solution.
4. Coverslips.
5. Bright-field inverted microscope.
6. Disposable microcentrifuge tubes (1.5 mL).
7. Pipettes (P-1000, P-200, and P-20).
8. Sterile disposable pipette tips (1–200 μL and 1 mL).
9. Sterile centrifuge tubes (15 and 50 mL).

2.15 Western Blotting Analysis

2.15.1 Sodium Dodecyl Sulfate-Polyacrylamide Gel Electrophoresis

1. Freshly prepared 10% (w/v) ammonium persulfate (APS).
2. Bis-acrylamide: 30% bis-acrylamide solution (stored at +4 °C).
3. Protein ladder.
4. Running buffer: prepare 1× by dissolving 3 g of 0.025 M Tris, 14.4 g of 0.192 M glycine, and 2 g SDS. Adjust the pH to 8.8, and then make up the volume to 1000 mL with ddH_2O.
5. 4× sample loading buffer: mix 2.5 mL 1 M Tris–HCl pH 6.8, 0.5 mL of ddH_2O, 1.0 g SDS, 0.8 mL 0.1% bromophenol blue, 4 mL 100% glycerol, and 2 mL β-mercaptoethanol (100% stock). Adjust the final volume to 10 mL with ddH_2O.
6. Resolving buffer: dissolve 36.35 g of 1.5 M Tris–HCl (pH 8.8) in 200 mL ddH_2O. Store the solution at 4 °C.
7. Stacking buffer: dissolve 12.1 g of 0.5 M Tris–HCl (pH 6.8) in 200 mL ddH_2O. Store the solution at 4 °C.
8. Dissolve 10 g of SDS in 100 mL of ddH_2O to get 10% (w/v). Store the SDS solution at room temperature.
9. N,N,N',N'-Tetramethyl-ethylenediamine (TEMED).
10. Western blot apparatus for running: glass plates, casting frames, combs, and casting racks.
11. Electrophoresis power supply.
12. Prepare the resolving gel percentage according to the following:

Solutions in total 20 mL	7.5%	10%	12.5%
Acrylamide-bis (30%)	5.0 mL	6.7 mL	6.25 mL
3 M Tris–HCl (pH 8.8)	5.0 mL	5.0 mL	3.75 mL
dH_2O	9.70 mL	8.0 mL	4.75 mL
TEMED	10.0 μL	10.0 μL	10.0 μL
10% SDS	200 μL	200 μL	200 μL
10% APS	100 μL	100 μL	100 μL

13. Prepare the stacking gel:

Solutions in total 10 mL	
Acrylamide-bis (30%)	1.0 mL
0.5 M Tris–HCl (pH 6.8)	2.5 mL
dH_2O	6.3 mL
TEMED	8.0 μL
10% SDS	100 μL
10% APS	80 μL

2.15.2 Protein Transfer and Detection

1. Polyvinylidene difluoride (PVDF) membrane.
2. Whatman filter papers.
3. Transfer buffer: prepare 1× by dissolving 3 g of 0.025 M Tris and 14.4 g of 0.192 M glycine. Adjust the pH to 8.0, and then make up the volume to 1000 mL with ddH_2O. Then mix 80 mL of transfer buffer with 20 mL methanol before use it.
4. Western blot apparatus for transfer: buffer tank, buffer tank lid, electrode assembly, blue cooling unit, holder cassette, and foam pads.
5. 1× TBS-T: dissolve 2.4 g of 1.5 M Tris–HCl and 29.29 g of NaCl in 800 mL ddH_2O, adjust the pH to 7.5 with HCl, and then make up the volume to 1000 mL after adding 0.5 mL of Tween 20.
6. Blocking buffer: 5 g of fat-free milk in 100 mL of TBS-T to get 5% (w/v) or 5 g of bovine serum albumin (BSA) in 100 mL of TBS-T to get 5% (w/v). Store the solution at +4 °C.
7. Primary antibody against the protein of interest. HuR, cyclin E (1:1000 dilution); caspase 9, PARP, and γH2AX (1:500); and β-actin (1:2000).

8. Enhanced chemiluminescent (ECL) reagent.
9. Horseradish peroxidase (HRP)-conjugated secondary antibody corresponding to the primary antibody.
10. A chemiluminescence imaging system.
11. GelQuant software.

2.15.3 Stripping and Reprobing

1. Freshly prepared mild stripping buffer: Mix equal volumes of H_2O_2 and PBS in 1:1 ratio.
2. Freshly prepared harsh stripping buffer: Mix 20 mL SDS 10%, 12.5 mL Tris–HCl, (pH 6.8; 0.5 M) 67.5 mL distilled water, and 0.8 mL β-mercaptoethanol under the fume hood.

2.16 Apoptotic Analysis

1. Harvested H1299 cells treated with Den-PEI-CDDP-HuR and Den-PEI-CDDP-HuR-FA.
2. Ice-cold 1× phosphate-buffered saline (PBS), pH 7.4.
3. Trypsin containing EDTA.
4. 1× annexin V and propidium iodide binding buffers.
5. FACSCalibur flow cytometer. Cell Quest software.
6. Deionized distilled water (dH_2O).
7. Disposable microcentrifuge tubes (1.5 mL).
8. Pipettes (P-1000, P-200, and P-20).
9. Sterile disposable pipette tips (1–200 μL and 1 mL).
10. Sterile centrifuge tubes (15 and 50 mL).

2.17 Quantitative Real-Time PCR (qRT-PCR)

2.17.1 Total RNA Isolation

1. Sterile disposable cell scraper.
2. 1× phosphate-buffered saline (PBS), pH 7.4.
3. TRIzol reagent.
4. DEPC-treated water.
5. Chloroform.
6. 100% isopropanol.
7. 70% ethanol.
8. Pipettes (P-1000, P-200, and P-20).
9. MultiMax Recovery Barrier pipette tips, (DNase-, RNase-, pyrogen-, human DNA-, and PCR inhibitor-free).
10. Sterile disposable serological pipettes (10, 5, and 1 mL).

2.17.2 Reverse Transcription (RT) and cDNA Synthesis

1. Total RNA samples.
2. Nuclease-free water.
3. Reverse transcriptase.
4. Reaction buffer containing dNTPs, magnesium chloride, and stabilizers.

5. Oligo (dt) primer.
6. Pipettes (P-1000, P-200, and P-20).
7. MultiMax Recovery Barrier pipette tips, (1–200 μL and 1 mL) DNase-, RNase-, pyrogen-, human DNA-, and PCR inhibitor-free.

2.17.3 qRT-PCR

1. Supermix, which contains double-stranded (dsDNA)-binding dye (SYBR Green), iTaq DNA polymerase, dNTPs, $MgCl_2$, Tris–HCl pH 8.4, KCl, fluorescein, and stabilizers.
2. Nuclease-free water.
3. Primers: forward and reverse were purchased from Integrated DNA Technologies.
4. Sterile microcentrifuge tubes.
5. Optical qPCR microplates or tubes.
6. cDNA template.

3 Methods

3.1 Maintaining Cell Lines and Seeding for Treatment of Cell Viability (H1299, A549, and MRC-9)

1. Inside a laminar flow hood, remove and discard the cell culture media from the culture flask or dish. Wash cells by adding phosphate-buffered saline (PBS), pH 7.4, to the side of the flask or dish (approximately 2 mL per 10 cm culture surface area). Gently shake the flask back-and-forth several times.
2. Add enough pre-warmed (37 °C) dissociation reagent trypsin-EDTA to the side of the flask (approximately 0.5 mL per 10 cm) to cover the cell layer, and gently rock the container to get complete coverage of the cell layer.
3. Incubate the culture vessel at room temperature for approximately 2 min. The incubation time varies with the cell line used.
4. Look for detachment of the cells under the microscope. If cells are less than 90% detached, increase the incubation time a few more minutes, checking for dissociation every 30 S. Tapping the flask expedites cell detachment.
5. Add pre-warmed (37 °C) culture medium (twice the volume used for the dissociation reagent), and then disperse the medium by pipetting over the cell layer surface several times.
6. Transfer the cells to a 15 mL conical tube, and then centrifuge at 200 × *g* for 5 min.
7. Discard the supernatant and suspended the cell pellet in 5 mL of pre-warmed culture medium. Remove an aliquot for counting.
8. Determine the total number of viable cells with a hemocytometer, by trypan blue exclusion, or count using an automated cell counter. If necessary, add growth medium to the cells to achieve the desired cell concentration, and recount the cells.

9. Calculate the volume needed to transfer 0.1 million cells from the suspended cells/well in 6-well plates in total volume of 2 mL culture medium. The number of seeded cells for treatment varies depends on the cell line used.
10. Add 1 mL of the suspended cells into a new flask containing 11 mL of pre-warmed culture medium for future experiments. Place the 6-well plates and the flask containing culture medium in the incubator. The flask cap should be loosened to allow proper gas exchange.
11. Allow the cells to attach and grow overnight, and then add the treatment.
12. Check for the confluence of the cells in the flask and change/replace the medium whenever the color turns pinkish yellow. After about 90% confluence, you can again subculture the cells.

3.2 Preparation of Den-PEI-CDDP-siRNA-FA Nanoparticles

1. Polyamidoamine (PAMAM) dendrimer (Den) generation 4.0 was purchased from Sigma-Aldrich. This dendrimer has 64 peripheral amine groups.
2. 20 mg of dendrimer was diluted with 2 mL of Tris–HCl (pH 7.4). We subsequently added 11.8 μg of polyethyleneimine (PEI) and bifunctional NHS-PEG-NHS (28 μg) cross-linker molecule to cross-link between the amine groups of the dendrimer and the PEI molecules.
3. The above reaction complexes were stirred for 24 h at room temperature.
4. The unreacted PEI and PEG molecules were removed by dialysis for 6 h against Tris–HCl.
5. The reaction mixture was kept in a dialysis tube (8-kDa MW cutoff cellulose membranes) and dialyzed against Tris–HCl pH 7.4 buffer for 6 h.
6. CDDP encapsulation into Den-PEI was performed through hydrolysis.
7. CDDP (15 mg) was dissolved in 10 mL of Milli-Q water at room temperature while stirring for 3 h.
8. 20 mg of the above Den-PEI nanoparticles was added into CDDP solution and stirred for 24 h at room temperature in the dark.
9. The unencapsulated CDDP was removed by dialysis, as mentioned in **step 5**.
10. The CDDP encapsulation efficiency was measured using orthophenylenediamine (OPDA) colorimetric assay (Subheading 3.8) by comparing the known CDDP concentrations stranded curve with the unbound concentration.

11. For targeting cell surface receptors, folic acid (FA) was conjugated to the surface of the amine groups of dendrimers in Den-PEI-CDDP nanoparticles using an FA-PEG-NHS cross-linker through amide bond.
12. 47 μg of Den in Den-PEI-CDDP was allowed to react at room temperature with different concentrations of folic acid-PEG-NHS linker (20, 40, 80, or 160 μg) in Tris–HCl (pH 7.4) buffer for 3 h.
13. The unbound folic acid molecules were removed by centrifugation for 15 min at 16,000 × *g*.
14. The conjugation efficiency was confirmed by FA absorbance at 280 nm.
15. Finally, siRNA was encapsulated with Den-PEI-CDDP-FA nanoparticles in Tris–HCl buffer.
16. Den-PEI-CDDP-FA nanoparticle was mixed with siRNA solution at a 10:1 nitrogen (dendrimer amine groups) to phosphate (nucleic acid phosphate) ratio (N:P, 10:1) in Tris–HCl pH 7.4 and incubated at room temperature for 20 min.
17. The formation of the Den-PEI-CDDP-siRNA-FA nanoparticles complex was confirmed by agarose gel electrophoresis assay (Subheading 3.7).

3.3 Zeta Potential Analysis

1. The surface charge of each step of conjugation for preparing Den-PEI-CDDP-siRNA-FA nanoparticles was measured using ZetaPALS.
2. All samples were dispersed in Tris–HCl (pH 7.4) buffer and vigorously mixed to avoid aggregates. (Note: the samples should be diluted enough for zeta potential measurement.)
3. The samples were measured for zeta potential in triplicate using a Brookhaven ZetaPALS analyzer.
4. The average zeta potential of each group was plotted in a graph.

3.4 Transmission Electron Microscopy (TEM) Imaging

1. The Den-PEI and Den-PEI-CDDP-siRNA-FA nanoparticle solution in deionized distilled water was filtered through a 400 nm syringe filter twice to eliminate any aggregates or dust particles.
2. A fraction from this sample was placed on a clean copper grid for 2 min to allow sedimentation over grid columns.
3. The leftover sample was carefully eliminated using a filter paper.
4. The sample was then stained using 2% phosphotungstic acid (PTA) for 2 min.
5. The stained samples were air dried and then directly used for TEM imaging with magnification of 50,000×.

3.5 CDDP Release Kinetics

1. CDDP-loaded Den-PEI-CDDP-siRNA and Den-PEI-CDDP-siRNA-FA (with known CDDP concentrations) were suspended in a mini dialysis tube (200 μL).
2. The tubes were then immersed in a 1.5 mL Eppendorf tube containing Tris–HCl (pH 7.4) and 10% FBS Tris–HCl.
3. Those tubes were kept in a shaker at 37 °C with 100 rpm.
4. At each predetermined time interval, 200 μL of sample was withdrawn from the 1.5 mL tube and replaced with fresh buffers in the respective tubes.
5. After 7 h of collection, the CDDP concentration was estimated in collected tubes using the OPDA colorimetric assay (Subheading 3.8) by measuring the absorbance of Pt-OPDA at 704 nm.
6. CDDP release was calculated as % cumulative release.
7. The drug release percentage was calculated by using the following equation:

% CDDP release = 100 (% of drug contents in nanoparticles is taken as 100) − [Amount of drug in 1 mL (μg/mL) × 50 (mL) × dilution factor] × 100.

3.6 siRNA Release Kinetics

1. siRNA-encapsulated Den-PEI-CDDP-siRNA and Den-PEI-CDDP-siRNA-FA nanoparticles were dispersed in 1.5 mL Eppendorf tubes containing Tris–HCl (pH 7.4) and 10% FBS Tris–HCl.
2. The above samples were incubated in a shaker (100 rpm) at 37 °C.
3. At predetermined time intervals, samples were centrifuged at 21,000 × *g* for 15 min, and the supernatant was collected.
4. The samples were again redispersed in the respective buffers.
5. After 24 h of collection, the collected samples were stained with 7.5 μL of PicoGreen reagent (as per the manufacturer's protocol).
6. The fluorescence intensity of PicoGreen-siRNA was measured with an Envision multiplate reader using excitation 485 nm and emission 535 nm wavelengths.
7. The cumulative percentage of siRNA release was calculated by comparison with standard siRNA controls.

3.7 Agarose Gel Electrophoresis

1. Samples are prepared by mixing siRNA with Den-PEI-CDDP and Den-PEI-CDDP-FA at a 10:1 N:P ratio in Tris–HCl (pH 7.4) and were kept for 20 min at room temperature.
2. Then, 10 μL of each sample was mixed with 10 μL of 2× sample loading buffer.

3. Agarose gel used for electrophoresis was 1.2% (0.6 g agarose in 50 mL TAE buffer).
4. 100-kbp ladder was loaded along with the sample in a separate well.
5. Free siRNA in loading buffer was used as control.
6. The gel was run up to 35 min at 50 V and was imaged in a UV-transilluminator imaging system.
7. The siRNA protected by nanoparticles remained in the well without migration.
8. For serum nuclease protection assay, Den-PEI-CDDP-siRNA nanoparticles were incubated in 10% FBS for predetermined time intervals (30 min, 1 h, and 3 h) and were loaded into the gel.
9. The siRNA-encapsulated samples showed no degradation, even after 3 h of serum incubation.

3.8 Ortho-phenylenediamine (OPDA) Assay

1. To estimate the CDDP concentration, the OPDA colorimetric assay was carried out. In this assay, the OPDA complexes with the Pt in CDDP to form a Pt-OPDA metal complex that gives a blue color. The concentration can be measured at the absorbance 704 nm, compared with a standard curve.
2. The different known concentrations of CDDP and unbound CDDP aliquots were taken in different individual tubes.
3. 70 μg of OPDA was added to each tube, and Tris–HCl buffer was added until the volume reached 500 μL in each tube.
4. 500 μL of *N*,*N*-dimethylformamide (DMF) was added to each tube, and tubes were incubated at 95 °C for 30 min.
5. Blue product formation indicates the formation of the Pt-OPDA complex.
6. The absorbance of the blue complex was measured using a DeNovix DS11 spectrophotometer at 704 nm.
7. We plotted the standard curve with known CDDP concentrations and estimated the unbound CDDP concentration using that standard curve.
8. Finally, we calculated the CDDP encapsulation using the following formula:

% Encapsulation efficiency = [(Initial amount of CDDP added − amount of CDDP left in the supernatant)/Initial amount of CDDP added] × 100.

3.9 Optimization of Folic Acid Concentration

1. For targeted delivery to FRA-expressing cells, the optimum FA concentration in Den-PEI-CDDP-siGLO-FA was determined by cell uptake assay.
2. The cell uptake efficiency was measured by fluorescence emitted by siGLO siRNA delivered by Den-PEI-CDDP-siGLO-FA.

3. Nanoparticles were modified using various amount of folic acid-PEG-NHS with folic acid amounts equivalent to 3.75, 7.5, 11, 18, 37, 56, or 75 μg/mL.
4. H1299 cells were seeded in 6-well plates for 24 h, with density of 0.1 million cells per well (Subheading 3.1).
5. After 24 h, H1299 cells were treated with Den-PEI-CDDP-siGLO-FA nanoparticles containing different FA concentrations, as mentioned above, by keeping a constant siGLO concentration (50 nM per well) in all of the wells after 1 h of serum starvation.
6. After 24 h of treatment, the cells were collected using trypsin-EDTA complex (Subheading 3.1), and we measured the fluorescence intensity using an Envision multiplate reader with excitation (557 nm) and emission (570 nm) wavelengths.
7. Then, we counted the cell numbers using the trypan blue exclusion method (Subheading 3.14) and normalized the fluorescence intensity for 10,000 cells.
8. From the normalization graph, the FA amount at which the highest fluorescence intensity obtained was determined as the optimal FA concentration in nanoparticles for targeted drug delivery.

3.10 Optimization of CDDP Concentration

1. To obtain the optimal combinatorial therapeutic effect, the concentration of CDDP was varied in the nanoparticles, while the siHuR concentration was kept constant.
2. The combinatorial Den-PEI-CDDP-siHuR-FA effect was tested for cell viability in H1299 cells.
3. CDDP was encapsulated into Den-PEI nanoparticles, as mentioned in Subheading 3.2.
4. H1299 cells (0.1 million) were seeded in 6-well plates (Subheading 3.1).
5. After 24 h of incubation, H1299 cells were treated with Den-PEI-CDDP nanoparticles with different CDDP concentrations of 10, 25, 40, 50, 60, 80, or 100 μM.
6. After 24 h of treatment, cells were harvested and counted using the trypan blue assay. The cell viability graph was plotted. The untreated cell count was considered 100% viable.
7. The IC_{20} concentration of CDDP was considered the optimal concentration for combined delivery.

3.11 Cell Uptake Study

1. H1299 cells were seeded onto glass coverslips at 50,000 cells/well and incubated for 24 h.
2. After incubation, the media was aspirated and the cells were washed with PBS.

3. Fluorescently labeled nanoparticles or fluorescent siRNA (siGLO)-containing nanoparticles (50 nM) were added onto the coverslips in 1 mL serum-free medium.
4. The cells were then incubated for 6 h.
5. The serum-free medium was replaced with serum-containing medium (10% FBS), and the cells were incubated.
6. Control wells contained no nanoparticles.
7. After 24 h, the cells were washed in PBS.
8. Then, the cells were stained with 7.5 μL of LysoTracker green endosomal staining solution for 3 h in the dark (Thermo-Fisher, LysoTracker® Green DND-26).
9. Then, the cells were washed with PBS three times, and the cells were fixed in 4% PFA for 20 min at room temperature.
10. The coverslips were again washed in PBS three times.
11. A 1 to 1000 dilution of DAPI nuclear dye was prepared in PBS and added onto coverslips. Coverslips were incubated at room temperature for 5 min.
12. The coverslips were then washed with PBS three times and mounted onto glass slides in mounting medium.
13. The sides of the coverslips were sealed using nail polish solution.
14. The cells were then ready for microscopy image capture using Nikon fluorescence microscopy. Fluorescence images of cells were acquired using a Nikon TiU microscope attached to a charge-coupled device (CCD) camera and imported into ImageJ analysis software (NIH).

3.12 Receptor-Specific Uptake Assay (Temperature-Dependent Receptor Uptake Study)

1. 0.1 million H1299 cells were seeded in two sets of 6-well plates (Subheading 3.1).
2. After 24 h of seeding, 1 h of serum starvation was conducted.
3. Den-PEI-CDDP-siGLO or Den-PEI-CDDP-siGLO-FA nanoparticles were added to two sets of plates with 50 nM siGLO concentrations.
4. Untreated cells were considered controls.
5. One set of plates was incubated at 37 °C for 4 h. The other set of plates was incubated at 4 °C for 4 h.
6. After 4 h of treatment, all groups of cells were collected with trypsin-EDTA and washed with PBS.
7. Then, we measured the fluorescence intensity using an Envision multiplate reader.

8. The measured fluorescence was normalized to 10,000 cells. The data were compared between 37 °C and 4 °C and between targeted and non-targeted nanoparticle treatment groups.

3.13 Receptor-Specific Uptake Assay (Receptor Blocking Uptake Study)

1. For the folate receptor blocking study, 0.1 million H1299 cells were seeded and incubated in media containing different exogenous FA concentrations.
2. The following media were used: regular RPMI-1640 (contains minimal FA), FA-free RPMI-1640, and RPMI-1640 medium with 1 mM of exogenous FA.
3. After 24 h of incubation, cells were starved for 1 h in serum-free medium.
4. Den-PEI-siGLO-FA nanoparticles containing 50 nM siGLO were then added to cells.
5. After 6 h of incubation, media was replaced with 10% serum-containing medium, and the incubation continued.
6. The cells were collected with trypsin-EDTA after 24 h of incubation.
7. The siGLO fluorescence was measured with an Envision multiplate reader, and the fluorescence intensity was normalized for 10,000 cells.
8. We plotted the graph by comparing fluorescence intensity between all groups.

3.14 Trypan Blue Exclusion Cell Viability Assay

1. This assay was performed to determine the cell density of cell suspension using a hemocytometer.
2. Prepare 0.4% solution of trypan blue in a buffered isotonic salt solution, pH 7.2–7.3 (i.e., phosphate-buffered saline).
3. Harvest the cells from a flask (Subheading 3.1) or from a 6-well plate (cells already treated) into a 15 mL conical tube.
4. Take an aliquot of the harvested cells.
5. Pipette 20 μL of an aliquot into a 1.5 mL centrifuge tube, and add 20 μL of 0.4% trypan blue stock solution. Mix it gently by pipetting.
6. Load 10 μL on each side of a hemocytometer, and examine it immediately under a microscope at low magnification (10×).
7. Count number of the bright cells (live cells), but not the blue cells (dead cells).
8. If you count more than 100 cells on each side of the hemocytometer, dilute the initial cell suspension, and repeat **steps 4–7**.

9. To calculate the number of viable cells per mL of culture, use the formula below. Remember to correct for the dilution factor.

   Number of viable cells × 10^4 × dilution factor × total volume of the cell suspension = cells/mL culture.

3.15 Western Blot Assay

3.15.1 Sodium Dodecyl Sulfate-Polyacrylamide Gel Electrophoresis

1. Wipe the glass plates with EtOH and mount them into the casting frames.
2. Add dH_2O to check for leakage.
3. Drain the dH_2O and dry using a Whatman 3M blotting paper.
4. Prepare the resolving gel according to the molecular weight of the protein to be detected. For example, to detect low molecular weight proteins, prepare a 12.5% gel; for high molecular weight protein, prepare 7.5% gel.
5. Mix the gel by pipetting and add 7.4 mL of the gel between the glass plates.
6. Add 300 μL of water-saturated isobutanol to remove the bubbles.
7. Let stand for 30–45 min until the gel polymerizes.
8. After 45 min, drain the isobutanol and wash with dH_2O to remove the remaining isobutanol.
9. Carefully remove excess dH_2O using a blotting paper, so as not to touch the gel.
10. Mix the stacking gel and add on top of the resolving gel.
11. Place the comb to make the wells (a 1.5 mm 10-well comb or 15-well comb can be used, depending on the number of wells required).
12. Let stand for 45 min until the gel polymerizes.
13. After 1 h, remove the combs carefully and gently wash with dH_2O to remove excess gel.
14. Drain the dH_2O carefully, without disturbing the gel.
15. Remove the glass plates from the gel stand and clamp it into the gel holder in the electrophoresis tank.
16. Pour 1× running buffer between the glass plates.
17. Mix the protein sample with lysing buffer (RIPA) according to the prepared chart to get equal loading concentrations of the proteins in a total of 20 μL. Then, add 20 μL of loading buffer and mix it with pipetting to get a total of 40 μL.
18. Cap the tubes, and then warm up the prepared 40-μL samples at 95 °C in a heat block for 10 min.
19. Centrifuge the samples briefly, then load 40 μL of the prepared samples into each well.

20. Mix 10 μL of protein ladder, 10 μL of lysis buffer (RIPA), and 20 μL of loading buffer to get a total of 40 μL, and then load it into one of the wells.
21. Run the gel at 55 V, monitoring the blue sample buffer line until it runs halfway off the gel, and then increase to 72 V.
22. Stop the gel once the blue line reaches the bottom of the gel.
23. Drain the running buffer and remove the slides from the stand.
24. Carefully open the two slides and remove the gel for transfer.
25. Transfer the proteins from the polyacrylamide gel to PVDF membrane.

3.15.2 Transferring Protein onto pvdf Membrane and Performing Detection

1. Add 1× transfer buffer in a dish and place the transfer cassette in the buffer.
2. Place a sponge pad on the cassette, followed by blotting filter paper(s), PVDF membrane (presoaked in methanol and then in 1× transfer buffer), polyacrylamide gel, 3M Whatman blotting paper, and sponge ("sandwiching" between two electrodes).
3. Put the transfer cassette into the cassette holder with the black side of the cassette toward the red side of the cassette holder (depending on the direction you are running the current).
4. Put the transfer cassette set-up in the container and pour 1× transfer buffer.
5. Put in a magnetic stirrer and an ice pack to keep the gel cool. Close the container.
6. Transfer the gel at 55 V for 1:30 h (depending on the molecular weight of the proteins we want to detect. Longer transfer may be required for high molecular weight proteins. Overnight transfer is preferred for proteins more than 180 kDa at 30 V in a cold room).
7. Following transfer, remove the stand and cut the PVDF membrane at the gel marks with a scalpel.
8. Place this membrane in a blocking dish and block with 5% fat-free milk in TBS-T for 1 h to block non-specific binding sites. After 1 h, drain the blocking buffer.
9. Add appropriate primary antibody diluted 1:500 to 1:1000 in 5% milk or BSA in TBS-T buffer. Leave the membrane on a shaker set at slow speed overnight at +4 °C.
10. The next morning, remove the BSA 5% containing the primary antibody. Add TBS-T blocking buffer to the membrane and wash the membrane three times for 5 min each time, placing it on a rotary shaker. Drain the washing buffer.

11. Add appropriate secondary antibody at a dilution of 1:5000 in 5% BSA or 5% fat-free milk blocking buffer.
12. Place on a rotary shaker set at slow speed and allow shaking for 1 h.
13. Wash the membrane as indicated in **step 11**. Drain the TBS-T.
14. Prepare the appropriate amount of chemiluminescence ECL reagent by mixing Detection Reagent 1 and Detection Reagent 2 in a ratio of 1:1. Add 1–2 mL of the mixed ECL reagent on a membrane for 3 min. Drain the ECL reagent and blot dry the membrane using a Whatman blotting paper.
15. Place the membrane in a chemiluminescence imaging system. Develop the membrane after exposure for few minutes to get multiple images to detect specific protein bands (always compare protein bands obtained with the protein ladder for the correctness in molecular weight of the desired protein).
16. The following antibodies were used in the study: HuR and cyclin E1 (1:1000), caspase 9, PARP, and γH2AX (1:500). β-Actin (1:2000) was used as an internal loading control.
17. The relative protein expression compared to β-actin was quantified using GelQuant software (Syngene).

3.15.3 Stripping and Reprobing

1. After obtaining the appropriate result, place the membrane in a dish previously used for washing the membrane.
2. Add 10 mL of stripping buffer to the membrane contained in the dish. Place the dish on a rotary shaker and shake for 30 min.
3. Drain the stripping buffer.
4. Add TBS-T buffer solution containing antibody to a loading protein, such as β-actin (1:1000).
5. Repeat the steps described in **steps 9–16** from Subheading 3.15.2 above.
6. Detection of a signal corresponding to the expected protein size for β-actin will allow the investigator to determine equal loading of protein samples and will serve as internal control.
7. Detection of the equal levels of β-actin will reflect true changes in the protein of interest.

3.16 Apoptotic Analysis

1. H1299 cells (0.1 million/well) were treated with Den-PEI-CDDP-HuR and Den-PEI-CDDP-HuR-FA groups with 10 μM CDDP and 100 nM HuR concentrations per well.
2. Untreated cells were considered controls.
3. After 72 h of treatment, the cells were harvested with trypsin-EDTA and washed twice with ice-cold PBS.

4. Then, cells were stained with 5 μL of Alexa Fluor 488-annexin V for 15 min at room temperature in the dark.
5. 1 μL of PI (1 mg/mL) solution was added.
6. The percentage of cells undergoing early apoptosis, late apoptosis, and necrosis was measured with a FACSCalibur flow cytometer using the Cell Quest software (BD Biosciences).
7. The percentage of early and late apoptotic cells was calculated as the percentage sum of annexin V-positive (Q2) and annexin V/PI-positive (Q3) cell population and was plotted as graph with different groups compared with controls.

3.17 Quantitative Real-Time PCR (qRT-PCR)

3.17.1 RNA Isolation

1. H1299 cells (0.1 million/well) and normal lung fibroblasts MRC9 (0.1 million/well) were treated with Den-PEI-CDDP, Den-PEI-HuR, or Den-PEI-CDDP-HuR nanoparticles, respectively. Untreated cells were considered controls.
2. Use concentrations of 10 μM CDDP and 100 nM siHuR per well.
3. After 72 h of treatment with nanoparticles, collect the cells with trypsin-EDTA and wash them with cold PBS.
4. Lyse the cell pellets in 1 mL TRIzol® per collected cells from three wells.
5. Incubate the TRIzol-containing lysed cells at room temperature for 5 min.
6. Complete dissociation of the nucleoprotein complex.
7. Add 0.2 mL of chloroform per 1 mL of TRIzol reagent used for homogenization. Cap the tubes securely.
8. Shake tubes vigorously by hand for 15 s.
9. Incubate for 2–3 min at room temperature.
10. Centrifuge the samples at 12,000 × *g* for 15 min at 4 °C.
11. The mixture separates into a lower red phenol-chloroform phase, an interphase, and a colorless upper aqueous phase. RNA exclusively remains in the aqueous phase, which is about 50% of the total volume.
12. Collect the aqueous phase of the sample in a new autoclaved microcentrifuge tube (1.5 mL).
13. Add 0.5 mL of 100% isopropanol to the aqueous phase.
14. Incubate and gently mix at room temperature for 10 min.
15. Centrifuge at 12,000 × *g* for 10 min at 4 °C.
16. The RNA forms a gel-like pellet at the bottom of the tube after centrifugation. Proceed to RNA wash.

17. Remove the supernatant from the tube, leaving the RNA pellet.
18. Wash the pellet with 1 mL of 75% ethanol.
19. Mix (tap) the samples briefly, and then centrifuge the tube at 7500 × *g* for 5 min at 4 °C. Discard the supernatant.
20. Air-dry the pellet at room temperature for 5 min. It is very important to prevent complete drying of the RNA pellet.
21. Add RNase-free water to resuspended pellet and let it dissolve for 10 min.
22. Quantify each sample using NanoDrop (DeNovix DS11).
23. Store at −80 °C until the RNA samples are used to make cDNA.

3.17.2 cDNA Synthesis Using Bio-Rad iScript Kit

1. Thaw the RNA samples and the cDNA synthesis kit containing 5× iScript reaction mix, iScript reverse transcriptase (Bio-Rad), and nuclease-free water on ice.
2. It is always recommended to quantify the RNA samples immediately prior to cDNA synthesis. RNA concentration might change due to degradation and/or poor storage.
3. Mix the RNA samples by tapping and centrifuge briefly.
4. Quantify each sample to determine the concentration and the quality of the RNA in triplicate, and then take the average.
5. RNA (1 μg) will be used for cDNA conversion in a 20 μL reaction volume. The reaction mixture will consist of 15 μL nuclease-free water, 4 μL 5× iScript reaction mix, and 1 μL iScript reverse transcriptase.
6. Incubate the complete reaction mixture in a thermal cycler using the following protocol:

   Priming for 5 min at 25 °C, reverse transcription for 20 min at 46 °C, RT inactivation for 1 min at 95 °C, and hold the reaction at 4 °C until tubes are removed.
7. Store the cDNA samples at −20 °C until further use.

3.17.3 Quantitative Real-Time Polymerase Chain Reaction (qRT-PCR)

1. Thaw cDNA samples, reverse and forward primers, iQ SYBR Green Supermix, and nuclease-free water.
2. Dilute the synthesized cDNA by ten times with RNase-free water.
3. Prepare 20 μL of reaction mix for each cDNA sample by mixing the following reagents in the same order: 6 μL RNase-free water, 0.5 μL forward primer (20 μM), 0.5 μL reverse primer (20 μM), water 4 μL iQ SYBR Green Supermix, and 3 μL diluted cDNA. Each reaction should be in triplicate to minimize pipetting error.

4. Load reaction in triplicate into PCR microplates and seal the reaction vessels. For best results, briefly spin the microplate or tubes in a microcentrifuge at low speed to remove any bubbles.
5. Run the reaction as follows: initial denaturation and enzyme activation at 95 °C for 3 min (1 cycle), denaturing 95 °C for 10 s, annealing at 60 °C for 10 s, extension at 72 °C for 30 s (denaturing, annealing, and denaturing cycles are repeated 40 times), and the Melt curve at 95 °C for 10 s (1 cycle).
6. The following HuR-specific forward and reverse oligonucleotide primers were used to prepare the reaction.

   Forward—5′ ATGAAGACCACATGGCCGAAGACT 3′

   Reverse—5′ TGTGGTCATGAGTCCTTCCACGAT 3′
7. The crossing threshold (Ct) value assessed by RT-PCR was noted for the transcripts and normalized with human GAPDH primers.

   Forward—5′ AGCCTCAAGATCATCAGCAATGCC 3′

   Reverse—5′ TGTGGTCATGAGTCCTTCCACGAT 3′
8. The CT values were calculated using Bio-Rad's CFX Maestro™ Software.
9. The HuR CT values were normalized with the GAPDH CT values.

4 Notes

1. Den nanoparticles when used without PEI modification exhibited poor transfection. Therefore, it is important to modify the nanoparticles with PEI to increase the transfection efficiency.
2. CDDP dissolved in 0.1 M NaCl solution showed less loading efficiency than did CDDP dissolved in water.
3. siRNA encapsulation into Den-PEI nanoparticles in the Tris–HCl (pH 7.4) showed better condensation than was observed in PBS (pH 7.4).
4. The PicoGreen assay was used for siRNA quantification because of its accuracy compared with UV (A260) measurement of siRNA.
5. The CDDP release kinetics was carried out in a dialysis tube, which was immersed in a 500 mL measuring cylinder containing the releasing buffer. Because of the large volume of the buffer, the released CDDP was highly diluted limiting CDDP detection by OPDA assay. Therefore, we recommend using a mini dialysis tube immersed in 1.5 mL of releasing medium for determining the drug release kinetics.

6. The growth rate of H1299 and A549 lung cancer cells is different. H1299 cells grow faster than A549 cells, and normal MRC9 lung fibroblast cells grow slower than H1299 and A549 cells. Therefore, for optimal confluency prior to the treatment, seeding density should be identified by careful titration.
7. Ensure the cells to be used are free of mycoplasma.
8. It is very important to record the name of the cell line with its passage number and the date of subculturing.
9. The recommended passage number of tumor cells and normal cells for use in experiments should not exceed 40 and 10, respectively.
10. While passaging the cell lines, washing is a necessary step to remove any traces of serum, calcium, and magnesium that would inhibit the action of the dissociation reagent. Remove and discard the PBS from the flask.
11. While developing polyvinylidene difluoride *(PVDF) membrane* to detect a protein, the incubation time might be adjusted depending on the ease of detecting the protein.

Acknowledgments

The work was supported in part by a grant received from the National Institutes of Health (NIH), R01 CA167516 (RR), an Institutional Development Award (IDeA) from the National Institute of General Medical Sciences (P20 GM103639) of the National Institutes of Health (RR & AM), VA MERIT grant 101B × 003420 (RR) and by funds received from the Stephenson Cancer Center Seed Grant (RR), Presbyterian Health Foundation Seed Grant (RR), Presbyterian Health Foundation Bridge Grant (RR), and Jim and Christy Everest Endowed Chair in Cancer Developmental Therapeutics (RR) at the University of Oklahoma Health Sciences Center. Rajagopal Ramesh is an Oklahoma TSET Research Scholar and holds the Jim and Christy Everest Endowed Chair in Cancer Developmental Therapeutics. The authors thank Ms. Kathy Kyler at the office of Vice President for Research, OUHSC, for editorial assistance.

References

1. Ma L, Kohli M, Smith A (2013) Nanoparticles for combination drug therapy. ACS Nano 7:9518–9525. https://doi.org/10.1021/nn405674m
2. Sonawane R, Harde H, Katariya M et al (2014) Solid lipid nanoparticles-loaded topical gel containing combination drugs: an approach to offset psoriasis. Expert Opin Drug Deliv 11:1833–1847. https://doi.org/10.1517/17425247.2014.938634
3. Cao P, Bae Y (2012) Polymer nanoparticulate drug delivery and combination cancer therapy. Future Oncol 8:1471–1480. https://doi.org/10.2217/fon.12.139

4. Kydd J, Jadia R, Velpurisiva P et al (2017) Targeting strategies for the combination treatment of cancer using drug delivery systems. Pharmaceutics 9:46. https://doi.org/10.3390/pharmaceutics9040046
5. Narsireddy A, Vijayashree K, Adimoolam MG et al (2015) Photosensitizer and peptide-conjugated PAMAM dendrimer for targeted in vivo photodynamic therapy. Int J Nanomedicine 10:6865–6878. https://doi.org/10.2147/IJN.S89474
6. Navath RS, Menjoge AR, Wang B et al (2010) Amino acid-functionalized dendrimers with heterobifunctional chemoselective peripheral groups for drug delivery applications. Biomacromolecules 11:1544–1563. https://doi.org/10.1021/bm100186b
7. Tekade RK, Kumar PV, Jain NK (2009) Dendrimers in oncology: an expanding horizon. Chem Rev 109:49–87. https://doi.org/10.1021/cr068212n
8. Xu Z, Kahr M, Walker KL et al (1994) Phenylacetylene dendrimers by the divergent, convergent, and double-stage convergent methods. J Am Chem Soc 116:4537–4550. https://doi.org/10.1021/ja00090a002
9. Grayson SM, Fréchet JM (2001) Convergent dendrons and dendrimers: from synthesis to applications. Chem Rev 101:3819–3868. https://doi.org/10.1021/cr990116h
10. Madaan K, Kumar S, Poonia N et al (2014) Dendrimers in drug delivery and targeting: drug-dendrimer interactions and toxicity issues. J Pharm Bioallied Sci 6:139–150. https://doi.org/10.4103/0975-7406.130965
11. Dufès C, Uchegbu IF, Schätzlein AG (2005) Dendrimers in gene delivery. Adv Drug Deliv Rev 57:2177–2202. https://doi.org/10.1016/j.addr.2005.09.017
12. Amreddy N, Babu A, Panneerselvam J et al (2017) Chemo-biologic combinatorial drug delivery using folate receptor-targeted dendrimer nanoparticles for lung cancer treatment. Nanomedicine 14:373–384. https://doi.org/10.1016/j.nano.2017.11.010

Chapter 21

Bioconjugated Gold Nanoparticles Enhance siRNA Delivery in Prostate Cancer Cells

Kamil Rahme, Jianfeng Guo, and Justin D. Holmes

Abstract

Here we describe a simple way to create a gold nanoparticle (AuNP)-based non-viral delivery system to deliver siRNA into prostate cancer cells. Therefore, positively charged polyethylenimine (PEI)-capped AuNPs were synthesized in water and further conjugated with the targeting ligand (folic acid) for folate receptors (AuNPs-PEI-FA). The AuNPs-PEI-FA could effectively complex small interfering RNA (siRNA) through electrostatic interaction. Flow cytometry displayed that AuNPs-PEI-FA could specifically deliver siRNA into LNCaP cells, a prostate cancer cell line overexpressing prostate-specific membrane antigen (PSMA) that exhibits a hydrolase enzymatic activity with a folate substrate. In contrast, internalization of siRNA into PC-3 cells, a prostate cancer cell line not expressing PSMA or folate receptors, was not achieved using AuNPs-PEI-FA.siRNA. Following endolysosomal escape, the AuNPs-PEI-FA-.siRNA formulation resulted in significant endogenous gene silencing when compared to the nontargeted formulation, suggesting the potential of AuNPs-PEI-FA for targeted delivery of therapeutic siRNAs in the treatment of prostate cancer.

Key words Gold nanoparticles, Targeting ligands, Receptor-mediated internalization, Non-viral siRNA delivery, Prostate cancer, Gene therapy

1 Introduction

Prostate cancer is one of the most leading causes of morbidity and mortality in men worldwide [1]. Gold nanoparticles (AuNPs) with size-/shape-dependent optical properties offered by the surface plasmon resonance (SPR) are now widely used in biomedical applications, including the contrast agents, photothermal agents, and radiosensitizers for cancer diagnosis and therapy [2, 3]. Furthermore, the affinity of AuNPs to strongly bind biomolecules containing thiol and/or amino groups makes them suitable as bioconjugated vehicles for selective drug delivery [4]. Small interfering RNA (siRNA) is known to selectively downregulate any gene implicated in the pathology of disease [5, 6]. Recent knowledge of molecular genetic data discovered by sequencing the whole

Lekha Dinesh Kumar (ed.), *RNA Interference and Cancer Therapy: Methods and Protocols*, Methods in Molecular Biology, vol. 1974, https://doi.org/10.1007/978-1-4939-9220-1_21, © Springer Science+Business Media, LLC, part of Springer Nature 2019

genome of cancer cells has resulted in significant changes in our understanding of the molecular pathogenesis of cancer and has facilitated the identification of a number of molecular targets against which RNAi has been developed [7, 8]. Indeed, delivery of siRNA using multifunctional nanoparticles (NPs) has been recently investigated to treat solid tumors and leukemia [9–11].

In this study, cationic bioconjugated AuNPs, namely, AuNPs-PEI-FA, were obtained using $HAuCl_4 \cdot 3H_2O$, ascorbic acid, polyethylenimine (PEI), and folic acids [folate, FA; in order to target the folate receptor (FR) overexpressed in prostate cancer cells]. Gel retardation showed that AuNP-PEI-FA could effectively complex siRNA through electrostatic interaction. Flow cytometry displayed that AuNPs-PEI-FA could specifically deliver siRNA into LNCaP cells. Consequently, the presence of the FA facilitated cell-specific receptor-mediated uptake, and the presence of PEI (a polycation with dramatic buffering capacity for "proton sponge" effect) induced endolysosomal escape of the siRNA subsequently achieving endogenous gene silencing in LNCaP cells.

2 Materials

Prepare all solutions using purified H_2O (resistivity ≈ 18.2 MΩ cm) as a solvent for the AuNPs synthesis. Clean all glassware used in the synthesis of AuNPs with aqua regia (three parts of concentrated HCl and one part of concentrated HNO_3); rinse with distilled water, ethanol, and acetone; and dry in the oven before use. Freshly prepared solutions were used (unless specified a stock solution). All siRNA solutions were prepared in RNase-free water following Sigma-Aldrich recommendations.

2.1 Nanoparticles Synthesis and Bioconjugation with Folic Acid

1. 1 mM Gold(III) chloride trihydrate ($HAuCl_4 \cdot 3H_2O$) ≥99.9% metal basis aqueous solution.
2. 100 mM L-ascorbic acid BioXtra, ≥99.0%, crystalline aqueous solution.
3. 2.3 mM PEI 2 kDa, average M_n ~ 1800 by GPC, average M_w ~ 2000 by LS aqueous solution stock at 4 °C.
4. Glass round flasks.
5. Magnetic stirrer.
6. Sonicator.
7. Folic Acid BioReagent, suitable for cell culture, suitable for insect cell culture, suitable for plant cell culture, ≥97%.
8. Dimethyl sulfoxide (DMSO) anhydrous ≥99.9%.
9. Trimethylamine ≥99.5%.

10. *N*-hydroxysuccinimide (NHS) 98%.
11. *N*,*N'*-Dicyclohexylcarbodiimide (DCC) for synthesis.
12. Whatman filter papers.
13. Schlenk line connected to Argon gas flow to operate under inert atmosphere.
14. Microbalance.
15. Centrifugator.

2.2 Preparation of AuNP.siRNA Complexes

1. AuNPs-PEI and AuNPs-PEI-FA were concentrated to 500 μg/mL by centrifugation (5000 rpm, 1677 × *g* for 15 min at RT).
2. Negative control siRNA; sense sequence 5′-UUC UCC GAA CGU GUC ACG U-3′; desalting purification.
3. *Fluorescein amidite* (FAM)-labeled siRNA; sense sequence 5′-UUC UCC GAA CGU GUC ACG U-3′, modified by 6-FAM on 5′ of sense sequence; desalting purification.
4. RelA siRNA; sense sequence 5′-CCA UCA ACU AUG AUG AGU U-3′; desalting purification.
5. Agarose (low gelling temperature, for molecular biology).
6. Tris/Borate/EDTA (TBE) buffer (BioReagent, suitable for electrophoresis, 10× concentrate).
7. SafeView.

2.3 Cell Culture

1. RPMI-1640 medium for LNCaP cells (with sodium bicarbonate, liquid, without L-glutamine, sterile-filtered, suitable for cell culture).
2. RPMI-1640 medium for PC-3 cells (with L-glutamine and sodium bicarbonate, liquid, sterile-filtered, suitable for cell culture).
3. LNCaP cells, a human prostate cancer cell line PC-3 cells, a human prostate cancer cell line.
4. 12- and 24-well cell culture plates.
5. Cell incubator.

2.4 Fluorescence-Activated Cell Sorting, Confocal Microscopy and In Vitro Gene Knockdown

1. Polystyrene Round-Bottom Tubes.
2. 6-well culture plates with glass bottom.
3. LysoTracker® Deep Red.
4. GenElute™ Mammalian Total RNA Miniprep Kit.
5. High-Capacity cDNA Reverse Transcription Kits.
6. Primers for RelA and GAPDH.

2.5 Characterization Equipments

1. UV-visible spectroscopy: CARY UV-visible spectrophotometer with a Xenon lamp (190–900 nm range, 0.5 nm resolution) was used to carry out optical absorption spectra.
2. Scanning electron microscopy (SEM): FEI 630NanoSEM equipped with an Oxford INCA energy dispersive X-ray (EDX) detector operating at 5 kV was used to image the AuNPs-PEI deposited from solution onto a Si wafer and air-dried prior to analysis.
3. Dynamic light scattering (DLS) and zeta potential measurements: Malvern Zetasizer Nano ZS was used to determine the size and charge distributions via dynamic light scattering (DLS) and electrophoretic mobility measurements, respectively.
4. Confocal Microscopy: Olympus FV 1000 microscope was used, and images were acquired and analyzed using Olympus FV10-ASW software.
5. Real-time quantitative PCR (qPCR): Gene expression was assessed by real-time quantitative PCR (qPCR) using the Applied Biosystems Real-Time PCR System (model 7300).

3 Methods

3.1 Synthesis of Folic Acid-Conjugated Gold Nanoparticles

3.1.1 Synthesis of PEI-Capped Gold Nanoparticles (AuNPs-PEI)

1. Add 1.42 mL of 2.3 mM PEI to 32.6 mL of 1 mM $HAuCl_4 \cdot 3H_2O$ solution with stirring at RT; the color of the solution changed from pale yellow to deep orange.
2. Add to the above solution (after about 1 min) 0.5 mL of a freshly prepared 100 mM L-ascorbic acid aqueous solution; the color changed immediately (~5 s) to red brownish (*see* **Note 1**).
3. Keep the solution stirring gently overnight at RT achieving AuNP-PEI (*see* **Note 2**) as shown by UV-visible spectroscopy and SEM shown in Fig. 1 (*see* **Note 3**).

3.1.2 Synthesis of Folic Acid Activated Ester (Folic Acid-NHS)

1. Dissolve 0.25 g (5.663×10^{-4} mol) of folic acid in 10 mL of dry DMSO.
2. Add to the above solution 0.125 mL (8.960×10^{-4} mol 1.5 equivalents) of triethylamine.
3. Sonicate the above solution for about 10 min, and further stir under Argon until folic acid is totally dissolved (*see* **Note 4**).
4. Add to the above homogeneous solution 0.13 g (11.295×10^{-4} mol, ~2 equivalents) of NHS and 0.235 g (11.38×10^{-4} mol, ~2 equivalents) of DCC.

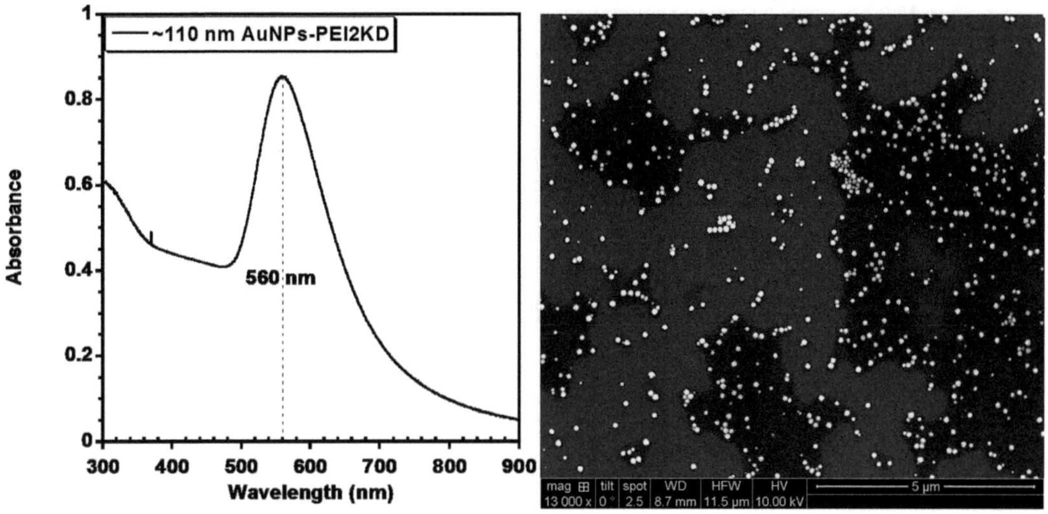

Fig. 1 UV-visible spectra of AuNPs-PEI (left) and SEM image (right) of AuNPs-PEI (bar = 5 μm)

5. Keep the mixture under stirring overnight in the dark at RT, to allow the formation of folic acid-NHS.
6. Filter the resulting solution to remove the dicyclohexylurea by-product (filtration is performed twice), and stock the final solution in the dark at room T (*see* **Note 5**).

3.1.3 Synthesis of AuNP-PEI-FA

1. Take 3 mL of AuNP-PEI, and add 30 μL of 25 mg/mL folic acid-NHS in DMSO. Keep the AuNP-PEI, folic acid-NHS under stirring in the dark at RT for 10 min.
2. Add 50 μL of 0.1 M NaOH to the solution, and leave it under stirring at RT for 48 h in the dark, achieving AuNP-PEI-folic acid (thereafter referred as AuNP-PEI-FA).
3. To purify the AuNP-PEI-FA solution centrifuge at 12,000 rpm (9660 × *g*) for 5 min, remove the supernatant containing unreacted folic acid-NHS.
4. The unreacted folic acid-NHS can be quantified by UV-vis (*see* **Note 6**).
5. The size and zeta potential distribution were determined by DLS and zeta potential measurements as shown in Fig. 2 (*see* **Note 7**).

3.2 Preparation of AuNP.siRNA Complexes

1. Solutions of 500 μg/mL AuNP-PEI and 500 μg/mL AuNP-PEI-FA were added to 0.264 μg/μL siRNA solutions at the mass ratio (MR) 20 of AuNPs to siRNA, followed by 1-h incubation with 400 rpm shaking at RT, in order to achieve the AuNP-PEI.siRNA and AuNP-PEI-FA.siRNA formulations.
2. The ability of AuNPs to complex siRNA was analyzed by gel retardation. Complexes of AuNPs and siRNA at different mass

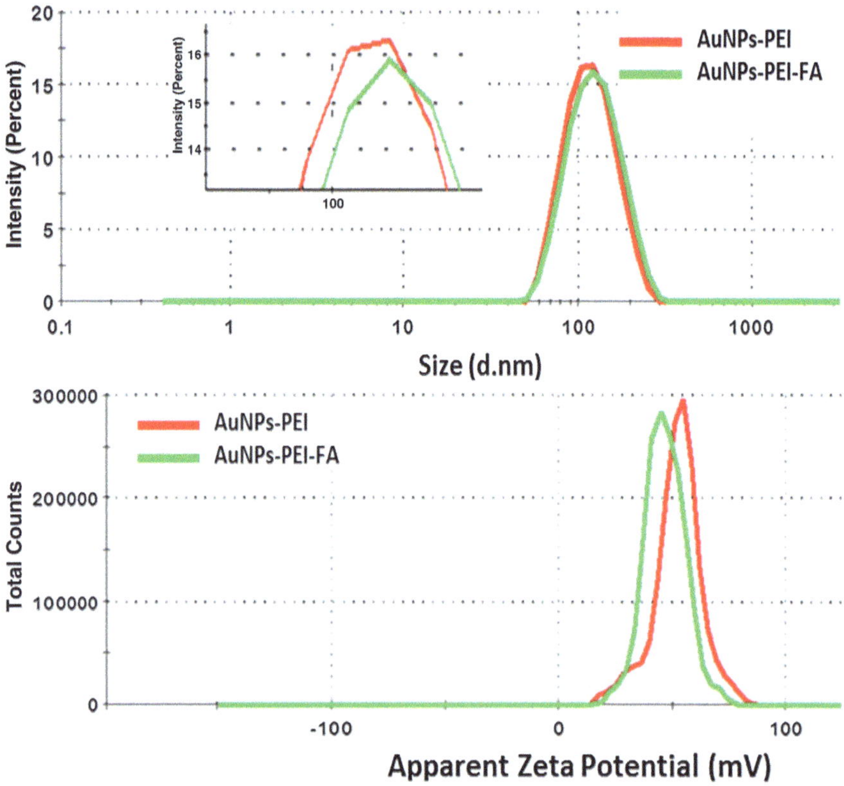

Fig. 2 Size distribution by intensity (up) of AuNPs-PEI (113 ± 1 nm, PDI = 0.07, *n* = 3) and AuNPs-PEI-FA (118 ± 1 nm, PDI = 0.09, *n* = 3); insert shows that the size is slightly increased; zeta potentials (down) of AuNPs-PEI (53 ± 1 mV, *n* = 3) and AuNPs-PEI-FA (46 ± 1 mV, *n* = 3) as obtained from DLS analysis. (Reproduced with permission from the Elsevier Copyright Clearance Center; J. Guo et al. *International Journal of Pharmaceutics 509 (2016) 16–27*)

ratios (μg AuNPs to μg siRNA, siRNA was fixed to 0.25 μg siRNA) were loaded onto 1% (w/v) agarose gels in 100 mL TBE buffer containing 6 μL SafeView.

3. Electrophoresis was performed at 120 V for 30 min, and the resulting gels were photographed under UV light (Fig. 3).

3.3 Cell Culture

1. LNCaP cells were maintained in RPMI-1640 medium supplemented with 10% FBS 50 units/mL penicillin, 50 μg/mL streptomycin, 2 mM L-glutamine, 10 mM HEPES, 7.2 mM D-glucose, and 1 mM sodium pyruvate.
2. PC-3 cells were maintained in RPMI-1640 medium supplemented with 10% FBS.
3. All cells were grown in the incubator at 37 °C with 5% CO_2 and 95% relative humidity.

3.4 Fluorescence-Activated Cell Sorting

1. 100,000 LNCaP cells and 100,000 PC-3 cells were, respectively, seeded per well in 24-well plates and incubated for 24 h under the growth conditions mentioned above.

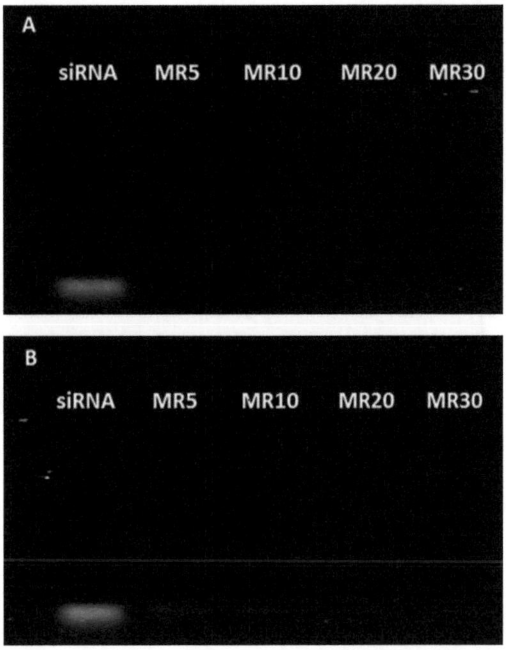

Fig. 3 Complexation of siRNA (0.25 μg) with (**a**) AuNPs-PEI and (**b**) AuNPs-PEI-FA at different mass ratios (MRs) using gel retardation (1% agarose gel at 120 mV for 30 min). (Reproduced with permission from the Elsevier Copyright Clearance Center; J. Guo et al. *International Journal of Pharmaceutics 509 (2016) 16–27*)

2. Cells were transfected with AuNP-PEI.siRNA (25 and 50 nM FAM siRNA) and AuNP-PEI-FA.sRNA (25 and 50 nM FAM siRNA) and incubated for 24 h in the growth medium mentioned above.
3. The following incubation cells were washed twice with PBS and trypsinized with trypsin-EDTA.
4. Cells were subsequently centrifuged (1000 rpm, 67 × *g* for 5 min) and resuspended in 1000 μL ice-cold PBS in polystyrene round-bottom tubes.
5. Ten thousand cells per sample were measured following the procedure outlined in Becton Dickinson FACScalibur manual. Fluorescein-positive cells (%) were displayed by Dot Plot (Fig. 4).

3.5 Confocal Microscopy

1. 300,000 LNCaP cells were seeded per well in 6-well culture plates with glass bottom. Cells were treated with the AuNP-PEI-FA formulation containing 50 nM FAM siRNA and incubated under the growth conditions (mentioned above) for 24 h.
2. In order to label late endosomes and lysosomes, 75 nM LysoTracker® Deep Red were added to cells for 30 min at 37 °C.
3. The medium was then replaced with fresh growth medium. The intracellular route was analyzed using the confocal microscopy (Fig. 5).

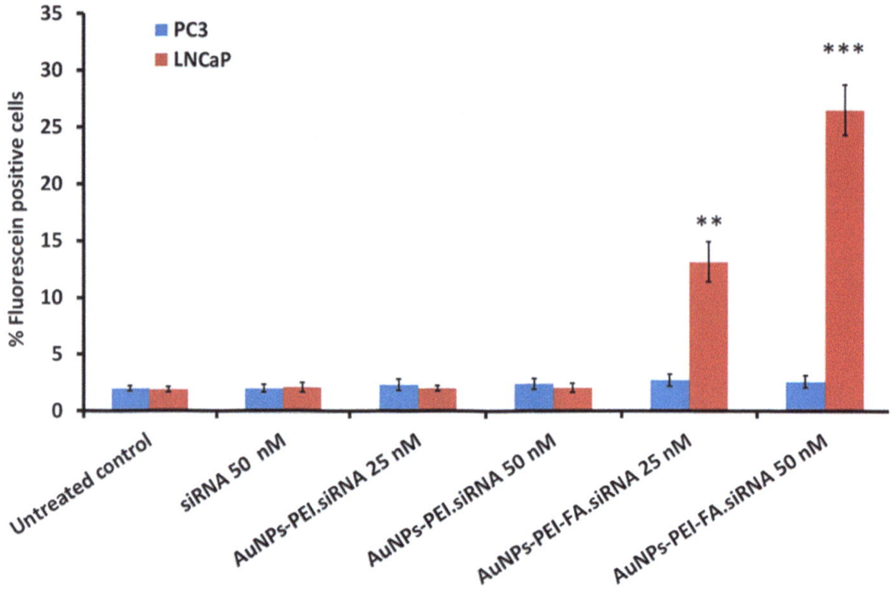

Fig. 4 Fluorescein-positive PC-3 and LNCaP cells (%, mean ± SD) after 24-h transfection of AuNPs-PEI-FA. FAMsiRNA (MR20, 50 nM siRNA) analyzed using flow cytometry (PC-3 = receptor-negative cells and LNCaP cells = receptor-positive cells). (**$P < 0.01$ and ***$P < 0.001$ compared to the untreated control). (Reproduced with permission from the Elsevier Copyright Clearance Center; J. Guo et al. *International Journal of Pharmaceutics 509 (2016) 16–27*)

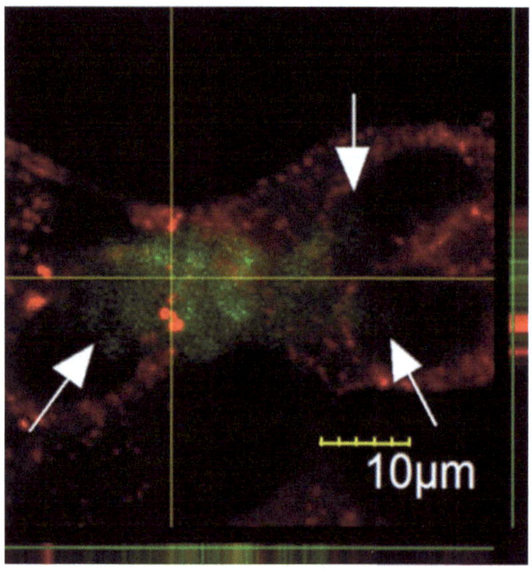

Fig. 5 A representative image showing intracellular distribution of AuNPs-PEI-FA formulation (MR20) containing FAM siRNA (50 nM) in LNCaP cells at 24-h post-transfection by confocal microscope. Late endosomes and lysosomes were labeled with 75 nM LysoTracker® Deep Red (Green = FAM siRNA, Red = LysoTracker® Deep Red). Most siRNAs were found in the cytoplasm. It is interesting to note that some FAM siRNAs were also found inside the nucleus as indicated by arrows

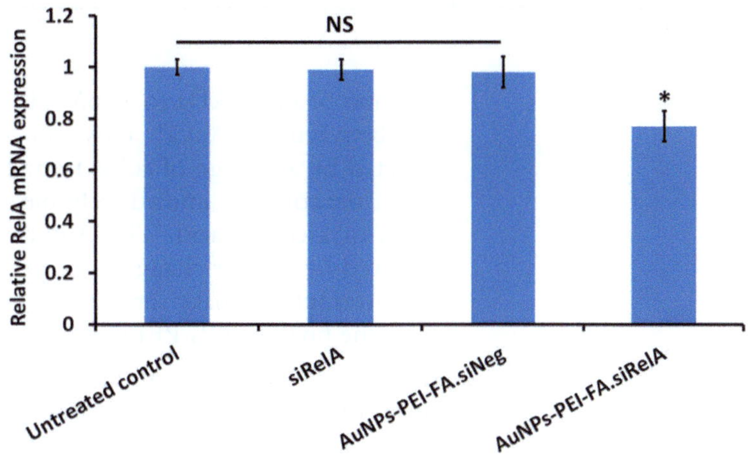

Fig. 6 Endogenous gene silencing in LNCaP cells by RelA siRNA (50 nM) formulated with AuNPs-PEI-FA (MR20) at 24-h post-transfection. The naked RelA siRNA (50 nM) and negative control siRNA (50 nM) formulated with AuNPs-PEI-FA (MR20) were used as controls (NS = no significance, $*P < 0.05$ compared to formulation containing negative control siRNA). (Mean ± SD, $n = 3$). (Reproduced with permission from the Elsevier Copyright Clearance Center; J. Guo et al. *International Journal of Pharmaceutics 509 (2016) 16–27*)

3.6 In Vitro Gene Silencing

1. 200,000 LNCaP cells were seeded per well in 12-well plates and incubated for 24 h under the growth conditions mentioned above.
2. Cells were then transfected by the AuNP-PEI-FA formulation (MR20) containing 50 nM RelA siRNA for 24 h. 50 nM naked RelA siRNA and the AuNPs-PEI-FA formulation containing 50 nM negative control siRNA were used as the negative controls.
3. The following incubation total RNA was isolated from the cells using the GenElute™ Mammalian Total RNA Miniprep Kit, according to the instructions supplied.
4. First-strand cDNA was generated from total RNA samples using high-capacity cDNA reverse transcription kits.
5. Gene expression was assessed by real-time quantitative PCR (qPCR) using the Applied Biosystems Real-Time PCR System. Assays were performed using appropriate primers for RelA and GAPDH. Amplification was carried out by 40 cycles of denaturation at 95 °C (15 s) and annealing at 60 °C (1 min). The quantitative level of each RelA mRNA was measured as a fluorescent signal corrected according to the signal for GAPDH mRNA using $2^{-\Delta\Delta}$ Ct method [12] (Fig. 6).

4 Notes

1. Addition of PEI 2 kDa to 1 mM $HAuCl_4 \cdot 3H_2O$ forms a complex between the $AuCl^{4-}$ ions and the amino groups in PEI detected by a change of color from pale yellow to orange. To reduce the complex, immediately inject an excess of ~1.55 equivalents of ascorbic acid. The AuNPs formation can be easily observed by a color change from yellow to red. Keep the stirring overnight to make sure that all gold is reduced and to stabilize the final AuNPs-PEI size.
2. Stirring a high concentration of 1 mM AuNPs overnight can cause some adsorption of the resulting AuNPs-PEI onto the wall of the round flask. Therefore, the adsorbed AuNPs might be recollected by sonication about 30 s.
3. The obtained AuNPs-PEI can be characterized by UV-vis spectroscopy (Fig. 1); the obtained AuNPs-PEI by the abovementioned method shows an absorption band at λ_{max} 560 nm at the origin of the red colloidal solution. This is the surface plasmon resonance band characteristic of AuNPs colloidal solutions. Furthermore, SEM analysis (Fig. 1) proves the formation of AuNPs with a diameter of ~110 nm as determined by *Image J* software.
4. To dissolve the 0.25 g of folic acid in DMSO, it was necessary to sonicate the solution for about 10 min followed by stirring for another 30 min. Folic acid needs to be manipulated in the dark as it is light sensitive.
5. The abovementioned synthesis of folic acid-NHS is known to yield approximately 80% γ-carboxyl group of folic acid, preserving the binding of activated folate to folate receptors [13]. The by-product dicyclohexylurea is insoluble in water and is separated by simple filtration. The solution of 25 mg/mL folic acid-NHS in DMSO can be used without further purification.
6. Folic acid activated with *N*-hydroxysuccinimide ester becomes ready to react with free amino group in PEI at a pH (7–8), to produce an amide linkage between the folic acid and AuNPs-PEI leading to AuNPs-PEI-FA. To purify the final AuNPs-PEI-FA solution centrifuge at 12,000 rpm (9660 × *g*) for 5 min, remove carefully the supernatant containing the unreacted folic acid-NHS, and run a UV-vis spectrum to quantify the folic acid attached to the AuNPs.
7. To confirm the formation of AuNP-PEI-FA, perform DLS/zeta potential analysis; which shows a very slight increase in the size distribution of AuNPs-PEI and a decrease in the value of the zeta potential confirming the successful attachment of FA (Fig. 2).

Acknowledgments

We acknowledge the financial support from Science Foundation Ireland and AMBER (Grant 12/RC/2278); the Irish Research Council, for a Government of Ireland Postdoctoral Fellowship (GOIPD/2013/150) and the Outstanding Youth Foundation from the Department of Science and Technology, Jilin Province (Project Number: 20170520046JH), to Jianfeng Guo; and the National Council for Scientific Research Lebanon (CNRS-L-GRP2015).

References

1. Siegel RL, Miller KD, Jemal A (2015) Cancer statistics. CA Cancer J Clin 65:5–29
2. WKNg V, Berti R, Lesage F, Kakkar A (2013) Gold: a versatile tool for in vivo imaging. J Mater Chem B 1:9–25
3. Jain S, Hirst DG, O'Sullivan JM (2012) Gold nanoparticles as novel agents for cancer therapy. Br J Radiol 85:101–113
4. Fratoddi I, Venditti I, Cametti C, Russo MV (2014) Gold nanoparticles and gold nanoparticle-conjugates for delivery of therapeutic molecules. Progress and challenges. J Mater Chem B 2:4204–4220
5. Guo J, Rahme K, Fitzgerald KA, Holmes JD, O'Driscoll CM (2015) Biomimetic gold nanocomplexes for gene knockdown: will gold deliver dividends for small interfering RNA nanomedicines? Nano Res 8:3111–3140
6. Wittrup A, Lieberman J (2015) Knocking down disease: a progress report on siRNA therapeutics. Nat Rev Genet 16:543–552
7. Fitzgerald KA, Evans JC, McCarthy J, Guo J, Prencipe M, Meghan K, Watson WR, O'Driscoll CM (2014) The role of transcription factors in prostate cancer and potential for future RNA interference therapy. Expert Opin Ther Targets 18:633–649
8. Guo J, McKenna SL, O'Dwyer ME, Cahill MR, O'Driscoll CM (2015) RNA interference for multiple myeloma therapy: targeting signal transduction pathways. Expert Opin Ther Targets 20:107–121
9. Strumberg D, Schultheis B, Traugott U, Vank C, Santel A, Keil O, Giese K, Kaufmann J, Drevs J (2012) Phase I clinical development of Atu027, a siRNA formulation targeting PKN3 in patients with advanced solid tumors. Int J Clin Pharmacol Ther 50:76–78
10. Tabernero J et al (2013) First-in-humans trial of an RNA interference therapeutic targeting VEGF and KSP in cancer patients with liver involvement. Cancer Discov 3:406–417
11. Guo J, Cahill MR, McKenna SL, O'Driscoll CM (2014) Biomimetic nanoparticles for siRNA delivery in the treatment of leukaemia. Biotechnol Adv 32:1396–1409
12. Schmittgen TD, Livak KJ (2008) Analyzing real-time PCR data by the comparative C(T) method. Nat Protoc 3:1101–1108
13. Van Steenis JH et al (2003) Preparation and characterization of folate-targeted PEG-coated pDMAEMA-based polyplexes. J Control Release 87:167–176

Chapter 22

Vaginal Suppositories with siRNA and Paclitaxel-Incorporated Solid Lipid Nanoparticles for Cervical Cancer: Preparation and In Vitro Evaluation

Gülay Büyükköroğlu, Behiye Şenel, and Evrim Yenilmez

Abstract

The objective of this study is to prepare vaginal suppository containing chemotherapeutic agent and genetic material that can be applied locally for cervical cancer. Cervical cancer is one of the most life-threatening types of cancer among women and is generally resistant to chemotherapy. Paclitaxel has been selected as chemotherapeutic agent, and siRNA that inhibits the Bcl-2 oncogene has been selected as the genetic material for simultaneous vaginal delivery. For this purpose, three different solid lipid nanoparticles (SLNs) were prepared that include Bcl-2 siRNA and paclitaxel and paclitaxel/Bcl-2 siRNA combination separately, and these SLN formulations were dispersed in vaginal suppositories prepared with PEG 6000. First, the physicochemical properties of SLNs, their cytotoxicities on HeLa cell lines, and the transfection ability of siRNA-incorporated SLN on the cells have been examined. Afterward, the release of SLNs from the three different vaginal suppositories prepared has been determined via horizontal diffusion chamber system. The loaded amount to the SLNs and release amount from suppositories of paclitaxel have been determined via HPLC, whereas stability, loading, and release amount of siRNA has been determined via gel retardation system and UV spectrophotometer.

Key words Bcl-2 siRNA, Paclitaxel, Cationic SLN, Vaginal suppository

1 Introduction

Cervical cancer can be due to various reasons and is the second most common type of cancer among gynecologic cancers [1]. It is characterized by uncontrolled cell growth that leads to cancer mass or tumor formation, one of the most important health problems today. This proliferation is caused by incapacity of cells to proceed to apoptosis as a result of mutations occurring in the DNA (activation of oncogenes, tumor suppressor gene inactivation, etc.) [2]. A variety of gene therapies have been developed to treat this type of cancer due to genetic disruption. In recent years, RNA interference has attracted attention among these methods [3]. In this technology, transmission of genetic information is

Lekha Dinesh Kumar (ed.), *RNA Interference and Cancer Therapy: Methods and Protocols*, Methods in Molecular Biology, vol. 1974, https://doi.org/10.1007/978-1-4939-9220-1_22, © Springer Science+Business Media, LLC, part of Springer Nature 2019

blocked by mRNA degradation or by causing a translational block thus downregulating the unwanted protein production which could have led to many genetic diseases [2, 4]. Many studies have shown that siRNAs/shRNAs could be successfully used for inhibition of angiogenesis and induction of apoptosis by inhibiting VEGF signaling pathway; for stopping tumor growth by inhibiting Wnt signaling, NF-κB pathways, EGFR, BCL-2, and Her-2/neu oncogenes; and also to increase radio- or chemosensitivity by inhibiting MDR1 [5, 6].

Various methods have been used to introduce siRNA in vitro and in vivo [7, 8]. Naked siRNAs are highly unstable in the serum, and also its penetration through lipid cellular membranes is ineffective given its half-life (~1 h) [9, 10]. Therefore, shRNAs in viral vectors (adenovirus, retrovirus, adeno-associated virus, etc.) [7, 11, 12] have been made use of for effective gene silencing. Since viral vectors are accompanied by its own drawback of reactivation of oncogenic properties, non-viral vectors like cationic lipids, liposomes, and other polymers were used for in vivo or in vitro transfer of siRNA/shRNAs to target cells [7, 13–16]. In recent studies, it was shown that paclitaxel (PTX) has a very high antitumoral effect and low systemic toxic effect on cervical cancer [17]. Traditional paclitaxel is administered through intravenous (IV) infusion. Due to its low water solubility, it is usually formulated with polyethoxy castor oil (Cremophor EL) and alcohol in a 1:1 ratio. Moreover, Cremophor EL has toxic effects and is incompatible with polyvinyl chloride (PVC) usually used in the IV application. These issues together give rise to clinically prepared dilutions which are unstable as well as exhibiting toxic characteristics [18]. It was reported that there is a need to develop localized and targeted paclitaxel delivery systems to reduce its toxicity and side effects [19]. From the early 1990s, many toxic drugs including doxorubicin, PTX, and retinoic acid were incorporated into SLN systems and tested [20]. SLNs are nanoparticulate systems composed of physiological and biocompatible lipids. These systems are in the colloidal size ranging between 10 and 1000 nm and can be loaded with both hydrophilic and lipophilic drugs [20, 21]. The other potential beneficial aspects of SLNs include controlled and sustained release of drugs, biodegradability and safety of carrier systems, and the potential for large scale-up and sterilization [20, 22]. Solid lipid nanoparticles have been widely used for the delivery of poorly soluble drugs, and for this reason, they are suitable delivery systems for lipophilic anticancer drugs such as PTX [23, 24]. The vagina is a region with a wide surface area and extensive vasculature thus high drug absorption capacity. However, it doesn't have a first pass effect unlike the liver and has low enzyme activity as compared to the gastrointestinal tract [25]. Hence it is a suitable place for administration of local and systemic medications as it is an alternative for parenteral drug administrations and convenience of self-administration by patients [26]. The drug delivery systems applied to the vagina are suppositories, tablets, bioadhesive tablets, capsules, solutions, emulsions, and suspensions and semisolid preparations such as cream gels and ointments, foams, active-ingredient-containing buffers, and rings [25–27]. Vaginal suppositories are usually single-dose preparations used with local purpose for antiseptic, anesthetic, antimicrobial, and contraceptive effects [28]. They are prepared by dispersing or dissolving one or more active compounds in an excipient which is water soluble or dispersed or melts at body temperature [29].

Therefore this study focuses on encapsulation of siRNA, paclitaxel, and siRNA/paclitaxel combination in SLN drug delivery system aiming for their use in cervical cancer treatment. Vaginal suppositories containing SLNs were prepared and evaluated in vitro [30].

2 Materials

2.1 Preparation of Formulations

1. Solid lipid: Stearoyl polyoxyl-32 glycerides (Gelucire 50/13).
2. Surfactants: Sorbitan trioleate (Span 85) and polyoxyethylene sorbitan monooleate (Tween 80).
3. Genetic materials: siRNA Bcl-2 and FITC (fluorescein isothiocyanate) conjugated siRNA.
4. Cationic lipid: *N*-(1-(2,3-dioleoyloxy)propyl)-*N*,*N*,*N*-trimethylammonium (DOTAP).
5. Anticancer drug: Paclitaxel (PTX) (from *Taxus yannanensis*, powder).
6. Dichloromethane (DCM).
7. Polyethylene glycol (PEG 6000).

2.2 Agarose Gel Electrophoresis

2.2.1 TBE Stock Solution

1. Prepare a stock solution of TBE 10× TBE (1 L): Dissolve (use magnetic stirrer) 108 g Tris and 55 g boric acid in 800 mL distilled water in a glass beaker (start stirring directly after addition of magnetic stirrer to the Tris/boric acid mix).
2. Add 40 mL 0.5 M EDTA (pH 8.0), and adjust volume to 1 L using 1 L graduated cylinder. Store at room temperature.
3. Prepare a stock solution of 0.5 M EDTA for 500 mL: Weigh 93.05 g EDTA, and dissolve in 200 mL deionized water (use magnetic stirrer before pH titration).
4. NaOH in fume hood to pH 8. Add NaOH until solution becomes clear. Adjust the volume to 500 mL with deionized water and store at 4 °C.

2.2.2 2% Gel Electrophoresis

1. Prepare sufficient 1× TBE electrophoresis buffer about your tank volume, and fill the tank with it.
2. Weigh 2 g agarose and add 100 mL 1× TBE buffer in Erlenmeyer flask. Heat this mix using a microwave oven to make the agarose completely dissolve in buffer (transparent) (*see* **Note 1**).
3. Let agarose solution cool down to about 45 °C (about when you can comfortably keep your hand on the flask), and add 8 μL ethidium bromide solution (5 mg/mL EtBr) to the dissolved agarose in the fume hood for avoiding of the respire EtBr, and mix.
4. Pour the agarose into a gel tray with the well comb in place.
5. Let it sit at room temperature for 20–30 min, until it has completely solidified.

2.2.3 Gel Electrophoresis Loading Buffer (6×)

1. Take 3 mL glycerol (%100) in 15 mL falcon tube.
2. Weigh 25 mg bromophenol blue, dissolve in 10 mL distilled water, and mix with glycerol.
3. Autoclave the mixture at 121 °C for 20 min, and store at room temperature.

2.3 Cell Culture Studies

HeLa (human cervical cancer adenocarcinoma) cell line was used for cell culture studies (*see* **Note 2**). All cell culture studies should be performed in laminar flow cabinet (*see* **Note 3**).

2.3.1 Growing of the Cells

1. Mix 445 mL Dulbecco's Modified Eagle Media (DMEM) that contained 4500 mg/L glucose, sodium bicarbonate, and phenol red with 50 mL FBS and 5 mL 100× streptomycin/penicillin, and filter from 0.2 μm filter (*see* **Note 4**).
2. Quickly thaw the HeLa cell line (2 mL) stored frozen in cold-resistant cryovials at −80 °C, in a water bath at 37 °C, and put the cell suspension into 20 mL of DMEM at 37 °C gently, and centrifuge at 134 × *g* for 5 min to remove the dimethyl sulfoxide (DMSO) from the medium (*see* **Note 5**).
3. Discard the supernatant at the end of centrifugation, and add 15 mL of DMEM (37 °C) to the pellet remaining.
4. Homogenize the cell suspension with a sterile disposable pipette.
5. Transfer the homogenized cell suspension to 75 cm^2 sterile flasks (cell culture container). Gently shake the flask orbital motion.
6. Write the name of the cell, the date of opening, and the number of passages onto the flask.
7. Put the flasks into an incubator at 37 °C in a fully humidified atmosphere containing 5% CO_2 and 95% air until the surfaces of the flask are completely covered by the cell.

2.3.2 Preparation of Cells for MTT Assay

1. Withdraw all the medium on the cells completely.
2. Add 15 mL of PBS to flasks, and aspirate all the solution from the flask to completely remove the medium from the cell surface.
3. After aspiration, add 2 mL of 0.25% trypsin/EDTA (*see* **Note 6**) to 75 cm^2 flasks for separation of the cell from the surface of the flask, and put the flasks in an incubator.
4. After about 2–3 min, the cells are checked in an inverted microscope. Cells must move in the solution.
5. After all cells are separated from the flask surface, add 20 mL of fresh DMEM (37 °C) to the flask to inhibit the trypsin.
6. Place all the contents in the flask into 50 mL Falcon Conical Centrifuge Tubes, and centrifuge at 134 × *g* for 5 min.
7. Discard the supernatant, add 10 mL of fresh DMEM (37 °C) on the remaining pellet and resuspend it to count the cells.

2.3.3 Cell Counting

Cell count was performed with hemocytometer. Briefly:

1. Wash the hemocytometer and coverslip with 70% alcohol and distilled water before use. Separate 100 μL medium containing cell suspension from 10 mL (**item** 7 of Subheading 2.3.2) to a microcentrifuge tube, and add the 100 μL of 0.4% tyrpan blue solution. Trypan blue stains dead cells but not viable cells.
2. Put the coverslip on hemocytometer. Hemocytometer has nine large squares, and each square has the same dimensions and contents 10^{-4} mL of suspension.
3. Take 10 μL cell solution with a pipette and put it the gap of between coverslip and hemocytometer, make sure that dyed cell suspension over the lines.
4. Place the hemocytometer under the inverted microscope and control the cells on the hemocytometer lines.
5. Count the total number of viable and nonviable cells in the five large squares shown in Fig. 1 with a 10× objective.
6. Determine the concentration (viable cells/mL) using the following equation (Eq. 1):

$$\text{Concentration} = \frac{\text{Average cells}}{\text{Square}} \times \text{Dilution factor} \times 10^4 \quad (1)$$

Example:
Counted total viable cell: 64
Counted total unviable cell: 8
Percentage of viable cells: 64/(64 + 8) × 100 = 88.89 = 88.89%
Average of cells/squares: 64/5 = 12.8

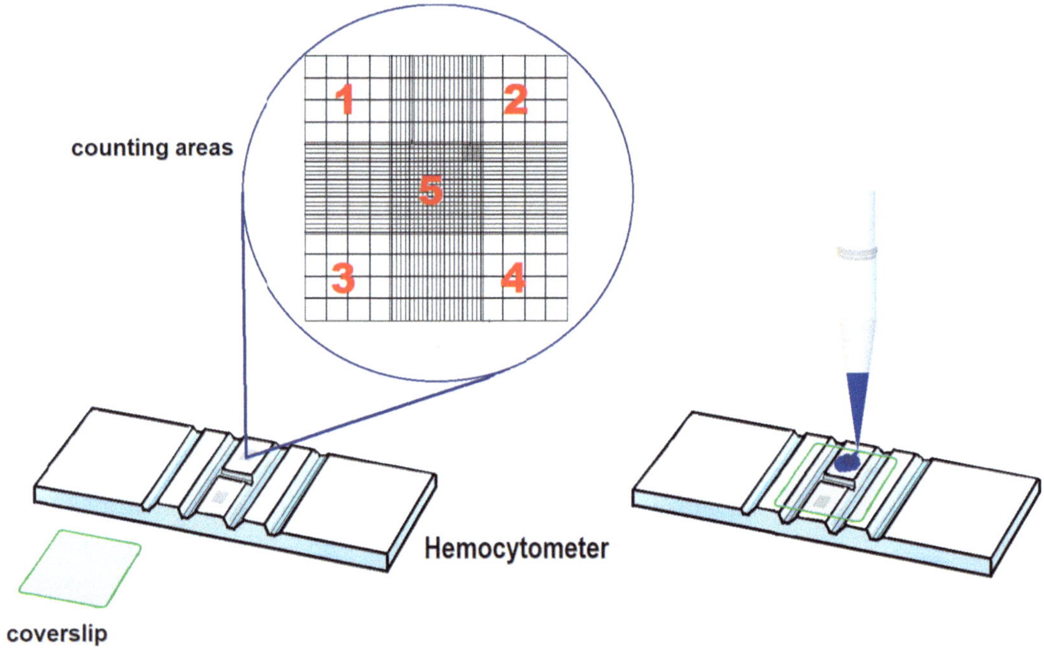

Fig. 1 Hemocytometer and counting areas

Dilution factor:

100 μL cell suspension + 100 μL tyrpan blue/100 μL cell suspension = 2

Concentration = $12.8 \times 2 \times 10^4$ = 256,000 cell/mL

Because our initial volume is 10 mL

$256{,}000 \times 10 = 2.56 \times 10^6$ cell/10 mL

2.3.4 Plates for MTT Studies

1. Prepare two 96 plates for the 7 samples that will be evaluated at 7 concentrations (*see* **Note** 7).
2. Dilue the cell suspensions to be approximately 2×10^4 cells for 100 μL as described at Subheading 2.3.3 and transfer 100 μL cell suspension to each well of 96 plate as Fig. 2.
3. To create a blank on the plate, do not put any cell suspension in a few wells on the plate.
4. Shut the lid of the plate, and put the plates in the incubators. Allow the cells to grow in the incubator at 37 ° C in a fully humidified atmosphere containing 5% CO_2 and 95% air for 1 day.
5. Following the 1-day reproduction period, remove the medium on the cells in the well.
6. Carefully, wash the wells with 1×PBS, one time.
7. The 96-well plate that contain the cells is ready for treatment with samples.

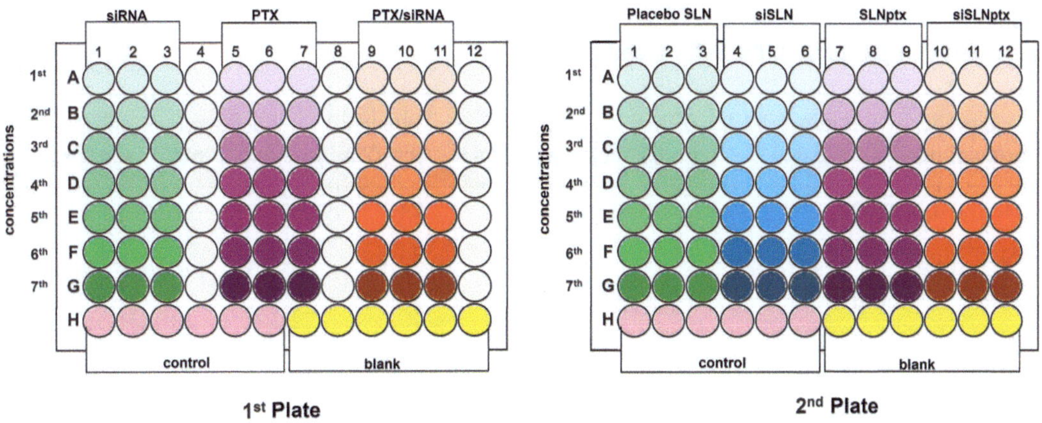

Fig. 2 Schematic illustration of cell seeding area on 96-well plate according to samples

2.3.5 Cell Storage

Unused subcultures were stored as cell stock at −80 ° C for a short period. The optimal freezing conditions depend on the cell line in use. For this:

1. Detach the cells gently from the flask above the procedure (Subheading 2.3.2) used during the subculture.
2. Resuspend the cells in a fresh 5 mL DMEM and count as in Subheading 2.3.3.
3. 100 μL DMSO was put into a sterile cryovial and stored at +4 °C until to use.
4. 1 mL medium containing 2–4 × 10^6/mL cell suspension was added into cryovial containing 100 μL DMSO; gently pipette.
5. Put the cryovials to +4 °C (~30 min), −20 °C (~1 h), and −80 °C (1 day), respectively.
6. Transfer the frozen cells to liquid nitrogen for a long period.

3 Methods

3.1 Preparation of SLN Formulations

SLNs were prepared using solvent emulsification/evaporation method according to the procedure described by Büyükköroğlu [26]. The ratios and code of the formulations are given in Table 1.

3.1.1 Preparation of Placebo SLN

1. For oily phase, dissolve 80 mg Gelucire 50/13 and 1 mg DOTAP with 1 mL DCM in a round bottom 5 mL glass flask.
2. For aqueous phase, dissolve 48 mg Tween 80 and 32 mg Span 85 (resulting a HLB value of 7.08) in 4 mL distilled and autoclaved (121 °C, 20 min) water (~pH 6.8) in a 15 mL falcon tube.
3. Add 2 mL aqueous phase to the oily phase in a round bottom 5 mL glass flask using micropipette.

Table 1
Codes and compositions of the SLNs prepared

Code of the formulations	Gelucire 50/13 (mg/2 mL)	DOTAP (mg/2 mL)	Span 85 (mg/2 mL)	Tween 80 (mg/2 mL)	Paclitaxel (mg/2 mL)	siRNA (μg/2 mL)
Placebo SLN	80	1	16	24	–	–
siSLN	80	1	16	24	–	125
SLNptx	80	1	16	24	4	–
siSLNptx	80	1	16	24	4	125

4. Sonicate the mixture for 1 min at 20% power using a probe sonicator in a temperature-controlled (2 °C) water bath.
5. Evaporate the emulsion under 20 rpm stirring and at about 150 mbar for 15–20 min in a water bath of 33 °C to remove the residual organic solvent using a Rotavapor.
6. Collect the formulation by measuring using micropipette from glass flask.
7. Determine how much water evaporated from formulation during evaporation step and complete the volume of formulation to 2 mL with water.
8. For homogenization of formulation, use a nano-homogenizer to obtain a homogeneous dispersion and also to reduce the particle size. For this purpose, transfer the formulation into "inlet reservoir" of homogenizers, apply high pressure (up to 40,000 psi or 2578 bar), and collect the formulation from "outlet reservoir" of homogenizers (Fig. 3).
9. Filter the obtained formulation using 0.2 μm non-pyrogenic filter in a laminar flow cabinet for sterilization, and store at 4 °C until use.

3.1.2 Preparation of SLNptx

1. For oily phase, dissolve 80 mg Gelucire 50/13, 1 mg DOTAP, and 4 mg PTX with 1 mL DCM in a round bottom 5 mL glass flask.
2. Apply Subheading 3.1.1, **steps 2–9**.

3.1.3 Preparation of siSLN

1. For oily phase, dissolve 80 mg Gelucire 50/13 and 1 mg DOTAP with 1 mL DCM in a round bottom 5 mL glass flask.
2. For aqueous phase, dissolve 24 mg Tween 80 and 16 mg Span 85 (resulting a HLB value of 7.08) in 1 mL distilled and autoclaved water in a 2 mL microcentrifuge tube.
3. Disperse the 125 μg siRNA in 1 mL water in another 2 mL microcentrifuge tube, and add on the aqueous phase.
4. Apply steps in Subheading 3.1.1 from **steps 3** to **9**.

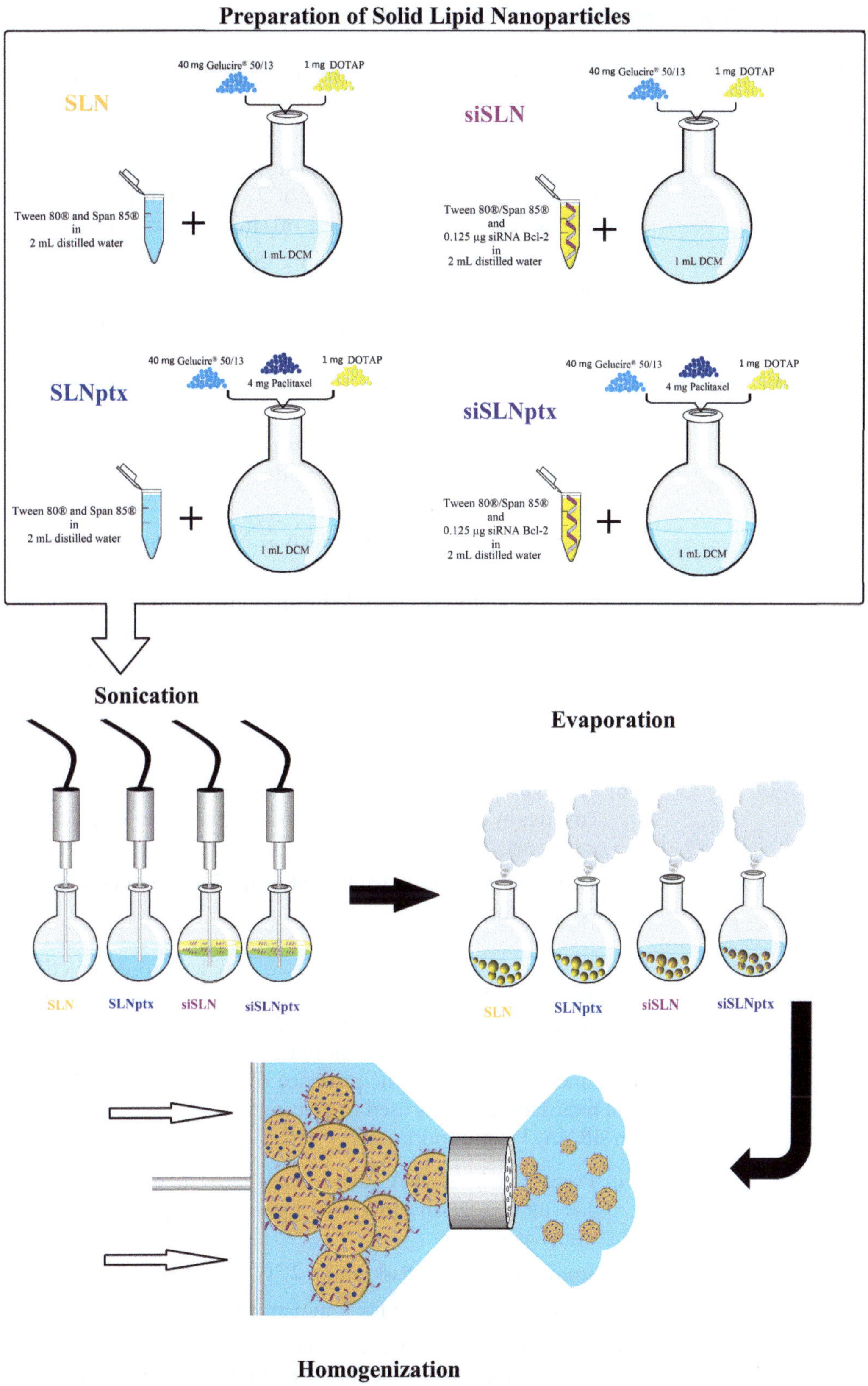

Fig. 3 Schematic illustration of preparation of SLN formulations [30]

3.1.4 Preparation of siSLNptx

1. For oily phase, dissolve 80 mg Gelucire 50/13, 1 mg DOTAP, and 4 mg PTX with 1 mL DCM in a round bottom 5 mL glass flask.
2. For aqueous phase, dissolve 24 mg Tween 80 and 16 mg Span 85 (resulting in a HLB value of 7.08) in 1 mL distilled and autoclaved water in another 2 mL microcentrifuge tube.
3. Disperse the 125 μg siRNA in 1 mL water and add the aqueous phase.
4. Apply steps in Subheading 3.1.1 from **steps 3** to **9**.

3.2 Characterization of SLNs

3.2.1 Determination of Particle Size and Zeta Potential

1. Use ultrapure distilled water adjusted to a constant conductivity of 50 μS using 0.9% NaCl.
2. Adjust the pH of this water to pH 7.4 using 0.1 M NaOH and pH 4.2 using 2% acetic acid solution prior to analyses at 25 °C [31].
3. Take the 30 μL samples of each SLN to microcentrifuge tube separately.
4. Add 1 mL adjusted distilled water onto samples, and disperse the SLNs in it.
5. Transfer these diluted dispersions to the Malvern Zeta capillary cuvettes, and determine average diameter and size distribution represented by the polydispersity index (PDI) using analyzer (Zetasizer NanoZS, Malvern Instruments, UK).
6. Analyze the Zeta potential using same sample in capillary cuvettes by the software based on the Helmholtz-Smoluchowski equation.
7. Repeat all analyses in triplicate for pH 7.4 and pH 4.2 at room temperature.

3.2.2 Determination of Sonication Effect on siRNA

Time and intensity of sonication used may lead to destruction in genetic material [32, 33]; therefore, the sonication time required for production of SLNs without damaging the siRNA and the sonication applied to siRNA should be determined in organic solvent and water without the presence of solid lipid and surfactant. Gel retardation assay is used to investigate the effects of sonication on siRNA. For this purpose:

1. Prepare the siRNA dispersion with a concentration of 125 μg/2 mL, and mix with 1 mL DCM in a round bottom 5 mL glass flask.
2. Immerse the glass flask into a 2 °C water bath.
3. Sonicate the mixture for 5 min at 20% power.
4. Stop the sonication every minute, and collect an aliquot of 20 μL sample from the mixture into a microcentrifuge tube.

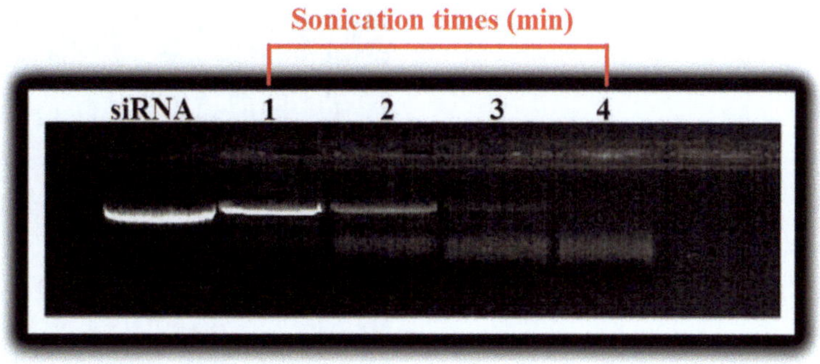

Fig. 4 Gel images of siRNA following sonication (siRNA 1.5 µg). Since siRNA was determined to be degraded by sonication at 20% power and beginning at the second minute, 1 min sonication has been found as appropriate time during SLN preparation

5. Allow the evaporation of DCM from samples at room temperature.
6. Add 5 µL 1× loading buffer on each sample and load on gel.
7. Electrophorese at 50 V for 2 h, and take the image using a gel documentation system (Fig. 4).

3.2.3 Extraction of siRNA from SLNs

To determine siRNA encapsulation efficiency and the quantification of formulations, an extraction method should be used to isolate siRNA from siSLN and siSLNptx formulations because encapsulated siRNA will not give any intensity on the gel image. For this purpose:

1. Transfer the 25 µL of the formulations into a microcentrifuge tube, and dilute with 100 µL water.
2. Add 1500 mL DCM ethanol (1:2) on the diluted dispersions to dissolve the lipid and PTX in the organic solvent and to precipitate siRNA.
3. Gently mix content of the tubes, and centrifuge at 134 × *g* for 30 min to precipitate siRNA.
4. Carefully discharge the supernatant, and add 500 mL ethanol on precipitate for removing the lipid residues from siRNA.
5. Evaporate the ethanol at room temperature, and disperse the siRNA in 25 µL water.
6. Determine the concentration of recovered siRNA using a nano-drop spectrophotometer at OD280 ($n = 6$).
7. In order to determine the effect of extraction on siRNA, load 20 µL sample of extract on the gel with 5 µL 1× loading buffer.
8. Electrophorese at 50 V for 2 h, and take the image using a gel documentation system (Fig. 5).

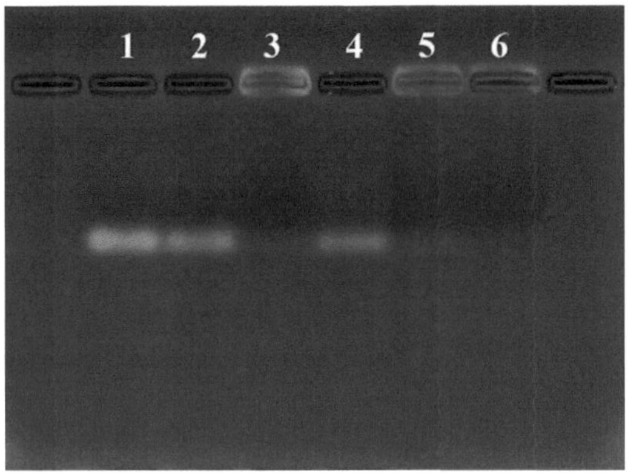

Fig. 5 Gel images of siRNA extracted from siSLN formulation (**1.** siRNA 1.5 μg, **2.** extracted siRNA from siSLN, **3.** siSLN, **4.** extracted siRNA from siSLNptx, **5.** siSLNptx, **6.** placebo SLN [30])

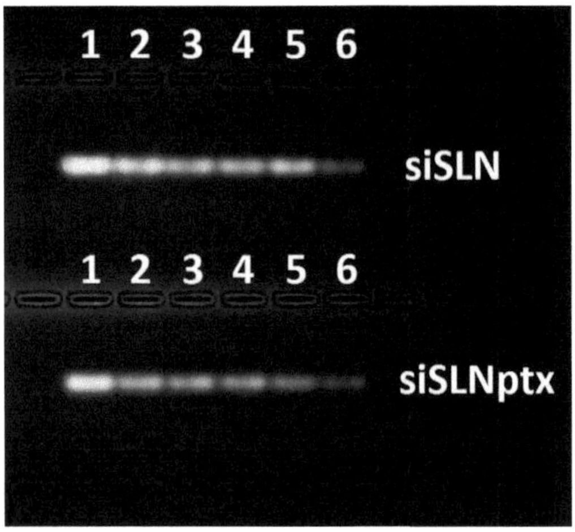

Fig. 6 Serum stability of siRNA after extraction from siSLN and siSLNptx formulations (**1.** siRNA, 1.5 μg; **2.** 3 h; **3.** 6 h; **4.** 12 h; **5.** 24 h; **6.** 48 h)

3.2.4 Evaluation Stability of siRNA in Serum

Protective character of the siRNA delivery system against serum components is of great importance for the maximum efficiency in cells [34]. Therefore, behavior of siSLN and siSLNptx should be examined in the culture medium containing serum components. For this purpose:

1. Take 25 μL of siSLN and siSLNptx separately in microtubes for 3, 6, 12, 24, and 48 h.
2. Add the 100 μL Dulbecco's Modified Eagle Medium (DMEM) which contained 10% fetal bovine serum (FBS).
3. Incubate mixtures at 37 °C for 3, 6, 12, 24, and 48 h, extract siRNA from mixtures using extraction method, load the extracted siRNAs on gel, and take image as described above (Fig. 6).

4. Also, determine the amount of recovered siRNA using a nano-drop spectrophotometer at OD280 ($n = 6$).

3.2.5 Determination of PTX Content with High-Performance Liquid Chromatography (HPLC)

A simple and validated HPLC method was adopted and developed for the determination of paclitaxel. The method was validated for accuracy, precision, and linearity with reference to the International Council for Harmonization guidelines. Validation of an analytical procedure is to demonstrate that it is suitable for its intended purpose. A validated analytical method ensures that it provides consistent, reliable, and accurate data (*see* **Note 8**).

1. Use a Shimadzu Shim-Pack CLC-ODS column with a diameter of 6.0 mm and length of 15.0 cm.
2. Prepare the mobile phase using a mixture of acetonitrile:water (50:50 v/v).
3. After mixing the solutions, in order to remove air bubbles, put the mixture into ultrasonic bath for 30 min.
4. Set the HPLC to 227 nm in order to analyze PTX with UV detection.
5. Set the flow rate of 1 mL/min.
6. Set the column temperature to 30 °C.
7. In order to examine the linearity of the study, dilute the paclitaxel in the range of 1–10 μL with the stock solution (50 μg/mL) in acetonitrile.
8. Take 20 μL from the stock solution, and complete to 1000 μL with acetonitrile into the vials.
9. In the following steps, 40, 60, 80, 100, 120, 140, 160, 180, and 200 μL of solutions were taken, respectively, and completed to 1000 μL with acetonitrile like in the **step 4**.
10. Pipette the aliquots of stock solutions into a series of 1 mL volumetric vials.
11. Close the vials and vortex for at least 1 min before analyzing with HPLC.
12. After setting the HPLC flow rate, column temperature, and UV nanometer, plot the program at least 1 h with the mobile phase.
13. Set the batch for the analyses.
14. Draw the calibration curves with the obtained results, and determine the correlation coefficient with statistical analyze program.
15. Construct the calibration curves by plotting peak area against the initial concentration of paclitaxel.
16. For the developed method in our lab, the linear regression analysis of the data gave the following equation:

$$y = 19{,}289x - 1503\left(r^2 = 0.9999\right). \quad (2)$$

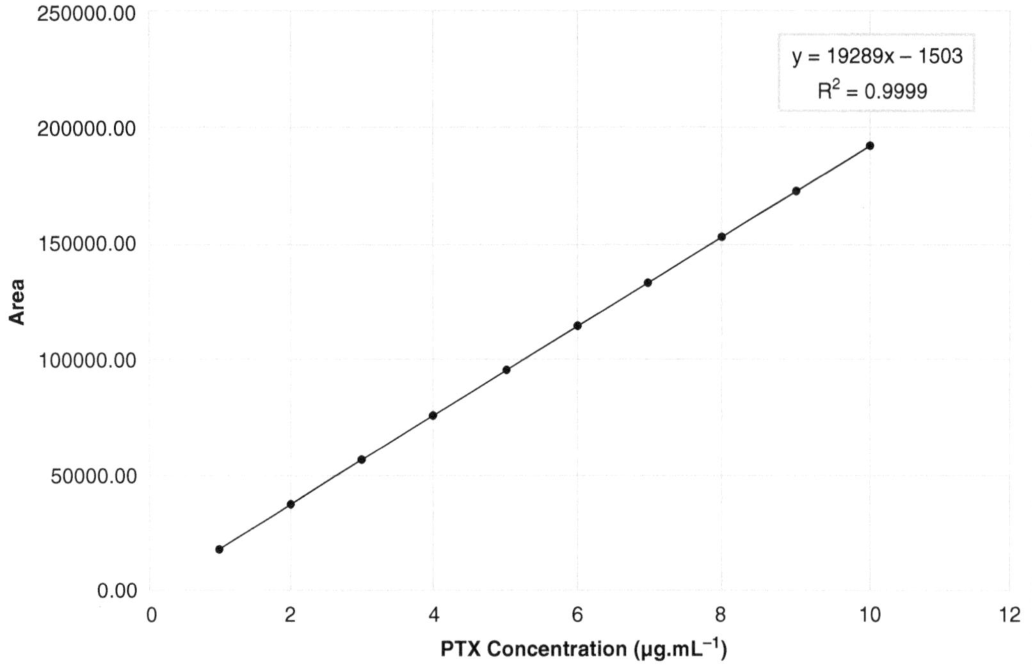

Fig. 7 High-pressure liquid chromatography standard curve of PTX

17. The method should have found to be linear and reproducible as Fig. 7.
18. In order to find the accuracy and precision of method, use six replicates of three different concentrations such as low-, medium-, and high-quality control samples.
19. Take 60, 120, and 180 μL of stock solutions into the vials, and complete to 1000 μL with acetonitrile.
20. Carry the analyses in three different days' time.
21. Using these steps, calculate the accuracy of the method. In our method, it was calculated as 100.4 ± 0.5 (n = 6). The intra- and inter-day precision values of the method were 1.5–2.0% for the 6 μg/mL concentration value and 0.90% and 0.92% for the 10 μg/mL concentration, respectively. The choice of method was made using a placebo formulation with no active ingredient from the placebo. Limit of detection (LOD) was calculated as 21 ng/mL, and the limit of quantitation (LOQ) was calculated as 62 ng/mL (*see* **Note 9**).

3.2.6 Determination of Loaded PTX Content from SLNs

1. In order to determine PTX amount in SLNptx and siSLNptx, take 500 μL of formulation and independently mix with methanol (1.5 mL) in 2 mL microtube (the same process must be applied on placebo SLN for control).

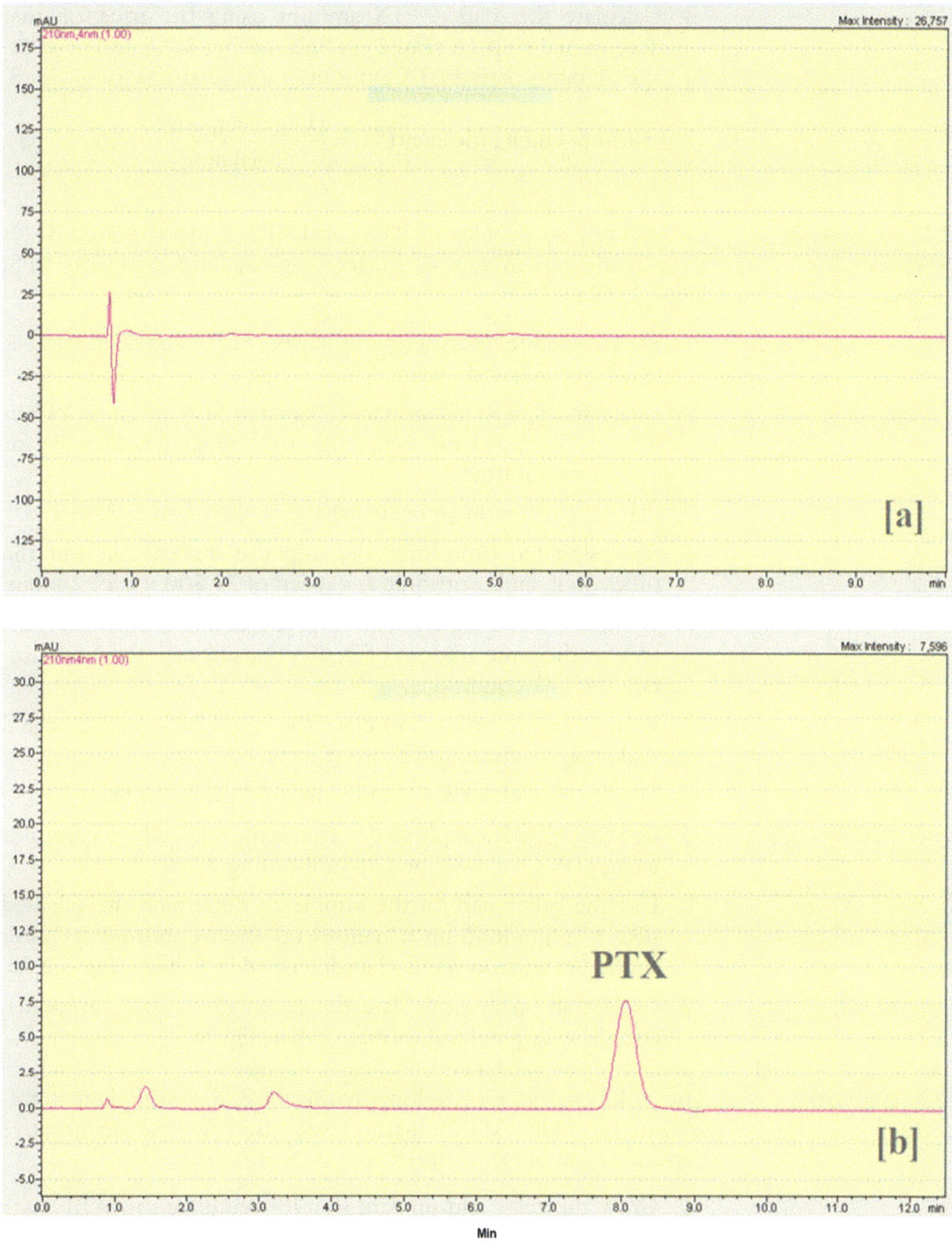

Fig. 8 Selectivity chromatograms for the PTX quantitation method (a = placebo SLN, b = pure PTX)

2. Filter the solution in step 1 through a 0.22 μm polyamide membrane filter.
3. Take 500 μL from filtrate to HPLC vial, add 500 μL mobile phase, and examine with the HPLC (Fig. 8).

4. Calculate the loaded PTX amount using the equation that determined **step 16** (Eq. 2) of Subheading 3.2.5 and percentage of incorporated PTX amount in formulations using Eq. 3.

$$\text{Incorporation efficiency}\,(\%) = \frac{\text{Drug incorporated}}{\text{Total drug}} \times 100 \qquad (3)$$

3.2.7 In Vitro Release of siRNA and PTX from SLNs

In vitro release profiles of PTX- and siRNA-incorporated SLNs are investigated in vaginal buffer (vaginal fluid simulant, VFS; pH 4.2) [35].

1. Determine the interval times such as 15, 30, and 45th, 1, 2, 4, 6, 12, 24, 32, and 48 h.
2. Disperse the 500 μL sample of the each SLN in 1 mL VFS at 37 ± 1 °C in a 2 mL centrifuge tube separately (n = 6) for every interval time.
3. Place the tubes on an orbital shaker at 120 rpm at 37 ± 1 °C.
4. At designated time intervals, stop the shaker, take out the tubes on it, and centrifuge at a speed of 25,300 × g for 20 min.
5. Withdraw 1 mL supernatant at predetermined time intervals, and for the same amount of fresh VFS, add each time to maintain the sink conditions.
6. Resuspend the pellet, and place the tubes back on the orbital shaker for continuous release studies.
7. Divide the withdrawal sample into 500 μL portions.
8. Use the 500 μL sample to determine the released PTX amount using HPLC as described in Subheading 3.2.6.
9. Use the other part of the sample to determine the released siRNA amount using a nano-drop spectrophotometer after applied extraction method as described in Subheading 3.2.3.
10. Collect all results, calculate the cumulative release amounts, and show as graphical format such as Fig. 9.

3.2.8 Cytotoxicity of SLNs

The colorimetric 3-(4,5-dimethylthiazol-2-yl)-2,5-diphenyltetrazolium bromide (MTT) method was used for the quantitative determination of cell cytotoxicity (*see* **Note 10**).

1. Grow the cells, and prepare the 96-well plate for MTT assay using descriptions from Subheadings 2.3.1–2.3.4.
2. Prepare every stock solution of formulations, calculate the amount of the formulation for each well, and multiply with 4.

   Examples

   <u>For first concentration of siSLNptx</u>

   0.5 μL formulation in 100 μL medium for one well

   Stock solution of formulations: 0.5 μL × 4 = 2 μL formulation in 398 μL medium.

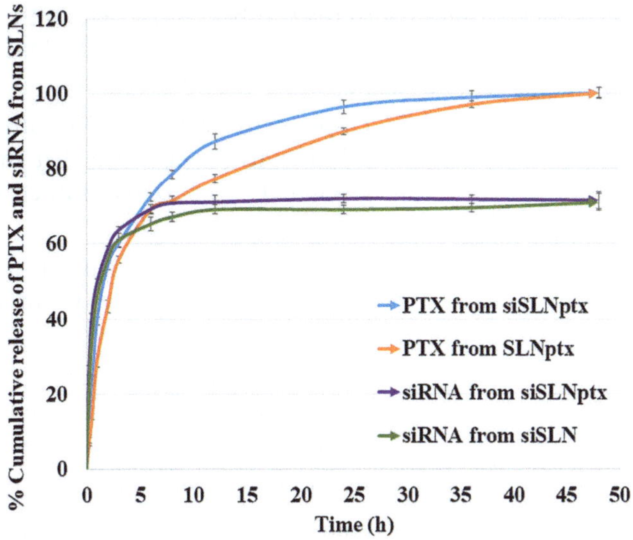

Fig. 9 The in vitro release profiles of siRNA and PTX from SLNs [30]

Table 2
Example for calculated dose of siSLNptx

Cons.	Formulation in 100 μL medium for one well (μL)	Formulation for four well (μL)	Medium for four well (μL)	PTX (μg) for one well	siRNA (ng) for one well
1	0.5	2	398	1	31.25
2	0.75	3	397	1.5	46.88
3	1	4	396	2	62.50
4	1.25	5	395	2.5	78.13
5	1.5	6	394	3	93.75
6	2	8	392	4	125.00
7	3	24	376	6	187.50

Amount of PTX: 4 mg PTX in 2000 μL siSLNptx formulation (Table 1).

1 μg PTX in 0.5 μL

Amount of siRNA: 0.125 mg siRNA in 2000 μL siSLNptx formulation.

31.25 ng siRNA in 0.5 μL

3. Apply 100 μL of these solutions separately to each three well as descripted in Fig. 1.
4. Weigh 10 mg PTX, and dissolve it in 100 μL DMSO. Make up to a volume of 1 mL with medium. Divide the stock solution to proportions that include quantity of PTX as given in Table 2

to microtubes, and make up to 100 μL with medium for each well. The final volume of DMSO should not exceed 0.1%; otherwise DMSO may show a cytotoxic effect.

5. Disperse the 2 μg siRNA with PBS to 100 μL. Divide this solution to proportions that included quantity of siRNA as given in Table 2 to microtubes, and complete with medium to 100 μL for each well.
6. Add the samples of formulations/PTX/siRNA onto the cell suspension with pipette. Change the pipette tips at each concentration change.
7. Add cell culture medium to six wells on the plate for forming control group as demonstrated in figure 2, this will be negative control.
8. Just add fresh cell culture medium to wells left as blank!
9. Shut the lid of plate, write the incubation time to be investigated, name the sample to be investigated, and note which concentration is put into which well.
10. Put the plates into incubator until the 48 h in the atmosphere containing 5% CO_2 and 95% air at 37 °C for evaluation of cell viability.
11. At the end of the desired period, remove the plates from the incubator, and aspirate the medium on cell in wells. Wash the wells with 1×PBS.
12. Add 20 μL MTT stain (5 mg/mL) to each well, and incubate for an additional 4 h at 37 °C so that the MTT stain could be converted into formazan salt by living cells.
13. At the end of 4 h, remove the MTT dye from wells, and add 200 μL DMSO to each well.
14. Put the wells in the incubator for an additional 30 min in order for DMSO to dissolve the formazan homogeneously.
15. After 30 min, analyze the absorbance of well with a microplate reader (spectrophotometer) at 570 nm wavelength (*see* **Note 11**).
16. You can see with the naked eye that the wells with living cells are purple. The wells are seen in lighter color depending on the increase of cell death.
17. The range of absorbance of formazan is from 400 to 650 nm with peak at 540 nm. You could use different wavelengths in this range; it will depend on your spectrophotometer.
18. A brief summary of the cytotoxicity study is shown in Fig. 10.
19. Receive average absorbance values of each concentration, and control and calculate % cell viability using Eq. 4.

$$\text{Cell viability\%} = \frac{\text{Average absorbance values}}{\text{Average absorbance values control}} \times 100 \qquad (4)$$

20. Collect all results and show a graphical format such as in Fig. 11.

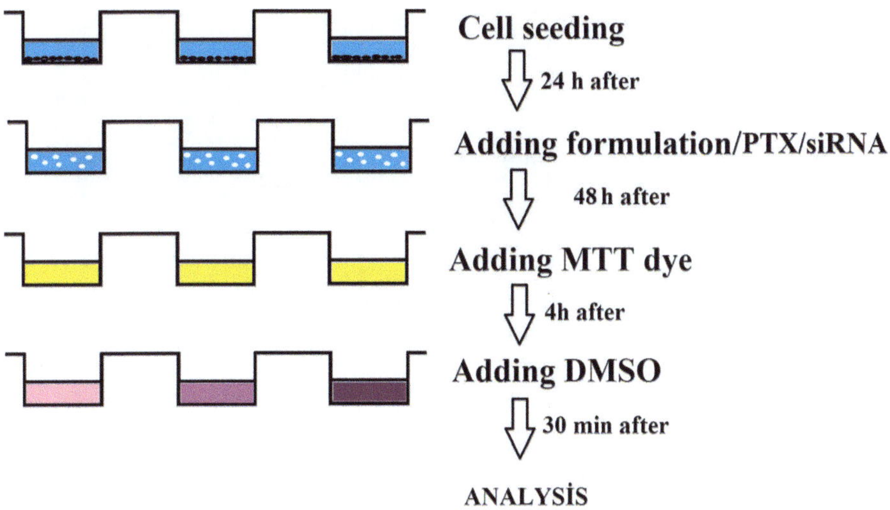

Fig. 10 Schematic illustration of MTT assay

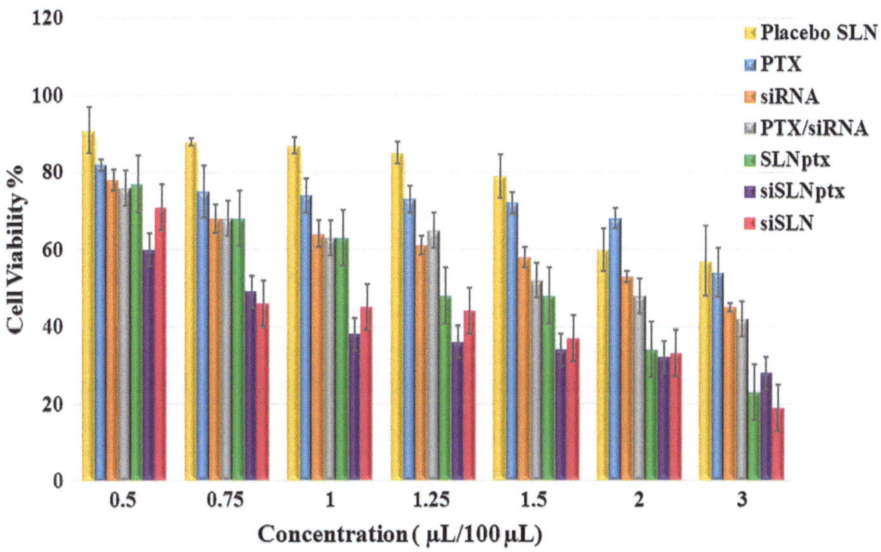

Fig. 11 Viability of Hela cells after being treated with PTX, siRNA, PTX/siRNA, and SLN formulation. Each formulation should be tested in seven replicates for 48 h

3.2.9 Cellular Uptake (Transfection Studies)

To determine the transfection efficacy of SLN, siSLN formulation was prepared using FITC-conjugated siRNA instead of Bcl-2 siRNA and named F-siSLN.

1. Seed the cells to 6-well cell culture plates with 2 mL medium containing 10% FBS without antibiotics, and incubate for 24 h at atmospheric environment containing 5% CO_2 and 95% air at 37 °C until 60–70% density.
2. At the end of 24 h, wash the wells with 1×PBS.
3. Dilute the 1.6 μL F-siSLN formulations that included 100 ng siRNA, mix gently with FBS/antibiotic-free fresh cell culture medium, and the final volume should be 100 μL.

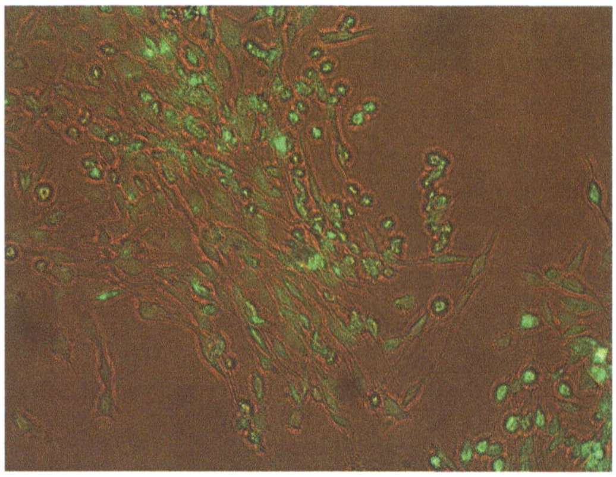

Fig. 12 Superimposed phase-contrast and fluorescence images of HeLa cells after being treated with F-siSLN for 24 h [30]

4. Add 100 μL of solution via dropwise to each well containing cell.
5. Shut the lid of the plate, and put in the incubator for 6 h.
6. At the end of 6 h, aspirate the medium from wells (optionally, medium may not be changed after 6 h).
7. Add fresh cell culture medium containing 20% FBS to the wells.
8. At the end of 48 hours incubation, visualize the transfection of siRNA under fluorescence microscope with a 10× or 20× or 40× objective (Fig. 12).
9. Count transfected and untransfected cells in various areas, and calculate the % transfection efficiency (TE) using Eq. 5.

$$\text{TE\%} = \frac{\text{Number of transfected cells}}{\text{Number of transfected cells} + \text{number of nontransfected cells}} \times 100 \quad (5)$$

3.3 Preparation of the Vaginal Suppositories

1. Take 1 mL SLN formulation into a round bottom 5-mL glass flask, and evaporate it under 20 rpm stirring and at about 100 mbar for 30 min in a water bath of 33 °C to reduce the volume to 250 μL.
2. Weigh the 1400 mg PEG 6000 in a 10 mL glass beaker, and melt it at 80 °C.
3. Add the 140 μL distilled water heated to the same temperature, and mix with melted PEG.
4. Following cooling of this mixture to 35 °C, add the reduced SLN dispersion on it, and mix using glass baguette.
5. Pour the mixture onto a greaseproof paper in droplets, and keep overnight to solidify at room temperature (Fig. 13).
6. Store at 4 °C until being used.

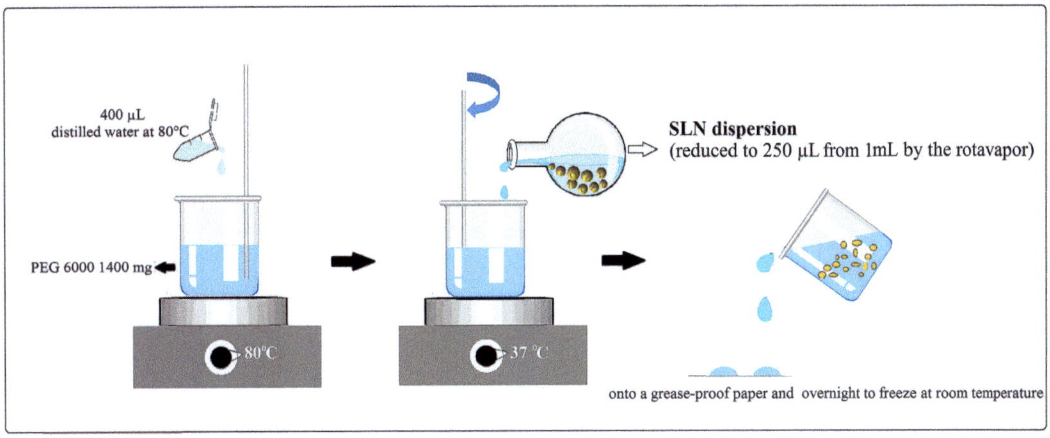

Fig. 13 Preparation of the vaginal suppositories [30]

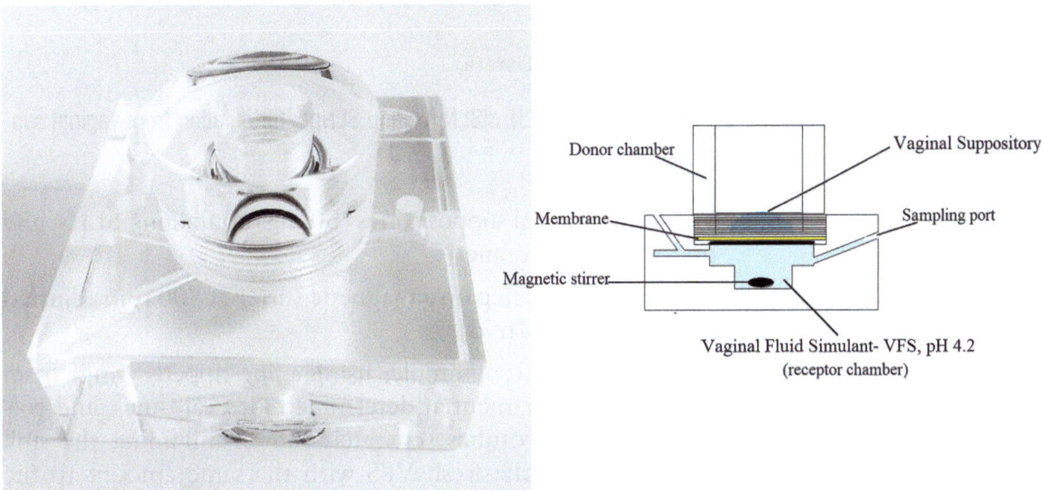

Fig. 14 Horizontal diffusion chamber and in vitro release from suppositories [30]

3.3.1 Determination of Release Profile of SLN Formulations from Vaginal Suppositories

Horizontal diffusion chamber system (HDCS) was used for the release of SLNs from suppositories. The receptor phase was VFS [35, 36], while the membrane was cellulose acetate with 0.45 μm pore size (Sartorius, Germany).

1. Soak the membranes into the buffer solution at room temperature for 1 h.
2. Determine the interval times such as 5, 15, and 30 min and 1, 2, 3, 4, and 6 h.
3. Fill up the receptor compartment with 3 mL VFS and put on a magnetic stirrer (0.3 × 0.5 cm).
4. Place the membrane on the horizontal diffusion chamber to be in contact with 1 cm^2 surface area and with no air bubbles remaining, and mount on the donor chamber of system as in Fig. 14.

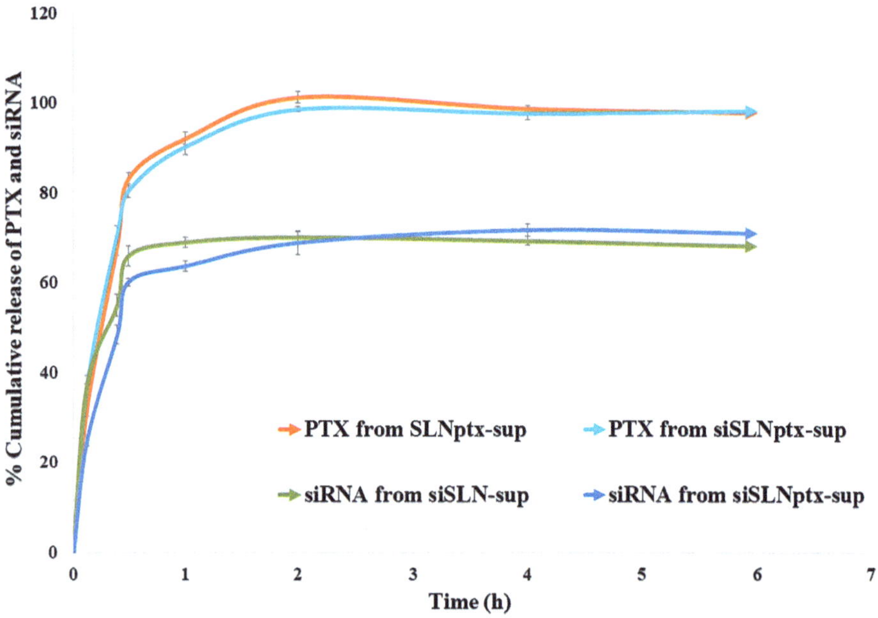

Fig. 15 In vitro time-dependent release profiles of siSLN, siSLNptx, and SLNptx formulation from vaginal suppositories (mean ± SE) [30]

5. Weigh the vaginal suppositories (50 mg) containing SLNs, and put on the membrane.
6. Start the stirring step on a magnetic stirrer at 700 rpm adjusted before to 37 °C (n=6).
7. Withdraw 250 μL of samples using a fine-tipped syringe from receptor compartment at determined interval times, and collect in the microtubes separately. For maintaining the sink condition, add the fresh VFS with the same amount to the compartment.
8. For determination of PTX content of samples, add 1 mL methanol to the samples, filter through 0.22 μm filter, and use 1 mL of the filtrate for the HPLC analyses as described in Subheading 3.2.6. Calculate the released PTX amount using Eq. 2.
9. For determination of siRNA content of samples, extract the siRNA from samples using extraction method **steps 2–6** in Subheading 3.2.3.
10. Create a graph of cumulative released PTX and siRNA as in Fig. 15.

4 Notes

1. You should not see any beads in the solution. If you have beads in the agarose, you will get distorted bands.
2. *HeLa cells* which are derived from immortal cervical cancer *cells are used in* cell culture studies.

3. *Laminar flow cabinet* is used to provide a sterile container and dust-free and particulate-free air with HEPA filters while running in experiments. The air in the chamber is taken into the cabinet and passed through a HEPA filter. The filtered air is passed through the experiment or devices and returns to the room with a correct flow. Therefore, sterile cabinets only protect internal experiments and materials. The preparation of materials used for cell culture, seeding of cell, and addition of formulations to cell on plate should be done in a laminar flow cabinet.
4. *DMEM (Dulbecco's Modified Eagle Media)* was used as a cell culture base medium in the culture. DMEM is the most frequently used nutrient component in somatic cell cultures today. DMEM has the necessary glucoside for feeding the cells, suitable osmolarity and pH for sustaining their vitality, and amino acids and vitamins necessary for their functioning. The pH should be around 7.4, and phenol red is used as pH indicator in most media. *Fetal bovine serum* (FBS) is a rich protein solution that is used to capture and proliferate cells. This protein solution contains hormones, enzymes, and growth factors that allow the cell to grow and multiply and intercellular matrix proteins that allow it to attach to the surface. The serum ratio of the media can vary depending on the cell culture medium and application. In a standard somatic cell culture, the serum ratio is 10%. When the solution is first opened, it should be aliquoted into small tubes (e.g., aliquot to 10, 20, and 50 mL falcon tubes) and taken as needed. The melted solutions should not be frozen again. *Penicillin and streptomycin* are antibiotics used to prevent bacterial contamination of culture. They have a good effective inhibition against gram-positive and gram-negative bacteria. In the cell culture study, antibiotic solution contains 10,000 units/mL of penicillin and 10,000 μg/mL of streptomycin were used at 100X concenration. When the solution is first opened, it should be aliquoted into small tubes (e.g., aliquot to 1 or 2 mL Eppendorf microcentrifuge tubes) and taken as needed. The melted solutions should not be frozen again.
5. *DMSO (dimethyl sulfoxide)* is a cryoprotectant that penetrates the cell. It has low molecular weight. Prior to the realization of the frost, due to the osmotic pressure difference, the liquid in the cell is displaced by the cryoprotectant materials, thereby reducing both cell volume changes and minimizing the formation of ice crystals within the cell, thereby minimizing damage to the cells during freezing.
6. *Trypsin/EDTA solution* is used to remove the cells from the flask. Trypsin is a serine protease-type enzyme that cleaves peptides from amino acids of lysine and arginine. In our laboratory, trypsin solution with 0.25% EDTA is used. It should not

be treated with cells for a long period of time as trypsin may damage cells over time. When the solution is first opened, it should be aliquoted into small tubes (e.g., aliquot to 1 or 2 mL Eppendorf microcentrifuge tubes), and take as needed. The melted solutions should not be frozen again.

7. *96-well plate* has many names like microplate, microwells, microtiter, and multiwell plates. The well plate is a flat plate. It looks like a tray with multiple wells. The bottom of the well can be round or flat. The flat bottom is used in cytotoxicity and transfection studies. There are 6, 12, 24, 48, and 384 well plates except for 96 wells for different studies. Generally, 96-well plates are used in cytotoxicity, whereas 6-, 12-, 24-, 48-, and 96-well plates can be used in transfection studies.
8. *High-performance liquid chromatography (HPLC)* is a modern form of liquid chromatography that uses small particle columns through which the mobile phase is pumped at high pressure. The separation of components depends on the extent of interaction between the solute component and the stationary phase. The component that has the lowest affinity for the stationary phase will elute first. HPLC is becoming a preferred method of analysis among various analytical methods for pharmaceuticals. HPLC methods provide rapid analysis, greater sensitivity, high resolution, easy sample recovery, and precise and reproducible results. The linearity of an analytical procedure is its ability (within a given range) to obtain test results which are directly proportional to the concentration (amount) of analyte in the sample. For HPLC strategies, the straight connection between identifier reaction (crest zone and tallness) and test fixation is resolved. The relationship can be exhibited specifically on tranquilize substance by weakening of standard stock or by partitioned weighing of the example segments, utilizing the proposed methods.
9. The proposed method does not require any laborious cleanup procedure before measurement. In addition, the method has wider linear dynamic range with good accuracy and precision.
10. *3-(4,5-dimethylthiazol-2-yl)-2,5-diphenyltetrazolium bromide (MTT) dye* is a substance that is actively passed to the cells and reduced in water-insoluble formazan which gives color by the reaction due to mitochondria. This method is based on the principle that mitochondrium can break down the tetrazolium ring of MTT dye in living cells. The result of fragmentation changed from pale yellow MTT colored to dark blue-violet formazan. While living and mitochondrial function intact cells stain purple, dead or mitochondrial dysfunctional cells are not stained. The MTT reduction ability of the cells is assessed as a measure of cell viability, and the dye density obtained as a result of MTT analysis is directly proportional to the number

of viable cells. These commercial dyes are stored at room temperature, while others are stored at +4 °C.

11. *Microplate readers* are instruments which are used to detect various biological and chemical samples in microtiter. The most common microplate formats used in research laboratories are 6-, 24-, 48-, 96-, and 384-well formats. In the MTT assay, 96-well plates are preferred. The absorbance, fluorescence intensity, and luminescence evaluation are common detection modes.

References

1. Urasa M, Darj E (2011) Knowledge of cervical cancer and screening practices of nurses at a regional hospital in Tanzania. Afr Health Sci 11(1):48–57
2. Volodko V, Gordon M, Salla M, Ghazaleh HA, Baksh S (2014) RASSF tumor suppressor gene family: biological functions and regulation. FEBS Lett 588:2671–2684
3. Zhang S, Zhao B, Jiang H, Wang B, Ma B (2007) Cationic lipids and polymers mediated vectors for delivery of siRNA. J Control Release 123:1–10
4. Xu C, Wang J (2015) Delivery systems for siRNA drug development in cancer therapy. Asian J Pharm Sci 10:1–12
5. Elbashir SM, Harborth J, Lendeckel W, Yalcin A, Weber K, Tuschl T (2001) Duplexes of 21-nucleotide RNAs mediate RNA interference in cultured mammalian cells. Nature 411:494–498
6. Tseng YC, Mozumdar S, Huang L (2009) Lipid-based systemic delivery of siRNA. Adv Drug Deliv Rev 61:721–731
7. Li CX, Parker A, Menocal E, Xiang S, Borodyansky L, Fruehauf JH (2006) Delivery of RNA interference. Cell Cycle 5(18):2103–2109
8. Takabatake Y, Isaka Y, Mizui M, Kawachi H, Takahara S, Imai E (2007) Chemically modified siRNA prolonged RNA interference in renal disease. Biochem Biophys Res Commun 363(2):432–437
9. Gary D, Puri N, Won YY (2007) Polymer-based siRNA delivery: perspectives on the fundamental and phenomenological distinctions from polymer-based DNA delivery. J Control Release 121:64–73
10. Reischl D, Zimmer A (2009) Drug delivery of siRNA therapeutics: potentials and limits of nanosystems. Nanomed Nanotechnol Biol Med 5(1):8–20
11. Gorbatyuk M, Justilien V, Liu J, Hauswirth WW, Lewin AS (2007) Suppression of mouse rhodopsin expression in vivo by AAV mediated siRNA delivery. VisionRes 47(9):1202–1208
12. Chen GX, Dong QH, Zhang JW, Fu FF, Xu ZF, Liang QY, Zheng S, Ding JY (2006) Knockdown of PLK1 mRNA by special siRNA. Yi Chuan 28(1):21–25
13. Sioud M, Sørensen DR (2003) Cationic liposome-mediated delivery of siRNAs in adult mice. Biochem Biophys Res Commun 312(4):1220–1225
14. Bouxsein NF, Mcallister CS, Ewert KK, Samuel CE, Safinya RC (2007) Structure and gene silencing activities of monovalent and pentavalent cationic lipid vectors complexed with siRNA. Biochemistry 46(16):4785–4792
15. Kataoka K, Itaka K, Nishiyama N, Yamasaki Y, Oishi M, Nagasaki Y (2005) Smart polymeric micelles as nanocarriers for oligonucleotides and siRNA delivery. Nucleic Acids Symp Ser 49(1):17–18
16. Katas H, Alpar HO (2006) Development and characterisation of chitosan nanoparticles for siRNA delivery. J Control Release 115(2):16–25
17. Savaese A, Cognetti F (2003) New drugs in the treatment of recurrent or metastatic cervical cancer. Crit Rev Oncol Hematol 48:323–327
18. Wu J, Liu Q, Lee RJ (2006) A folate receptor-targeted liposomal formulation for paclitaxel. Int J Pharm 316:148–153
19. Thu TH, Dass CR, Dunstan DE (2008) Injectable chitosan hydrogels for localized cancer therapy. J Control Release 126:205–216
20. Chimmiri P, Rajalakshmi R, Mahitha B, Ramesh G, Noor Ahmed VH (2012) Solid lipid nanoparticles: a novel carrier for cancer therapy. Int J Pharm Biol Sci 3(3):405–413
21. Serpea L, Catalano MG, Cavalli R, Ugazio E, Bosco O, Canaparo R, Muntoni E, Frairia R, Gasco MR, Eandi M, Zara GP (2004) Cytotoxicity of anticancer drugs incorporated in solid lipid nanoparticles on HT-29 colorectal cancer cell line. Eur J Pharm Biopharm 58:673–680

22. Shah R, Eldridge D, Palombo E, Harding I (2014) Optimisation and stability assessment of solid lipid nanoparticles using particle size and zeta potential. J Phys Sci 25(1):59–75
23. Banerjee I, De K, Mukherjee D, Dey D, Chattopadhyay S, Mukherjee M, Mandal M, Bandyopadhyay AK, Gupta A, Ganguly S, Misra M (2016) Paclitaxel-loaded solid lipid nanoparticles modified with Tyr-3-octreotide for enhanced antiangiogenic and anti-glioma therapy. Acta Biomater 38:69–81
24. Yu YH, Kim E, Park DE, Shim G, Lee S, Kim YB, Kim CW, Oha YK (2012) Cationic solid lipid nanoparticles for co-delivery of paclitaxel and siRNA. Eur J Pharm Biopharm 80: 268–273
25. Pavelić Z, Skalko-Basnet N, Filipović-Grcić J, Martinac A, Jalsenjak I (2005) Development and in vitro evaluation of a liposomal vaginal delivery system for acyclovir. J Control Release 106:34–43
26. Hussain A, Ahsan F (2005) The vagina as a route for systemic drug delivery. J Control Release 103:301–313
27. Dobaria N, Mashru R, Vadia NH (2007) Vaginal drug delivery systems: a review of current status. East Cent Afr J Pharm Sci 10:3–13
28. Choudhury A, Das S, Kar M (2011) A review on novelty and potentiality of vaginal drug delivery. Int J PharmTech Res 3(2): 1033–1044
29. Haas S, Woerdenbag H, Sznitowska M (2015) Rectal and vaginal. In: Bouwman-Boer Y, Fenton-May V, Brun PL (eds) Practical pharmaceutics: an international guideline for the preparation, care and use of medicinal products. Springer International Publishing AG, Cham, pp 190–225
30. Büyükköroğlu G, Şenel B, Başaran E, Yenilmez E, Yazan Y (2016) Preparation and in vitro evaluation of vaginal formulations including siRNA and paclitaxel-loaded SLNs for cervical cancer. Eur J Pharm Biopharm 109:174–183
31. Büyükköroğlu G, Yazan EY, Öner AF (2015) Preparation and physicochemical characterizations of solid lipid nanoparticles containing DOTAP® for DNA delivery. Turk J Chem 39:1012–1024
32. Gvili K, Benny O, Danino D, Machluf M (2006) Poly(D,L-lactide-coglycolide acid) nanoparticles for DNA delivery: waiving preparation complexity and increasing efficiency. Biopolymers 85:379–391
33. Mehnert W, Mäder K (2001) Solid lipid nanoparticles: production, characterization and applications. Adv Drug Deliv Rev 47(2–3): 165–196
34. Şenel B, Büyükköroğlu G, Yazan Y (2015) Solid lipid and chitosan particulate systems for delivery of siRNA. Pharmazie 8:698–705
35. Neves J, Amaral MH, Bahia MF (2008) Performance of an in vitro mucoadhesion testing method for vaginal semisolids: influence of different testing conditions and instrumental parameters. Eur J Pharm Biopharm 69: 622–632
36. Owen DH, Katz DF (1999) A vaginal fluid simulant. Contraception 59:91–95

Chapter 23

The Use of Star Polymer Nanoparticles for the Delivery of siRNA to Mouse Orthotopic Pancreatic Tumor Models

Joshua A. McCarroll, George Sharbeen, Maria Kavallaris, and Phoebe A. Phillips

Abstract

Pancreatic cancer is a lethal malignancy which is refractory to most chemotherapy drugs. Recent landmark studies have shed new light on the complex genetic heterogeneity of pancreatic cancer and provide an opportunity to utilize "precision-based medicines" to target genes based on the genetic profile of an individual's tumor to increase the efficiency of chemotherapy and decrease tumor growth and metastases. Gene-silencing drugs in the form of short-interfering RNA (siRNA) have the potential to play an important role in precision medicine for pancreatic cancer by silencing the expression of genes including those considered difficult to inhibit (undruggable) using chemical agents. However, before siRNA can reach its clinical potential a delivery vehicle is needed to carry siRNA across the cell membrane and into the cytoplasm of the cell. Herein, we detail the methods required to use star polymer nanoparticles to deliver siRNA to pancreatic tumors in an orthotopic pancreatic cancer mouse model to silence the expression of an "undruggable" gene (βIII-tubulin) that regulates pancreatic cancer growth and chemosensitivity.

Key words Pancreatic cancer, Nanoparticles, Short-interfering RNA (siRNA), Star polymers

1 Introduction

Pancreatic cancer is predicted to be the second-leading cause of cancer death by 2030 [1]. Despite intensive treatment the 5-year survival rate for pancreatic cancer remains at less than 8% [2]. This dismal prognosis is in part due to chemotherapy drug resistance and metastases [3]. Recent large-scale genome sequencing of pancreatic tumors has significantly increased our understanding of this complex disease as well as identified a host of new potential therapeutic gene targets [4–6]. However, some of these genes have no known chemical inhibitors or are considered undruggable.

Joshua A. McCarroll and George Sharbeen contributed equally to this chapter.

Lekha Dinesh Kumar (ed.), *RNA Interference and Cancer Therapy: Methods and Protocols*, Methods in Molecular Biology, vol. 1974, https://doi.org/10.1007/978-1-4939-9220-1_23, © Springer Science+Business Media, LLC, part of Springer Nature 2019

Gene-silencing therapy has the potential to solve this problem with use of short-interfering RNA (siRNA), which can be designed to selectively target any gene of choice and inhibit its expression with high potency [7]. siRNA therapies are in development to treat a wide range of different diseases including cancer. For example, the siRNA therapeutic drug "patisiran" recently produced positive results in a phase III clinical trial (clinicaltrials.gov ID: NCT02939820) for patients suffering with a rare inherited and often fatal disease TTR amyloidosis (ATTR). This drug uses siRNA packaged inside a nanoparticle to selectively inhibit the expression of a gene (mutated transthyretin, TTR) that causes ATTR [8]. However, before siRNA therapies can be applied to the clinic to treat pancreatic cancer a delivery vehicle is required to traverse the characteristic dense fibrotic extracellular matrix and highly disrupted vascular blood supply [9].

Non-viral nanoparticles can be engineered with precision to self-assemble or package siRNA and deliver it to different cell types. Nanoparticles can be generated to have physical and chemical properties which make them an ideal delivery vehicle for therapeutic siRNA for the treatment of cancer, namely, (1) they can package and deliver multiple siRNAs to selectively inhibit a cocktail of tumor-promoting genes simultaneously based on a patients tumor profile (a precision medicine tool); (2) have a size and surface charge which allows them to take advantage of the enhanced permeability and retention effect and accumulate in solid tumors [10]; (3) have targeting moieties (peptides, antibodies, or antibody fragments) attached to their surface that allows for active targeting of siRNAs to tumor cells; and (4) improve tumor drug bioavailability by having PEG lipids coating their surface which helps to minimize binding to serum proteins and uptake by circulating immune cells. Our research program recently developed star-shaped polymeric nanoparticles for the delivery of siRNA to pancreatic cancer cells in vitro and in vivo [11]. Star polymers are cost-effective to produce, can self-assemble siRNA via a simple mixing process, and retain a small size (<50 nm) when complexed to siRNA [11]. Importantly, we demonstrated that star polymer nanoparticles can deliver therapeutic siRNA to pancreatic tumors in a clinically relevant orthotopic pancreatic cancer mouse model and silence βIII-tubulin [11], which is undruggable using chemical inhibitors and plays an important role in promoting pancreatic tumor growth and chemoresistance [12]. βIII-Tubulin is overexpressed in pancreatic tumors leading to poor outcome in patient survival and promotes pancreatic tumor growth and chemotherapy resistance [12–14]. We have previously shown that inhibition of βIII-tubulin using gene-silencing agents (shRNA) decreases pancreatic tumor growth and metastases in vivo [12]. However, due to βIII-tubulin's close amino acid sequence homology with other microtubule proteins, the design of chemical inhibitors for βIII-tubulin is problematic. Therefore, βIII-tubulin is an ideal target for gene-silencing nanomedicines.

In this chapter we describe the methodology required to use star polymer nanoparticles to self-assemble siRNA and deliver it to orthotopic mouse pancreatic tumors to silence βIII-tubulin expression.

2 Materials

2.1 Orthotopic Pancreatic Cancer Mouse Model

1. Human pancreatic cancer cell lines stably expressing luciferase. These cells can be generated in the laboratory by transfection using lipid-based nanoparticles or viral transduction of a luciferase reporter construct. In this study, we used human MiaPaCa-2 pancreatic cancer cells which were transfected with 2 μg of pGL4.50 [Luc2/CMV/Hygro] luciferase reporter construct using Lipofectamine 2000. Clones with stable integration of the construct were isolated by culture in hygromycin B.
2. Dulbecco Modified Eagle's Medium (DMEM; 4.5 g/L glucose, 110 mg/mL sodium pyruvate) supplemented with 10% fetal bovine serum and 2 mM L-glutamine.
3. 0.25% Trypsin/EDTA.
4. Sterile phosphate buffered saline (PBS).
5. Eight-week-old immune-deficient mice (athymic nude BalbC mice) housed under pathogen-free conditions and fed standard chow and water ad libitum (*see* **Note 1**).
6. Growth factor-reduced Matrigel.
7. Stinger research anaesthetic gas machine with attached DEE Clean Air filter, oxygen tank, and nose piece.
8. Isoflurane.
9. Analgesia; buprenorphine hydrochloride.
10. Safil 5/0 polyglycolic acid, braided, coated, round-needle absorbable sutures.
11. Autoclip® 9 mm stainless steel wound clips, wound clip applicator, and remover.
12. Fine scissors, delicate straight (11 cm length).
13. Graefe forceps, serrated, slight curve, and straight (0.8 × 0.7 mm tip, 10 cm length).
14. Olsen-Hegar needle holder with suture cutter (2 mm serrated tip, 14 cm length).
15. 27 gauge needle attached to 0.5 mL syringe and 29 gauge needle attached to 0.5 and 1 mL syringe.
16. Dulbecco's phosphate buffered saline.

17. Betadine.
18. Alcohol swabs.
19. Sterile gauze (BSN Medical (Aust) Pty Ltd, VIC, Australia).
20. Sterile saline (0.9% NaCl) solution.
21. Heat pad.
22. Clean cage/s with fresh food and drinking water.
23. Organ harvest (endpoint analysis):
    (a) Tissue-Tek® optimum cutting temperature (OCT) compound and plastic OCT cassettes.
    (b) 4% paraformaldehyde in Dulbecco's phosphate buffered saline, aliquoted in 15 mL (6–10 mL aliquot) and 50 mL (20–30 mL aliquot) falcon tubes for collection of tissues (*see* **Note 2**).
    (c) 1.5 mL sterile Eppendorf microtubes.
    (d) Disposable scalpels with blades.
    (e) Sterile 60 mm plastic petri dishes.
    (f) Dry ice.
    (g) Liquid nitrogen.
    (h) 70% ethanol.
    (i) IVIS Spectrum CT Imaging System with the Living Image Software for luminescent and fluorescent imaging.
    (j) Firefly D-luciferin potassium salt (diluted to 15 mg/mL in Dulbecco's phosphate buffered saline and sterile-filtered using a 0.22 μm syringe filter).

2.2 Bioluminescent Imaging of Orthotopic Pancreatic Tumors

1. IVIS Spectrum CT Imaging System with Living Image Software. Refer to operator's manual for equipment instructions and analyses of bioluminescent signal using the region of interest (ROI) method.
2. Stinger research anaesthetic gas machine with attached DEE Clean Air filter, oxygen tank, induction chamber, and IVIS Spectrum CT Imaging System.
3. Firefly D-luciferin potassium salt diluted to 15 mg/mL in Dulbecco's phosphate buffered saline and sterile-filtered using 0.22 μm syringe filter.
4. 29 gauge needle attached to 0.5 mL syringe.

2.3 Star Polymer-siRNA Assembly

1. RNA-/DNase-free ultrapure sterile water (500 mL).
2. Stock solution of star polymer (2–4 mg/mL) dissolved in sterile-filtered DNA-/RNase-free ultrapure water and syringe filtered using a 0.45 μm sterile filter.

3. siRNAs [non-specific control, TUBB3 (βIII-tubulin), non-specific control labeled with Dy647 (Cy5™ alternate) fluorescent dye at the 5′ end of the sense strand].

   TUBB3 siRNA: sense strand: *5′-GUACGUGCCU CGAGCCAUUUU-3′*; antisense strand: *5′-UUCAUGCA CGGAGAGCUCGGUAA-3′*.

   Non-silencing control siRNA: sense strand: *5′-GCUAUG GCUGAAUACAAAUUUU-3′*; antisense strand: *5′-UUCG AUACCGACUUAUGUUUAA-3′*.

   Fluorescent siRNA: sense strand: *5′-(DY647)GCUAUG GCUGAAUACAAAUUUU-3′*; antisense strand: *3′-UUCGAUACCGACUUAUGUUUAA-5′*.

   All siRNAs are in vivo purification quality [siRNAs processed by counter-ion (Na^+) exchange. Desalted, sterile-filtered, and endotoxin tested] 100–500 nmol (Millennium Science).
4. 5× siRNA dilution buffer.
5. 0.2 μm syringe filter.
6. 5 mL syringe.
7. Sterile 1.5 mL Eppendorf microtubes.

2.4 Star Polymer-siRNA In Vivo Biodistribution

1. IVIS Spectrum CT Imaging System with Living Image Software.
2. Stinger research anaesthetic gas machine with attached DEE Clean Air filter, oxygen tank, induction chamber, and IVIS Spectrum CT Imaging System.
3. Star polymer complexed to fluorescently labeled siRNA, as per Subheading 2.3.

2.5 Histology of Orthotopic Pancreatic Tumors: Immunofluorescence and Immunohistochemistry

1. Superfrost™ plus microscope slides.
2. Immunofluorescence:
   (a) Dulbecco's phosphate buffered saline and Dulbecco's phosphate buffered saline Tween solutions (0.05% Tween-20 in Dulbecco's phosphate buffered saline).
   (b) Goat serum.
   (c) CD31 antibody.
   (d) Goat anti-rabbit-Alexa Fluor 488.
   (e) ProLong™ Gold antifade with DAPI mounting medium.
   (f) Fluorescence confocal microscope.
3. Immunohistochemistry:
   (a) Xylene.

(b) Goat, rabbit, or mouse serum (depending on secondary antibody).

(c) Antigen-retrieval citrate buffer (10 mM citrate, 0.05% Tween-20, pH 6).

(d) Peroxidase blocking solution (1% H_2O_2, 3.3% methanol).

(e) Mouse-on-mouse staining kit.

(f) Primary antibody against target of interest. We used anti-βIII-tubulin (clone TUJ1, mouse monoclonal IgG2a; Biolegend, CA, USA) and an IgG2a isotype control.

(g) Vectastain ABC Kit.

(h) ImmPACT™ DAB substrate.

(i) Mayer's hematoxylin.

(j) Aperio ScanScope XT Slide Scanner and ImageScope software.

(k) PAP pen.

(l) EUKITT mounting medium.

2.6 Pancreatic Tumor Protein Extraction and Western Blotting

1. TissueRuptor and TissueRuptor disposable Probes.
2. Handheld sonicator with probe.
3. Radioimmunoprecipitation assay buffer (RIPA) containing 1% protease inhibitor cocktail, 1 mM PMSF, and 1% phosphatase inhibitor cocktail.
4. Refrigerated benchtop microfuge.
5. Bio-Rad gel electrophoresis tank.
6. Laemmli loading buffer:

   3.55 mL MilliQ water.

   1.25 mL 0.5 M Tris–HCl, pH 6.8.

   2.5 mL glycerol.

   2.0 mL 10% (w/v) SDS.

   0.2 mL 0.5% (w/v) bromophenol blue.

   Add 50 μL β-mercaptoethanol to 950 μL of loading buffer just before addition to sample.
7. Discontinuous SDS-PAGE Gel (4–10%):

   (a) *Resolving gel (10%):*

   4.1 mL MilliQ water.

   3.3 mL 30% degassed acrylamide/bis—stored at 4 °C.

   2.5 mL 1.5 M Tris–HCl, pH 8.8.

   0.1 mL 10% SDS.

Add 50 μL 10% APS and 5 μL TEMED, and swirl gently to initiate setting. Immediately after pouring, gently overlay a layer of water and allow to polymerize for 45 min–1 h.

(b) *Stacking gel (4%):*

6.1 mL MilliQ water.

1.3 mL 30% degassed acrylamide/bis—stored at 4 °C.

2.5 mL 0.5 M Tris–HCl, pH 6.8.

0.1 mL 10% SDS.

Add 50 μL 10% APS and 10 μL TEMED, and swirl gently to initiate setting. Pour to top of thinner plate, then insert comb immediately, and allow to polymerize for 30–45 min.

8. 10× SDS Running buffer:

   30.3 g Tris base.

   144.0 g glycine.

   10.0 g SDS.

   Make up to 1 L with MilliQ H_2O. Dilute 1:10 for working stock.

9. 10× transfer buffer:

   30.3 g Tris base.

   144 g glycine.

   Dissolve in 1 L MilliQ water. To make working stock, mix 100 mL 10× transfer buffer + 200 mL methanol + 700 mL MilliQ water.

10. Dulbecco's phosphate buffered saline + 0.1% (v/v) TWEEN-20 (DPBST):

    0.5 mL TWEEN-20.

    495 mL Dulbecco's phosphate buffered saline.

    500 mL total.

11. DPBST + 5% (w/v) skim milk (blocking buffer):

    2.5 g skim milk powder per 50 mL DPBST.

12. DPBST + 5% (w/v) BSA:

    2.5 g BSA per 50 mL DPBST.

13. Nitrocellulose membrane.
14. Whatman 3mm paper 10 cm × 10 cm.
15. Fiber pads and transfer cassette.
16. LAS4000 scanner and ImageQuant TL software.
17. Pierce™ BCA Protein Assay Kit.

3 Methods

3.1 Orthotopic Pancreatic Cancer Mouse Model

3.1.1 Day Prior to Orthotopic Implantation of Human Pancreatic Cancer Cells

1. Clean and autoclave all surgical equipment.
2. Ensure human pancreatic cancer cells are in their log phase of growth and are 70–80% confluent.
3. Allow growth factor-reduced Matrigel to thaw overnight in an ice-filled container, placed in a 4 °C fridge. Place 27 gauge needle syringes and 1000 μL pipette tips at −20 °C overnight. These should be transferred to ice on the day of surgery.
4. Prepare 20 μg/mL buprenorphine (also known as Temgesic) in sterile saline. This will be administered subcutaneously to the mice postsurgery at 0.1 mg/kg of mouse.
5. Weigh all mice within 24 h of surgery to calculate amount of analgesic to be administered.

3.1.2 Pancreatic Cancer Cell Preparation

1. Lift MiaPaCa-2 cells from their cell culture flasks using 0.25% Trypsin/0.9 mM EDTA solution.
2. Neutralize the Trypsin with complete cell culture growth medium, and then immediately pellet cells at 335 × *g* for 5 min at room temperature using a benchtop centrifuge.
3. Resuspend the cells in culture media, and take an aliquot to perform cell counts using a hemocytometer or automated cell counter. 1×10^6 MiaPaCa-2 cells will be required per mouse. To account for dead volume in the syringe and loss of volume due to pipetting, ensure that you have at least twice the number of cells that you need per mouse, that is, at least 2×10^6 MiaPaCa-2 cells per mouse. This number may have to be adjusted for other human pancreatic cancer cell lines.
4. Transfer 2×10^6 MiaPaCa-2 cells per mouse to a separate 15–50 mL sterile falcon tube and pellet as described above in **step 2**.
5. Aspirate culture supernatant, and place the cell pellet on ice. Gently flick the pellet to loosen the cells.
6. Using a 1000 μL pipette with pre-chilled 1000 μL tips, gently resuspend the cell pellet in 50 μL growth factor-reduced Matrigel per 1×10^6 MiaPaCa-2 cells, on ice.

3.1.3 Orthotopic Implantation of Human Pancreatic Cancer Cells into the Mouse Pancreas

1. Ensure the anaesthetic machine and surgical area is ready (Fig. 1; *see* **Note 3**). Before starting surgeries check:
   (a) The DEE Clean Air filter and ensure their weight has not increased more than 50 g relative to start weight.
   (b) Liquid isoflurane in the anaesthetic machine is full.
   (c) Oxygen tank is full, switched on and flowing through the anaesthetic machine to the nose piece. The flow rate should be set to 0.8 LPM.
   (d) Switch on the heating pad and maintain at 37 °C.

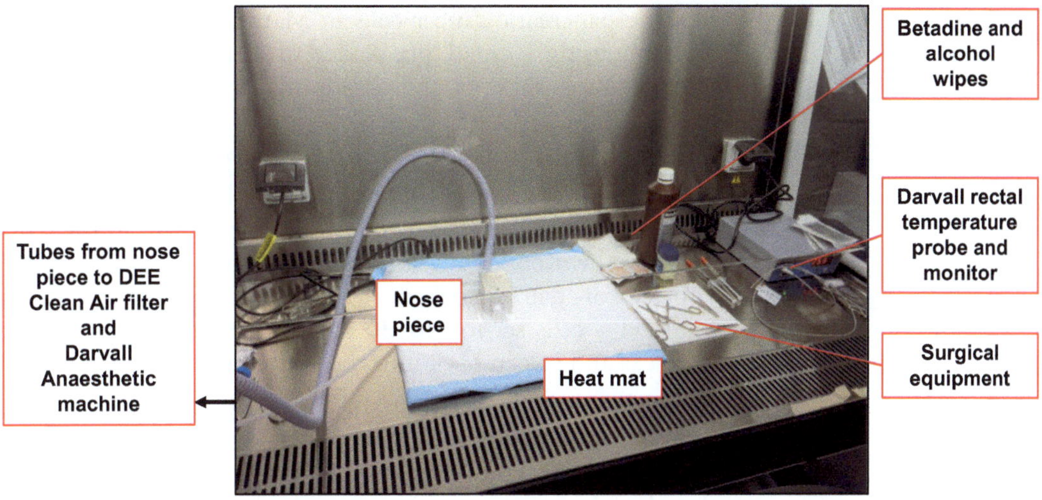

Fig. 1 Diagram illustrating the experimental setup required for orthotopic implantation of human pancreatic cancer cells into the mouse pancreas. For example, set up of biosafety cabinet for surgical implantation of human pancreatic cancer cells into the mouse pancreas

2. Transfer a mouse to the surgical area, and while scruffing the back of the neck, place its nose into the nose piece of the anaesthetic machine. Induce anaesthesia with 4% isoflurane, and then maintain at 3%. Gently dab ophthalmic ointment on the eyes.
3. Carefully roll the mouse on to its back, and tape down the hind legs, and then wipe the abdomen of the mouse with betadine on sterile gauze. Clean off the betadine with an alcohol swab—the mouse is now ready for the first incision.
4. Using a pair of forceps, lift the skin on the left side of the mouse abdomen approximately 2 cm below the bottom of the rib cage, and make a shallow cut with a pair of surgical scissors.
5. Extend the incision in a straight line approximately 1.5 cm in length upward and toward the rib cage, being careful not to damage the peritoneal membrane.
6. Using a pair of forceps, lift the exposed peritoneal membrane slightly, making sure not to pick up any underlying organs. Make a small initial cut in the peritoneal membrane.
7. Carefully slide the pair of scissors under the membrane, through the initial cut, and make a straight incision up toward the rib cage, to expose the head of the spleen and attached tail of the pancreas.
8. Using a pair of forceps, carefully locate the pancreas (cream-colored organ, resembling fat and attached to the head of spleen), and gently pull a small amount of the tail of the pancreas with attached spleen out through the peritoneal incision.

9. Mix the pancreatic cancer cells in Matrigel on ice, by gently pipetting using 1000 μL pre-chilled tips. Slowly aspirate 50 μL of the cell mix in a pre-chilled 0.5 mL syringe with 27 gauge needle.
10. Using a pair of forceps, gently grasp the tail of the pancreas, and inject the cell mixture. The cells should not leak from the injection site and should form a solid bubble within the pancreas. In the event of a leak, immediately dab the leakage with a piece of gauze soaked in betadine. Let the injected Matrigel + cells set for 20–30 s, and then carefully push the pancreas and spleen back into the abdominal cavity.
11. Close the peritoneal membrane using an absorbable suture, and perform a continuous stitch. Ensure the suture is double-knotted at the ends and that the peritoneal membrane is closed.
12. Close the skin by applying two Autoclip® 9 mm staples using a wound clip applicator.
13. Carefully remove the tape from the hind legs, and roll the mouse on to its abdomen, ensuring the nose does not leave the nose piece.
14. Inject 0.1 mg buprenorphine/kg mouse subcutaneously into the right flank using a 0.5 mL syringe attached to a 29 gauge needle.
15. Inject 1 mL warm sterile saline subcutaneously above the neck using a 1 mL syringe attached to a 29 gauge needle.
16. Remove the mouse from anaesthetic, and place in a clean cage under a heat lamp. Allow the mouse to recover under the heating lamp for 15 min. Once the mouse has recovered, return the cage to the appropriate animal facility cage rack (*see* **Note 4**).

3.2 *In Vivo* Bioluminescent Imaging

1. Weigh mice within 24 h of imaging, to calculate administration volume of D-luciferin.
2. Switch on and initialize the IVIS Spectrum CT Imaging System, using the linked Living Image Software according to the manufacturer's instructions. This will require training by your relevant custodian/facility. Initialization will ensure the platform in the imaging system is pre-warmed to 37 °C and the camera is precooled and ready for image acquisition.
3. Prepare anaesthetic apparatus as per Subheading 3.1. This should be set up so that anaesthetic flows to the plastic induction chamber and to the nosepieces in the Spectrum Imaging System.
4. Allow 15 mg/mL D-luciferin to reach room temperature, and then inject each mouse intraperitoneally with 10 μL of 15 mg/

mL D-luciferin per gram of body weight, using a 0.5 mL syringe attached to a 29 gauge needle.

5. Transfer the injected mice to the plastic induction chamber, and switch on isoflurane to 4% for induction and then 3% for maintenance.
6. Once mice are anaesthetized, carefully transfer them to the nosepieces in the spectrum imaging system. The mice should be on their backs, and slightly rolled to their right side, so as to raise the pancreatic tumor side toward the camera.
7. Begin acquiring photographs under the "luminescence" setting, with "medium" binning, F/Stop = 1 and 10 s exposure. To image all the five mice at once, select the "D" field of view (*see* **Note 5**).
8. Acquire photographs within 8 min of D-luciferin injection, at 2 min intervals, until the luminescent signal for the tumor plateaus. The signal can be monitored using the "ROI" tool to select tumor regions and "photons/s" setting to track the change in luminescence over the time course.
9. Once the time course is completed, use Living Image Software to extract the "total flux" (photons/s) per tumor per time point. Plot these readings against time, and select the maximum luminescence in the plateau as the weekly reading for that tumor (Fig. 2a).
10. Repeat this weekly until the endpoint of the model, based on the requirements of your animal ethics approval. Plot the weekly readings throughout the entire model (Fig. 2b) to determine the optimal time to administer nanoparticle-siRNA for biodistribution, gene knockdown, or therapeutic studies (*see* **Note 6**).

3.3 Star Polymer-siRNA Assembly and Administration

3.3.1 Formulation of Star Polymer-siRNA Nanocomplex

1. Prepare stock solutions for siRNAs by resuspending in 1× siRNA buffer (dilute 5× siRNA buffer using sterile DNA-/RNA-free water). The concentration of stock siRNA can range from 200 to 400 μM (2–4 μg/μL, depends on total quantity of siRNA ordered and amount of resuspension buffer used to dilute the siRNA). Mix the solution using a 1000 μL pipette, and rotate the tube slowly at room temperature for 20 min to ensure that the siRNA is fully resuspended in the solution. Place the siRNA into sterile 0.5 mL Eppendorf tubes as small aliquots (50–100 μL) to prevent repeated freeze-thawing and store at −80 °C.
2. Prepare stock solution of star polymer in DNA-/RNase-free water at concentrations ranging from 1 to 4 mg/mL. Filter sterilize the stock solution using a 5 mL plastic syringe attached to a 0.45 μm sterile filter. The stock solution can be stored at 4 °C for up to 12 months without any loss of activity.

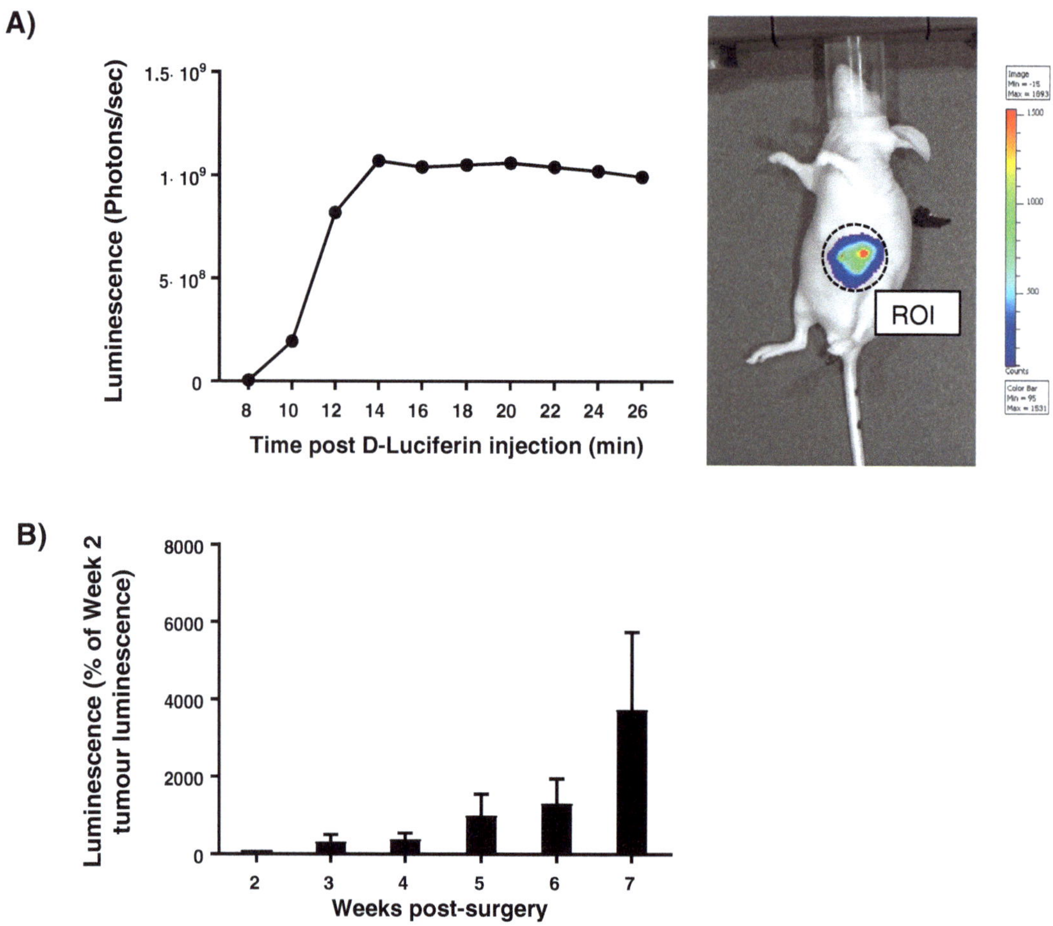

Fig. 2 Bioluminescent monitoring of tumor growth in vivo. (**a**) Kinetic curve for in vivo luminescence up to 26 min postinjection of D-luciferin. The right panel shows a representative luminescent photograph from an IVIS Spectrum CT Imaging System with scale bar and the placement of an ROI to measure luminescence (photons/s). The maximum luminescence in the plateau is used to prepare the graph in (**b**). (**b**) Tumor growth kinetics for MiaPaCa-2 orthotopic pancreatic tumors based on weekly in vivo tumor bioluminescence imaging

3.3.2 Preparation of Star-siRNA Nanocomplex for Assessment of siRNA Biodistribution In Vivo

1. Prepare a stock solution of siRNA-labeled with Dy647 fluorescent dye at 2.5 μg/μL in 1× siRNA buffer.
2. Prepare a stock solution of star polymer at 1.6 mg/mL in RNA-/DNA-free sterile water.
3. In a 1 mL sterile DNA-/RNase-free Eppendorf tube, add 16 μL of siRNA to 34 μL of sterile DNA-/RNA-free water or 1× HEPES buffer, and mix using a 100 μL pipette. Store the tube on ice.
4. In a separate sterile DNA-/RNase-free Eppendorf tube, add 100 μL of the star polymer stock solution, and mix using a 100 μL pipette and store on ice.

5. Using a 200 μL pipette slowly, add 50 μL of the siRNA solution to 100 μL of the star polymer solution. Mix several times using a 200 μL pipette, and allow the solution to complex at room temperature for 5–10 min.
6. Draw up 150 μL of the star-siRNA nanocomplex using a 0.5 mL syringe attached to a 27 gauge needle and administer into the tail vein of mice (*see* **Note 7**). Total amount of star-siRNA delivered is 160 μg star polymer-40 μg siRNA (4:1 w/w ratio) (*see* **Note 8**).

3.3.3 Preparation of Star-siRNA Nanocomplex for Assessment of Gene-Silencing Activity in Pancreatic Tumors In Vivo

1. Prepare stock solutions of control (nontargeting) siRNA and target gene siRNA (2–4 μg/μL). Prepare stock solution of star polymer nanoparticles as described above in Subheading 3.3.2. For our studies we used siRNA designed against green fluorescent protein (GFP) as a control and TUBB3 (βIII-tubulin) as target gene (*see* Subheading 2.3 for sequence details). siRNA is complexed to star polymer and administered to mice using the same protocol as described above in Subheading 3.3.1 (*see* **Note 9**).
2. Administer star polymer-siRNA nanocomplexes via the tail vein using the same procedure as described above in Subheading 3.3.2. Star polymer-siRNA complexes are administered once a day for 3 days. Twenty-four hours after the final injection, the mice are humanely killed and pancreatic tumors collected for histological and biochemical analysis.

3.3.4 Organ Harvest to Assess siRNA Biodistribution and siRNA Gene-Silencing Activity

1. Humanely sacrifice mice according to your approved institutional protocol and guidelines.
2. Pin the euthanized mouse onto a foam dissection board, and sterilize the skin by wiping with 70% ethanol.
3. Make incisions in the skin/peritoneal membranes to generate two flaps that open to expose the internal organs (Fig. 3a).

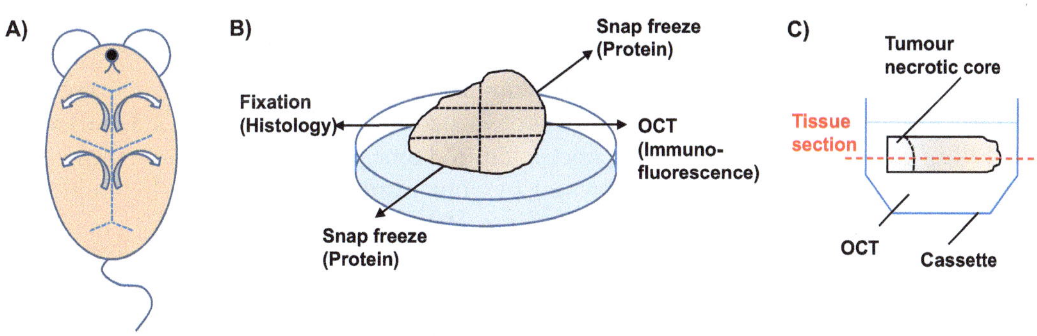

Fig. 3 Schematic diagram illustrating the dissection of orthotopic pancreatic tumors for histological and biochemical analysis. (**a**) Dissection of mice for pancreatic tumor/organ collection. Dotted lines indicate sites for incision on the skin and peritoneal membranes, which are then pulled back as four flaps to expose organs. (**b**) Dissection of orthotopic pancreatic tumors for multiple downstream histological and biochemical analyses. (**c**) Example orientation of tumor tissue in OCT cassette for frozen sectioning

4. Using a pair of forceps and scissors, locate and carefully cut out the following organs in order, and then place into Dulbecco's phosphate buffered saline in a 6-well plastic tissue culture plate: pancreatic tumor, spleen, intestines, liver, and kidneys. Also cut open the diaphragm/chest cavity, and remove the heart and lungs.
5. Proceed to ex vivo fluorescence imaging of the organs (described in Subheading 3.4) to measure siRNA biodistribution. Otherwise, proceed to **step 6**.
6. Dissect the pancreatic tumor in a clean 60 mm petri dish as shown in Fig. 3b. Prepare tissues for downstream analyses as follows:
   (a) *Protein extraction*: Snap-freeze a piece of tumor in a 1.5 mL microtube using liquid nitrogen. Transfer to dry ice when frozen, and store long term at −80 °C.
   (b) *Immunohistochemistry*: Transfer all organs into a tube with 4% fresh paraformaldehyde. Store at 4 °C overnight, and then rinse organs with Dulbecco's phosphate buffered saline and transfer into 70% ethanol. Store at 4 °C.
   (c) *Immunofluorescence:* Pre-fill plastic tissue cassettes with OCT freezing compound, providing enough depth to completely submerge a piece of tumor. Carefully submerge and position the tumor into OCT so that the cross section through the center of the tumor faces down. This is to ensure that sections cut for subsequent immunofluorescence analysis go across the cross section of the tumor rather than through it (Fig. 3c). Once the correct position has been obtained, snap-freeze the sample in OCT by touching the bottom of the cassette to liquid nitrogen until frozen through. Transfer to dry ice when frozen, and store long term at −80 °C.

3.4 Ex Vivo Fluorescence Imaging of Organs and Pancreatic Tumors to Detect Fluorescent siRNA

1. Initialize the IVIS Spectrum Imaging System as described in Subheading 3.2. Oxygen and anaesthetic are not required.
2. After systemic administration of star polymer-fluorescently labeled siRNA as described above in Subheading 3.3, humanely euthanize mice at desired time points postinjection, using protocols approved by your institutional animal ethics and care committee (*see* **Note 10**).
3. Dissect organs and transfer to Dulbecco's phosphate buffered saline as described above in Subheading 3.1. Discard the stomach/intestines as autofluorescence from digested food can interfere with the fluorescent signal. The gall bladder can be left on the liver and used as a positive control, as it produces a focused autofluorescent signal.

4. Select the "Imaging Wizard" on the IVIS spectrum for acquisition, and choose the "fluorescence/epiluminescence" settings.
5. Use the drop-down menu to select the fluorophore to be imaged. Ideally use fluorophores that emit far-red wavelengths to avoid interference from organ autofluorescence. We and others have successfully used Dy647 (Cy5 equivalent; [11]) dye to detect siRNA biodistribution in vivo (*see* **Note 11**).
6. Select "autoexposure" and the desired field of view "A–D." Use a field of view that encompasses all organs in a single photo.
7. Place organs onto a clean, nonreflective piece of black cardboard. Pipette Dulbecco's phosphate buffered saline onto organs to keep them from drying during the imaging process (*see* **Note 12**).
8. Acquire fluorescence images of the pancreatic tumor and/or organs (Fig. 4a), then immediately transfer whole organs into OCT freezing medium, and snap-freeze the samples as described above in Subheading 3.1.
9. Use the Living Image Software to calculate the fluorescence (radiant efficiency) from each organ using the "ROI" tool.
10. Subtract the fluorescence from organs imaged from the control mice (injected with PBS or 0.9% saline) as background to obtain the final fluorescence reading.

3.5 Immunofluorescence Imaging of Pancreatic Tumors to Examine siRNA Localization

1. Prepare 10 μm thick frozen sections of pancreatic tumors on Superfrost™ plus microscope slides.
2. Allow the sections to dry at room temperature for 10 min.
3. Wash sections with Dulbecco's phosphate buffered saline for 5 min on a benchtop orbital rocker.
4. Fix the tissue with fresh 4% paraformaldehyde in Dulbecco's phosphate buffered saline for 10 min at room temperature.
5. Wash tissue section three times in Dulbecco's phosphate buffered saline as described in **step 3**.
6. Permeabilize tissue section with 0.2% Triton X-100 in Dulbecco's phosphate buffered saline for 5 min at room temperature.
7. Wash the tissue section once in Dulbecco's phosphate buffered saline as described in **step 3**.
8. Carefully draw around the tissue section with a PAP pen.
9. Block tissue sections within the borders of the PAP pen using 10% goat serum in DPBS for 30 min at room temperature.
10. Pour off the blocking solution, and pipette anti-CD31 antibody diluted 1:200 in blocking solution (Dulbecco's phos-

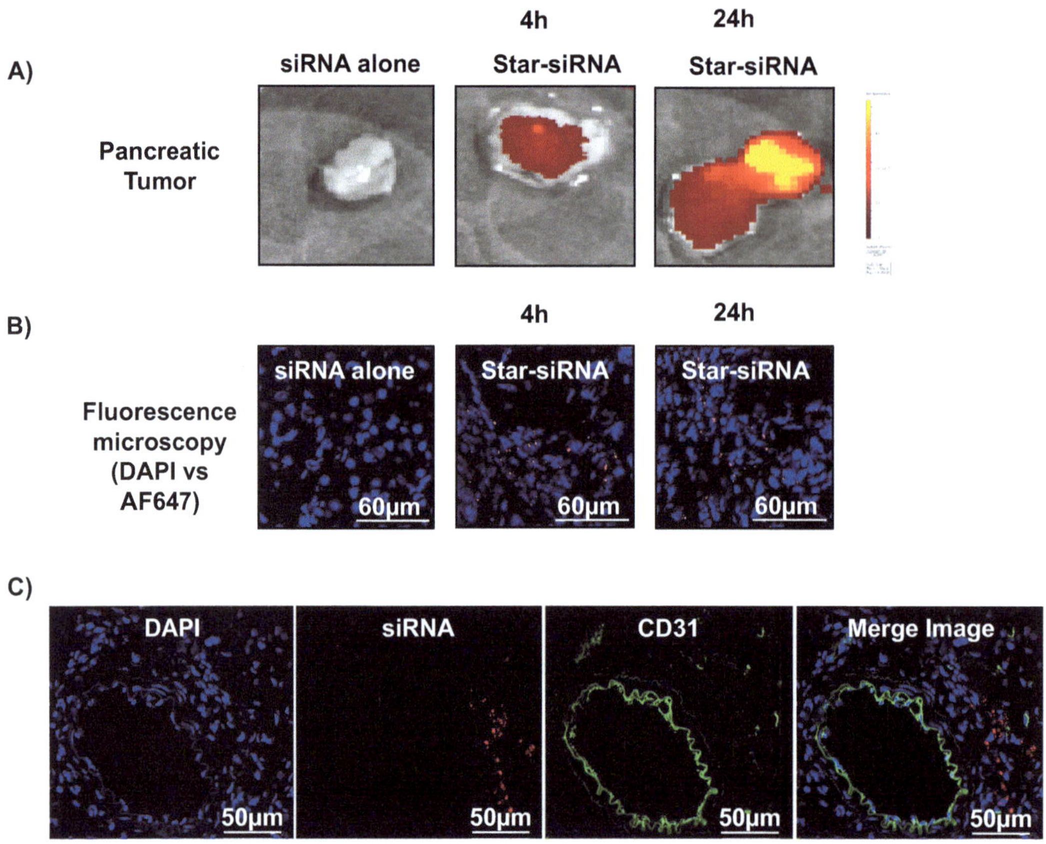

Fig. 4 Biodistribution of star polymers-siRNA in vivo. (**a**) Example of ex vivo fluorescence images of star polymer+siRNA-Dy647 (star+siRNA) uptake in orthotopic pancreatic tumors (MiaPaCa-2; top panels) versus siRNA-Dy647 alone (siRNA alone; control), 4 h and 24 h post I.V. injection of complexes. (**b**) Corresponding fluorescence confocal microscopy for Dy647 fluorescence (red) emitted from the siRNA versus DAPI (nuclear blue stain) in tumor tissue sections. (**c**) Immunofluorescent stain for CD31 (blood vessel marker, green), with DAPI counterstain (blue) in orthotopic pancreatic tumor tissue section with star+siRNA-Dy647 (red) uptake. This approach can be used to check for extravasation of nanoparticles within tumors. Reprinted (adapted) with permission from Teo et al., Biomacromolecules, 2016, 17,2337–51. Copyright (2016) American Chemical Society

phate buffered saline + 10% goat serum) within the borders of the PAP pen. Incubate at 4 °C overnight in a humidified dark box (*see* **Note 13**).

11. Wash the tissue section in Dulbecco's phosphate buffered saline as described in **step 3**.
12. Pipette goat anti-rabbit-AF488 secondary antibody diluted to 4 µg/mL in diluent (Dulbecco's phosphate buffered saline + 10% goat serum) within the borders of the PAP pen. Incubate for 1 h at room temperature in a humidified dark box.
13. Wash tissue sections twice in Dulbecco's phosphate buffered saline as described in **step 3**.

14. Mount the tissue sections with ProLong™ Gold antifade with DAPI mounting medium and a cover slip. Alternatively, sections can be immediately mounted in this way, without CD31 staining, to visualize general siRNA uptake in the tumor (Fig. 4b).
15. Seal the edges of the cover slip with nail polish, and visualize on a confocal fluorescence microscope: DAPI excitation = 358 nm, emission = 461 nm; CD31-AF488 excitation = 490 nm, emission = 525 nm; and star polymer-siRNA-Dy647 excitation = can use Alexa Fluor 647 or Cy5 emission/excitation channels. *See* Fig. 4c for an example of these images.

3.6 Assessment of Gene-Silencing Activity in Orthotopic Pancreatic Tumors

3.6.1 Immunohistochemistry

1. Embed paraformaldehyde-fixed tumor tissue into paraffin using standard protocols.
2. Prepare 5 μm thick sections on Superfrost™ plus glass microscope slides.
3. Incubate slides at 60 °C in a laboratory oven for 30 min.
4. Remove slides, and then immediately deparaffinize/re-hydrate tissue sections using standard protocol as described below (all incubations are performed at room temperature):
   (a) Incubate sections in 2× washes of xylene for 5 min each.
   (b) Incubate sections in 2× washes of 100% ethanol for 5 min each.
   (c) Incubate sections in 2× washes of 95% ethanol for 5 min each.
   (d) Incubate sections in 2× washes of 70% ethanol for 5 min each.
   (e) Wash sections in tap water for 5 min.
5. *Antigen retrieval*:
   (a) Pour 400–500 mL of fresh antigen retrieval (citrate) buffer into a 2 L glass beaker, and then place the slides into the beaker. Make sure that the solution covers the slides.
   (b) Cover the beaker with two layers of gladwrap, place holes in the top of the wrap, and place the beaker into a microwave oven. Microwave the solution for a maximum of 4 min or until the solution turns cloudy. It is important to not let the solution boil.
   (c) Replace the cover with two layers of aluminum foil, put some holes on the top of the foil, and place the beaker inside a 104 °C preheated laboratory oven for 15 min.
   (d) Remove the beaker from the oven, and allow to cool at room temperature for 1 h.

(e) Once the solution has cooled, wash the slides in 2× changes of MilliQ water for 5 min.

6. *Block endogenous peroxidase activity:*

   (a) Incubate sections in 3% hydrogen peroxide in MilliQ water for 5 min.

   (b) Wash sections twice in Dulbecco's phosphate buffered saline for 2 min each.

7. Transfer the slides to a plastic humidified chamber, and then carefully dry around the tissue sections and draw a circle around them using a PAP pen.

8. *Block non-specific IgG receptors in tissue sections:*

   (a) Incubate sections at room temperature for 1 h in working solution of mouse-on-mouse kit IgG blocking reagent (provided with kit; prepare mouse-on-mouse kit IgG blocking reagent by adding 9 μL mouse-on-mouse kit IgG blocking reagent per 250 μL Dulbecco's phosphate buffered saline). Mouse-on-mouse kit should only be used when the primary antibody has been raised in a mouse.

   (b) Pour off IgG block solution, and wash sections twice for 2 min in Dulbecco's phosphate buffered saline.

9. *Primary antibody stain:*

   (a) Incubate tissue sections for 5 min in mouse-on-mouse kit diluent (provided with kit).

   (b) Pour off mouse-on-mouse kit diluent, and then pipette anti-βIII-tubulin antibody or mouse IgG2A isotype control antibody diluted 1:50 in 1× mouse-on-mouse kit diluent, onto the tissue sections, within the boundaries of the PAP pen. Incubate at room temperature for 60 min (*see* **Note 14**).

   (c) 20 min into the incubation, prepare the ABC complex (Vectastain ABC kit): Add one drop of reagent A into 2.5 mL Dulbecco's phosphate buffered saline, then add one drop of reagent B, mix well, and allow to form a complex at room temperature for 30 min.

   (d) Remove the primary antibody, and wash sections twice in Dulbecco's phosphate buffered saline for 2 min each.

10. *Secondary antibody stain:*

   (a) Pipette mouse-on-mouse kit biotinylated anti-mouse IgG reagent (provided with kit; dilute 1:250 in mouse-on-mouse kit diluent) onto the tissue sections, within the boundaries of the PAP pen. Incubate at room temperature for 10 min.

(b) Wash sections twice for 2 min each in Dulbecco's phosphate buffered saline.

11. *Tertiary stain:*

 (a) Incubate sections with 100 μL ABC complex at room temperature for 5 min.

 (b) Wash sections twice for 2 min each in Dulbecco's phosphate buffered saline.

12. *Visualization of target protein:*

 (a) Just before use, prepare the DAB substrate solution (DAB kit from DAKO; Add one drop of DAB concentrate into 1 mL of DAB buffer), and mix well by pipetting.

 (b) Pipette the substrate onto the tissue sections, within the boundaries of the PAP pen. Incubate for 2–3 min or until the tissue section starts turning brown in color (*see* **Note 15**).

 (c) Immediately stop the DAB reaction by placing all slides into tap water at the same time.

13. *Counterstain using Mayer's hematoxylin solution:*

 (a) Stain tissue sections with the hematoxylin solution at room temperature for 60 s.

 (b) Immediately rinse all slides with tap water to remove excess reagent.

14. Remove slides, then immediately dehydrate tissue sections as follows (all at room temperature):

 (a) Incubate sections in 1× wash of 70% ethanol 5 min.

 (b) Incubate sections in 1× wash of 95% ethanol for 5 min.

 (c) Incubate sections in 2× washes of 100% ethanol for 5 min each.

 (d) Incubate sections in 2× washes of xylene for 5 min each.

15. Apply mounting media to tissue sections and coverslip. Allow to dry in a fume cabinet for at least 30 min. Samples can be imaged immediately after this or stored at room temperature. Positive DAB signal will be brown, with blue hematoxylin counterstain for nucleus (Fig. 5a).

3.6.2 Western Blotting

1. Transfer snap-frozen pancreatic tumor tissue samples in 1.5 mL microtubes to dry ice and label a 2 mL sterile microtube per tissue sample.

2. Place pancreas tissue into a 60 mm cell culture petri dish placed on dry ice using a sterile pair of forceps. Dice pancreatic tissue sample into smaller pieces (2 mm × 2 mm) in the dish using a sterile scalpel.

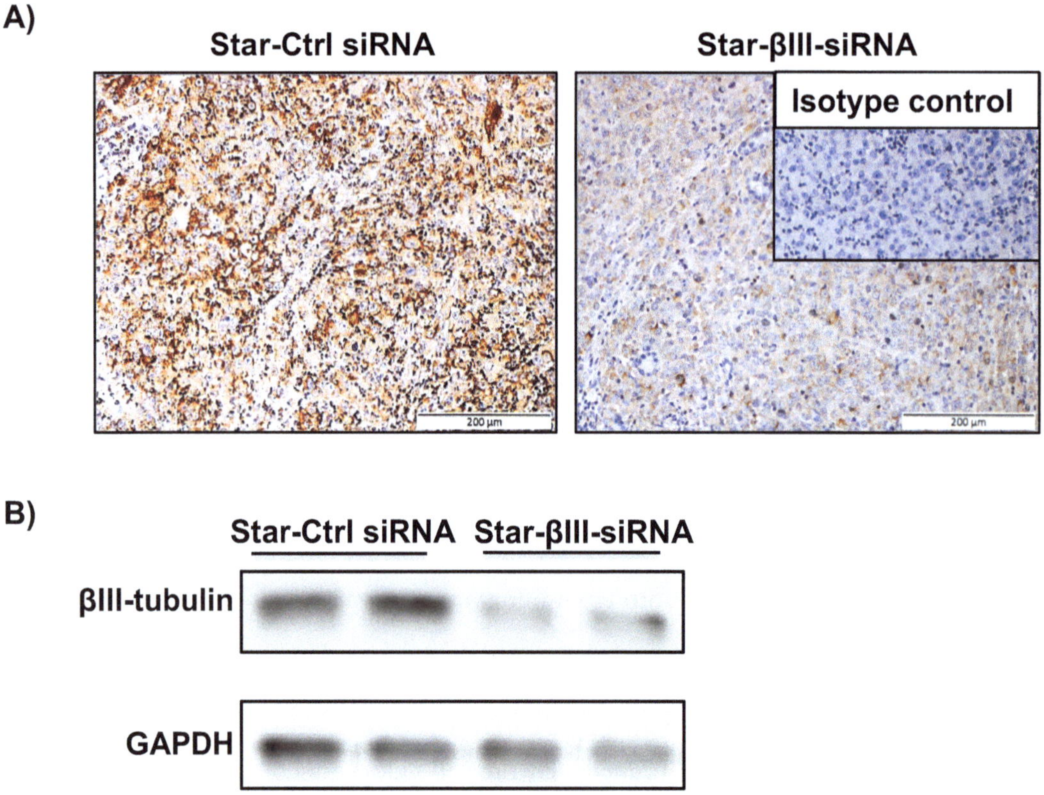

Fig. 5 Knockdown of βIII-tubulin in mouse orthotopic pancreatic tumors via systemic delivery of star polymer+siRNA. (**a**) Immunohistochemistry and (**b**) Western blot for βIII-tubulin protein, in orthotopic pancreatic tumors from mice injected with 4 mg/kg star polymer-βIII-tubulin siRNA (Star-βIII-siRNA) or star polymer-control siRNA (Star-Ctrl siRNA) 24 h after the final injection (3× daily injections total). Inset in (**b**) shows isotype control. GAPDH was used as a protein loading control in (**b**). Reprinted (adapted) with permission from Teo et al., Biomacromolecules, 2016, 17,2337–51. Copyright (2016) American Chemical Society

3. Use the lid of the petri dish, and pre-chilled on dry ice, to weigh 10–15 mg of tissue. Transfer these pieces of tissue immediately to a 2 mL microtube on dry ice.
4. Prepare 1 mL ice-cold 1× RIPA cell lysis buffer per sample by diluting 10× RIPA buffer (cell signaling) in MilliQ water and supplementing with 1% protease inhibitor cocktail (Sigma), 1% phosphatase inhibitor cocktail (Sigma), and 1 mM phenylmethyl sulfoxide (PMSF; Sigma) (*see* **Note 16**).
5. Add 1 mL ice-cold complete RIPA buffer per sample, one sample at a time, then immediately homogenize using a TissueRuptor with a sterile TissueRuptor probe. Samples must be kept on ice during and after homogenization.
6. Once the tissue has been homogenized, further disrupt any remaining small pieces of tissue using a sonicator probe.
7. Incubate the samples on ice for 30 min.

8. Centrifuge the tissue suspension at 14,000 × *g* for 10 min in a precooled microcentrifuge at 4 °C.
9. Transfer the supernatant into a new cold 1.5 mL microtube.
10. Measure the concentration of protein samples by performing a BCA protein assay according to manufacturer's instructions.
11. Prepare 10 μg of protein sample for SDS-PAGE by mixing 1:1 with Laemmli loading buffer and heating at 95 °C for 5 min (*see* **Note 17**).
12. Electrophorese proteins on a 10% discontinuous SDS-PAGE gel at 100 V for 1 h, in a Bio-Rad gel electrophoresis tank according to the manufacturer's instructions.
13. Transfer electrophoresed proteins to a nitrocellulose membrane at 100 V for 1 h, using the Bio-Rad wet transfer system (Bio-Rad gel electrophoresis tank, plastic cassette, fiber pads, 3MM Whatmann paper) according to the manufacturer's instructions.
14. Once samples have been transferred on to the nitrocellulose membrane, use a clean scalpel blade to cut the membrane between 37 kDa (GAPDH) and 50 kDa (βIII-tubulin).
15. Place the two pieces of membrane in blocking buffer on a rocker for 1 h at room temperature.
16. Add the primary antibody (Anti-βIII-tubulin or anti-GAPDH diluted 1:1000 and 1:50,000 in blocking buffer, respectively). Incubate on a rocker at 4 °C overnight.
17. Wash the membrane 3× in DPBST for 5 min each.
18. Add secondary antibody diluted in blocking buffer (anti-mouse-HRP, 1:2000). Incubate on a rocker for 1 h at room temperature.
19. Rinse the membrane 3× in DPBST for 5 min each. Perform a final rinse for 5 min in Dulbecco's phosphate buffered saline.
20. Add ECL reagent directly to the membrane for 30 s, and then visualize luminescence on a LAS4000 imaging system. Quantify band intensity using the ROI tool in ImageQuant TL software (Fig. 5b) *See* **Note 18**.

4 Notes

1. It is important to obtain relevant approvals from your appropriate institutional Animal Ethics Committee before conducting any experiments involving animals.
2. 16% paraformaldehyde can be purchased from ProSciTech and diluted to 4% in Dulbecco's phosphate buffered saline. This should be prepared on the day of harvest.

3. The anaesthetic machine will require training from the appropriate custodian and should be used according to manufacturer's instructions.
4. Buprenorphine must be readministered as per subheading 3.1.3 **step 14** every 6–12 h for the first 24 h, then as required based on any signs of pain. Staples can be removed 7 days postsurgery.
5. Note that exposure and binning can be raised if the image counts are below 600 or lowered if the software detects saturating pixels.
6. In vivo bioluminescent imaging can be used to determine the tumor growth kinetics of each pancreatic cancer cell line. This information can then be used to select the ideal time during tumor growth to administer star polymer-siRNA nanocomplexes, based on tumor size and growth phase. For example, in our previous study using MiaPaCa-2 cells [11], our time point for nanoparticle injection was 5 weeks post-implant, based on a pilot model in which orthotopic MiaPaCa-2 tumor growth was monitored weekly using in vivo bioluminescent imaging.
7. To account for volume loss in the syringe when you draw up the nanocomplex, we multiply the total volume of siRNA and star nanoparticle by 1.1. This extra volume will account for any volume loss and allow you to easily draw up 150 μL of the nanocomplex for systemic administration into the tail vein of mice.
8. The amount of nanoparticle needed to fully complex siRNA was determined using in vitro gel retardation assays [11]. If using another nanoparticle system, the amount of nanoparticle needed may be different and should be optimized using in vitro siRNA-nanoparticle binding assays. Amounts of siRNA and star nanoparticle delivered to mice in Subheading 3.3.2 were based on a 20 g mouse.
9. To assess siRNA biodistribution in vivo, the experiential plan must include for a group of mice injected with Dulbecco's phosphate buffered saline or 0.9% saline (no nanoparticle or siRNA) to control for background fluorescence from isolated organs.
10. A series of time points following administration of star polymer-fluorescent siRNA should be assessed to examine the early and late biodistribution profile of siRNA in vivo. In our study we chose to look at siRNA biodistribution 1 h, 4 h, and 24 h postinjection [11].
11. While the software will select peak excitation and emission wavelengths for the selected fluorophore, multiple emission channels can also be manually selected. This can be advantageous as there may be a suboptimal emission wavelength that has cleaner background fluorescence.

12. Livers should be imaged separately to all other organs as they will have the brightest fluorescence, especially at the earliest time points. The high accumulation of nanoparticle-siRNA in the liver is common for most polymeric nanoparticles.
13. It is important to determine whether siRNA localized with the pancreatic tumor is able to extravasate from tumor vessels into the surrounding tumor cells. We assess this by performing immunofluorescence using CD31 antibody to label tumor blood vessels.
14. The staining immunohistochemistry procedure described may have to be optimized for your selected target protein. In this study we detected βIII-tubulin expression in mouse orthotopic pancreatic tumor sections using a mouse-on-mouse staining kit.
15. The DAB exposure time may have to be optimized, to avoid saturating the DAB signal and masking protein knockdown.
16. All pancreatic tumor tissue samples must be kept on dry ice until RIPA cell lysis buffer is added. Protease and phosphatase inhibitors should only be added to the RIPA cell lysis buffer just before use.
17. The amount of protein loaded onto the gel may have to be optimized for each new target protein. In this study we optimized the amount of protein and antibody staining condition to detect βIII-tubulin expression in mouse pancreatic tumors.
18. Final considerations: Although this protocol details the methodology for silencing βIII-tubulin expression in orthotopic mouse pancreatic tumors using star polymer nanoparticles complexed with siRNA, it can also be used to target other genes in pancreatic tumors using siRNA. Our methodology above evaluates gene-silencing activity using immunohistochemistry and Western blot to analyze protein expression; however we also routinely evaluate gene expression using qPCR [11]. The described methodology may also prove highly useful for delivering siRNA to orthotopic pancreatic tumors using other nanoparticle systems which have chemical or physical properties similar to star polymers.

Acknowledgments

This work was supported by grants from the National Health and Medical Research Council (NHMRC; P. A. Phillips, J. McCarroll, M. Kavallaris; APP1024895), Cancer Council New South Wales (P. A. Phillips, J. McCarroll, M. Kavallaris), Cure Cancer Australia Foundation Grant (P. A. Phillips), Cancer Institute NSW Fellowship (J. McCarroll, G. Sharbeen), NHMRC CDF Fellowship (P. A.

Phillips, APP1024896). M. Kavallaris is also supported by funding from National Health and Medical Research Council (NHMRC) Program Grant (APP1091261), Cancer Council NSW Program Grant (PG16-01), Australian Research Council Centre of Excellence in Convergent Bio-Nano Science and Technology (project number CE140100036) and an NHMRC Principal Research Fellowship (APP1119152). Joshua A. McCarroll and George Sharbeen contributed equally to this work.

Conflict of interest disclosure: The authors declare no competing financial interest.

References

1. Rahib L, Smith BD, Aizenberg R, Rosenzweig AB, Fleshman JM, Matrisian LM (2014) Projecting cancer incidence and deaths to 2030: the unexpected burden of thyroid, liver, and pancreas cancers in the United States. Cancer Res 74(11):2913–2921. https://doi.org/10.1158/0008-5472.CAN-14-0155
2. Siegel RL, Miller KD, Jemal A (2018) Cancer statistics, 2018. CA Cancer J Clin. https://doi.org/10.3322/caac.21442
3. McCarroll JA, Naim S, Sharbeen G, Russia N, Lee J, Kavallaris M, Goldstein D, Phillips PA (2014) Role of pancreatic stellate cells in chemoresistance in pancreatic cancer. Front Physiol 5:141. https://doi.org/10.3389/fphys.2014.00141
4. Bailey P, Chang DK, Nones K, Johns AL, Patch AM, Gingras MC, Miller DK, Christ AN, Bruxner TJ, Quinn MC, Nourse C, Murtaugh LC, Harliwong I, Idrisoglu S, Manning S, Nourbakhsh E, Wani S, Fink L, Holmes O, Chin V, Anderson MJ, Kazakoff S, Leonard C, Newell F, Waddell N, Wood S, Xu Q, Wilson PJ, Cloonan N, Kassahn KS, Taylor D, Quek K, Robertson A, Pantano L, Mincarelli L, Sanchez LN, Evers L, Wu J, Pinese M, Cowley MJ, Jones MD, Colvin EK, Nagrial AM, Humphrey ES, Chantrill LA, Mawson A, Humphris J, Chou A, Pajic M, Scarlett CJ, Pinho AV, Giry-Laterriere M, Rooman I, Samra JS, Kench JG, Lovell JA, Merrett ND, Toon CW, Epari K, Nguyen NQ, Barbour A, Zeps N, Moran-Jones K, Jamieson NB, Graham JS, Duthie F, Oien K, Hair J, Grutzmann R, Maitra A, Iacobuzio-Donahue CA, Wolfgang CL, Morgan RA, Lawlor RT, Corbo V, Bassi C, Rusev B, Capelli P, Salvia R, Tortora G, Mukhopadhyay D, Petersen GM, Australian Pancreatic Cancer Genome Initiative, Munzy DM, Fisher WE, Karim SA, Eshleman JR, Hruban RH, Pilarsky C, Morton JP, Sansom OJ, Scarpa A, Musgrove EA, Bailey UM, Hofmann O, Sutherland RL, Wheeler DA, Gill AJ, Gibbs RA, Pearson JV, Waddell N, Biankin AV, Grimmond SM (2016) Genomic analyses identify molecular subtypes of pancreatic cancer. Nature 531(7592):47–52. https://doi.org/10.1038/nature16965
5. Biankin AV, Waddell N, Kassahn KS, Gingras MC, Muthuswamy LB, Johns AL, Miller DK, Wilson PJ, Patch AM, Wu J, Chang DK, Cowley MJ, Gardiner BB, Song S, Harliwong I, Idrisoglu S, Nourse C, Nourbakhsh E, Manning S, Wani S, Gongora M, Pajic M, Scarlett CJ, Gill AJ, Pinho AV, Rooman I, Anderson M, Holmes O, Leonard C, Taylor D, Wood S, Xu Q, Nones K, Fink JL, Christ A, Bruxner T, Cloonan N, Kolle G, Newell F, Pinese M, Mead RS, Humphris JL, Kaplan W, Jones MD, Colvin EK, Nagrial AM, Humphrey ES, Chou A, Chin VT, Chantrill LA, Mawson A, Samra JS, Kench JG, Lovell JA, Daly RJ, Merrett ND, Toon C, Epari K, Nguyen NQ, Barbour A, Zeps N, Australian Pancreatic Cancer Genome Initiative, Kakkar N, Zhao F, Wu YQ, Wang M, Muzny DM, Fisher WE, Brunicardi FC, Hodges SE, Reid JG, Drummond J, Chang K, Han Y, Lewis LR, Dinh H, Buhay CJ, Beck T, Timms L, Sam M, Begley K, Brown A, Pai D, Panchal A, Buchner N, De Borja R, Denroche RE, Yung CK, Serra S, Onetto N, Mukhopadhyay D, Tsao MS, Shaw PA, Petersen GM, Gallinger S, Hruban RH, Maitra A, Iacobuzio-Donahue CA, Schulick RD, Wolfgang CL, Morgan RA, Lawlor RT, Capelli P, Corbo V, Scardoni M, Tortora G, Tempero MA, Mann KM, Jenkins NA, Perez-Mancera PA, Adams DJ, Largaespada DA, Wessels LF, Rust AG, Stein LD, Tuveson DA, Copeland NG, Musgrove EA, Scarpa A, Eshleman JR, Hudson TJ, Sutherland RL, Wheeler DA, Pearson JV,

McPherson JD, Gibbs RA, Grimmond SM (2012) Pancreatic cancer genomes reveal aberrations in axon guidance pathway genes. Nature 491(7424):399–405. https://doi.org/10.1038/nature11547

6. Waddell N, Pajic M, Patch AM, Chang DK, Kassahn KS, Bailey P, Johns AL, Miller D, Nones K, Quek K, Quinn MC, Robertson AJ, Fadlullah MZ, Bruxner TJ, Christ AN, Harliwong I, Idrisoglu S, Manning S, Nourse C, Nourbakhsh E, Wani S, Wilson PJ, Markham E, Cloonan N, Anderson MJ, Fink JL, Holmes O, Kazakoff SH, Leonard C, Newell F, Poudel B, Song S, Taylor D, Waddell N, Wood S, Xu Q, Wu J, Pinese M, Cowley MJ, Lee HC, Jones MD, Nagrial AM, Humphris J, Chantrill LA, Chin V, Steinmann AM, Mawson A, Humphrey ES, Colvin EK, Chou A, Scarlett CJ, Pinho AV, Giry-Laterriere M, Rooman I, Samra JS, Kench JG, Pettitt JA, Merrett ND, Toon C, Epari K, Nguyen NQ, Barbour A, Zeps N, Jamieson NB, Graham JS, Niclou SP, Bjerkvig R, Grutzmann R, Aust D, Hruban RH, Maitra A, Iacobuzio-Donahue CA, Wolfgang CL, Morgan RA, Lawlor RT, Corbo V, Bassi C, Falconi M, Zamboni G, Tortora G, Tempero MA, Australian Pancreatic Cancer Genome Initiative, Gill AJ, Eshleman JR, Pilarsky C, Scarpa A, Musgrove EA, Pearson JV, Biankin AV, Grimmond SM (2015) Whole genomes redefine the mutational landscape of pancreatic cancer. Nature 518(7540):495–501. https://doi.org/10.1038/nature14169
7. McCarroll J, Teo J, Boyer C, Goldstein D, Kavallaris M, Phillips PA (2014) Potential applications of nanotechnology for the diagnosis and treatment of pancreatic cancer. Front Physiol 5:2. https://doi.org/10.3389/fphys.2014.00002
8. Galant NJ, Westermark P, Higaki JN, Chakrabartty A (2017) Transthyretin amyloidosis: an under-recognized neuropathy and cardiomyopathy. Clin Sci (Lond) 131(5):395–409. https://doi.org/10.1042/CS20160413
9. Phillips P (2012) Pancreatic stellate cells and fibrosis. In: Grippo PJ, Munshi HG (eds) Pancreatic cancer and tumor microenvironment. Transworld Research Network, Trivandrum, India
10. Matsumura Y, Maeda H (1986) A new concept for macromolecular therapeutics in cancer chemotherapy: mechanism of tumoritropic accumulation of proteins and the antitumor agent smancs. Cancer Res 46(12 Pt 1):6387–6392
11. Teo J, McCarroll JA, Boyer C, Youkhana J, Sagnella SM, Duong HT, Liu J, Sharbeen G, Goldstein D, Davis TP, Kavallaris M, Phillips PA (2016) A rationally optimized nanoparticle system for the delivery of RNA interference therapeutics into pancreatic tumors in vivo. Biomacromolecules 17(7):2337–2351. https://doi.org/10.1021/acs.biomac.6b00185
12. McCarroll JA, Sharbeen G, Liu J, Youkhana J, Goldstein D, McCarthy N, Limbri LF, Dischl D, Ceyhan GO, Erkan M, Johns AL, Biankin AV, Kavallaris M, Phillips PA (2015) betaIII-tubulin: a novel mediator of chemoresistance and metastases in pancreatic cancer. Oncotarget 6(4):2235–2249. https://doi.org/10.18632/oncotarget.2946
13. Kato A, Naiki-Ito A, Naitoh I, Hayashi K, Nakazawa T, Shimizu S, Nishi Y, Okumura F, Tadahisa I, Takada H, Kondo H, Yoshida M, Takahashi S, Joh T (2018) The absence of class III beta-tubulin is predictive of a favorable response to nab-paclitaxel and gemcitabine in patients with unresectable pancreatic ductal adenocarcinoma. Hum Pathol. https://doi.org/10.1016/j.humpath.2018.01.009
14. Lee KM, Cao D, Itami A, Pour PM, Hruban RH, Maitra A, Ouellette MM (2007) Class III beta-tubulin, a marker of resistance to paclitaxel, is overexpressed in pancreatic ductal adenocarcinoma and intraepithelial neoplasia. Histopathology 51(4):539–546. https://doi.org/10.1111/j.1365-2559.2007.02792.x

Chapter 24

Microfluidic Assembly of siRNA-Loaded Micelleplexes for Tumor Targeting in an Orthotopic Model of Ovarian Cancer

Daniel P. Feldmann, Steven Jones, Kirk Douglas, Anthony F. Shields, and Olivia M. Merkel

Abstract

The use of cationic polymers to interact with negatively charged siRNA via charge complexation to form polyelectrolyte complexes has been widely studied ever since the 1998 report on RNA interference. These polyelectrolyte complex formulations aim to overcome the many pitfalls associated with the use of RNA interference as a potential cancer therapy. The triblock copolymer polyethylenimine-polycaprolactone-polyethylene glycol (PEI-PCL-PEG) contains the cation PEI and has been shown to be an efficient carrier capable of complexing with nucleic acids for gene delivery. This copolymer system also allows for targeting moieties to be linked to the micelleplex, thereby exploiting overexpressed receptors (such as the folate receptor) located within tumors. Additionally, we demonstrated recently that microfluidic mixing of PEI-PCL-PEG nanoparticles allows for the rapid, scaled-up production of micelleplexes while maintaining small and uniform particle distributions. The preparation of small and reproducible particles is imperative for clinical translation of nanomedicine and for tumor targeting via systemic administration. Furthermore, to enable tracing of its deposition in vivo after its administration, micelleplexes can be radiolabeled. To assess tumor targeting over time, the noninvasive imaging technique single-photon emission computed tomography (SPECT) offers the ability to examine the same subject at multiple time points and generate biodistribution profiles. Since the biodistribution and tumor targeting of the therapeutic load of micelleplexes is of foremost interest, we recently described an approach to modify siRNA with a DTPA (diethylenetriaminepentaacetic acid) chelator. Herein, we explain the details of encapsulating indium-labeled siRNA via microfluidic mixing in PEI-PCL-PEG nanoparticles with a folic acid targeting ligand for assessment of their in vivo tumor targeting in an orthotopic ovarian cancer model.

Key words siRNA delivery, SPECT imaging, Indium labeling, Microfluidic mixing, Triblock copolymer, Folate receptor targeting

1 Introduction

After the landmark 1998 study published in Nature by Mello et al. that provided the first demonstration of RNA interference (RNAi), the use of siRNA as a therapeutic molecule has attracted significant attention as an alternative cancer treatment. However due to its

Lekha Dinesh Kumar (ed.), *RNA Interference and Cancer Therapy: Methods and Protocols*, Methods in Molecular Biology, vol. 1974, https://doi.org/10.1007/978-1-4939-9220-1_24, © Springer Science+Business Media, LLC, part of Springer Nature 2019

intrinsic properties, delivery of siRNA requires a carrier to ensure efficient therapeutic effects in vivo [1–4]. Therefore, the use of specialized delivery vehicles is imperative in order for therapeutic siRNA to be effective. Along with the benefit of shielding “naked” siRNA, the use of polymeric nanoparticles as a carrier also adds the ability to package multiple payloads, increases siRNA’s circulation time, and allows for tumor-specific targeting through surface functionalization [5]. Previously, the triblock copolymer polyethylenimine-polycaprolactone-polyethylene glycol (PEI-PCL-PEG) has been reported to spontaneously assemble with siRNA to form “micelleplexes” that are able to mediate RNAi in vivo and in vitro [6–8]. However, due to the chaotic nature of the self-assembly process, it is well documented that polyelectrolyte complexes have broad size distributions that limit their ability to be produced in large-scale batch reactors [9]. One strategy to address this limitation involves the use of microfluidic devices to increase colloidal stability by controlling the interaction of cationic polymer with anionic nucleic acid [10]. This strategy has proven useful in increasing siRNA complexation and uptake of the micelleplexes, which leads to a more efficient gene knockdown in vivo [11].

Therapeutic nanoparticle administration routes include intravenous, transdermal, pulmonary, and intraocular [12]. For therapeutic nanoparticles intended for cancer therapy, intravenous delivery is hypothesized to enable nanoparticles to accumulate preferentially in the tumor after extravasation from the bloodstream due to the enhanced permeability and retention (EPR) effect [13]. Intravenous administration, however, is not without its own hurdles to be overcome. Parameters such as the nanoparticle’s deposition, distribution, and elimination are key components to the success of any therapeutic.

With over 85% of all ovarian tumors overexpressing folate receptor alpha (FRα), the receptor has become an attractive candidate for targeted therapy of ovarian cancer [14]. To date, numerous groups have exploited folate receptors to deliver a targeted payload to the cancer cells of interest by attaching folic acid to the backbone of the drug itself or onto the delivery vehicle [15–17]. Subsequently, this approach has been applied in nanomedicine as a way to selectively deliver nanoformulated payloads to target cells, thereby reducing its uptake in healthy cells and decreasing off target toxicity. Additionally, attaching folic acid onto nanoparticles serves as active targeting of the carrier to the tumor site, therefore increasing the overall amount of payload reaching the intended target.

To assess the accumulation of nanoparticles in the organ or tissue of interest, SPECT imaging can be used. SPECT imaging is often employed to detect radioactive species within the body [18]. Furthermore, siRNA has the ability to be radioactively labeled, and this radioactive tag can be used as a tracer to track where the siRNA travels throughout the bloodstream and its deposition inside the body. Specifically, SPECT imaging can illustrate whether siRNA and polymeric carrier disassemble in circulation before reaching the tumor site, allows monitoring of biodistribution, and finally its route of elimination [19, 20]. Therefore, this approach provides useful information about the parameters that are crucial for intravenous delivery. Due to the need to overcome these hurdles for a successful treatment, siRNA imaging techniques are needed.

This chapter outlines the technique to label siRNA with a DTPA chelator and its subsequent chelation of ^{111}Indium. We further describe the microfluidic mixing of the radioactively labeled siRNA with folic acid conjugated triblock copolymer PEI-PCL-PEG to be used as a tracer in an orthotopic model of ovarian cancer following intravenous and intraperitoneal injection of the resulting micelleplexes via SPECT imaging.

2 Materials

2.1 siRNA Formulation

Due to their increased stability, high activity, and ability to be covalently modified, 2′-*O*-methylated 25/27mer DsiRNA14 [21] targeting EGFP was used and is recommended for use. For coupling of DTPA, amine-labeled siRNA is recommended. Here, we used a duplex with an amino-hexyl modification at the 5-prime of the antisense strand:

siEGFP: sense: 5′-pACCCUGAAGUUCAUCUGCACCACdCdG

Antisense: 3′-mAmCmUGmGGmACmUUmCAmAGmUAmGAmCGUGGUGGC-$C_6H_{12}NH_2$

whereby p denotes a phosphate residue, capital letters are ribonucleotides, d denotes 2′-deoxyribonucleotides, and m denotes 2′-*O*-methylribonucleotides.

2.2 Covalent Modification of siRNA with p-SCN-Bn- DTPA

1. siEGFP—see above.
2. p-SCN-Bn-DTPA (Macrocyclics, Plano, TX).
3. 0.1 M $NaHCO_3$ in DEPC water—filtered through a 0.22 μm filter before use.
4. 2 M NaOAc in DEPC water—filtered through a 0.22 μm filter before use.
5. Dried DMSO (about 3 mL).
6. 0.22 μm filter.
7. 2 mL centrifuge tube.
8. Metal spatula wrapped in parafilm—used to weigh out p-SCN-Bn-DTPA (*see* **Note 1**).
9. Aluminum foil.
10. Vortex mixer.

2.3 Precipitation of the siRNA-DTPA Complex

1. 2 M NaOAc in RNase-free water—filtered through a 0.22 μm filter before use.
2. Absolute ethanol—filtered through a 0.22 μm filter before use.
3. 15 mL conical tube.

2.4 Isolation of the siRNA-DTPA Complex

1. Ultracentrifuge (*see* **Note 2**).
2. Lysis buffer from Absolutely RNA miRNA Kit.
3. 2 M NaOAc in RNase-free water—filtered through a 0.22 μm filter before use.
4. Absolute ethanol—filtered through a 0.22 μm filter before use.
5. RNeasy Midi Kit (10) columns.

2.5 siRNA-DTPA Purification

1. Centrifuge.
2. Low-salt buffer from the Absolutely RNA miRNA Kit.
3. RNase-free water.
4. Sterile 2 mL collection centrifuge tubes.

2.6 siRNA Concentration Measurement

1. Spectrophotometer (e.g., Nanodrop™ 2000c).
2. RNase-free water.
3. 0.5 mL centrifuge tubes.
4. Parafilm.
5. Dry heat bath set to 94 °C.
6. Timer.

2.7 DTPA Concentration Measurement

1. Stock solution of the yttrium(III)-arsenazo III complex containing 5 μM arsenazo(III) and 1.6 μM yttrium(III) chloride (Acros Organics, New Jersey, US) in a 0.15 M NaOAc buffer at pH 4.
2. Stock solution of 0.123 mM DTPA dissolved in DI-H_2O with three molar equivalents of NaOH (*see* **Note 3**).
3. UV-Vis spectrophotometer.
4. UV-Vis disposable cuvette.

2.8 Indium Labeling

1. Radioactive Indium(III) chloride.
2. GE Healthcare Disposable PD-10 Desalting Columns.
3. RNase-free water.
4. Scintillation vials (make and model to fit gamma counter).
5. Gamma counter (e.g., Packard Tricarb 2910TR).
6. Spectrophotometer (e.g., Nanodrop™ 2000c).

2.9 Establishment of Orthotopic Model of Ovarian Cancer

1. Syringes, 1 mL.
2. 25-gauge sterile needle.
3. Sterile PBS buffer.
4. Mice (e.g., 6-week-old female nu/nu nude mice).
5. Folic acid-deficient diet.
6. Human ovarian cancer cell line (e.g., SKOV-3/LUC cells).

2.10 Monitoring Tumor Progression via Bioluminescence Imaging

1. Stock solution of D-luciferin in sterile PBS (e.g., 15 mg/mL).
2. Syringes, 1 mL.
3. 25-gauge sterile needle.
4. Bioluminescence imager (e.g., In-Vivo Extreme II).
5. Anesthetic agent (e.g., 3% isoflurane).

2.11 Microfluidic Preparation of Indium-Labeled siRNA Micelleplexes

1. 70% isopropyl alcohol.
2. 5% glucose solution—filtered through a 0.22 μm filter before use.
3. Polyethylenimine-*graft*-polycaprolactone-*block*-polyethylene glycol (PEI-g-PCL-b-PEG) triblock copolymer dissolved in water and sterile filtered to yield desired polymer concentration based upon the PEI 25 kDa content (*see* **Note 4**).
4. Luer Lock disposable syringes.
5. Syringe pump (e.g., KD Scientific 220 syringe pump).
6. Fluorinated ethylene propylene tubing (1/16″ OD × 0.25 mm ID).
7. Dolomite micromixer chip with hydrophobic coating.
8. Sterile collection tubes.

2.12 SPECT/CT Animal Imaging

1. Sterile syringes.
2. Sterile indium-labeled siRNA micelleplexes.
3. Mouse anesthesia.
4. SPECT imaging device and mouse cradle.

2.13 Pharmaco kinetics

1. Microhematocrit blood collection tubes.
2. Gauze sponges.
3. General anesthetic agents and equipment.
4. Scintillation vials (make and model to fit gamma counter).
5. Gamma counter (e.g., Packard Tricarb 2910TR).

2.14 Biodistribution of Micelleplexes

1. Surgery table and surgical instruments.
2. Scintillation vials (make and model to fit gamma counter).
3. Gamma counter (e.g., Packard Tricarb 2910TR).

3 Methods

3.1 React siRNA with p-SCN-Bn- DTPA

1. Weigh out 5.11 mg of double-stranded siRNA in a 2 mL centrifuge tube, and dissolve it in 100 μL of RNase-free water (*see* **Note 5**).
2. To the 2 mL tube, add 100 μL of filtered 0.1 M $NaHCO_3$.
3. Next, weigh out 9.76 mg of p-SCN-Bn-DTPA, and dissolve it in 540 μL of dry DMSO (*see* **Notes 1** and **6**).
4. Add the 540 μL of the DTPA to siRNA mixture. The new total volume should be 740 μL (*see* **Note** 7).
5. Vortex the solution thoroughly, and incubate for 6 h. Agitate the solution every 30 min.

3.2 Precipitation of the siRNA-DTPA Complex

1. Add 74 μL of filtered 2 M Na-acetate to the mixture (10% of the total amount of mixture).
2. Transfer the mixture to a 15 mL conical tube.
3. Add filtered absolute ethanol so that the final concentration is 80% v/v.
4. Freeze solution overnight at −80 °C.

3.3 Isolation of the siRNA-DTPA Complex

1. Centrifuge the sample for 30 min at 12,000 × *g* in an ultracentrifuge (*see* **Note 2**).
2. Discard the supernatant.
3. Add 2.5 mL of lysis buffer from "Absolutely RNA miRNA Kit."
4. Vortex the solution.
5. Add 250 μL of filtered 2 M Na-Acetate.
6. Add 7.25 mL of filtered absolute ethanol for a total of 10 mL (*see* **Note 8**).
7. Vortex the solution and equally distribute the 10 mL onto five RNeasy Midi Kit Qiagen Columns (*see* **Note 9**).

3.4 siRNA-DTPA Purification

1. Centrifuge the columns at 4500 × *g* for 5 min, discard the flow through.
2. To each column, add 200 μL of the low-salt buffer from the "Absolutely RNA miRNA Kit."
3. Centrifuge the solution at 4500 × *g* for 2 min, discard the flow through.

4. Repeat **steps 2** and **3**.
5. To dry the column, spin them down at 5000 × *g* for 5 min.
6. Transfer the columns to a new collection tube, and add 200 μL of 60 °C hot RNase-free water.
7. Centrifuge the solution at 5000 × *g* for 5 min to collect the purified siRNA-DTPA.
8. Add 100 μL of 60 °C hot RNase-free water and centrifuge at 5000 × *g* for 5 min.
9. Combine the flow through from all of the columns into one sterile 2 mL tube.

3.5 siRNA Concentration Measurement

1. Measure the siRNA concentration on a spectrophotometer (e.g., Nanodrop 2000c). Use RNase-free water as your blank.
2. Under the hood, dilute the siRNA to a desired concentration, and aliquot into 0.5 mL sterile tubes (*see* **Note 10**).
3. Sterilize siRNA-DTPA with a 0.22 μm syringe filter.
4. Parafilm each tube, and anneal the siRNA at 94 °C for exactly 2 min.
5. Let the samples cool down to room temperature.
6. Freeze the samples, and keep frozen until needed.

3.6 DTPA Concentration Measurement

1. Using a UV-Vis spectrophotometer, create the standard curve for DTPA concentrations.
2. Pipette 3 mL of the Y(III)-arsenazo III complex stock solution into a cuvette and read (652 nM) this as the blank.
3. Add 5 μL of the stock DTPA solution to the cuvette, gently mix, and read the solution again (*see* **Note 11**).
4. Add another 5 μL of the stock DTPA solution, read, and repeat.
5. Do this until you have generated enough points for your standard curve (each new data point will have an additional 5 μL added into the cuvette).
6. Once all the standards have been made and read on the spectrophotometer, discard the solution inside the cuvette, and put 3 mL of fresh Y(III)-arsenazo III complex stock solution into the cuvette (*see* **Note 12**).
7. Add 5 μL of your siRNA-DTPA sample into the cuvette and take the measurement (652 nM).
8. Plot the standard curve for the DTPA concentrations versus absorbance and insert a linear line of best fit (Fig. 1).
9. Using the equation yielded from the line of best fit, plot the absorbance value obtained from your sample measurement, and plug that into the *Y*-value of the equation in order to solve for *X*.

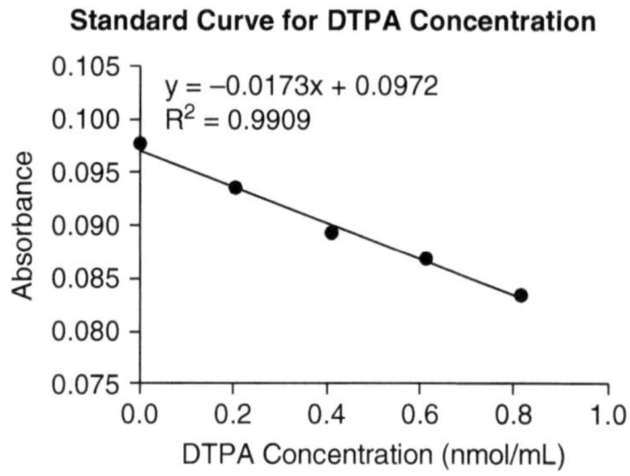

Fig. 1 Standard curve for DTPA concentration. Scatter plot obtained from creating the standard curve of the DTPA concentrations. The amount of DTPA added to the cuvette is on the *X*-axis, and Absorbance at 652 nm is on the *Y*-axis. From here, concentration of the DTPA inside the siRNA-DTPA mixture can be obtained

10. The *X*-value obtained will be the concentration of DTPA in your sample.
11. Now that the DTPA and siRNA concentrations have been found for the siRNA-DTPA mixture, calculate the molar amounts of the siRNA and DTPA within your sample. Use this to determine the molar equivalency of the siRNA and DTPA (*see* **Note 13**).

3.7 Indium Labeling and Purification

1. React radioactive $^{111}InCl_3$ with DTPA-siRNA. In the example shown below, 116.9 MBq $^{111}InCl_3$ were reacted with 15 nmol DTPA- siRNA. Incubate for 2 min at 94 °C, and allow solution to cool down to room temperature.
2. Equilibrate a PD-10 column with RNase-free water by washing it with 25 mL.
3. Prepare 24 scintillation vials in a rack and label them from 1 to 24.
4. Place vial 1 underneath the PD-10 column, and start adding the siRNA-indium mixture to the column slowly.
5. Collect 13 drops in the first vial as fraction 1 and then move on to the next vial. Collect 13 drops per fraction. Once the complete volume of the siRNA-indium mixture is applied to the column, add RNase-free water. Collect 24 fractions.
6. Close the scintillation vials, and measure the counts per minute (CPM) in every vial using a gamma counter.
7. Plot the CPM versus the fraction number (Fig. 2).
8. Determine the siRNA concentration in the peak fraction using a Nanodrop spectrophotometer (*see* **Note 14**).

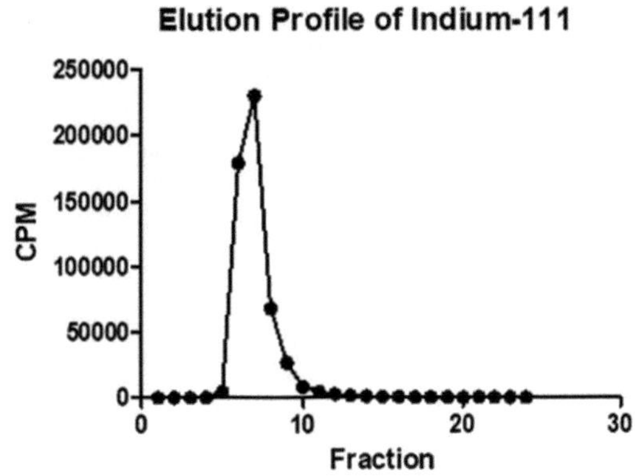

Fig. 2 Elution profile of ^{111}In. Scatter plot obtained from purifying and eluting ^{111}In-labeled siRNA over a PD-10 column. The radioactivity as measured in counts per minute (CPM) are shown on the *Y*-axis as a function of the fraction eluted on the *X*-axis. A clear peak is shown in fraction 7

3.8 Establishment of Orthotopic Model of Ovarian Cancer

1. Culture and expand luciferase (LUC)-expressing ovarian cancer cells per established protocols.
2. Harvest cells, and inject each mouse with 1×10^6 cells resuspended in phosphate-buffered saline intraperitoneally (*see* **Note 15**).
3. Four weeks post-injection, place all mice on folic acid-deficient diet (*see* **Note 16**).
4. Monitor mice and observe tumor growth for a total of 6 weeks after cancer cell inoculation.

3.9 Monitoring Tumor Progression via Bioluminescence Imaging

1. Two weeks after cancer cell inoculation, inject mice with LUC-expressing tumors intraperitoneally with 100 μL of a freshly prepared 15 mg/mL D-Luciferin stock solution in PBS per 10 g of their body weight.
2. After 10 min, sedate mice with 3% isoflurane (*see* **Note 17**).
3. Place mice inside imager cradle and take BLI images with a 3-min exposure under high sensitivity and aperture of the lens set at an f-stop of 1.1.
4. Simultaneously, take X-ray images with standard 1.2-s exposure.
5. Repeat this procedure every 2 weeks until the 6-week time point has been met (Fig. 3).

3.10 Microfluidic Preparation of Indium-Labeled siRNA Micelleplexes

1. Fill two syringes with 70% isopropyl alcohol, and pump the solution through the micromixer chip at 1 mL/min. Repeat this wash step with sterile 5% glucose solution.
2. After determining the optimal N/P ratio to be used, dilute PEI-g-PCL-b-PEG triblock copolymer and indium-labeled

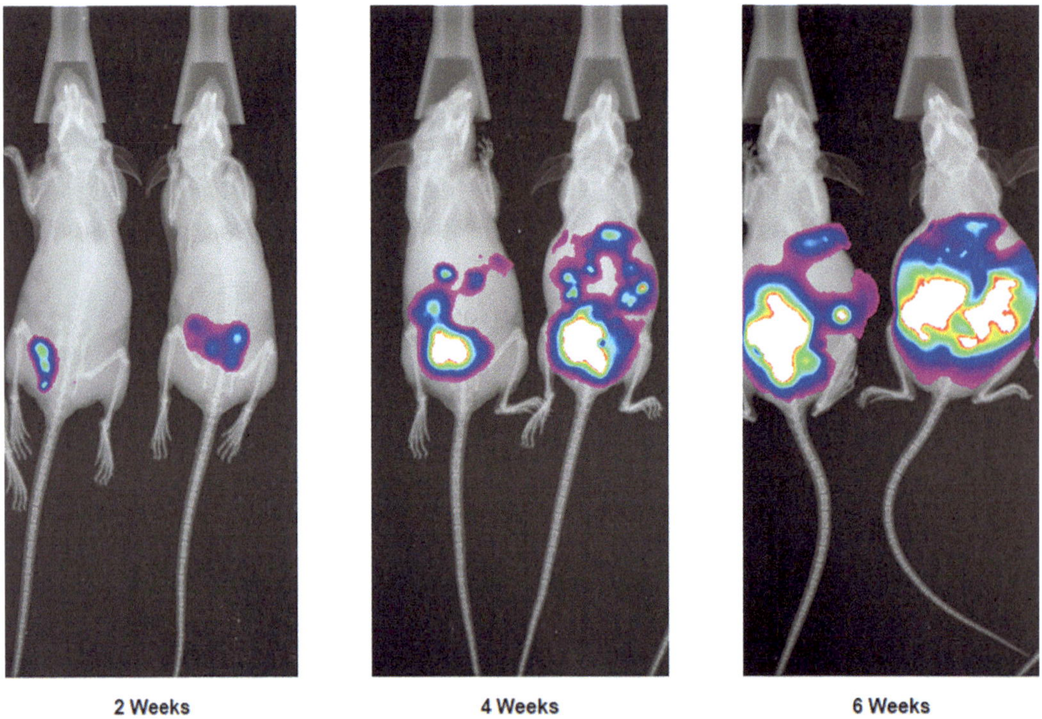

Fig. 3 Assessment of tumor growth via bioluminescence. Luciferase activity is shown predominantly in the primary tumor as well as in metastases

siRNA with sterile 5% glucose to desired concentration. In the example below, micelleplexes were prepared with 2 nmol siRNA per animal which was equivalent to approximately 3 MBq per animal (*see* **Note 18**).

3. Load diluted polymer and siRNA solution into separate syringes, and ensure that their plungers align (*see* **Note 19**).
4. Load each syringe into the syringe pump, and attach them to their corresponding female Luer adapter/FEP tubing.
5. Ensure that the syringe pump is set to the correct syringe diameter, and set the flow rate to 0.5 mL/min.
6. After steady-state flow is reached within the micromixer chip, collect assembled micelleplex solution in a separate sterile collection tube.
7. Following collection of formed micelleplexes, refill syringes with 70% isopropyl alcohol, and wash micromixer chip to avoid blockages forming during chip storage.

3.11 SPECT/CT Animal Imaging

1. Anesthetize the animals and administer the indium-labeled siRNA micelleplexes. In the example below, intravenous tail vein injection and intraperitoneal injection were chosen.

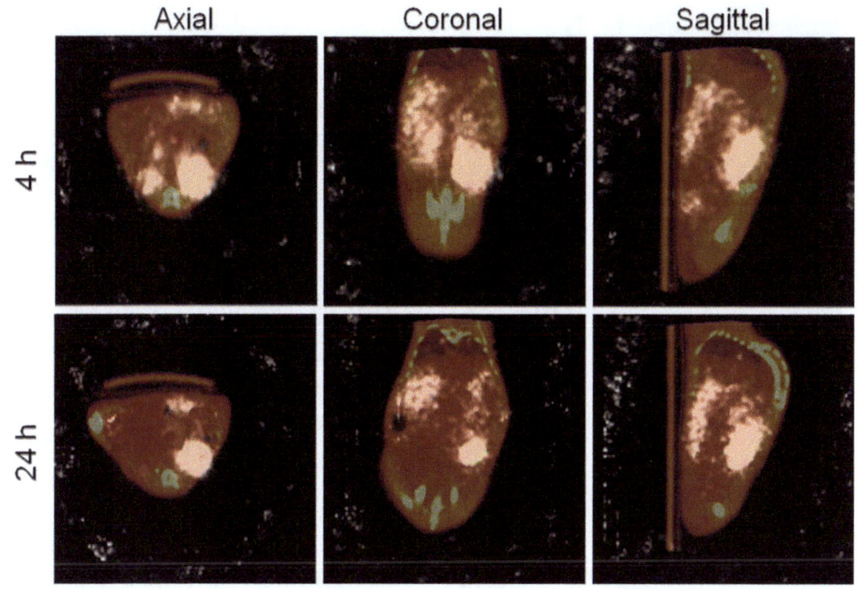

Fig. 4 In vivo SPECT/CT images. Biodistribution analysis of nude mice 4 and 24 h post intraperitoneal injection of PEI-PCL-PEG micelleplexes loaded with ^{111}In-labeled siRNA

2. Place the animals, one after the other, into the SPECT mouse cradle, and start the 360° imaging program (Fig. 4) (*see* **Note 20**).

3.12 Pharmacokinetic Modeling

1. Immediately following administration of micelleplexes, collect a 25 μL blood sample via retro orbital collection at various time points with a microhematocrit blood tube. In the example below, blood sampling was done at 1, 3, 5, 15, 30, 60, and 120 min following both administration routes.
2. Transfer blood samples into separate scintillation vials.
3. Close the scintillation vials, and measure the counts per minute (CPM) in every vial using a gamma counter.
4. Record CPM of each blood sample for every time point.
5. To determine the percent injected dose per mL of blood, the CPMs were multiplied by 40 to account for an estimated total volume of blood in the mouse to be 2 mL.
6. Following calculations, the new CPM value was fit to a fresh ^{111}In standard curve to determine the percent injected dose of ^{111}In in the blood sample.
7. Once the values were obtained, the PK values were plotted on a logarithmic scale and fit to a two-compartment model (*see* **Note 21**).

3.13 Biodistribution of Micelleplexes

1. Twenty-four hours after the administration of micelleplexes, sacrifice mice in order to harvest organs.

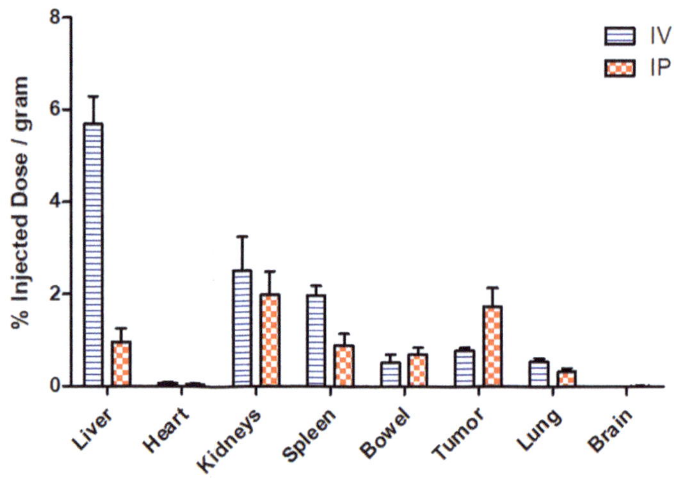

Fig. 5 In vivo biodistribution analysis of nude mice. Biodistribution analysis of ^{111}In after intraperitoneal and intravenous injection of PEI-PCL-PEG micelleplexes loaded with ^{111}In-labeled siRNA. Organs were harvested 24 h post injection and read under gamma scintillation counting

2. Dissect the liver, kidneys, lungs, brain, spleen, bowels, and tumors.
3. Record the weight of each organ individually, and place into separate scintillation vials.
4. Close the scintillation vials and measure the counts per minute (CPM) in every vial using a gamma counter.
5. To determine the percent injected dose per gram of tissue, the CPMs were adjusted for activity to account for decay.
6. Following adjustment, the CPM of the tissue was divided by the CPM value of the total injected ^{111}In. (Note: The CPM of the injected amount was determined by fitting the value into a fresh standard curve of ^{111}In).
7. The new value was divided by the weight of each organ to determine the percent injected dose per gram of tissue (Fig. 5) (*see* **Note 22**).

4 Notes

1. Wrap the metal spatula in parafilm so the DTPA does not complex to the metal ions from the spatula.
2. You should get a nice visible white pellet at the bottom of the 15 mL conical tube. A regular centrifuge that reaches a speed of 12,000 × *g* may as well be used.

3. You should first dissolve the DTPA in DMSO before diluting in the DI-H_2O with NaOH. Make sure the DMSO is at least diluted out by a factor of 1:100.
4. Quantification of 25 kDa PEI can be achieved via a TNBS (2,4,6-trinitrobenzene sulfonic acid) assay [22].
5. The siRNA used here had a MW of 17,950.36 g/mol. Therefore, we used 0.285 μmol of siRNA.
6. Total p-SCN-Bn-DTPA (MW = 649.9 g/mol) is 15.02 μmol.
7. The solution turned cloudy upon the addition of the DTPA to the siRNA solution.
8. Upon addition of the ethanol, the solution should turn slightly cloudy again.
9. You should put roughly 2 mL into each column since each column can only retain 1 mg RNA and a limited volume. If you add too much, you may lose siRNA during the purification steps.
10. To make calculations easier in the future, dilute the siRNA to either 100 μM or 50 μM. Aliquot the samples into small portions to prevent several freeze-thaw cycles.
11. Mix the samples by gently pipetting up and down within the cuvette. Be careful not to create any bubbles.
12. Make sure you rinse out the cuvette very well. When you read the fresh 3 mL of the complex solution, verify that the values are in line with the previous measurements.
13. Since each siRNA strand has only one amine group for DTPA to complex to, if performed correctly, your ratio should be approximately a 1:1 molar equivalence of DTPA and siRNA. If the ratio of DTPA per siRNA is higher than 1:1, residual-free DTPA was not removed during the purification.
14. It may be necessary to combine two or more peak fractions based on the CPM values and RNA concentrations. If free DTPA is present in the siRNA solution when it is radiolabeled, a second small peak will appear around fraction 12, and free Indium appears around fraction 20.
15. Make sure to harvest ovarian cancer cells when they are less than 80–90% confluent. This ensures that the cells are in a period of exponential growth and will increase tumor engraftment success rate.
16. By placing mice on a folic acid-deficient diet, it reduces their serum folate to a level near that of human serum and increases the folate receptor alpha status of the ovarian cancer cells.
17. Isoflurane was used during imaging to keep the mice sedated.
18. Minimum volume of both polymer and siRNA solution must be ≥200 μL.

19. When loading solutions into syringes, use a microliter pipettor that has pipette tips smaller than the diameter of the Luer tip to avoid sealing the syringe barrel. In addition, retract the syringe plunger while simultaneously ejecting the polymer or siRNA solution from the pipettor to help draw up solution.
20. The imaging procedure can be repeated at any given time. The half-life of ^{111}In is 2.6 days, and a significant amount of siRNA is typically excreted renally or even hepatically. Therefore, imaging at time points later than 48 h can become challenging.
21. Nanoparticles tend to accumulate in the liver as a deep compartment. Therefore, a two-compartment model usually fits best.
22. Make sure to collect the primary tumor first, and separate any surrounding metastatic tumors. These can then be measured separately and combined during the analysis.

Acknowledgments

Research reported in this publication was supported by the Wayne State University Nano Incubator grant to Olivia Merkel, the Ruth L. Kirschstein National Research Award T32-CA009531 fellowship to Steven Jones, and the Wayne State School of Medicine GRA support of Daniel Feldmann.

References

1. Whitehead KA, Langer R, Anderson DG (2009) Knocking down barriers: advances in siRNA delivery. Nat Rev Drug Discov 8(2):129–138. https://doi.org/10.1038/nrd2742
2. Dickerson EB, Blackburn WH, Smith MH, Kapa LB, Lyon LA, McDonald JF (2010) Chemosensitization of cancer cells by siRNA using targeted nanogel delivery. BMC Cancer 10:10. https://doi.org/10.1186/1471-2407-10-10
3. Wagner E (2012) Polymers for siRNA delivery: inspired by viruses to be targeted, dynamic, and precise. Acc Chem Res 45(7):1005–1013. https://doi.org/10.1021/ar2002232
4. Merkel OM, Zheng M, Debus H, Kissel T (2012) Pulmonary gene delivery using polymeric nonviral vectors. Bioconjug Chem 23(1):3–20. https://doi.org/10.1021/bc200296q
5. Vlahov IR, Leamon CP (2012) Engineering folate-drug conjugates to target cancer: from chemistry to clinic. Bioconjug Chem 23(7):1357–1369. https://doi.org/10.1021/bc2005522
6. Jones SK, Lizzio V, Merkel OM (2016) Folate receptor targeted delivery of siRNA and paclitaxel to ovarian cancer cells via folate conjugated triblock copolymer to overcome TLR4 driven chemotherapy resistance. Biomacromolecules 17(1):76–87. https://doi.org/10.1021/acs.biomac.5b01189
7. Endres T, Zheng M, Beck-Broichsitter M, Kissel T (2012) Lyophilised ready-to-use formulations of PEG-PCL-PEI nano-carriers for siRNA delivery. Int J Pharm 428(1–2):121–124. https://doi.org/10.1016/j.ijpharm.2012.03.003
8. Endres T, Zheng M, Kilic A, Turowska A, Beck-Broichsitter M, Renz H, Merkel OM, Kissel T (2014) Amphiphilic biodegradable PEG-PCL-PEI triblock copolymers for FRET-capable in vitro and in vivo delivery of siRNA and quantum dots. Mol Pharm 11(4):1273–1281. https://doi.org/10.1021/mp400744a

9. Trierweiler LF, Trierweiler JO (2011) Industrial production of polymeric nanoparticles: alternatives and economic analysis. In: Beck R, Guterres S, Pohlmann A (eds) Nanocosmetics and nanomedicines. Springer, New York, pp 123–138
10. Hou S, Ziebacz N, Wieczorek SA, Kalwarczyk E, Sashuk V, Kalwarczyk T, Kaminski TS, Holyst R (2011) Formation and structure of PEI/DNA complexes: quantitative analysis. Soft Matter 7(15):6967–6972
11. Feldmann DP, Xie Y, Jones SK, Yu D, Moszczynska A, Merkel OM (2017) The impact of microfluidic mixing of triblock micelleplexes on in vitro/in vivo gene silencing and intracellular trafficking. Nanotechnology 28(22):224001. https://doi.org/10.1088/1361-6528/aa6d15
12. Watts JK, Corey DR (2010) Clinical status of duplex RNA. Bioorg Med Chem Lett 20(11):3203–3207. https://doi.org/10.1016/j.bmcl.2010.03.109
13. Bertrand N, Wu J, Xu X, Kamaly N, Farokhzad OC (2014) Cancer nanotechnology: the impact of passive and active targeting in the era of modern cancer biology. Adv Drug Deliv Rev 66:2–25. https://doi.org/10.1016/j.addr.2013.11.009
14. Kalli KR, Oberg AL, Keeney GL, Christianson TJ, Low PS, Knutson KL, Hartmann LC (2008) Folate receptor alpha as a tumor target in epithelial ovarian cancer. Gynecol Oncol 108(3):619–626. https://doi.org/10.1016/j.ygyno.2007.11.020
15. Mohammadi M, Li Y, Abebe DG, Xie Y, Kandil R, Kraus T, Gomez-Lopez N, Fujiwara T, Merkel OM (2016) Folate receptor targeted three-layered micelles and hydrogels for gene delivery to activated macrophages. J Control Release 244(Pt B):269–279. https://doi.org/10.1016/j.jconrel.2016.08.020
16. Esmaeili F, Ghahremani MH, Ostad SN, Atyabi F, Seyedabadi M, Malekshahi MR, Amini M, Dinarvand R (2008) Folate-receptor-targeted delivery of docetaxel nanoparticles prepared by PLGA-PEG-folate conjugate. J Drug Target 16(5):415–423. https://doi.org/10.1080/10611860802088630
17. Reddy JA, Dorton R, Bloomfield A, Nelson M, Vetzel M, Guan J, Leamon CP (2014) Rational combination therapy of vintafolide (EC145) with commonly used chemotherapeutic drugs. Clin Cancer Res 20(8):2104–2114. https://doi.org/10.1158/1078-0432.CCR-13-2423
18. Renette T, Librizzi D, Endres T, Merkel O, Beck-Broichsitter M, Bege N, Petersen H, Curdy C, Kissel T (2012) Poly(ethylene carbonate) nanoparticles as carrier system for chemotherapy showing prolonged in vivo circulation and anti-tumor efficacy. Macromol Biosci 12(7):970–978. https://doi.org/10.1002/mabi.201100499
19. Pippin CG, Parker TA, McMurry TJ, Brechbiel MW (1992) Spectrophotometric method for the determination of a bifunctional DTPA ligand in DTPA-monoclonal antibody conjugates. Bioconjug Chem 3(4):342–345
20. Merkel OM, Librizzi D, Pfestroff A, Schurrat T, Behe M, Kissel T (2009) In vivo SPECT and real-time gamma camera imaging of biodistribution and pharmacokinetics of siRNA delivery using an optimized radiolabeling and purification procedure. Bioconjug Chem 20(1):174–182. https://doi.org/10.1021/bc800408g
21. Rose SD, Kim DH, Amarzguioui M, Heidel JD, Collingwood MA, Davis ME, Rossi JJ, Behlke MA (2005) Functional polarity is introduced by Dicer processing of short substrate RNAs. Nucleic Acids Res 33(13):4140–4156. https://doi.org/10.1093/nar/gki732
22. Snyder SL, Sobocinski PZ (1975) An improved 2,4,6-trinitrobenzenesulfonic acid method for the determination of amines. Anal Biochem 64(1):284–288. https://doi.org/10.1016/0003-2697(75)90431-5

Chapter 25

Synthesis, Physicochemical, and Biological Evaluation of Spherical Nucleic Acids for RNAi-Based Therapy in Glioblastoma

Serena Tommasini-Ghelfi, Andrew Lee, Chad A. Mirkin, and Alexander H. Stegh

Abstract

Spherical nucleic acids (SNAs), an emerging class of gene-regulatory nanotherapeutics, typically consist of a nanoparticle core densely functionalized with a shell of radially oriented small interfering RNA (siRNA) oligonucleotides, microRNA (miRNA) mimics, or antagonists. The unique three-dimensional SNA structure regardless of core type (e.g., gold or lipids) confers heightened resistance to nuclease-mediated degradation and accounts for robust cell entry in the absence of auxiliary transfection vehicles. In murine models of glioblastoma (GBM), the most aggressive and prevalent form of malignant brain cancers, systemically administered siRNA or miRNA-conjugated SNAs penetrated blood-brain and blood-tumor barriers and robustly reduced tumor progression. Here, we describe methods for the synthesis and physicochemical and biological characterization of SNA gene silencing effects in glioma cells in vitro and in patient-derived xenograft models in vivo.

Key words RNA interference, Biotherapeutic gene silencing, Spherical nucleic acids, Nanotherapeutic

1 Introduction

Glioblastoma (GBM) is the most lethal form of malignant brain tumors with a median patient survival of only 14–16 months after diagnosis. GBM progression is rapid, and the development of therapy-resistant recurrent disease is nearly universal [1]. Functional oncogenomic investigations continue to refine the molecular basis of GBM tumor progression and the evolution of therapy resistance. The majority of genes identified as contributing to the pathogenesis of GBM, however, represent nonenzymatic activities with unknown modi operandi, which are considered "undruggable" using conventional small molecule therapeutics.

RNA interference (RNAi)-based biotherapeutic gene silencing emerged as a promising strategy to target the genetic basis of cancer,

Lekha Dinesh Kumar (ed.), *RNA Interference and Cancer Therapy: Methods and Protocols*, Methods in Molecular Biology, vol. 1974, https://doi.org/10.1007/978-1-4939-9220-1_25,

by blocking the translation of mRNA and thus reducing the expression of pathological protein. Small interfering RNAs (siRNAs) are generated by cleavage of long double-stranded (ds) RNAs into ~20 nucleotide-containing siRNAs by the enzyme Dicer. Unwinding of siRNAs into two single-stranded (ss) RNAs, incorporation of the guide strand into the RNA-induced silencing complex (RISC), and binding of siRNAs to complementary mRNA trigger the degradation of endogenous mRNA by Argonaute, the catalytic component of the RISC complex [2]. MicroRNAs (miRNAs) are transcribed by RNA polymerase, processed within the cell nucleus into pri-miRNA and subsequently into pre-miRNA, which upon export into the cytoplasm are cleaved by Dicer and converted into a miRNA duplex. Only one strand is typically incorporated into RISC and interacts with the target mRNA. miRNA-mediated gene silencing occurs via mRNA degradation, deadenylation or translational repression [3].

Due to the negative charge of the RNA backbone, siRNA and miRNA oligonucleotides require delivery systems to overcome negatively charged membranes and to counteract electrostatic repulsion. With physico-biological properties significantly dependent on particle size, charge, and hydrophobicity [4], certain nanomaterials at the submicron scale have features that make them ideal carrier systems for RNAi delivery. For example, spherical nucleic acid (SNA) nanoparticle conjugates, which are nanostructures consisting of densely packed, highly oriented DNA, siRNA, or miRNA oligonucleotides surrounding an inorganic (e.g., gold) or organic (e.g., liposomal) nanoparticle core, are a platform for gene silencing. Figure 1 shows a schematic of the various SNA design elements [5, 6]. SNAs are actively recognized by scavenger receptors and behave as single-entity agents capable of transfection and gene regulation without the need for auxiliary carriers. They are stable in physiological environments, resist nuclease degradation, and result in efficient and persistent gene knockdown in cells and cancerous tissues without triggering a significant immune response [7] and off-target effects [8]. Here we describe siRNA- and miRNA-based SNA conjugates, which were designed to carry tumor-suppressive miR-182 sequences that target the caspase and p53 inhibitor Bcl2-like 12 (Bcl2L12). This knocks down the expression of DNA repair enzyme O^6-methylguanine DNA methyltransferase (MGMT), which removes DNA alkylation adducts and antagonizes antitumor effect of alkylating agents [9–11]. siRNA- or miRNA-conjugated SNAs robustly penetrated GBM cells and reduced the expression of cancer-causing genes via RNAi. In vivo, systemically administered SNAs cross the blood-brain and blood-tumor barriers in mice with intracranial GBM xenograft that pervasively and selectively infiltrate tumor tissue. Selective accumulation in tumor but not normal brain elements is likely due to an "enhanced permeability and retention" (EPR) effect, i.e., the increased accumulation of nanomaterials in tumors due to abnormal architecture and function of tumor blood vessels. The first versions

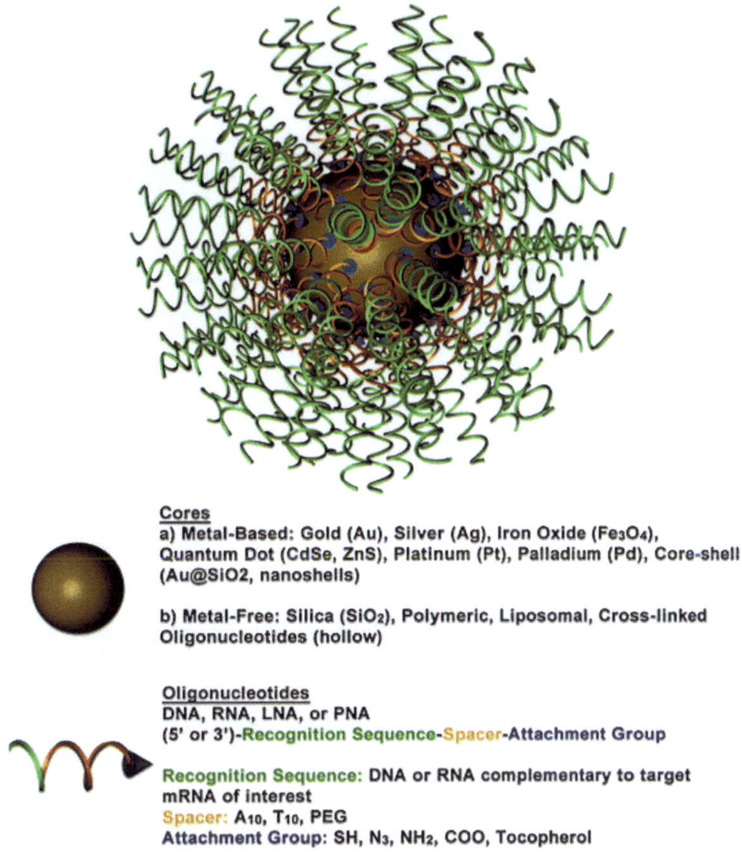

Fig. 1 A 3D drawing of a spherical nucleic acid (SNA). SNAs consist of densely functionalized and highly oriented nucleic acids on the surface of a nanoparticle. Cores can be metal based (a) or metal free (b). Nucleic acids are formed of a recognition sequence, a spacer, and an attachment group. Adapted with permission from Cutler et al. Copyright 2012. American Chemical Society [5]

of SNAs designed to target the Bcl2L12 GBM oncogene have entered clinical trials for treating GBM in humans (NCT03020017).

To determine the gene-regulatory activity of SNAs in derivative xenogeneic tumors in real time, we developed tumor models that stably co-express optical reporters for a near-infrared fluorescent protein (iRFP670) fused to the gene of interest and luciferase, the latter used for noninvasive bioluminescence imaging as readout for tumor volume (Fig. 2). Upon indication of tumor progression, established by fluorescence and bioluminescence monitoring, mice were administered gene-specific siRNA-SNAs by systemic intravenous administration. Tumor luminescence and fluorescence values were obtained subsequent to treatment and compared against pre-treatment readings. Relative protein knockdown was assessed by comparing the fluorescence to luminescence ratios of siMGMT-SNA-treated to siCo-SNA-treated mice (Fig. 3).

Here, we describe experimental procedures to generate and characterize gold-based SNAs functionalized with tumor suppres-

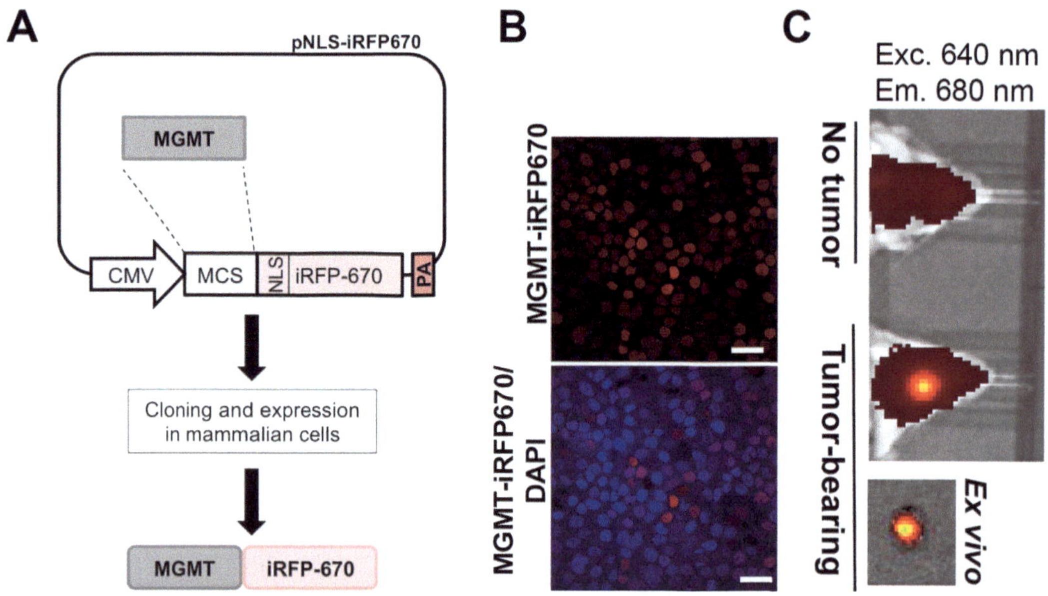

Fig. 2 Generation of MGMT-iRFP670-expressing cells and derivative orthotopic grafts. (**a**) Schematic of the expression plasmid encoding the MGMT-iRFP670 fusion product. (**b**) Confocal microscopy images of $SF767^{MGMT\text{-}iRFP670}$ glioma cells. Scale bar, 30 μm. (**c**) IVIS imaging of SCID mice bearing SF767 and $SF767^{MGMT\text{-}iRFP670}$ cell orthotopic GBM xenografts. Adapted with permission from Sita et al. Copyright 2017. PNAS [10]

sive siRNA or miRNA and co-conjugated with fluorochromes and contrast-enhancing agents, i.e., gadolinium (III), which allow for assessment of SNA cell and tissue distribution via fluorescence and magnetic resonance imaging. We provide detailed methodology for the quantitative assessment of SNA cell and intratumoral accumulation and for SNA knockdown efficacy in advanced optical dual reporter xenograft models.

2 Materials

2.1 Synthesis and Physicochemical Characterization of siRNA- and miRNA-Functionalized SNAs

2.1.1 Synthesis of 13-nm, Citrate-Capped Gold Nanoparticles

1. $HAuCl_4 \cdot 3H_2O$.
2. Trisodium citrate dihydrate.
3. Hydrochloric acid and nitric acid (aqua regia is a 3:1 mixture of HCl and HNO_3).
4. 0.2 μm filter.
5. Diethylpyrocarbonate (DEPC).
6. Iodine solution (0.16 M I_2, 1.0 M KI).
7. Sodium borohydride, in a freshly prepared aqueous solution (2.0 M in 0.05 M NaOH).
8. Dithiothreitol (DTT; 0.3 M aqueous solution).
9. Sodium dihydrogen phosphate (NaH_2PO_4, 1.0 M aqueous solution).

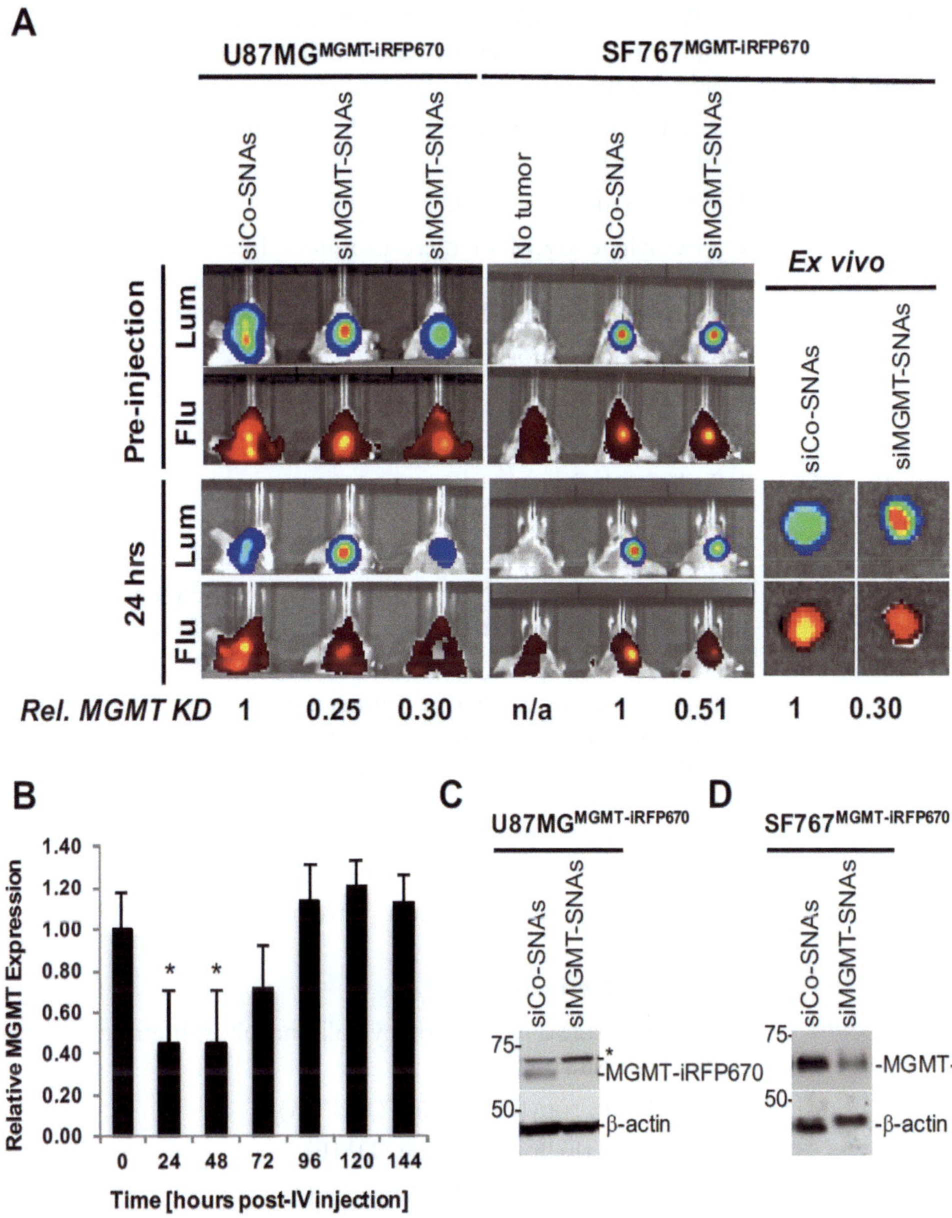

Fig. 3 siMGMT-SNAs silence MGMT-iRFP670 expression in intracranial xenografts. (**a**) Fluorescence (Flu) and luminescence (Lum) images of mice bearing U87MG$^{MGMT-iRFP670}$ or SF767$^{MGMT-iRFP670}$ xenografts before and 24 h after injection with siCo-SNAs or siMGMT-SNAs (1.4 mg/kg). (**b**) Longitudinal assessment of relative MGMT expression in intracranial tumors in response to siMGMT-SNA treatment. Asterisk (*) indicates $p < 0.02$ for difference between relative MGMT expression at 0 h time point compared to 24 and 48 h time points ($n = 8$–28 mice per time point). (**c** and **d**) Western blots of MGMT-iRFP670 in U87MG$^{MGMT-iRFP670}$ and SF767$^{MGMT-iRFP670}$ tumor lysates 48 h after injection with siCo-SNAs or siMGMT-SNAs. Asterisk, unspecific band. Adapted with permission from Sita et al. Copyright 2017. PNAS [10]

2.1.2 Generation of Gene-Regulatory SNAs by Adsorption of siRNA or miRNA to Gold Nanoparticles

1. siRNA or miRNA oligonucleotides, in which one of the oligonucleotides is chemically functionalized with a 3′ spacer disulfide group.
2. Duplex buffer (30 mM HEPES, 100 mM potassium acetate, pH = 7.5).
3. NaCl solution (5.0 M).
4. Triethylene glycol 11-mercaptoundecyl ether (0.005 M solution in ethanol); alternatives are hexaethylene glycol 11-mercaptoundecyl ether and thiol-functionalized PEG.
5. Wash buffer (phosphate-buffered saline).
6. Urea (8.0 M).
7. Quanti-iT Oligreen reagent.
8. Iodine solution.

2.1.3 Cy5- and Cy5.5-Functionalized SNAs

1. Same materials as those described in Subheading 2.1.2 but with siRNA or miRNA for which the sense sequence oligonucleotide is synthesized with a 5′-Cy5 or Cy5.5.

2.1.4 Gd(III)-Functionalized SNAs

1. Same materials as that in Subheading 2.1.2 but with single-stranded DNA oligonucleotides that are chemically modified with a 3′-spacer disulfide group and five internal dT-alkyne nucleotides: 5′-TT(T*)TT(T*)TT(T*)TT(T*)TT(T*)T-3′-propyl-disulfide, where the (T*) nucleotide is a derivative of dT with alkyne functional groups.
2. Tri-*tert*-butyl 1,4,7,10-tetraazacyclododecane-1,4,7,10-tetraacetate.
3. 3-Azido-1-propanamine.
4. Benzotriazole-1-yl-oxy-tris-pyrrolidino-phosphonium hexafluorophosphate (PyBOP).
5. *N*,*N*-Diisopropylethylamine (DIPEA).
6. *N*,*N*-Dimethylformammide (DMF).
7. Aqueous saturated $NaHCO_3$.
8. Ethyl acetate.
9. $MgSO_4$.
10. Trifluoroacetic acid.
11. Triethylsilane.
12. $GdCl_3{\cdot}6H_2O$.
13. NaOH.
14. Acetonitrile.
15. Tris(3-hydroxypropyltriazolylmethyl)amine.
16. Sodium ascorbate.
17. Copper(II) sulfate.

2.2 In Vitro Testing of SNAs in Glioma Cells

2.2.1 Assessment of Cellular Uptake by ICP-MS

1. Cell lines: transformed GBM cells, e.g., U87MG (obtained from the American Type Tissue Collection, ATCC).
2. DMEM 1× with 4.5 g/L glucose, L-glutamine, and sodium pyruvate (Corning) and supplemented with 10% fetal bovine serum (Life Technologies) and 0.5% Pen-Strep (Life Technologies).
3. siRNA- or miRNA-conjugated SNAs.
4. Oven.
5. HCl.
6. Nitric acid.
7. Indium.
8. Thermo Fisher iCAPTM Q ICP-MS.

2.2.2 Validation of RNAi Triggered by SNA Treatment by 5′ RNA Ligase-Mediated Rapid Amplification of cDNA Ends Polymerase Chain Reaction (5′ RLM-RACE-PCR)

1. Cell lines: transformed GBM cells, e.g., U87MG (ATCC).
2. DMEM 1× with 4.5 g/L glucose, L-glutamine, and sodium pyruvate (Corning) and supplemented with 10% fetal bovine serum (Life Technologies) and 0.5% Pen-Strep (Life Technologies).
3. siRNA- or miRNA-conjugated SNAs.
4. RNeasy kit (Qiagen).
5. Phenol/chloroform.
6. Ethanol.
7. SuperScript III reverse transcriptase kit.
8. GeneRacer™ kit (Life Technologies).
9. 3′ reverse primers specific for the gene of interest.
10. Gene-specific nested primer for the gene of interest.
11. Gel extraction kit (Promega).
12. Agarose.

2.3 Evaluation of SNA Effect In Vivo: Real-Time Assessment of Intratumoral Gene Knockdown Using a Dual Luciferase/Near-Infrared (NIR) Fluorescence Reporter Xenograft Model

2.3.1 Lentiviral Infection of Luciferase-Expressing Glioma Cells with an iRFP670-Gene-of-Interest cDNA Fusion Construct

1. Cell lines: Luciferase-expressing transformed GBM cells or patient-derived glioma-initiating cells (GICs), HEK293T.
2. DMEM 1× with 4.5 g/L glucose, L-glutamine, and sodium pyruvate and supplemented with 10% fetal bovine serum and 0.5% Pen-Strep.
3. DMEM/F12 50:50 supplemented with L-glutamine for growth of GICs.
4. Supplements for GIC medium: 0.5% Pen-Strep, B27, N2, human epidermal growth factor (hEGF), human fibroblast growth factor (hFGF), leukemia inhibitory factor h(LIF), and GlutaMAX. Plasmids: pNLS-iRFP670 plasmid or piRFP670 plasmid.
5. Plasmids for lentiviral expression construct: CSII-CMV-MCS-IRES2-Venus, pMD2.G, and psPAX2.

6. Restriction enzymes of choice and relevant buffers.
7. Calf intestinal alkaline phosphatase (CIP).
8. cDNA-specific primers containing restriction enzyme recognition sites.
9. Agarose.
11. Gel extraction kit.
10. T4 ligation and buffers.
11. E. cloni® 5-alpha chemically competent cells.
12. SOC medium or recovery medium supplied with the competent cells.
13. Agar plates with the appropriate antibiotic.
14. LB broth.
15. Miniprep kit.
16. Maxiprep kit.
17. Lipofectamine 2000.
18. Antibody specific for the detection of the protein of interest via Western blot.

2.3.2 Orthotopic Implantation of iRFP670 Fusion Protein Expressing Glioma Cells

1. Female CB17 SCID mice (6–8 weeks old).
2. Isoflurane.
3. Shaving cream.
4. Anesthetic chamber.
5. Meloxicam.
6. Stereotactic frame.
7. Betadine.
8. Microsurgical drill.
9. Hamilton syringe.
10. Sutures.
11. Luciferin.
12. Insulin syringes.

2.4 Assessment of Intratumoral SNA Accumulation

2.4.1 Semiquantitative Assessment of Intratumoral Target Protein Knockdown

1. Insulin syringes.
2. Heat pad.
3. SNAs.
4. Luciferin.
5. Anesthetic chamber and isoflurane.

2.4.2 Magnetic Resonance Imaging of Tumor Brain Resected from Mice Treated with Gd(III)-SNAs

1. 4% Paraformaldehyde.
2. Fomblin.
3. 25 mm quadrature birdcage RF coil.
4. 9.4T Bruker BioSpec MRI system (Bruker BioSpin, Billerica, MA).

2.4.3 Assessment of Intratumoral Accumulation by Silver Staining

Silver staining amplifies the size of nanoparticles via the deposition of silver on gold and allows for visualization of gold presence by optical microscopy.

1. 10% formalin.
2. Ethanol.
3. Silver Enhancement Kit for Light and Electron Microscopy (Ted Pella, Inc.).
4. Mayer's Hematoxylin.
5. VectaMount AQ.

2.4.4 Quantification of Intratumoral Gold Content by ICP-MS

1. Oven.
2. HCl.
3. Nitric acid.
4. Matrix solution (2% HCl and 2% HNO_3).
5. Indium.
6. Thermo Fisher iCAPTM Q ICP-MS.

3 Methods

3.1 Synthesis and Physicochemical Characterization of siRNA- and miRNA-Functionalized SNAs

3.1.1 Synthesis of 13-nm, Citrate-Capped Gold Nanoparticles

1. Rinse a three-necked 1-L round bottom flask, glass stoppers, stir bar, and reflux condenser with aqua regia and water, and dry. Load the flask onto a heating mantle on top of a stir plate.
2. Add 0.193 g of $HAuCl_4 \cdot 3H_2O$ to the flask, and add 500 mL of water to dissolve; the solution will be yellow. Stopper two of the three necks of the flask, and into the third (central) position, insert the reflux condenser, and connect to a water supply. Stir the solution and heat to a full reflux.
3. Into a 50-mL plastic tube, add 0.57 g of trisodium citrate and 10 mL of water to dissolve.
4. When the solution of $HAuCl_4$ reaches a boil, remove one of the glass stoppers, and pour the trisodium citrate solution into the three-necked flask. The reaction between citrate and $HAuCl_4$ will cause an evolution in the color of the solution from yellow to black to wine read.

5. Turn off the heat, and let the solution cool to room temperature. Filter the solution through a 0.2 μm membrane, and store in a glass bottle.
6. This procedure will generate ~500 mL of 13-nm gold nanoparticles (citrate capped) at an approximate concentration of 13 nM (in gold nanoparticles). The concentration of gold nanoparticles can be estimated by analyzing a tenfold diluted sample of the gold nanoparticles by absorbance spectroscopy. The localized surface plasmon resonance should lead to a maximum absorbance at 517 nm, and the concentration of gold nanoparticles can be estimated from the absorbance at 517 nm, using an extinction coefficient of 2.7×10^8 M^{-1} cm^{-1}.
7. To ensure sterility and to deactivate any RNAses, treat the solution of gold nanoparticles by adding 0.5 mL of DEPC, stirring overnight, and then autoclaving the solution.

3.1.2 Generation of Gene-Regulatory SNAs by Adsorption of siRNA or miRNA to Gold Nanoparticles

1. Obtain chemically functionalized siRNA oligonucleotides designed to target a specific oncogene and a non-targeting control sequence. The oligonucleotides with the "sense" sequence must be chemically modified with a 3′-propyl-disulfide group (accessible by solid-phase synthesis with a solid support functionalized with disulfide groups). The oligonucleotide with the antisense sequence can be prepared without chemical modification to add disulfide or thiol functionality.
2. Generate 15 nmol of duplex siRNA (where the sense sequence oligonucleotide is chemically functionalized at the 3′-end with a propyl disulfide group), by adding 15 nmol each of the sense sequence oligonucleotide and antisense sequence oligonucleotide (*see* **Note 1**). Dissolve into 100 μL of duplex buffer. Heat to 90 °C for 10 min, and gradually cool to room temperature and then to 4 °C.
3. Add 15 nmol of duplex siRNA to 10 mL of citrate-capped gold nanoparticles. Sonicate and add NaCl to a concentration of 0.1 M, by adding 0.3 mL of a 5 M NaCl stock. Allow the initial adsorption of disulfide-modified siRNA to take place for >4 h.
4. Add 0.3 mL of NaCl stock to the reaction mixture, to increase the NaCl concentration to 0.2 M, and wait for 4 h; then add another 0.3 mL of NaCl stock, to increase the NaCl concentration to 0.3 M, and wait for an overnight period (*see* **Note 2**).
5. Passivate the surfaces of the gold nanoparticles by adding triethylene glycol 11-mercaptoundecyl ether (hexaethylene glycol 11-mercaptoundecyl ether and thiol-functionalized PEG are alternatives) to the reaction mixture, to a concentration of 30 μM (e.g., 90 μL of a 0.005 M stock should be added to a 15 mL reaction volume). Allow the alkanethiols to adsorb for 1 h.

6. To purify the siRNA and alkanethiol-functionalized gold nanoparticles from excess siRNA and alkanethiol, pipette the reaction mixture into centrifuge tubes (those that are rated for centrifugation at 21,000 × *g*). Centrifuge for 20 min to sediment the gold nanoparticles to the bottom of the tubes. Remove the supernatant; add phosphate-buffered saline (PBS) to resuspend and wash the SNAs. Repeat the centrifugation and wash steps two more times. Finally, resuspend the SNAs to a total volume of ~1.5 mL of PBS, and store at 4 °C. This procedure should generate a stock of SNAs in PBS at a gold nanoparticle concentration that is ≥100 nM; dilute an aliquot of the stock by 100-fold, and analyze by absorbance spectroscopy to experimentally estimate the gold nanoparticle concentration (*see* **Note 3**).
7. For the quantification of the average number of antisense oligonucleotides per gold nanoparticle, take an aliquot of gold nanoparticles (2×10^{-12} mol), and dilute by >10-fold into a solution of 8.0 M urea (e.g., the dilution of 20 μL of a 100 nM SNA stock into 1.0 mL of 8.0 M urea). Heat to 45 °C for 20 min; these steps cause the dissociation of the duplex siRNA and the release of the antisense oligonucleotides into solution (*see* **Note 4**). Centrifuge the mixture at 21,000 × *g* for 20 min; remove and analyze the supernatant solution. Add an aliquot of the supernatant solution to the Quant-iT Oligreen reagent, to a total volume of 150 μL (*see* **Note 5**).
8. Prepare calibration standards by diluting aliquots taken from a stock solution of antisense oligonucleotides into 8.0 M urea and treating with the Quant-iT Oligreen reagent; volumes of 150 μL with oligonucleotide concentration of ~0.2–2.0×10^{-8} M typically generate a linear dependence of fluorescence with concentration.
9. Analyze the fluorescence (λ_{ex} = 485, λ_{em} = 515 nm) of the calibration standards, and estimate the concentration of antisense oligonucleotides released into solution by the treatment of SNAs with urea.
10. For the quantification of the average number of sense oligonucleotides per gold nanoparticle, wash the pelleted gold nanoparticles from **step** 7 with urea solution (8.0 M), and resuspend in 200 μL of water. Remove 20 μL for analysis (2×10^{-13} mol of gold nanoparticles), and place in a reaction tube. Add to the gold nanoparticle sample 10 μL of iodine solution, and allow for 5 min for the reaction to complete. Add 10 μL of NaH_2PO_4 and mix. Prepare a 1:5 mixture of 2 M $NaBH_4$:0.3 M DTT (e.g., 0.2 and 0.8 mL). To reduce the iodine and Au ions, add 40 μL of $NaBH_4$:DTT mixture. Mix and allow the reaction 5 min to complete; the solution will bubble and turn from brown-yellow to colorless. Sonicate the mixture and centrifuge at 21,000 × *g* for 10 min. Separate the supernatant solution

(~80 μL) from the solids, and analyze by diluting fourfold into a 96 well plate with the Quant-iT Oligreen reagent (e.g., 50 μL of mixture to a total volume containing the Quant-iT Oligreen reagent). The fluorescence signal (λ_{ex} = 485, λ_{em} = 515 nm) from these samples must be analyzed with a standard curve generated from reaction mixtures containing citrate-capped gold nanoparticles (~2 × 10^{-13} mol) and sense sequence oligonucleotides (0.4–4.0 × 10^{-11} mol).

3.1.3 Cy5- and Cy5.5-Functionalized SNAs

1. SNAs with Cy5 or Cy5.5 fluorochromes attached to the siRNA can be prepared by following the same steps to generate SNAs described in Subheading 3.1.2 but with the use of Cy5- or Cy5.5-conjugated oligonucleotides.
2. Obtain or synthesize Cy5- or Cy5.5-conjugated sense sequence oligonucleotides. These can be prepared by manual coupling of Cy5 or Cy5.5 phosphoramidite reagents to controlled pore glass (CPG) that has been used in the solid-phase synthesis of the sense strand oligonucleotide using UltraMild monomers (Glen Research is a provider), prior to cleavage and deprotection. Following coupling, the Cy5- or Cy5.5-conjugated molecules can be cleaved from the solid support by treatment with 30% ammonium hydroxide at room temperature for 2 h, and 2′-silyl-deprotected with trimethylamine trihydrofluoride, and purified by reverse phase HPLC.
3. Generate duplex siRNA consisting of the Cy5- or Cy5.5-conjugated sense oligonucleotide and antisense oligonucleotide, using the procedure described in **step 2** of Subheading 3.1.2.

3.1.4 Gd(III)-Functionalized SNAs

1. SNAs with Gd(III) contrast-enhancing agent can be prepared by following the same steps to generate SNAs described in Subheading 2.1.2 but with the use of single-stranded, chemically modified DNA, in place of chemically modified siRNA or miRNA (*see* **Note 6**).
2. Obtain single-stranded DNA oligonucleotides with two types of chemical modifications: (1) 3′-propyl disulfide groups (for subsequent attachment to gold nanoparticle surfaces) and (2) chemically modified nucleotides with alkyne groups (e.g., through the use of C8-alkyne-dT-CE phosphoramidites), for subsequent conjugation to N_3-functionalized Gd(III) chelates through click reactions.
3. The synthesis of the N_3-functionalized Gd(III) chelate through a two reaction sequence is described in detail in Mastarone et al. [12]. The first is the reaction of tri-*tert*-butyl 1,4,7,10-tetraazacyclododecane-1,4,7,10-tetraacetate (0.6 mmol) in 5 mL of DMF with 3-azido-1-propanamine (1.0 mmol), PyBOP (0.7 mmol), and DIPEA (0.9 mmol). Stir the mixture for 4 h, and then add saturated aqueous $NaHCO_3$ (20 mL). Add 50 mL

of ethyl acetate and the mixture to a separatory funnel, and shake to extract. Collect the ethyl acetate portion and reserve. Repeat the addition of 50 mL of ethyl acetate and shaking and collection of the ethyl acetate portion two more times. Combine the three portions of ethyl acetate and add $MgSO_4$; remove the $MgSO_4$ by filtration, and collect the ethyl acetate solution; remove the ethyl acetate by rotary evaporation. This reaction and purification generates the product in crude form, 1-(3-((3-azidopropyl)amino)-2-oxoethyl)-4,7,10-tris (carboxymethyl)-1,4,7,10-tetraazacyclododecane.

4. The second reaction is the deprotection and metalation of the product from **step 3**. Deprotect the product from **step 3** by adding trifluoroacetic acid (6 mL), triethysilane (0.1 mL) and water (0.25 mL), and stir overnight. Evaporate the mixture, and dissolve in water containing $GdCl_3{\cdot}6H_2O$ (0.744 mmol), and adjust the pH to 6.5 with NaOH; stir and heat to 50 °C. Monitor the pH; the reaction is complete when the pH stops falling. Purify the product by reverse phase HPLC using a C18 column and gradient elution with water:acetonitrile, using a ramp from 100:0 to 57.5:42.5 over 25 min, followed by a ramp to 0:100 over 5 min. This procedure generates the N_3-functionalized Gd(III) chelate in pure form.
5. To conjugate N_3-functionalized Gd(III) to 5′-TT(T*)TT(T*)TT(T*)TT(T*)TT(T*)T-3′-propyl-disulfide, prepare a reaction mixture that is 1.0 mL of an aqueous solution of NaCl (0.2 M), tris(3-hydroxypropyltriazolylmethyl)amine (2.0 mM), sodium ascorbate (2.0 mM), and copper(II) sulfate (0.4 mM), containing 10 mg of N_3-functionalized Gd(III) chelate. Add to this solution the T*-containing oligonucleotide (1.0 mmol), and allow for reaction at room temperature for 2 h. Purify the click reaction-modified oligonucleotide by size-exclusion chromatography (e.g., through the use of a NAP-10 column), followed by a final purification through RP-HPLC with a semi-prep reverse phase (C18) column.
6. Follow **steps 3–5** of Subheading 3.1.2, with the use of Gd(III)-conjugated single-stranded DNA oligonucleotides in place of siRNA.

3.2 In Vitro Testing of SNAs in Glioma Cells

3.2.1 Assessment of Cellular Uptake by ICP-MS

1. Plate U87MG in a 96 well plate, and allow adhering for 24 h.
2. Treat cells with 5 nM SNAs (based on gold content) or the desired RNA concentration for 24–48 h.
3. Resuspend cells in 1 mL concentrated $HCl{:}HNO_3$ (3:1) overnight at 65 °C.
4. Dilute samples and standards with known gold concentrations in 5 mL of matrix solution (2% HCl and 2% HNO_3) and 5 μL of 1 ppb indium as an internal standard.
5. Analyze on a Thermo Fisher iCAPTM Q ICP-MS.

3.2.2 Validation of RNAi Triggered by SNA Treatment by 5′ RNA Ligase-Mediated Rapid Amplification of cDNA Ends Polymerase Chain Reaction (5′ RLM-RACE-PCR)

1. Use the GeneRacer™ kit (Life Technologies) with some modifications as detailed below.
2. Figure 4 illustrates the overall concept and steps of a 5′RLM-RACE-PCR, which are described below.
3. Treat cells with 5 nM SNAs (based on gold content) or the desired RNA concentration for 48 h, and harvest the total RNA with the RNeasy kit (Qiagen).
4. Extract the RNA with phenol/chloroform, precipitate with ethanol, and use SuperScriptIII to reverse transcribe it using a 3′ primer specific for the gene of interest.
5. For the 5′RACE reaction, use 1.5–6.5 μg of total harvested RNA, and ligate it to 250 ng GeneRacer™ Oligo for 1 h at 37 °C.
6. The reverse transcription is for 50 min at 55 °C followed by heat inactivation at 70 °C for 15 min.
7. Take 2 μL from the previous reaction, and use it for first-round PCR with Gene Racer™ 5′ Primer and a 3′ reverse primer specific for the gene of interest. The cycling conditions recommended by the manufacturer are 94 °C for 2 min (1 cycle), 94 °C for 30 s, 72 °C for 1 min (5 cycles), 94 °C for 30 s, 60 °C for 30 s, 72 °C for 1 min (25 cycles), and 72 °C for 10 min (1 cycle).
8. Perform a second round of nested PCR with the GeneRacer 5′ Nested Primer and a gene-specific nested primer using the following cycle program: 95 °C for 2 min (1 cycle), 95 °C for 30 s, 60 °C for 30 s, 72 °C for 10 min (40 cycles), and 72 °C for 10 min (1 cycle) (*see* **Note 7**).
9. Separate the PCR products on a 3% agarose gel, cut out the band, purify DNA using the gel extraction kit, and sequence DNA to determine nucleotide identity.

3.3 Evaluation of SNA Effect In Vivo: Real-Time Assessment of Intratumoral Gene Knockdown Using a Dual Luciferase/Near-Infrared (NIR) Fluorescence Reporter Xenograft Model

3.3.1 Lentiviral Infection of Luciferase-Expressing Glioma Cells with an iRFP670-Gene-of-Interest cDNA Fusion Construct

1. Amplify the cDNA of the gene of interest using primers containing restriction enzyme recognition sites.
2. Run the product on a gel to check that the bands are the expected size, cut the bands, and extract them from the gel using the gel extraction kit.
3. Digest the PCR products and the plasmid with the corresponding restriction enzymes, overnight at 37 °C. Add 1 μL of CIP to the plasmid and incubate for 1 h at 37 °C.
4. Run the digested products and plasmid on a gel, excise the bands, and purify them using the gel extraction kit.
5. Set up a ligation using an optimized ratio of insert and vector, and incubate at 16°C for 3 h for sticky ends, and overnight for blunt ends. Set up an empty vector ligation (no insert) to check for self-ligation.

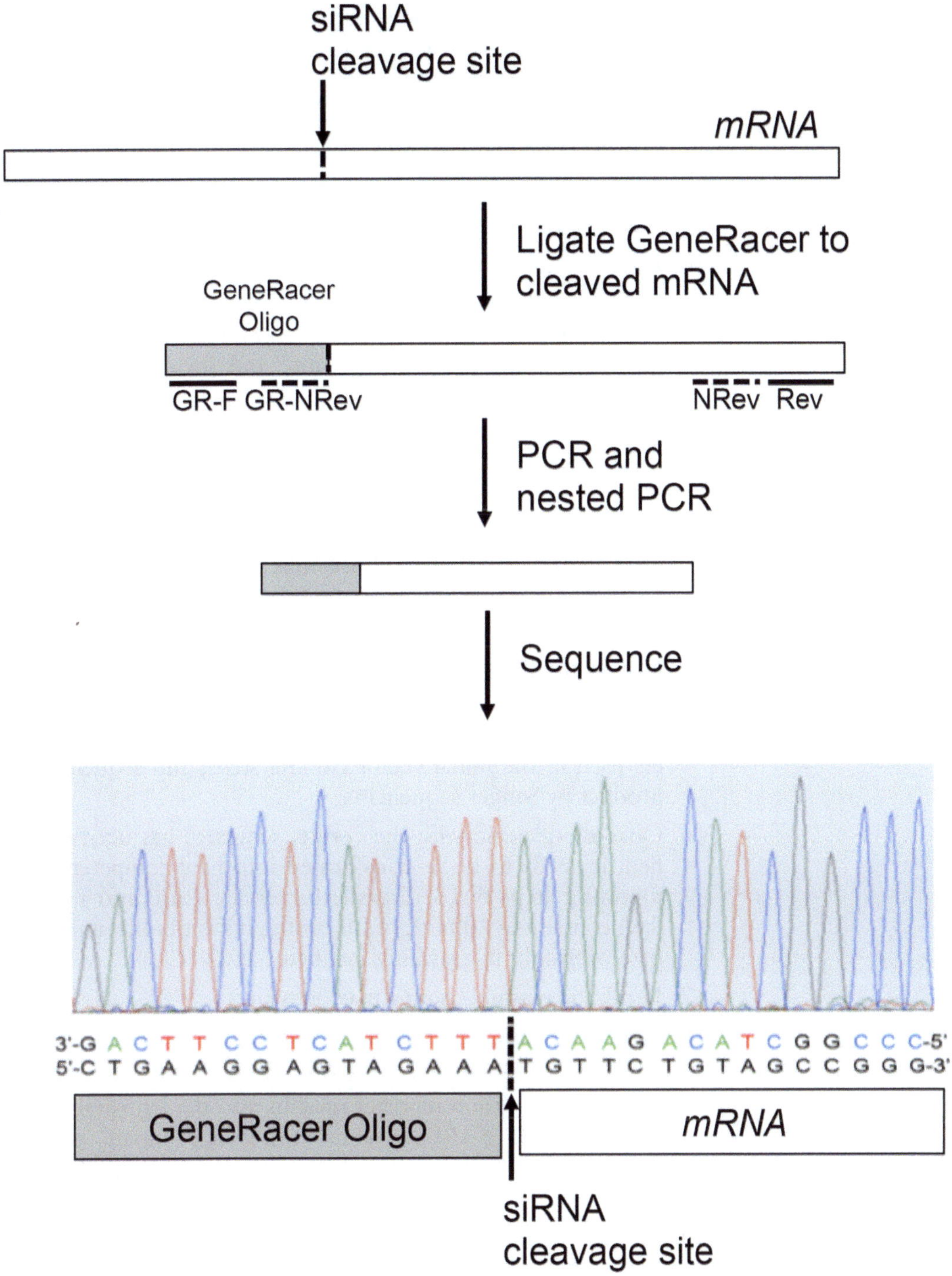

Fig. 4 Schematic of a 5′RNA ligase-mediated rapid amplification of cDNA ends (5′RLM-RACE)-PCR. GR-F, GeneRacer forward primer; GR-NF, GeneRacer nested forward primer; Rev, gene-specific reverse primer; NRev, gene-specific nested reverse primer

6. For vector transformation, add 4 μL of ligation reaction mixture to 50 μL of E. cloni® 5-alpha chemically competent cells. Stir briefly with pipette tip.
7. Incubate on ice for 30 min; heat shock cells by placing them in a 42 °C water bath for 30 s.
8. Incubate the cells on ice for 2 min.
9. Add 950 μL of SOC medium to the cells.
10. Place the tubes in a shaking incubator at 250 rpm for 1 h at 37 °C; meanwhile pre-warm the agar plates at 37 °C.
11. Plate 100 μL of transformed cells on nutrient agar plates containing the appropriate antibiotic, and incubate the plates overnight at 37 °C.
12. Next day check for the presence of colonies. If colonies are present, set up 5–10 culture tubes containing LB broth and the appropriate antibiotic for selection.
13. Use a pipette tip to gently touch one colony at the time, and eject the tip inside the culture tube (if preparing five tubes, pick five colonies). Incubate the tubes in a shaking incubator overnight.
14. Extract the plasmids from each individual tube using a miniprep kit.
15. Check the presence of the insert by PCR (using the same primers used in the initial vector cloning step), and sequence the product by Sanger sequencing.
16. Once the plasmid with the correct sequence has been identified, it needs to be re-transformed into new competent cells following **steps 6–13**. Transfer bacterial culture into a bigger flask containing 200 mL of LB broth plus antibiotic, and incubate overnight in a shaking incubator at 37 °C.
17. Purify the plasmid using a maxiprep kit.
18. Figure 2a shows the schematic of the pNLS-iRFP670 vector encoding for an MGMT-iRFP670 fusion protein.
19. Subclone the fusion reporter protein into the lentiviral vector CSII-CMV-MCS-IRES2-Venus, using appropriate primers and restriction enzymes and following procedure outlined in **steps 1–17**.
20. Produce the lentivirus by co-transfecting HEK293T cells with the CMV-MCS-IRES2-Venus plasmid containing the iRFP670-fusion cDNA together with additional plasmids encoding for other components of the virus (i.e., pMD2.G and psPAX2) using Lipofectamine 2000.
21. Venus-positive cells will be enriched using FACS-based cell sorting.

22. Figure 2b shows results from a confocal immunofluorescence microcopy analysis of glioma cells stably expressing an MGMT-iRFP670 fusion protein.
23. Verify the presence of the fusion protein by western blot using antibodies specific for iRFP670 or the protein of interest.

3.3.2 Orthotopic Implantation of iRFP670 Fusion Protein Expressing Glioma Cells

1. Trypsinize and resuspend the cells (with or without the expression of luciferase and/or iRFP670) in 3 μL of Hank's Balanced Salt Solution per mouse (*see* **Note 8**).
2. Cover the area between the ears of the mice with shaving cream to remove the hair. The cream can be left for 1 min and be removed with a wet cloth (*see* **Note 9**).
3. Place the mice into the anesthetic chamber containing isoflurane.
4. Once the animals are anesthetized, administer 50 μL of meloxicam, and place them in the stereotactic frame making sure they stay under anesthesia.
5. Clean the surgical area three times by rubbing with a cotton bud dipped in betadine and wiping with an ethanol prep.
6. Make an incision making sure the membrane between the skin and the skull is also open.
7. Make a 0.7 mm burr hole in the skull with a microsurgical drill, 2 mm lateral right of the sagittal suture and 0.5 mm posterior of bregma.
8. Administer the cells through a Hamilton syringe, inserted 3 mm into the brain. Inject the cells over a period of 3 min.
9. Remove the needle, close the incision with a suture, and place the mice in a recovery cage on a heated pad.
10. Twenty-four hours after surgery, administer another dose of meloxicam.
11. Verify tumor formation by noninvasive bioluminescence imaging using in vivo imaging system (IVIS) apparatus.
12. Administer luciferin at a dose of 200 mg/kg.
13. Place the mice into the anesthetic chamber containing isoflurane.
14. After 10 min postinjection, acquire the image using the bioluminescence settings on the IVIS machine (Time set on Auto, medium binning, *see* **Note 10**).
15. To also acquire the iRFP signal, choose fluorescence, and apply the spectral unmixing settings in order to isolate the fluorophore-specific signal.
16. Figure 2c shows representative near-infrared fluorescence images of mice harboring a xenografted tumor that are engineered to express an MGMT-iRFG670 fusion protein.

17. Figure 3 shows representative bioluminescence and near-infrared fluorescence images of mice harboring tumors that express an iRFP670-MGMT fusion protein; mice were treated with control and MGMT-specific SNAs, and the fluorescence to bioluminescence ratio was assessed.

3.4 Assessment of Intratumoral SNA Accumulation

3.4.1 Semiquantitative Assessment of Intratumoral Target Protein Knockdown

1. To demonstrate intratumoral knockdown efficacy in vivo, inject SNAs conjugated with siRNA for the protein of interest, or control SNAs via the tail vein of the mice. For this step, keep the mice on a heat pad to facilitate vasodilation.
2. The administered RNA concentration chosen for prior studies was 1.4 mg of RNA per kg of mouse [9].
3. 12–48 h post-SNA administration, check for iRFP expression (and therefore fusion protein knockdown) and luciferase activity by IVIS.
4. Administer luciferin at a dose of 200 mg/kg, and place mice into the anesthetic chamber containing isoflurane.
5. Ten minutes postinjection, acquire the image using the bioluminescence settings on the IVIS machine (Time set on Auto, medium binning).
6. To acquire the iRFP signal, choose fluorescence at the appropriate wave length, and apply the spectral unmixing settings in order to isolate the fluorophore-specific signal.
7. When analyzing the data using the Living Image software, the level of knockdown is calculated using iRFP expression relative to luminescence expression (i.e., tumor volume/size).
8. When assessing the therapeutic efficacy of SNAs, administer multiple SNA dosages, and assess tumor size periodically using IVIS as described above.

3.4.2 Magnetic Resonance Imaging of Tumor Brain Resected from Mice Treated with Gd(III)-SNAs

1. Inoculate tumor cells as described via orthotopic injection into mouse brain, and administer Gd(III)-SNAs via tail vein injection as described.
2. Excise the brain and fix in 4% paraformaldehyde for 24–48 h.
3. Prior to MR imaging, rinse and immerse in Fomblin.
4. Position the sample in a 25 mm quadrature birdcage RF coil, and image on a 9.4T Bruker BioSpec MRI system with 12 cm diameter gradient insert.
5. Acquire localizer images.
6. Image the brain using a T1-weighted three-dimensional gradient echo sequence (FLASH) with repetition time (TR) of 15 ms, echo time (TE) of 6 ms, flip angle of 30°, 64 averages, matrix size 120 × 100 × 100, and field of view (FOV) 1.5 × 1.25 × 1.25 cm^3, resulting in 0.125 mm^3 isotropic resolution (total scan time: 2 h 40 min).

7. Acquire images in a coronal or sagittal orientation, and reslice them using JIM software (Java Imaging Matrix) to an axial orientation.
8. Perform 3D reconstruction of the distribution of Gd(III)-SNA by first segmenting the brain contour and the enhanced area of the 3D FLASH images using a semiautomated contouring method in JIM.
9. Use the regions of interest to create a stack of masked images for animation and rendering in Maya 2011 using Mental Ray.

3.4.3 Assessment of Intratumoral Accumulation by Silver Staining

1. Fix brain samples in 10% formalin for 72 h.
2. Store them in 70% ethanol until they are paraffin-embedded and sectioned.
3. Deparaffinize samples in xylene (10 min), and hydrate through a series of ethanol (100%, 90%, 70%, 50%; 3 min each) and water (3 min × 3 times) washes.
4. Perform silver enhancement using the Silver Enhancement Kit for Light and Electron Microscopy.
5. Counterstain the samples with Mayer's Hematoxylin, mount them with VectaMount AQ aqueous mounting medium, place the coverslip, and dry them before visualization at the microscope.

3.4.4 Quantification of Intratumoral Gold Content by ICP-MS

1. Dry the tissues in an oven overnight at 50 °C.
2. Weigh the tissues, and digest them with 1 mL of concentrated $HCl:HNO_3$ (3:1) overnight at 65 °C.
3. Dilute digested samples and standards with known gold concentrations in 5 mL of matrix solution and 5 μL of 1 ppb indium as an internal standard.
4. Analyze on a Thermo Fisher iCAPTM Q ICP-MS.

4 Notes

1. Evaporate to dryness if the volumes of these reagents exceed 10 μL.
2. The increasing concentration of NaCl increases adsorption and surface density of siRNA on the surfaces of the gold nanoparticles.
3. An alternative technique, suited for preparation of SNAs at scale, is the removal of unconjugated siRNA and NaCl by dialysis or by tangential flow filtration with the use of 100-kDa cutoff membranes.
4. The thiol-terminated sense oligonucleotides remain immobilized on the gold nanoparticles.

5. A dilution factor of 5 of the supernatant solution is a good starting point.
6. It is important to note that these SNAs, because of their use of DNA oligonucleotides, and chemically modified nucleotides at internal positions of the oligonucleotide, will not have the ability of the siRNA- or miRNA-functionalized SNAs to perform gene silencing.
7. Of note, first- and second-round PCR reactions might require optimization of cycle conditions.
8. The optimal number of cells required for tumor formation needs to be determined empirically for each cell line.
9. If the hair removal is done 24 h before the surgery, it gives less irritation of the area and therefore less stress for the animals.
10. A kinetic experiment should be performed in order to determine the peak of luminescence after luciferin injection. Typically, this time is between 10 and 15 min.

References

1. Stupp R, Mason WP, van den Bent MJ, Weller M, Fisher B, Taphoorn MJ, Belanger K, Brandes AA, Marosi C, Bogdahn U, Curschmann J, Janzer RC, Ludwin SK, Gorlia T, Allgeier A, Lacombe D, Cairncross JG, Eisenhauer E, Mirimanoff RO, European Organisation for Research and Treatment of Cancer Brain Tumor and Radiotherapy Groups, National Cancer Institute of Canada Clinical Trials Group (2005) Radiotherapy plus concomitant and adjuvant temozolomide for glioblastoma. N Engl J Med 352(10):987–996. https://doi.org/10.1056/NEJMoa043330
2. McManus MT, Sharp PA (2002) Gene silencing in mammals by small interfering RNAs. Nat Rev Genet 3(10):737–747. https://doi.org/10.1038/nrg908
3. Djuranovic S, Nahvi A, Green R (2012) miRNA-mediated gene silencing by translational repression followed by mRNA deadenylation and decay. Science 336(6078):237–240. https://doi.org/10.1126/science.1215691
4. Singh R, Lillard JW Jr (2009) Nanoparticle-based targeted drug delivery. Exp Mol Pathol 86(3):215–223. https://doi.org/10.1016/j.yexmp.2008.12.004
5. Cutler JI, Auyeung E, Mirkin CA (2012) Spherical nucleic acids. J Am Chem Soc 134(3):1376–1391. https://doi.org/10.1021/ja209351u
6. Banga RJ, Chernyak N, Narayan SP, Nguyen ST, Mirkin CA (2014) Liposomal spherical nucleic acids. J Am Chem Soc 136(28):9866–9869. https://doi.org/10.1021/ja504845f
7. Massich MD, Giljohann DA, Seferos DS, Ludlow LE, Horvath CM, Mirkin CA (2009) Regulating immune response using polyvalent nucleic acid-gold nanoparticle conjugates. Mol Pharm 6(6):1934–1940. https://doi.org/10.1021/mp900172m
8. Zheng D, Giljohann DA, Chen DL, Massich MD, Wang XQ, Iordanov H, Mirkin CA, Paller AS (2012) Topical delivery of siRNA-based spherical nucleic acid nanoparticle conjugates for gene regulation. Proc Natl Acad Sci U S A 109(30):11975–11980. https://doi.org/10.1073/pnas.1118425109
9. Jensen SA, Day ES, Ko CH, Hurley LA, Luciano JP, Kouri FM, Merkel TJ, Luthi AJ, Patel PC, Cutler JI, Daniel WL, Scott AW, Rotz MW, Meade TJ, Giljohann DA, Mirkin CA, Stegh AH (2013) Spherical nucleic acid nanoparticle conjugates as an RNAi-based therapy for glioblastoma. Sci Transl Med 5(209):209ra152. https://doi.org/10.1126/scitranslmed.3006839
10. Sita TL, Kouri FM, Hurley LA, Merkel TJ, Chalastanis A, May JL, Ghelfi ST, Cole LE, Cayton TC, Barnaby SN, Sprangers AJ, Savalia N, James CD, Lee A, Mirkin CA, Stegh AH (2017) Dual bioluminescence and near-infrared fluorescence monitoring to evaluate spherical nucleic acid nanoconjugate activity in vivo. Proc Natl Acad Sci U S A 114(16):4129–4134. https://doi.org/10.1073/pnas.1702736114

11. Kouri FM, Hurley LA, Daniel WL, Day ES, Hua Y, Hao L, Peng CY, Merkel TJ, Queisser MA, Ritner C, Zhang H, James CD, Sznajder JI, Chin L, Giljohann DA, Kessler JA, Peter ME, Mirkin CA, Stegh AH (2015) miR-182 integrates apoptosis, growth, and differentiation programs in glioblastoma. Genes Dev 29(7):732–745. https://doi.org/10.1101/gad.257394.114

12. Mastarone DJ, Harrison VS, Eckermann AL, Parigi G, Luchinat C, Meade TJ (2011) A modular system for the synthesis of multiplexed magnetic resonance probes. J Am Chem Soc 133(14):5329–5337. https://doi.org/10.1021/ja1099616

Chapter 26

Folate-Decorated Polyamidoamine Dendrimer Nanoparticles for Head and Neck Cancer Gene Therapy

Leyuan Xu and Hu Yang

Abstract

Gene delivery systems have been developed on the basis of dendrimers and many other types of nanoparticle carriers, but few have been developed for head and neck squamous cell carcinomas (HNSCC). Herein, we describe the design and synthesis of fluorescently labeled, folic acid-decorated polyamidoamine (PAMAM) generation 4 (G4) dendrimer conjugates for HNSCC-targeted gene delivery. This delivery system comprises a dendrimer as the carrier that is conjugated with folic acid (FA) as HNSCC targeting moiety and imaging agents fluorescein isothiocyanate (FITC) or IRDye 800CW (NIR) for in vitro trafficking or bioimaging, respectively. By complexing with plasmid or siRNA, G4-FA/plasmid (or siRNA) significantly enhances gene transfection or knockdown efficiency in HNSCC cells. In a mouse xenograft model of HNSCC, this versatile G4-FA vector shows high biocompatibility, tumor targeting, high uptake, and sustained retention, making it a suitable platform for HNSCC gene therapy.

Key words Dendrimer, Targeted delivery, Gene therapy, Folate receptor, Head and neck cancer

1 Introduction

Head and neck squamous cell carcinoma (HNSCC) is the most common malignancy of this region, and it ranks the sixth most common cancer worldwide [1, 2]. Treatment of HNSCC often requires multimodal intervention including surgery, chemotherapy, and radiation; however, their impact on improvement of 5-year mortality rate of patients with HNSCC is still minimal [3]. Chemotherapy is often given as a combination of docetaxel, cisplatin and fluorouracil to treat HNSCC. These anticancer drugs interfere with DNA replication and cell division and consequently lead to cellular demise. However, incomplete tumor response and development of acquired drug resistance commonly occur and lead to systemic toxicity and disease relapse [4].

A paradigm shift in design of anticancer therapeutics has been brought about by RNA interference (RNAi) [5]. Gene therapy has become a rapidly emerging field for cancer the therapeutics;

Lekha Dinesh Kumar (ed.), *RNA Interference and Cancer Therapy: Methods and Protocols*, Methods in Molecular Biology, vol. 1974, https://doi.org/10.1007/978-1-4939-9220-1_26,

however, not much success has been reported for HNSCC. The challenge of gene therapy is the off-target effects and insufficient delivery of genetic materials. Increased understanding of molecular events involved in HNSCC initiation and progression helps reduce off-target effects, whereas developing safe targeted non-viral vectors can improve gene delivery efficiency by lowering immunogenicity, protecting gene from degradation, reducing non-specific absorption, enhancing cellular uptake and specificity of HNSCC, and facilitating transfection [6–10].

Polyamidoamine (PAMAM) dendrimers possess a highly branched, three-dimensional architecture with very low polydispersity and high functionality [5, 11, 12]. PAMAM dendrimers with primary amines on the surface are appealing vectors for gene delivery because of their inherent functions [13, 14]. Folate receptor alpha (FRα) is typically expressed at very low levels in normal tissues. However, to meet the high folate demand of rapidly dividing cells in many cancers including HNSCC, FRα is expressed at significantly elevated levels [15, 16]. Consistently, we have shown that FRα is highly expressed in HN4 and HN6 cells (derived from human primary HNSCC) as well as HN12 cells (derived from a human HNSCC lymph node metastasis) as shown in Fig. 1 [17]. Hence, FRα can be targeted for developing active targeting therapies for HNSCC [18–20]. Given the scarcity of non-viral vectors for gene therapy of HNSCC and limited clinical outcomes, it is highly desirable to develop and evaluate vectors under the context of HNSCC.

Herein, we describe the protocols for synthesis of fluorescence-labeled folic acid-anchored PAMAM generation 4 dendrimer conjugates for HNSCC-targeted gene delivery. The dendrimer conjugates consist of three functional components: (1) folic acid (FA), to specifically target folate receptor (FRα) overexpressing HNSCC cells; (2) imaging probe of fluorescein isothiocyanate (FITC) or IRDye 800CW (NIR) for in vitro trafficking or bioimaging, respectively; and (3) PAMAM fourth generation (G4) dendrimer to complex with plasmid or siRNA to enhance gene delivery in HNSCC cells and a mouse xenograft model of HNSCC as shown in Fig. 2. This versatile G4-FA conjugate exhibited (1) facilitating the cellular uptake of plasmids in a FRα-dependent manner; (2) selectively delivering plasmids to FRα-highly expressing HN12 cells other than FRα-low expressing U87 cells; (3) competing with free FA for the same binding site, i.e., FRα, on the surface of HN12 cells; (4) enhancing gene transfection/knockdown efficiency; (5) prolonging its retention in the HN12 xenograft tumor following intratumoral injection; and (6) locally delivering siRNA against vascular endothelial growth factor A (siVEGFA) to profoundly inhibit HN12 xenograft tumor growth by significantly reducing tumor angiogenesis [17, 21]. Taken together, G4-FA shows high biocompatibility, tumor targeting, uptake, and sustained retention properties, making it a suitable platform for HNSCC gene therapy.

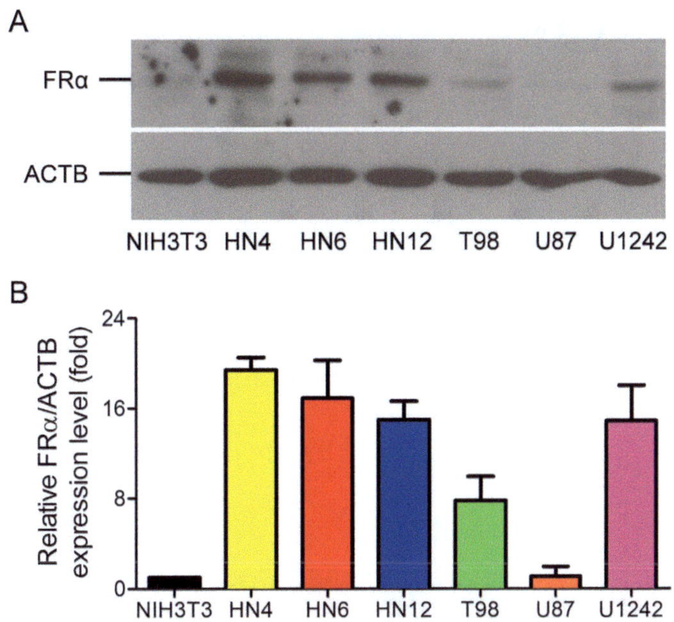

Fig. 1 Endogenous expression levels of folate receptor α (FRα) as determined by Western blot analysis conducted on four independent occasions ($n = 4$). A representative image is shown (reproduced from [17])

2 Materials

2.1 Chemicals

1. Diaminobutane (DAB) core polyamidoamine (PAMAM) dendrimer generation 4 (technical grade).
2. Dimethylformamide (DMF).
3. Dimethyl sulfoxide (DMSO).
4. Folic acid (FA).
5. Trifluoroacetic acid (TFA).
6. 1-ethyl-3-[3-dimethylaminopropyl] carbodiimide hydrochloride (EDC).
7. *N*,*N*-diisopropylethylamine (DIPEA).
8. Fluorescein isothiocyanate (FITC).
9. IRDye 800CW NHS ester (simply referred to as NIR).
10. 4′,6-diamidino-2-phenylindole (DAPI).
11. Phosphate-buffered saline (PBS).
12. Acetonitrile (ACN).
13. Formalin solution (10 wt.% in H_2O).

2.2 Biologicals

1. HN12 cells derived from an HNSCC lymph node metastasis.
2. Yellow fluorescent protein (YFP)-expressing HN12 cells (HN12-YFP).

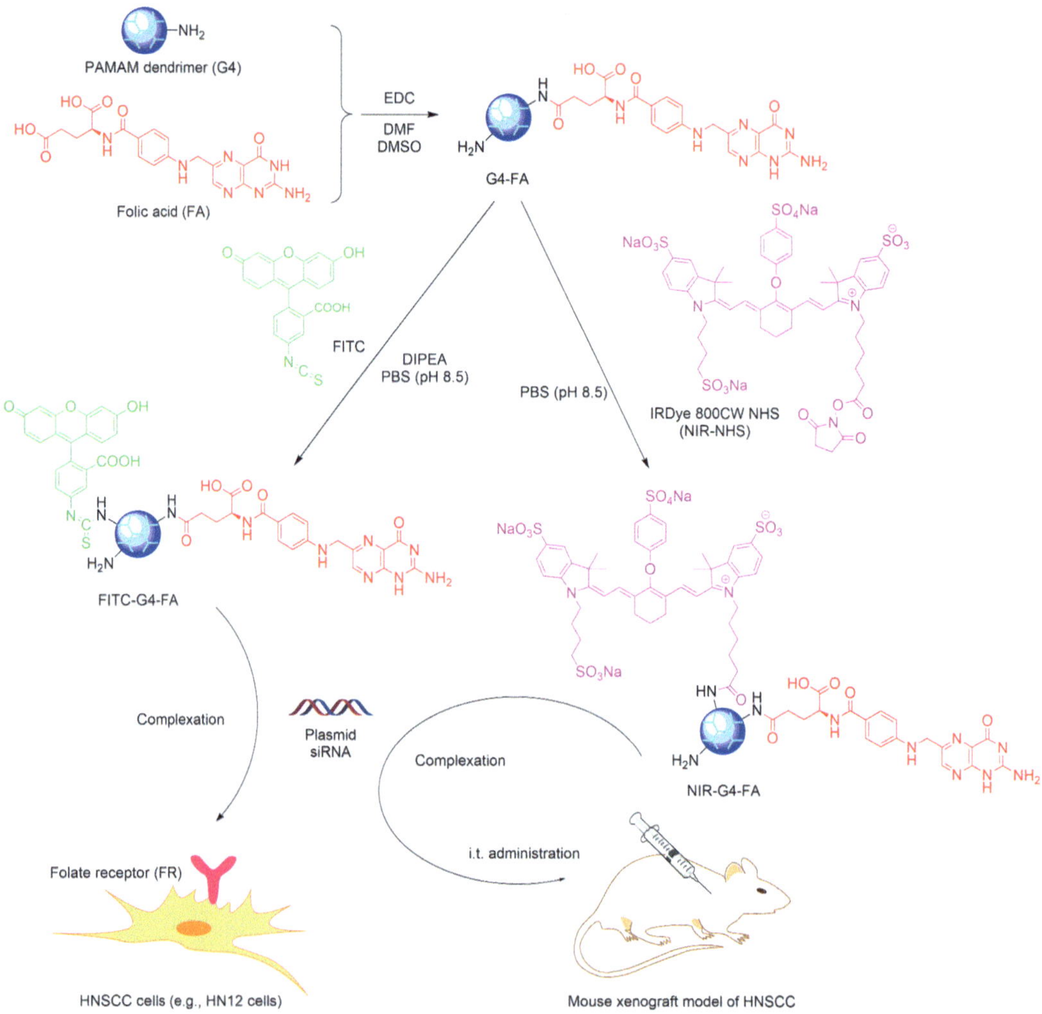

Fig. 2 Synthesis of folic acid-decorated PAMAM dendrimer G4-FA, FITC-labeled G4-FA (FITC-G4-FA) for in vitro transfection, and near-infrared dye-labeled G4-FA (NIR-G4-FA) for in vivo targeted gene therapy

3. U87 cells.
4. Dulbecco's Modified Eagle's Medium (DMEM).
5. Cosmic calf serum.
6. Penicillin and streptomycin.
7. Athymic nude mice.
8. pMAX-GFP plasmid (pGFP).
9. Label IT® plasmid delivery control Cy3® (referred to as Cy3-plasmid).
10. Duplex siRNA against VEGFA (siVEGFA)
    siVEGFA sense strand: 5′-GGAGUACCCUGAUGAGAU CdTdT-3′.
    siVEGFA antisense strand: 5′-GAUCUCAUCAGGGUA CUCCdTdT-3′.

2.3 Equipment

1. Lyophilizer.
2. Reverse-phase high performance liquid chromatography (RP-HPLC) system (Waters).
3. Proton nuclear magnetic resonance (^1H NMR) (Bruker).
4. Spectrophotometer (Thermo Fisher Scientific).
5. Dynamic light scattering (DLS) (Malvern).
6. Flow cytometer (Millipore Sigma).
7. Fluorescent microscope (Nikon).
8. Pearl Trilogy small animal imaging system (Li-COR Biotechnology).

3 Methods

3.1 Synthesis of Folic Acid (FA)-Anchored PAMAM G4 Dendrimer Conjugates

3.1.1 Synthesis of Folic Acid (FA)-Anchored PAMAM G4 Dendrimer (G4-FA) Conjugate

1. Dry 1 mL of 10 wt.% PAMAM-G4-NH_2 dendrimer solution in methanol to obtain 100 mg (7.03 μmol) of PAMAM-G4-NH_2 in a vial.
2. Dissolve 100 mg of PAMAM-G4-NH_2 in 50 mL of DI water.
3. Dissolve 18.6 mg (42.2 μmol) of FA in a mixture of 12 mL of DMF and 4 mL of DMSO for 1 h, and then add it dropwise to 50 mL of PAMAM-G4-NH_2.
4. Stir the reaction mixture vigorously at room temperature for 2 days.
5. Purify the reaction mixture by dialysis against DI water using dialysis tubing with MWCO of 7 kDa for 2 days.
6. Dry the resultant G4-FA conjugate by lyophilization, and characterize it by UV-Vis spectrometry, RP-HPLC, ^1H NMR, and DLS (described in **Note 1**).

3.1.2 Synthesis of FITC-Labeled G4-FA (FITC-G4-FA) Conjugate

1. Dissolve 10 mg (0.63 μmol) of G4-FA conjugate in 10 mL of PBS (pH 8.5) in a vial.
2. Dissolve 1.23 mg (3.16 μmol) of FITC in 2 mL of DMSO, and then add it dropwise to 10 mL of G4-FA conjugates along with 0.82 mg (6.32 μmol) of DIPEA.
3. Stir the reaction mixture continuously in the dark at room temperature for 1 day.
4. Purify the reaction mixture by dialysis against DI water using dialysis tubing with MWCO of 7 kDa in the dark for 2 days.
5. Dry the resultant FITC-G4-FA conjugate by lyophilization, and characterize it by UV-Vis spectrometry, RP-HPLC, ^1H NMR, and DLS.

3.1.3 Synthesis of NIR-Labeled G4-FA (NIR-G4-FA) Conjugate

1. Dissolve 10 mg (0.63 μmol) of G4-FA conjugate in 10 mL of PBS (pH 8.5) in a vial.
2. Dissolve 1.48 mg (1.26 μmol) of NIR-NHS ester in 2 mL of PBS, and then add it dropwise to 10 mL of G4-FA conjugates along with 0.82 mg (6.32 μmol) of DIPEA.
3. Stir the reaction mixture continuously in the dark at room temperature for 1 day.
4. Purify the reaction mixture by dialysis against DI water using dialysis tubing with MWCO of 7 kDa in the dark for 2 days.
5. Dry the resultant NIR-G4-FA conjugate by lyophilization, and characterize it by UV-Vis spectrometry, RP-HPLC, ^{1}H NMR, and DLS.

3.2 Complexation of G4-FA Conjugate with Plasmid or siRNA

3.2.1 Preparation of G4-FA/Plasmid Polyplex

1. Pre-dissolve 5 μg of G4-FA or FITC-G4-FA conjugate in 300 μL of DMEM.
2. Resuspend 1 μg of pGFP plasmid or Cy3-plasmid in 200 μL of DMEM.
3. Mix 300 μL of dendrimer-containing DMEM with 200 μL of plasmid-containing DMEM.
4. Homogenize the mixture for 10 s by vortexing.
5. Equilibrate the mixture for 30 min at room temperature.
6. Adjust the volume to 3 mL with addition of 2.5 mL of the complete medium.
7. Use the prepared polyplexes immediately (described in **Note 2**).

3.2.2 Preparation of G4-FA/siRNA Polyplex

1. Pre-dissolve 10 μg of G4-FA conjugate in 300 μL of DMEM.
2. Resuspend 2 μg of siVEGFA or siGFP in 200 μL of DMEM.
3. Mix 300 μL of dendrimer-containing DMEM with 200 μL of siRNA-containing DMEM.
4. Homogenize the mixture for 10 s by vortexing.
5. Equilibrate the mixture for 30 min at room temperature.
6. Adjust the volume to 3 mL with addition of 2.5 mL of the complete medium.
7. Use the prepared polyplexes immediately (described in **Note 2**).

3.3 In Vitro Trafficking of FITC-G4-FA Conjugate and Polyplex

3.3.1 Head and Neck Cancer Cell Line Culture

1. Culture HN12 (or HN12-YFP) cells in the complete medium (described in **Note 3**) at 37 °C in 90% air/10% CO_2.

3.3.2 Determination of Cellular Uptake Kinetics of G4-FA Conjugate and Polyplex by Flow Cytometry

1. Seed HN12 cells in the 60-mm dishes at a density of 50,000 cells/dish.
2. Allow the cells to attach overnight.
3. Incubate the cells with FITC-G4-FA conjugate (5 μg/3 mL) or FITC-G4-FA/Cy3-labeled plasmid polyplex (5 μg/1 μg/3 mL) at 37 °C for 0, 1, 2, 6, and 24 h.
4. Wash the cells with PBS three times at the end of each treatment.
5. Resuspend the cells with trypsinization for 5–10 min.
6. Neutralize trypsin by adding equal volume of the complete medium.
7. Collect the cells by centrifugation at 180 × *g* at room temperature for 10 min.
8. Resuspend the cells with PBS for flow cytometry analysis.

3.3.3 Determination of Cellular Uptake Kinetics of G4-FA Conjugate and Polyplex by Fluorescence Imaging

1. Fix the cells with 4% formaldehyde at room temperature for 20 min at the end of each treatment same as those in flow cytometry assessment.
2. Wash the cells with PBS three times.
3. Permeate the cells with 0.1% Triton X-100 for 5 min.
4. Wash the cells with PBS three times.
5. Counter stain the cell nuclei with DAPI for 5 min.
6. Wash the cells with PBS three times and image the cells by fluorescence microscope and confocal microscope as shown in Fig. 3.

3.4 Assessment of Folate Receptor Targeting of G4-FA Conjugate

3.4.1 Competitive Inhibition Assay

1. Seed HN12 cells in the 60-mm dishes at a density of 50,000 cells/dish.
2. Allow the cells to attach overnight.
3. Incubate the cells with FITC-G4-FA conjugate (5 μg/3 mL) or FITC-G4-FA/Cy3-labeled plasmid polyplex (5 μg/1 μg/3 mL) in the absence or presence of free FA (0.5 mg/mL) at 37 °C for 1 or 2 h (described in **Note 4**).
4. Wash the cells with PBS three times at the end of each treatment.
5. Resuspend the cells for flow cytometry analysis (as described in Subheading 3.3.2) or fix, permeate, and counterstain the cells with DAPI for fluorescence microscope imaging (as described in Subheading 3.3.3).

3.4.2 $FR\alpha^{high}/FR\alpha^{low}$ Co-culture Model Assessment

1. Seed equal numbers of HN12-YFP cells and U87 cells in a 6-well plate at a density of 10,000 cells/well.
2. Allow the cells to attach overnight.

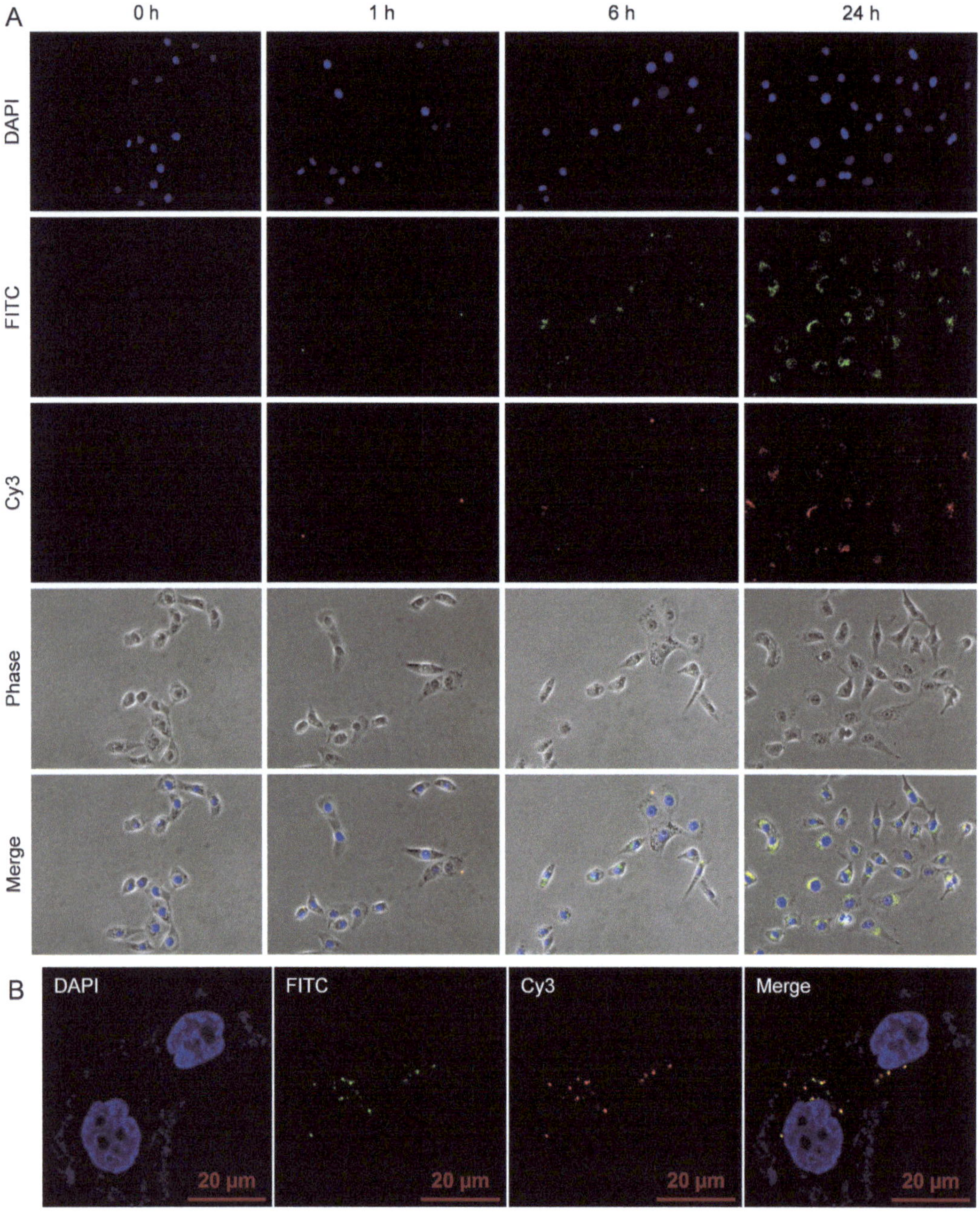

Fig. 3 Cellular uptake and intracellular distribution of FITC-G4-FA/Cy3-plasmid polyplexes in HN12 cells. (**a**) Cellular uptake of FITC-G4-FA/Cy3-plasmid polyplexes in HN12 cells over a period of 24 h was monitored with fluorescence microscopy (200×). (**b**) Intracellular distribution of FITC-G4-FA/Cy3-plasmid polyplexes in HN12 cells at 24-h incubation illustrated in the confocal images (630×) (reproduced from [17])

3. Incubate the cells with G4-FA/Cy3-labeled plasmid polyplex (5 μg/1 μg/3 mL) at 37 °C for 6 and 24 h.
4. Wash the cells with PBS three times at the end of each treatment.
5. Fix, permeate, and counter stain the cells with DAPI for fluorescence microscope imaging (as described in Subheading 3.3.3) as shown in Fig. 4.

3.5 Evaluation of G4-FA Conjugate-Mediated Transfection Efficiency

3.5.1 Assessment of Gene Transfection Efficiency

1. Seed HN12 cells in a 6-well plate at a density of 10,000 cells/well.
2. Allow the cells to attach overnight.
3. Incubate the cells with G4-FA/pGFP polyplex (5 μg/1 μg/3 mL) at 37 °C for 2 days, followed by another 2-day culture in the fresh complete medium.
4. Remove the spent media at the end of each treatment.
5. Harvest the cells for protein analysis of GFP by Western blotting.
6. Resuspend the cells for flow cytometry analysis (as described in Subheading 3.3.3).
7. Fix, permeate, and counterstain the cells with DAPI for fluorescence microscope imaging (as described in Subheading 3.3.3).
8. Determine the cell viability by WST-1 cell proliferation assay following the manufacturer's protocol as shown in Fig. 5.

3.5.2 Assessment of Gene Knockdown Efficiency

1. Seed HN12 cells in a 6-well plate at a density of 20,000 cells/well.
2. Allow the cells to attach overnight.
3. Incubate the cells with G4-FA/siVEGFA polyplex (10 μg/2 μg/3 mL) at 37 °C for 2 days, followed by another 2-day culture in the fresh complete medium.
4. Collect the spent media at the end of each treatment and determine released VEGFA level by ELISA.
5. Harvest the cells for RNA analysis of VEGFA by quantitative PCR (qPCR).

3.6 Assessment of Biodistribution of NIR-G4-FA Conjugate

3.6.1 Establishment of Mouse Xenograft Model of Head and Neck Cancer

1. Culture NH-12 cells (or NH12-YFP cells) in 175 mm^2 flasks.
2. Wash the cells with PBS.
3. Resuspend the cells with trypsinization for 5–10 min.
4. Neutralize trypsin by adding equal volume of the complete medium.
5. Collect the cells by centrifugation at 180 × g at room temperature for 10 min.

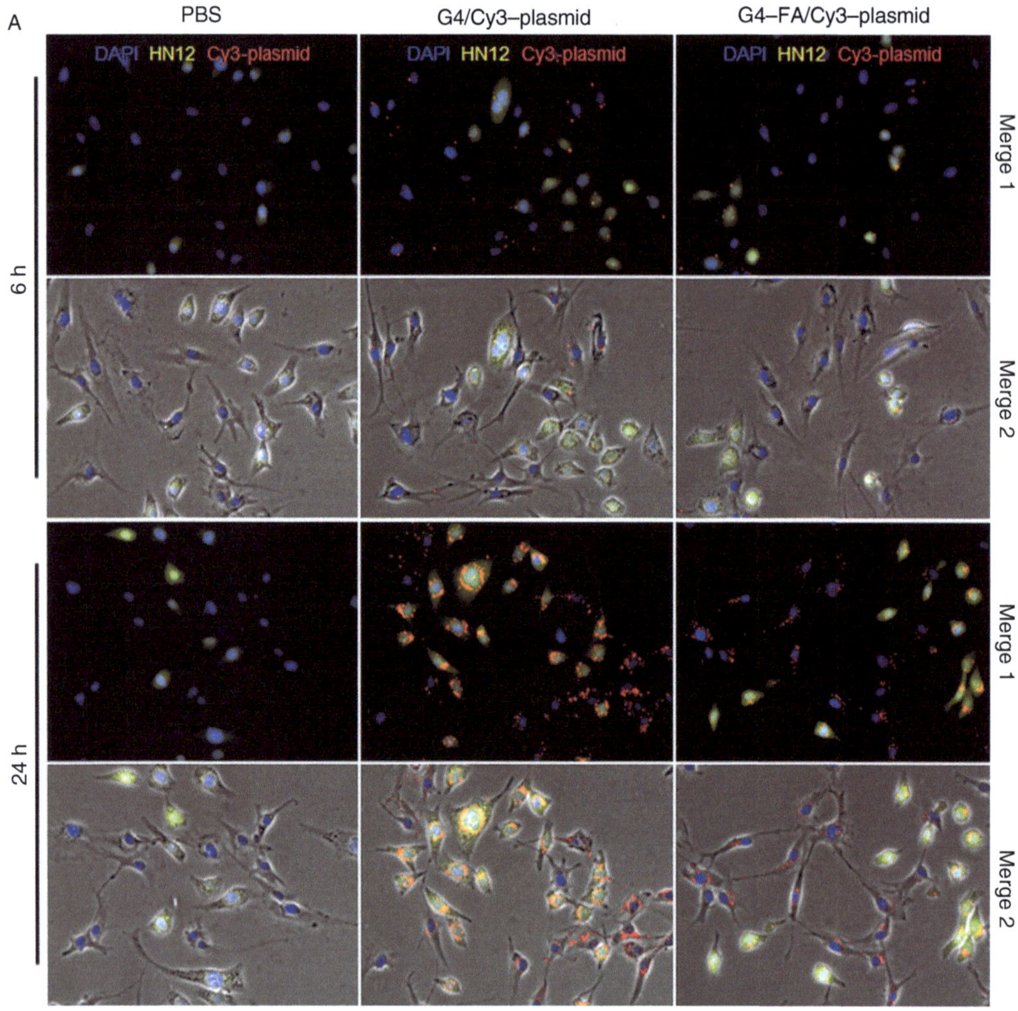

Fig. 4 G4-FA promotes preferential cellular uptake of plasmid in folate receptor-high cells. Cellular uptake of polyplexes in an HN12-YFP/U87 (FR-high/FR-low) co-culture model monitored by using fluorescence microscopy. Blue, DAPI; yellow, HN12-YFP cells; and red, Cy3-plasmid (reproduced from [17])

6. Count the cell numbers.
7. Resuspend the cells with PBS at 20 × 10^6 cells/mL.
8. Inject 0.25 mL of the cell suspension (5 × 10^6 cells/mouse) subcutaneously into the right flank of 4-week-old female athymic nude mice.
9. Monitor the tumor size and mouse body weight.

3.6.2 Assessment of Biodistribution of NIR-G4-FA Conjugate

1. Inject NIR-G4-FA intratumorally (i.t.) to the HN12 tumor-bearing mice at 8 days after xenograft.
2. Measure the tumor size (calculated by the equation described in **Note 5**) and mouse body weight every other day.

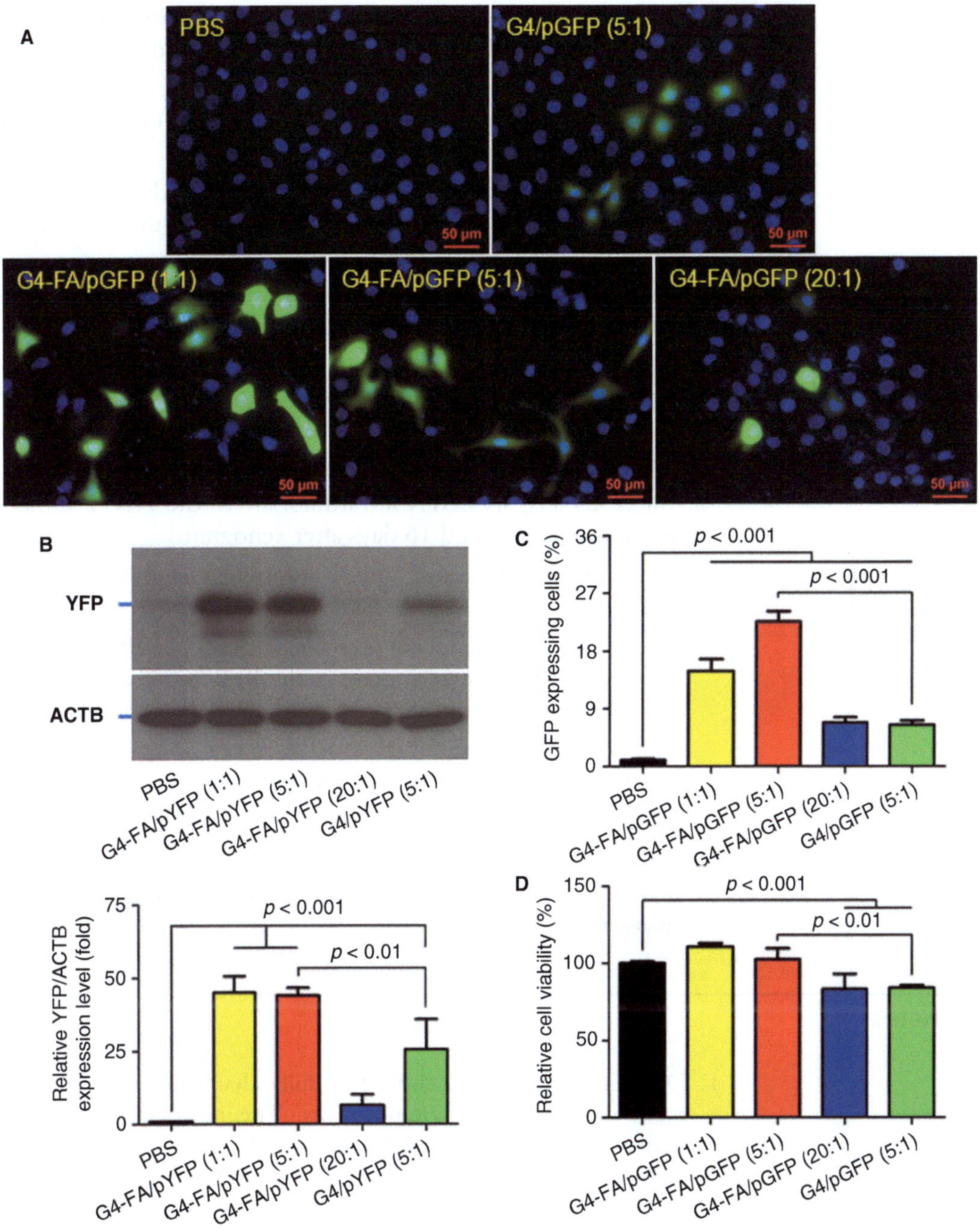

Fig. 5 Enhanced gene transfection in HN12 cells by G4-FA. (**a**) Fluorescence images of HN12 cells receiving different treatments. (**b**) Quantification of YFP expression levels by using western blot analysis. (**c**) Quantification of GFP-expressing cell populations as determined by flow cytometry. (**d**) Post-transfection cell viability assessment as determined by WST-1 assay (reproduced from [17])

3. Image the whole-body fluorescence using a Pearl Trilogy small animal imaging system at 1 h, 1, 3, 7, 14, and 21 days after NIR-G4-FA injection.
4. Record and quantify the fluorescence intensity of ROI (tumor region).
5. Collect and image all the major organs including the brain, heart, lung, liver, spleen, kidneys, and tumor at the end of each treatment.
6. Record and quantify the fluorescence intensity of each organ and normalize it to the weight as shown in Fig. 6.

3.7 Assessment of Antitumor Efficacy of FITC-G4-FA/siVEGFA Polyplex

1. Inject 0.25 mL of HN12-YFP cell suspension (5×10^6 cells/mouse) subcutaneously into the right flank of 4-week-old female athymic nude mice.
2. Inject G4-FA/siVEGFA intratumorally to the HN12 tumor-bearing mice at 8 and 16 days after xenograft.
3. Measure the tumor size and mouse body weight every other day.
4. Image the tumor yellow fluorescence using IVIS 200 system to monitor tumor size and quantify the YFP fluorescence intensity of tumor region at 8, 16, and 24 days after xenograft (described in **Note 6**) as shown in Fig. 7.
5. Sacrifice the mice at 26 days after xenograft.
6. Collect and weigh the tumor.
7. Immerse the tumor into 10% formalin for histology (H&E staining) and immunohistochemistry (IHC staining of CD31) assessment.

4 Notes

1. Dendrimer derivatives should be fully characterized for assessment of purity and compositions. The purity of dendrimer derivatives is determined by using RP-HPLC. TFA can be included in the mobile phase as a counterion to neutralize dendrimer surface charges. The mobile phase of H_2O:ACN:TFA (750:250:0.38, v/v/v) is freshly made, degased, and eluted at a flow rate of 1 mL/min to separate dendrimer derivatives from free FA at 275 nm (and FITC at 485 nm). The area under the curve (AUC) of G4-FA and free FA is determined, and the purity of G4-FA is calculated by using the equation:

$$\text{Purity} = \text{AUC}_{\text{G4-FA}} / \left(\text{AUC}_{\text{G4-FA}} + \text{AUC}_{\text{FA}}\right) \times 100\%$$

The composition of dendrimer derivatives is determined by using ^{1}H NMR.

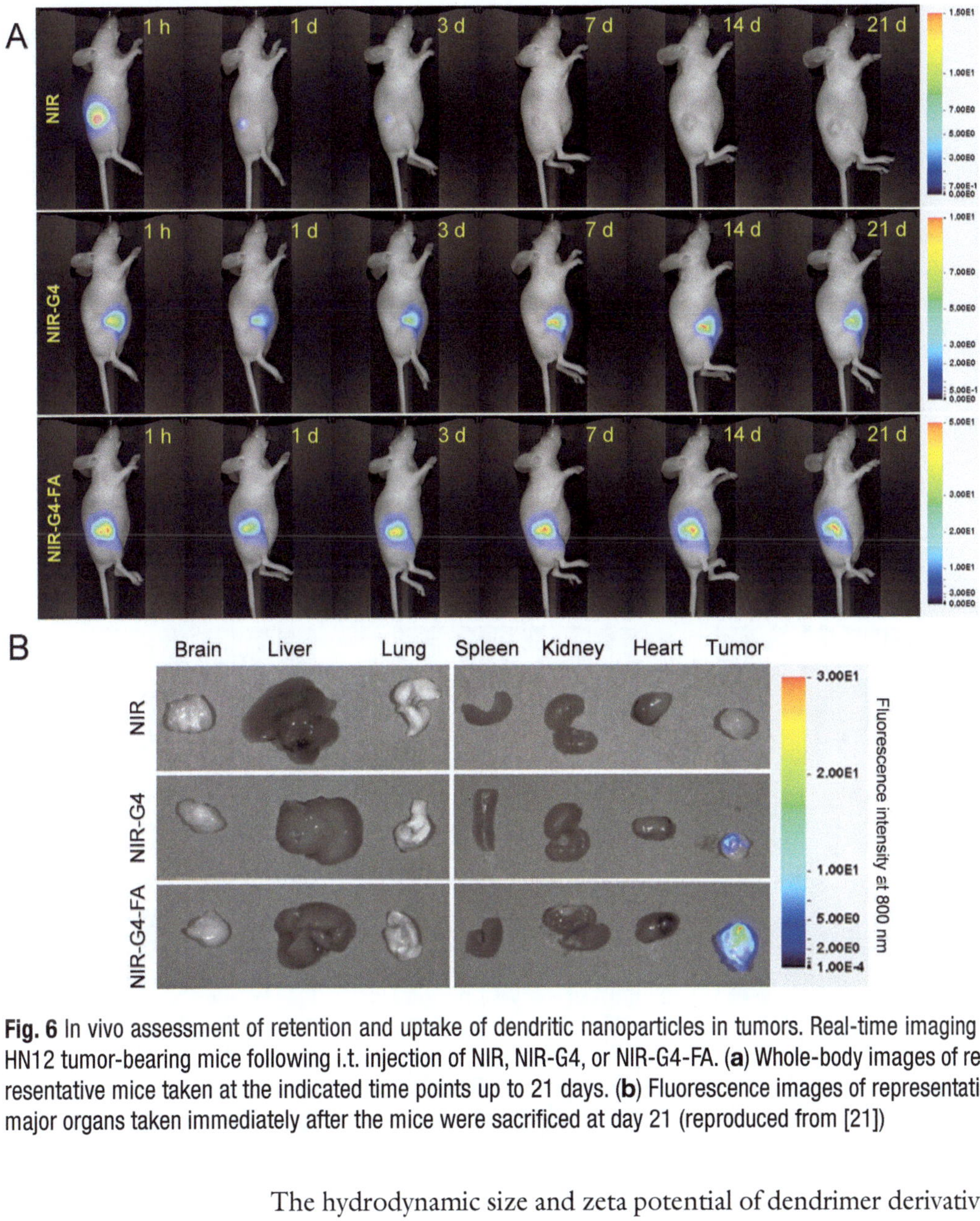

Fig. 6 In vivo assessment of retention and uptake of dendritic nanoparticles in tumors. Real-time imaging of HN12 tumor-bearing mice following i.t. injection of NIR, NIR-G4, or NIR-G4-FA. (**a**) Whole-body images of representative mice taken at the indicated time points up to 21 days. (**b**) Fluorescence images of representative major organs taken immediately after the mice were sacrificed at day 21 (reproduced from [21])

The hydrodynamic size and zeta potential of dendrimer derivatives are determined by using DLS. Dendrimer derivatives are resuspended and equilibrated in the PBS (filtered through 20 nm filter) at 0.5 mg/mL at room temperature prior to the measurement.

The number of FA and FITC molecules coupled to each dendrimer is quantified by using UV-Vis spectrophotometer at 275 nm and 485 nm, respectively.

2. Naïve, FITC-labeled, and NIR-labeled PAMAM G4 dendrimer (G4, FITC-G4, or NIR-G4) at the same weight concentrations

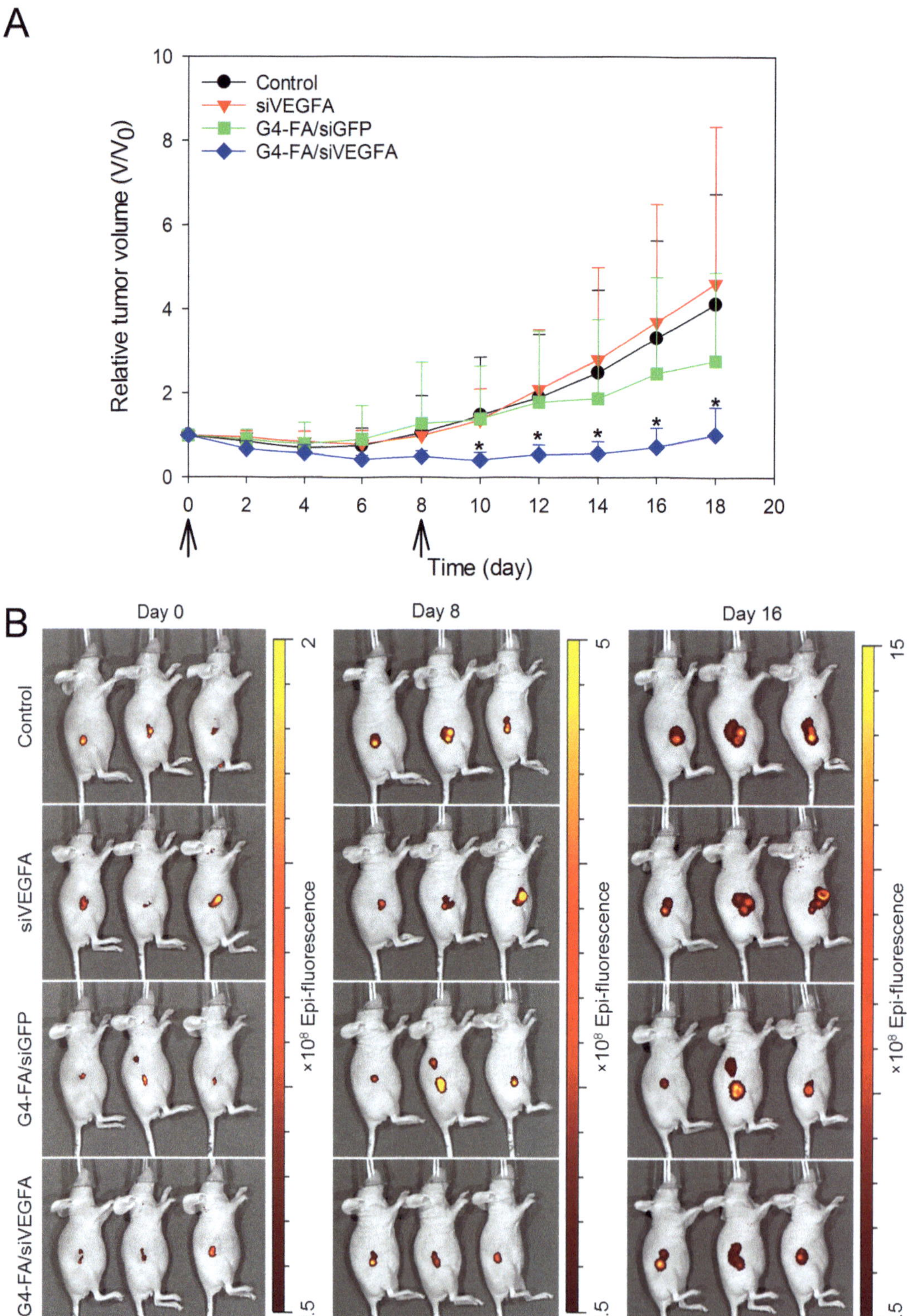

Fig. 7 In vivo assessment of antitumor effects. HN12-YFP tumor-bearing mice received one dose of PBS, siVEGFA, G4-FA/siGFP, or G4-FA/siVEGFA at day 0 and a second dose at day 8 via i.t. injection. The experiment was terminated at day 18. (**a**) Tumor volume measured at indicated time points. Arrows indicate the days when the mice were inoculated. $*p < 0.05$ versus the control ($n = 6$). (**b**) Semiquantitative in vivo assessment of antitumor effects by noninvasive imaging of tumors in the HN12-YFP tumor-bearing mice. Whole-body fluorescence images of the mice in the two-dose experiment at day 0, day 8, and day 16 (reproduced from [21])

as G4-FA, FITC-G4-FA, or NIR-G4-FA are used as non-FR-targeted controls for all the experiments, respectively. PBS is used as negative control. Free siVEGFA alone is used as free drug control for the in vivo assessments. Duplex siRNA against GFP (siGFP) is used as a negative control for siVEGFA knock-down assessment in vitro and in vivo.

siGFP sense strand: 5′-GCACGACUUCUUCAAGUCCd TdT-3′.

siGFP antisense strand: 5′-GGACUUGAAGAAGUCGUGC dTdT-3′ [22].

3. Complete medium is DMEM supplied with 10% cosmic calf serum, 100 units/mL of penicillin, and 100 μg/mL of streptomycin.
4. Four approaches can be used to assess the FRα-mediated endocytosis/transfection using G4-FA. First, free FA (0–2 mg/mL) is used as a competitive inhibitor for cellular uptake of G4-FA in FRα-highly expressing HN12 cells. Second, the cellular uptake kinetics and transfection efficiency of G4-FA are compared to those of naïve G4 in HN12 cells. Third, the cellular uptake and transfection efficiency of G4-FA in HN12 cells and FRα-highly expressing T98 cells are compared to those in FRα-low expressing U87 cells. Fourth, the cellular uptake of G4-FA and naïve G4 is determined in HN12-YFP cells co-cultured with U87 cells at the same cell density.
5. Calculation of tumor volume (V) by using the equation:

$$V = \text{Length} \times \text{Width}^2 / 2$$

Where, the width is smaller than the length [23].

6. Quantification of tumor YFP fluorescence intensity can be used to semiquantitatively assess HN12-YFP tumor growth in addition to the tumor size measurement.

References

1. Harrington K, Alvarez-Vallina L, Crittenden M, Gough M, Chong H, Diaz RM, Vassaux G, Lemoine N, Vile R (2002) Cells as vehicles for cancer gene therapy: the missing link between targeted vectors and systemic delivery? Hum Gene Ther 13:1263–1280
2. Brunotto M, Zarate AM, Bono A, Barra JL, Berra S (2014) Risk genes in head and neck cancer: a systematic review and meta-analysis of last 5 years. Oral Oncol 50:178–188
3. Thomas SM, Grandis JR (2009) The current state of head and neck cancer gene therapy. Hum Gene Ther 20:1565–1575
4. Dobbelstein M, Moll U (2014) Targeting tumour-supportive cellular machineries in anticancer drug development. Nat Rev Drug Discov 13:179–196
5. Xu L, Yeudall WA, Yang H (2013) Dendrimer-based RNA interference delivery for cancer therapy. In: Scholz C, Kressler J (eds) Tailored polymer architectures for pharmaceutical and biomedical applications, vol 1135. American Chemical Society, Washington, DC, pp 197–213
6. Ameyar-Zazoua M, Guasconi V, Ait-Si-Ali S (2005) siRNA as a route to new cancer therapies. Expert Opin Biol Ther 5:221–224
7. Cabanillas R, Rodrigo JP, Astudillo A, Dominguez F, Suarez C, Chiara MD (2007) P53 expression in squamous cell carcinomas of

the supraglottic larynx and its lymph node metastases: new results for an old question. Cancer 109:1791–1798
8. Cardinali M, Jakus J, Shah S, Ensley JF, Robbins KC, Yeudall WA (1998) p21(WAF1/Cip1) retards the growth of human squamous cell carcinomas in vivo. Oral Oncol 34:211–218
9. Ndoye A, Dolivet G, Hogset A, Leroux A, Fifre A, Erbacher P, Berg K, Behr JP, Guillemin F, Merlin JL (2006) Eradication of p53-mutated head and neck squamous cell carcinoma xenografts using nonviral p53 gene therapy and photochemical internalization. Mol Ther 13:1156–1162
10. Morris JC, Wildner O (2000) Therapy of head and neck squamous cell carcinoma with an oncolytic adenovirus expressing HSV-tk. Mol Ther 1:56–62
11. Xu L, Zhang H, Wu Y (2014) Dendrimer advances for the central nervous system delivery of therapeutics. ACS Chem Neurosci 5:2–13
12. Yang H (2010) Nanoparticle-mediated brain-specific drug delivery, imaging, and diagnosis. Pharm Res 27:1759–1771
13. Kesharwani P, Iyer AK (2015) Recent advances in dendrimer-based nanovectors for tumor-targeted drug and gene delivery. Drug Discov Today 20(5):536–547
14. Lee CC, MacKay JA, Frechet JM, Szoka FC (2005) Designing dendrimers for biological applications. Nat Biotechnol 23:1517–1526
15. Xu L, Bai Q, Zhang X, Yang H (2017) Folate-mediated chemotherapy and diagnostics: an updated review and outlook. J Control Release 252:73–82
16. Chen C, Ke J, Zhou XE, Yi W, Brunzelle JS, Li J, Yong EL, Xu HE, Melcher K (2013) Structural basis for molecular recognition of folic acid by folate receptors. Nature 500:486–489
17. Xu L, Kittrell S, Yeudall WA, Yang H (2016) Folic acid-decorated polyamidoamine dendrimer mediates selective uptake and high expression of genes in head and neck cancer cells. Nanomedicine (Lond) 11:2959–2973
18. Parker N, Turk MJ, Westrick E, Lewis JD, Low PS, Leamon CP (2005) Folate receptor expression in carcinomas and normal tissues determined by a quantitative radioligand binding assay. Anal Biochem 338:284–293
19. Elnakat H, Ratnam M (2004) Distribution, functionality and gene regulation of folate receptor isoforms: implications in targeted therapy. Adv Drug Deliv Rev 56:1067–1084
20. Ross JF, Chaudhuri PK, Ratnam M (1994) Differential regulation of folate receptor isoforms in normal and malignant tissues in vivo and in established cell lines. Physiologic and clinical implications. Cancer 73:2432–2443
21. Xu L, Yeudall WA, Yang H (2017) Folic acid-decorated polyamidoamine dendrimer exhibits high tumor uptake and sustained highly localized retention in solid tumors: its utility for local siRNA delivery. Acta Biomater 57:251–261
22. Kim SH, Jeong JH, Kim TI, Kim SW, Bull DA (2009) VEGF siRNA delivery system using arginine-grafted bioreducible poly(disulfide amine). Mol Pharm 6:718–726
23. Guo S, Lin CM, Xu Z, Miao L, Wang Y, Huang L (2014) Co-delivery of cisplatin and rapamycin for enhanced anticancer therapy through synergistic effects and microenvironment modulation. ACS Nano 8:4996–5009

INDEX

Lekha Dinesh Kumar (ed.), *RNA Interference and Cancer Therapy: Methods and Protocols*, Methods in Molecular Biology, vol. 1974, https://doi.org/10.1007/978-1-4939-9220-1, © Springer Science+Business Media, LLC, part of Springer Nature 2019

D

E

F

G

H

I

O

P

Q

R

S

MIX
Papier aus verantwortungsvollen Quellen
Paper from responsible sources
FSC® C105338

If you have any concerns about our products,
you can contact us on
ProductSafety@springernature.com

In case Publisher is established outside the EU,
the EU authorized representative is:
Springer Nature Customer Service Center GmbH
Europaplatz 3, 69115 Heidelberg, Germany

Printed by Libri Plureos GmbH
in Hamburg, Germany